AF389419

Advances in Nanomaterials for Photovoltaic Applications

Advances in Nanomaterials for Photovoltaic Applications

Editor

Vlad-Andrei Antohe

MDPI • Basel • Beijing • Wuhan • Barcelona • Belgrade • Manchester • Tokyo • Cluj • Tianjin

Editor
Vlad-Andrei Antohe
University of Bucharest
Romania

Editorial Office
MDPI
St. Alban-Anlage 66
4052 Basel, Switzerland

This is a reprint of articles from the Special Issue published online in the open access journal *Nanomaterials* (ISSN 2079-4991) (available at: https://www.mdpi.com/journal/nanomaterials/special_issues/nano_photovoltaics).

For citation purposes, cite each article independently as indicated on the article page online and as indicated below:

LastName, A.A.; LastName, B.B.; LastName, C.C. Article Title. *Journal Name* **Year**, *Volume Number*, Page Range.

ISBN 978-3-0365-7050-1 (Hbk)
ISBN 978-3-0365-7051-8 (PDF)

Contents

About the Editor

Vlad-Andrei Antohe

Vlad-Andrei ANTOHE received in 2002 a Bachelor's degree (B.Sc.) in Physics Education, and in 2005 graduated with a Master diploma (M.Sc.) in Physical Electronics, from the Faculty of Physics, University of Bucharest, Romania. He also obtained in 2003 an Engineering degree (M.Eng.) in Electronics from the Faculty of Electronics and Telecommunications, Polytechnic University of Bucharest, Romania. Subsequently, in 2012 he finished a Ph.D. program in Applied Engineering Sciences, the diploma being awarded by Catholic University of Louvain, Belgium, where he continued his Post-Doctoral studies until 2016. Since then, he has been an Associate Professor within Faculty of Physics, University of Bucharest, Romania, as well as a Scientific Collaborator within Institute of Condensed Matter and Nanosciences, Catholic University of Louvain, Belgium.

Besides teaching activities with undergraduate and graduate students, his main research interests are in the areas of materials science and nanotechnology. His particular focus is on the development and investigation of nanostructured materials and low-dimensional architectures, with the aim of generating novel structural arrangements tailored to specific desired properties. Mr. Vlad-Andrei ANTOHE also has expertise in the fabrication and characterization of electronic and optoelectronic devices based on inorganic, organic, or hybrid organic/inorganic nanomaterials, such as photovoltaic cells, sensors and biosensors, magnetic media for various applications, as well as micro-batteries and micro-supercapacitors.

Assoc. Prof. Ph.D. Eng. Vlad-Andrei ANTOHE (Web of Science ID: D-2158-2012; Scopus ID: 8725853100) co-authored around 60 scientific publications, of which more than 15 ISI research papers are published in journals with a high Impact Factor (IF), ranging approximately from 5 to 32.

Preface to "Advances in Nanomaterials for Photovoltaic Applications"

This Special Issue of Nanomaterials, entitled: "Advances in Nanomaterials for Photovoltaic Applications", includes theoretical and experimental studies reporting on the innovative processing and characterization of materials and nanostructured materials engineered for photovoltaic applications. In particular, the book presents a collection including an editorial and twelve high-quality articles, among which are two comprehensive reviews, one communication letter and nine original research papers. These works exhaustively cover state-of-the-art aspects of the four essential constitutive building-blocks of a photovoltaic element: the main absorber component, the "window" layer, the electron- and hole-transporter media, as well as the back- and top-electrode charge carrier collectors.

I am very grateful to all authors who have contributed to this Special Issue of Nanomaterials, as well as to the reviewers for their excellent input into raising the overall scientific rigorousness and quality of each work covered in this book. Ultimately, I would like to express my appreciation towards the editorial team of Nanomaterials and to send my special thanks to the Section Managing Editor, Ms. Andreea-Beatrice GÎRBACIU, for constantly providing me with consistent administrative and technical support.

It is my hope that scientists the world over will enjoy this Special Issue and will fruitfully enrich their knowledge on this topic as they read through its content.

Vlad-Andrei Antohe

Editor

nanomaterials

Editorial

Advances in Nanomaterials for Photovoltaic Applications

Vlad-Andrei Antohe [1,2]

[1] Research and Development Center for Materials and Electronic & Optoelectronic Devices (MDEO), Faculty of Physics, University of Bucharest, 077125 Măgurele, Ilfov, Romania; vlad.antohe@fizica.unibuc.ro

[2] Institute of Condensed Matter and Nanosciences (IMCN), Université catholique de Louvain (UCLouvain), B-1348 Louvain-la-Neuve, Wallonie, Belgium; vlad.antohe@uclouvain.be

Citation: Antohe, V.-A. Advances in Nanomaterials for Photovoltaic Applications. *Nanomaterials* **2022**, *12*, 3702. https://doi.org/10.3390/nano12203702

Received: 17 October 2022
Accepted: 19 October 2022
Published: 21 October 2022

Publisher's Note: MDPI stays neutral with regard to jurisdictional claims in published maps and institutional affiliations.

The development of novel nanomaterials became a subject of intensive research, due to high market needs for innovative applications in virtually all aspects of life. In particular, the evolution of photovoltaic (PV) devices encountered a great scientific progress in the last few years, mainly because solar power has great potential to cover society needs in the context of energy crisis the world is facing nowadays. In this case, the researchers' efforts are especially directed lately into designing solar cell architectures with improved photo-conversion efficiency, while mainly employing environmentally-friendly materials and inherently maintaining low manufacturing costs. In this context, I am really delighted to observe the large diversity of topics chased by the authors in this Special Issue of *"Nanomaterials"*, dealing practically with all the four essential sub-systems of a PV cell, i.e., (i) photo-absorber component, (ii) "window" layer, (iii) electrons- and holes-transporter media, as well as (iv) charge collecting electrodes.

Basically, the most important component in a PV structure, whatever it is the generation it belongs to, is the main photo-absorber, as this material must be typically engineered to gather photons from as wide region of solar spectrum as possible. In this sense, there are several papers focused onto studying novel materials typically prepared by low-cost techniques, used as photoactive layers to design cheaper and more efficient PV devices. For instance, it is worth mentioning here the comprehensive review of S. Gedi et al. debating with the physical properties of a relatively new class of materials relying onto tin-based binary sulfides (Sn_xS_y), exclusively synthesized by simple chemical bath deposition (CBD) approaches [1]. Besides, as the authors present, these materials are ecologically-friendly, abundant on earth and they could be successfully employed not only in solar energy photo-conversion devices, but in other applications to generate "clean" energy, such as photocatalysis or thermoelectricity. Very good performances of the second generation' solar cells based on copper indium gallium selenide (CIGS) thin films exclusively grown by electrochemical pathways are also demonstrated elsewhere [2], as well as the work on ultra-thin noncrystalline cadmium telluride (CdTe) films prepared by the novel blade coating technique that promises to offer a reliable solution for low-cost and large-scale fabrication of solar cells based on semiconducting nanocrystals [3]. Not in the least, good results were also obtained while testing other main absorber materials for PV cells like antimony selenoiodide (SbSeI) thin films [4], or by implementing specific innovative fabrication technologies such as engagement of aluminium arsenide (AlAs) capping layers onto indium arsenide/galium arsenide (InAs/GaAs) quantum dot (QD) structures for solar cells [5]. The latter strategy was proven to be effective on improving the photovoltaic efficiency, when compared to the reference traditional QD-based solar cells.

Subsequently, the properties of the "window" layer in a PV device are in most of the cases greatly responsible for the overall wideness of the photoactive region of the cell. Noteworthy, J. Lee et al. pointed out that the efficiency of the chalcogenide solar cells (i.e., specifically the CIGS-based) is highly-dependent on the quality of the "window" layer and especially on its thickness [6]. Although the application of such ultra-thin "window" layers is typically limited by the capabilities of the deposition method, the authors demonstrated that by replacing the commonly employed sputtering method with the atomic layer deposition (ALD) technique, ultra-thin zinc oxide (A-ZnO) layers (i.e., with a thickness of only 12 nm) could be achieved, featuring superior properties to act as "window" layers and to determine thus better photovoltaic performances when compared to those PV cells based on sputtered intrinsic zinc oxide (i-ZnO) "window" layers. According to the authors, the ALD process could be also advantageously engaged into the large-scale mass production of CIGS-based solar cells. In contrast, other researchers demonstrated that zinc selenide (ZnSe) thin films prepared by radio-frequency (RF) magnetron sputtering under optimized specific conditions could possess excellent physical properties to successfully act as environmentally-friendly "window" materials for solar cells, helping thus at diminishing the amount of cadmium (Cd) still commonly used for the second generation of solar cells relying on cadmium telluride (CdTe) as main absorber [7].

The transport of the photo-generated charge carriers towards the cell's electrodes play also an important role in the overall photovoltaic response. For this reason, a great interest of the scientific community is dedicated to this process, as these materials should exhibit good compatibility with the neighbouring components of the cell and most importantly, they must feature an energetic band diagram structure with adequate LUMO (lowest unoccupied molecular orbital) and HOMO (highest occupied molecular orbital) levels that facilitate the transport of electrons and holes towards the cathode and anode electric contacts of the PV cell, respectively. In this sense, the thorough studies onto complexes based on titanium dioxide (TiO_2) acting as electron transporter layer (ETL) materials must be acknowledged [8,9], as well as the exhaustive review on hole transporter layer (HTL) materials for organic PV cells by C. Anrango-Camacho et al. [10].

Ultimately, the collection of the photo-generated charge carriers substantially depend on the physical properties of the back- and top-electrodes of the cell. Obviously, at least one of these electrodes must combine a high electrical conductivity with an excellent transparency in the visible region of the electromagnetic spectrum. These two antagonistic properties are sometimes hard to reconcile, motivating the important work dedicated to the study of novel transparent conducting oxide (TCO) materials for PV cells. Herein, a new composite relying on indium-zinc-tin oxide (IZTO) was successfully engaged as a transparent electrode, the authors demonstrating its potential when used within construction of an ultra-flexible organic PV cell used for powering human wearable devices [11]. Other novel TCO materials with superior physical properties have been also studied by L. Hrostea et al. [12]. In this case, the authors deposited oxide/metal/oxide multi-layers as alternative to the well-studied indium-doped tin oxide (ITO), onto glass and especially plastic substrates, hence owning to a great potential for flexible photovoltaic devices.

It is obvious now that a great dedication of studying in details each building-block of the PV device is essential, as the morphological, compositional, structural, optical and photo-electrical properties of each of the constituting material have to be all-together well-harmonized in order to get the expected increase in the performance of a solar cell, whatever its type and generation.

Funding: The author acknowledges financial support received from the "Executive Unit for Financing Higher Education, Research, Development and Innovation" (UEFISCDI, Romania) through the grant: TE 115/2020 (PN-III-P1-1.1-TE-2019-0868).

Data Availability Statement: Data might be available from the authors of the cited papers.

Conflicts of Interest: The author declares no conflict of interest.

References

1. Gedi, S.; Reddy, V.R.M.; Kotte, T.R.R.; Park, C.; Kim, W.K. Fundamental Aspects and Comprehensive Review on Physical Properties of Chemically Grown Tin-Based Binary Sulfides. *Nanomaterials* **2021**, *11*, 1955. [CrossRef] [PubMed]
2. Saeed, M.; González Peña, O.I. Mass Transfer Study on Improved Chemistry for Electrodeposition of Copper Indium Gallium Selenide (CIGS) Compound for Photovoltaics Applications. *Nanomaterials* **2021**, *11*, 1222. [CrossRef] [PubMed]

3. Xiao, K.; Huang, Q.; Luo, J.; Tang, H.; Xu, A.; Wang, P.; Ren, H.; Qin, D.; Xu, W.; Wang, D. Efficient Nanocrystal Photovoltaics via Blade Coating Active Layer. *Nanomaterials* **2021**, *11*, 1522. [CrossRef] [PubMed]
4. Choi, Y.C.; Jung, K.W. One-Step Solution Deposition of Antimony Selenoiodide Films via Precursor Engineering for Lead-Free Solar Cell Applications. *Nanomaterials* **2021**, *11*, 3206. [CrossRef] [PubMed]
5. Ruiz, N.; Fernández, D.; Stanojević, L.; Ben, T.; Flores, S.; Braza, V.; Carro, A.G.; Luna, E.; Ulloa, J.M.; González, D. Suppressing the Effect of the Wetting Layer through AlAs Capping in InAs/GaAs QD Structures for Solar Cells Applications. *Nanomaterials* **2022**, *12*, 1368. [CrossRef] [PubMed]
6. Lee, J.; Jeon, D.H.; Hwang, D.K.; Yang, K.J.; Kang, J.K.; Sung, S.J.; Park, H.; Kim, D.H. Atomic Layer Deposition of Ultrathin ZnO Films for Hybrid Window Layers for Cu(In$_X$,Ga$_{1-X}$)Se$_2$ Solar Cells. *Nanomaterials* **2021**, *11*, 2779. [CrossRef] [PubMed]
7. Toma, O.; Antohe, V.A.; Panaitescu, A.M.; Iftimie, S.; Răduţă, A.M.; Radu, A.; Ion, L.; Antohe, Ş. Effect of RF Power on the Physical Properties of Sputtered ZnSe Nanostructured Thin Films for Photovoltaic Applications. *Nanomaterials* **2021**, *11*, 2841. [CrossRef] [PubMed]
8. Lee, J.; Kim, J.; Kim, C.S.; Jo, S. Compact SnO$_2$/Mesoporous TiO$_2$ Bilayer Electron Transport Layer for Perovskite Solar Cells Fabricated at Low Process Temperature. *Nanomaterials* **2022**, *12*, 718. [CrossRef] [PubMed]
9. Hsu, C.H.; Chen, K.T.; Lin, L.Y.; Wu, W.Y.; Liang, L.S.; Gao, P.; Qiu, Y.; Zhang, X.Y.; Huang, P.H.; Lien, S.Y.; et al. Tantalum-Doped TiO$_2$ Prepared by Atomic Layer Deposition and its Application in Perovskite Solar Cells. *Nanomaterials* **2021**, *11*, 1504. [CrossRef] [PubMed]
10. Anrango-Camacho, C.; Pavón-Ipiales, K.; Frontana-Uribe, B.A.; Palma-Cando, A. Recent Advances in Hole-Transporting Layers for Organic Solar Cells. *Nanomaterials* **2022**, *12*, 443. [CrossRef] [PubMed]
11. Choi, J.Y.; Park, I.P.; Heo, S.W. Ultra-Flexible Organic Solar Cell Based on Indium-Zinc-Tin Oxide Transparent Electrode for Power Source of Wearable Devices. *Nanomaterials* **2021**, *11*, 2633. [CrossRef] [PubMed]
12. Hrostea, L.; Lisnic, P.; Mallet, R.; Leontie, L.; Girtan, M. Studies on the Physical Properties of TiO$_2$:Nb/Ag/TiO$_2$:Nb and NiO/Ag/NiO Three-Layer Structures on Glass and Plastic Substrates as Transparent Conductive Electrodes for Solar Cells. *Nanomaterials* **2021**, *11*, 1416. [CrossRef] [PubMed]

Review

Fundamental Aspects and Comprehensive Review on Physical Properties of Chemically Grown Tin-Based Binary Sulfides

Sreedevi Gedi [1], Vasudeva Reddy Minnam Reddy [1,*], Tulasi Ramakrishna Reddy Kotte [2], Chinho Park [1] and Woo Kyoung Kim [1,*]

[1] School of Chemical Engineering, Yeungnam University, 280 Daehak-ro, Gyeongsan 38541, Korea; drsrvi9@gmail.com (S.G.); chpark@ynu.ac.kr (C.P.)

[2] Department of Physics, Sri Venkateswasra University, Tirupati 517 502, India; ktrkreddy@gmail.com

* Correspondence: drmvasudr9@gmail.com (V.R.M.R.); wkim@ynu.ac.kr (W.K.K.)

Abstract: The rapid research progress in tin-based binary sulfides (Sn_xS_y = o-SnS, c-SnS, SnS_2, and Sn_2S_3) by the solution process has opened a new path not only for photovoltaics to generate clean energy at ultra-low costs but also for photocatalytic and thermoelectric applications. Fascinated by their prosperous developments, a fundamental understanding of the Sn_xS_y thin film growth with respect to the deposition parameters is necessary to enhance the film quality and device performance. Therefore, the present review article initially delivers all-inclusive information such as structural characteristics, optical characteristics, and electrical characteristics of Sn_xS_y. Next, an overview of the chemical bath deposition of Sn_xS_y thin films and the influence of each deposition parameter on the growth and physical properties of Sn_xS_y are interestingly outlined.

Keywords: o-SnS; c-SnS; SnS_2; Sn_2S_3; CBD; solar cells

Citation: Gedi, S.; Minnam Reddy, V.R.; Kotte, T.R.R.; Park, C.; Kim, W.K. Fundamental Aspects and Comprehensive Review on Physical Properties of Chemically Grown Tin-Based Binary Sulfides. *Nanomaterials* **2021**, *11*, 1955. https://doi.org/10.3390/nano11081955

Academic Editor: Vlad Andrei Antohe

Received: 20 June 2021
Accepted: 26 July 2021
Published: 29 July 2021

Publisher's Note: MDPI stays neutral with regard to jurisdictional claims in published maps and institutional affiliations.

1. Introduction

To make a significant contribution to the energy needs of society with low production cost, thin film photovoltaic (TFPV) technology has been developed. Currently, the CdTe/CdS and Cu(In,Ga)Se$_2$(CIGS)/CdS heterojunction TFPV technologies have received worldwide attention because these solar cells achieved record efficiencies of 22.1% [1] and 23.35% [2], respectively. However, their wider impact is hindered due to major concerns raised on the presence of harmful elements (Cd and Se) and scarcity of constituent elements (Te, In and Ga). In view of that, considerable efforts have been made to develop environmentally friendly absorbers and buffers that are free from the aforementioned toxic and inadequacy elements. Along this path, the tin-based binary sulfides (Sn_xS_y) such as tin monosulfide (orthorhombic (ORT)-SnS and cubic (CUB)-SnS), tin disulfide (SnS_2), and tin sesquisulfide (Sn_2S_3) have drawn much attention because these are abundant, inexpensive, and nontoxic [3]. According to the merits of Sn_xS_y (see Table 1), the o-SnS, c-SnS, and Sn_2S_3 are strongly expected as potential and alternative absorbers to the conventional CdTe and CIGS, and SnS_2 is expected as an appropriate and alternative buffer to the regular CdS. Furthermore, their simple composition and promising physical properties made them suitable for other applications such as photocatalytic, thermoelectric, etc. (see Figure 1). Among Sn_xS_y, o-SnS, SnS_2, and Sn_2S_3 occur naturally, whereas c-SnS was synthesized in the laboratory. The historical information and the applications of Sn_xS_y available in the literature are presented in Table 2.

The deposition of Sn_xS_y in thin film form became prominent owing to their wide applications. The selection of deposition technique along with the growth conditions is critical because the properties of Sn_xS_y thin films are susceptible to their growth method. Sn_xS_y films should be prepared by low-cost techniques such as solution processes to further reduce the production cost of TFPV devices. The preparation of Sn_xS_y thin films via chemical methods, especially by chemical bath deposition(CBD), includes a slightly low cost and is, in fact, unchallenging on the preparation side. In addition, CBD provides several experimental flexibilities such as non-vacuum thin-film deposition, a wide selection of various substrates, easy doping of elements, and room temperature film growth, which are suitable for the large-scale fabrication of flexible devices for industrial applications.

The deposition of Sn_xS_y thin films using CBD is relatively new, and their process–property relationships must be understood for the desired application. Further, the formation of single-phase o-SnS, c-SnS, SnS_2, and Sn_2S_3 thin films in CBD is highly dependent on preparative conditions. In addition, identification and separation of the o-SnS, c-SnS, SnS_2, and Sn_2S_3 phases are also critical criteria. However, there is a lack of comprehensive studies on the optimization of growth parameters until now. Therefore, extensive research studies on the growth, deposition mechanism, and preparative parameters that affect phase separation and the physical properties of Sn_xS_y thin films are crucial to a successful device design in the production of clean energy.

In this scenario, the present article provides an overview of the bulk properties of Sn_xS_y and a comprehensive review of the deposition, growth mechanism, and effect of growth parameters on the physical properties of Sn_xS_y thin films. According to the authors' knowledge, this is the first review of Sn_xS_y thin films by CBD.

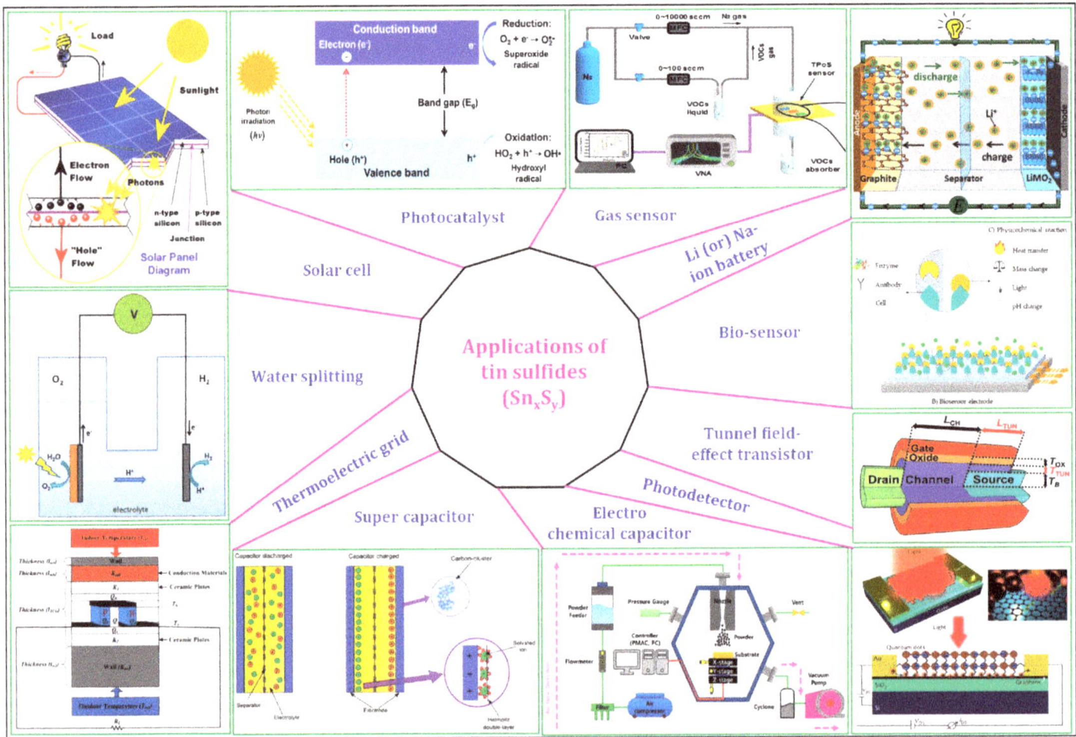

Figure 1. Applications of Sn_xS_y [4–12]. Solar cell (o-SnS, c-SnS, Sn_2S_3 as light absorber and SnS_2 as buffer); photodetector (o-SnS, c-SnS, Sn_2S_3 as light absorber and SnS_2 as buffer); Li- and Na-ion batteries (o-SnS, c-SnS, and SnS_2 as anode materials); gas- and bio sensors (o-SnS, c-SnS, and SnS_2 as sensing materials); tunnel field-effect transistors (TFET) (o-SnS, c-SnS, and SnS_2 as top or back gates); electrochemical and super capacitors (o-SnS, c-SnS, and SnS_2 as electrode materials); capacitor; thermoelectrics (o-SnS, c-SnS, and Sn_2S_3 as grids); and water-splitting (o-SnS, c-SnS, and SnS_2 as photocathodes).

Table 1. The advantages of Sn_xS_y compared to the conventional absorbers (CdTe, CIGS, and CZTS) and the buffer (CdS).

Characteristics	PV Absorbers						PV Buffers	
	CdTe	CIGS	CZTS	o-SnS	c-SnS	Sn_2S_3	CdS	SnS_2
Earth abundance	No	No	Yes	Yes	Yes	Yes	Yes	Yes
Eco-friendly	No	No	Yes	Yes	Yes	Yes	No	Yes
Band gap (eV)	1.45–1.5 eV [13]	1.1–1.5 [14]	1.0–1.5 [15]	1.16–1.79 [16–23]	1.64–1.75 [24–29]	0.95–2.03 [30–34]	2.35–2.50 [14,35]	2.04–3.30 [36–41]
Absorption coefficient	$>10^4$	10^5	$>10^4$	10^5	10^5	10^4	–	–
Conductivity type	p-type	p-type	p-type	p-/n-type	p-type	p-/n-type	n-type	n-type
Carrier density (cm^{-3})	10^{14}–10^{17} [35]	10^{12}–10^{18} [14]	10^{16}–10^{18} [42]	10^{11}–10^{18} [43–49]	10^{11}–10^{18} [29,50,51]	10^{14}–10^{16} [45,52,53]	10^{12}–10^{18} [14,35]	10^{13}–10^{17} [54–56]
Structure	Zinc blend [13]	Chalcopyrite [14]	Kesterite [57]	Orthorhombic [58,59]	Cubic [60]	Orthorhombic [51]	Hexagonal [35]	Hexagonal [61]
Maximum theoretical efficiency (%)	~29	~29	31 [62]	31 [63]	>25 [64]	–	–	–

Table 2. The historical information and the applications of Sn_xS_y.

Tin Sulfides	Mineral Form [65–67]	Appearance [68]	Other Names	Discovered/Reported [69,70]	Applications [24,71–91]
o-SnS	Herzenbergite	Black color with dark red–brown internal reflections.	Kolbeckine	Reported by Ramdohr from the Maria-Teresa mine (Oruro, Bolivia) in 1934.	PV, photodetectors [24], photocatalysts [71], water splitting [72], supercapacitors [83], field-effect transistors [85], sodium-ion and lithium-ion batteries [86,87], gas sensors [88], biosensors [89] thermoelectric [90], and electro chemical capacitors [91].
SnS_2	Berndtite	Pale yellow with intense brownish to yellow–orange internal reflections.	Mosaic gold	Discovered at the Stiepelmann mine in Arandis, Namibia, as described by Ramdohr in 1935.	PV, photocatalysts [73], water splitting [74], supercapacitors [75], field-effect transistors [76], lithium-ion and sodium-ion batteries [77,78], gas sensors [79], thin film diodes [80], and high-speed photodetectors [81].
Sn_2S_3	Ottemanite	Gray with orange–brown internal reflections.	-	Reported by Moh from the Cerro de Potosi mine (Bolivia) in 1964.	PV, optoelectronic [82], thermoelectric and IR detectors [84].

2. Physical Properties of o-SnS, c-SnS, SnS_2, and Sn_2S_3

The physical properties such as structural, optical, and electrical properties of Sn_xS_y can significantly influence the device's performance. The crystal structure of a material can influence its optical and electrical properties, which can affect the material-related device performance [92]. Understanding and obtaining knowledge on the electronic band structure and optical characteristics of Sn_xS_y is essential before using them for device applications because the main optical parameter, band gap energy (E_g), is very sensitive to the crystal structure and defects, which directly influences the performance of a PV device. Electrical characteristics such as conduction type, resistivity, carrier concentration, and mobility of Sn_xS_y play a key role in achieving high-performance photovoltaic devices. These electrical properties critically depend on the formation of defects in Sn_xS_y. Therefore, a good understanding of the physical properties is required for the development of effective devices.

2.1. Crystal Structure and Structural Characteristics

In o-SnS, c-SnS, SnS_2, and Sn_2S_3, Sn exhibits '2+' (divalent Sn(II)) or/and '4+' (tetravalent Sn(IV)) oxidation states (Table 3). The capability of Sn to elect different oxidation states is the origin of structural diversity/different structures (Figure 2a) of resulting compounds [93]. o-SnS (α-SnS, space group, Pnma-D_{16}^{2h}) and c-SnS (π-SnS, space group, P2$_1$3) are the two polymorphic forms of SnS, resulting from distortions in the crystal lattice depending on growth conditions. The o-SnS consists of a double layer of Sn and S atoms (two-dimensional SnS sheets) with zigzagged chains in which each Sn atom bonds to two S atoms in the b–c plane of the layer with a bond length of 2.671 Å and one additional S atom at a short bond length of 2.633 Å perpendicular to the plane (along a-axis) in the layer stack. The interlayer bond length of Sn-Sn atoms is 3.48 Å, and the distance between the layer stacks is 2.79 Å. The lone pair of $5s^2$ in the Sn atom occupies the fourth coordination site and weakens interaction along the b-axis. In the case of c-SnS, each Sn atom bonds with three nearest S atoms at 2.7 Å and forms a trigonal pyramidal environment, with the Sn-S bond at the trigonal base and the $5s^2$ lone pair pointing toward the apex. The stereo chemical activity of Sn(II) $5s^2$ lone pair creates a highly distorted internal structure of c-SnS. The local coordination in c-SnS is similar to o-SnS; however, a three-dimensional network is formed by a covalent bond [60]. Additionally, SnS also exhibits some of other polymorphs [94] such as, β-SnS (formed at T > 880 K), γ-SnS [95], δ-SnS [96,97], RS-SnS (rock salt-SnS, Fm$\bar{3}$m) [98]. One key point is that c-SnS and RS-SnS belong to the cubic crystal system. Further, c-SnS is a simple cubic lattice type, and RS-SnS is a face-centered cubic lattice (important note: the structure of c-SnS was incorrectly assigned as ZB-SnS previously in the literature. To avoid confusion in the literature, the readers should replace ZB-SnS with c-SnS in previous literature). More details can be found in the literature clarifying these assignments [28,93,99,100]. Furthermore, the weak interactions between distorted lone pair of $5s^2$ in Sn and neighboring S form the metastable SnS crystals or polymorphs of SnS [101].

Next, the SnS_2 adopts a layered hexagonal structure with P$\bar{3}$ml space group, similar to the structure of the CdI_2 system. In this structure, the layers are arranged in the b–c plane, which is perpendicular to the a-axis, and each layer is composed of S, Sn, and S atomic parallel monolayers. Each Sn atom forms bonds in an octahedral environment with six S atoms, similar in rutile SnO_2 structure [93]. It has a symmetric edge-sharing Sn(IV)S6 octahedral with 2D planes that are separated by weak van der Waals interaction between 3.6 Å distant S atoms [61]. Finally, the Sn_2S_3 exhibits an orthorhombic crystal structure similar to o-SnS with the same space group of Pnma. It contains tetravalent (4+) and divalent (2+) Sn atoms in equal proportions, and they form Sn(IV)S$_6$ octahedral 1D chains covered by Sn(II) tetrahedral [51]. The lone pair in Sn(II) occupies one coordination site as in the ground state form of o-SnS. These lone pairs are responsible for weak interchain interactions. The optimized theoretical lattice parameters along with experimental values for all these Sn_xS_y phases are presented in Table 3. The structural characterization of Sn_xS_y thin films is generally performed by X-ray diffraction (XRD). The standard XRD patterns of o-SnS, c-SnS, SnS_2, and Sn_2S_3 (Figure 2b) showed the highest intensity for peaks located at 2θ of: 31.53°, 26.63°/30.84°, 15.03°, and 21.49°, arising from diffraction from (111), (222)/(400), (001), and (130) planes, respectively [25,28,102–104].

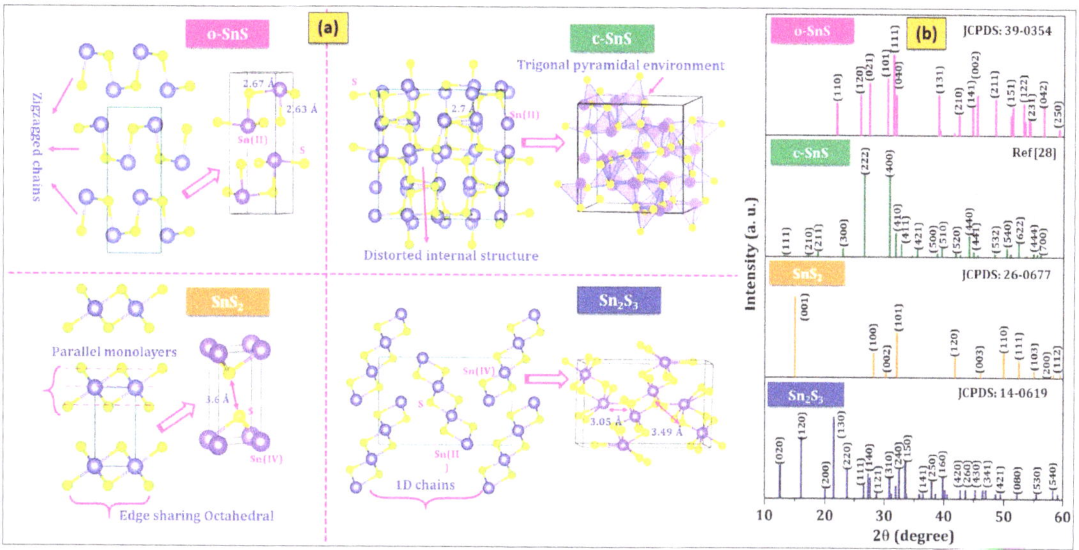

Figure 2. (**a**). Crystal structures of ground state Sn_xS_y forms (reprinted with permission [51] © 2017, Royal Society of Chemistry) and (**b**) standard powder diffraction patterns for Sn_xS_y.

The o-SnS, c-SnS, SnS_2, and Sn_2S_3 thin films have similarities in XRD patterns (Figure 3a). Thus, clear differentiation of one phase to another is difficult by using diffraction analysis alone. In this respect, it is preferable to identify these phases in pure form using a complementary method such as Raman spectroscopy because the Raman spectrum is sensitive to mainly crystal quality, structural symmetry, and strength of the chemical bond between atoms [105]. The o-SnS has 21 optical vibrational modes with the irreducible representation of $\Gamma = 4A_g + 2B_{1g} + 4B_{2g} + 2B_{3g} + 2A_u + 4B_{1u} + 2B_{2u} + 4B_{3u}$ [106,107]. In these phonons, two are inactive ($2A_u$), seven are infrared-active ($3B_{1u}$, $3B_{3u}$, and $1B_{2u}$), and twelve are Raman-active ($4A_g$, $2B_{1g}$, $4B_{2g}$, and $2B_{3g}$). In the case of c-SnS, there are 189 optic branches, and they can be reduced to 3 in the form of $\Gamma = 16A + 16E + 47T$ [51]. Next, the SnS_2 has six vibrational modes with the irreducible representation, $\Gamma = A_{1g} + E_g + 2A_{2u} + 2E_u$ [108]. The six optic modes are divided into two Raman-active modes (A_{1g} and E_g), two infrared-active (A_{2u} and E_u), and two acoustic modes (A_{2u} and E_u). Additionally, the Sn_2S_3 has 57 optic modes with reduced form of $\Gamma = 10A_g + 5A_u + 5B_{1g} + 9B_{1u} + 10B_{2g} + 4B_{2u} + 5B_{3g} + 9B_{3u}$ [51,109]. The simulated Raman spectra [51] (Figure 3b) clearly showed the significant differences in frequencies and spectral intensities because the o-SnS, c-SnS, SnS_2, and Sn_2S_3 phases have differences in structure and bonding. The Raman spectrum of the o-SnS showed three prominent peaks at 160 cm^{-1} (narrow mode, B_{2g}), 189 cm^{-1} (highest intensity mode, A_g), and 220 cm^{-1} (narrow mode, A_g). The spectrum also showed a weak A_g mode at approximately 92 cm^{-1}, which has a narrow line width up to room temperature. In the case of c-SnS, there are three strong A phonon modes at 174 cm^{-1}, 187 cm^{-1}, and 202 cm^{-1} and two prominent E modes at 166 cm^{-1} and 183 cm^{-1} along with a group of weak modes in between the ranges of 50–125 cm^{-1} and 200–250 cm^{-1}. Next, the SnS_2 showed a single and strong mode at 305 cm^{-1} (A_g), which has a constant line width with the temperature. Furthermore, the Sn_2S_3 showed a significantly high-intensity A_g mode at 291 cm^{-1} and a moderate-intensity A_g mode at 300 cm^{-1} with a narrow line width. Its spectrum also showed weak modes at 182 cm^{-1}, 210 cm^{-1}, 226 cm^{-1}, 244 cm^{-1}, and 252 cm^{-1}, which are observed at high temperatures. Notably, the Raman mode of Sn_2S_3 (307 cm^{-1}) overlaps marginally with the active mode of SnS_2 (310 cm^{-1}). However, the phase can be easily identified based on the band width of modes. Sn_2S_3 has a band width that is significantly greater than that of SnS_2 (Figure 3b). From the experimental Raman spectra (Figure 3c), o-SnS, c-SnS, SnS_2, and Sn_2S_3 films showed Raman active modes at 93 cm^{-1}, 161 cm^{-1}, 192 cm^{-1}, and 218 cm^{-1} [110]; 59 cm^{-1}, 71 cm^{-1}, 90 cm^{-1}, 112 cm^{-1}, 123 cm^{-1}, 176 cm^{-1}, 192 cm^{-1}, 202 cm^{-1}, and 202 cm^{-1} [111]; 224 cm^{-1} and 310 cm^{-1} [112]; and 61 cm^{-1}, 91 cm^{-1}, 179 cm^{-1}, 220 cm^{-1}, and 307 cm^{-1} [113], respectively, which matched well with theoretically calculated data. The Raman results suggested that o-SnS, c-SnS, SnS_2, and Sn_2S_3 phases have distinct modes. Moreover, S-rich impurity phases can easily be found in SnS due to the sharp Raman mode of SnS_2 (310 cm^{-1}).

2.2. Electronic Band Structure and Optical Characteristics

According to the electronic band structures of Sn_xS_y (Figure 3d), the valence band maxima (VBM) of o-SnS, c-SnS, and Sn_2S_3 are formed mostly of S 3p and Sn 5s hybrid states with a tiny contribution from Sn 5p states, whereas SnS_2 is primarily composed of S 3p orbitals. The conduction band minimum (CBM) of SnS_2 and Sn_2S_3 is formed by the Sn 5s bands, whereas SnS is mainly composed of Sn 5p orbitals. The VBM of SnS_2 is lower than those of o-SnS, c-SnS, and Sn_2S_3; however, the CBMs of all o-SnS, c-SnS, SnS_2, and Sn_2S_3 are almost aligned. The partial hybridizations with S 3s and Sn 5p states result in SnS polymorphs due to the change in density of state.

On the other hand, the band-edge positions deviated from the special points in the reciprocal space except for the CBM of Sn_2S_3. However, they fall between the Brillouin zone center and the zone boundaries [64,114]. From the ab initio band-structure calculations and Kohn–Sham density-functional theory, o-SnS and c-SnS exhibited the indirect and direct energy gaps of 1.6 eV and 1.8 eV; and 1.72 eV and 1.74 eV, respectively, which is due to an inherent error in calculating band structure [58]. The energy difference between direct and indirect band gaps is small; thus, the change in the nature of the band gap could be due to the effect of temperature through thermal expansion and electron-phonon coupling [64]. According to the band-structure calculations from the Hartree−Fock exchange HSE06 functional technique, o-SnS, SnS_2, and Sn_2S_3 showed indirect energy gaps of 1.11 eV, 2.24 eV, and 1.09 eV, respectively [115]. The optical characterization of Sn_xS_y thin films is generally studied by a UV-Vis-NIR spectrometer. Although theoretical calculations showed the indirect band gap energy of Sn_xS_y, most of the experimental studies (Figure 3e) proved that Sn_xS_y thin films have direct band gap energies, with the following ranges (Tables 4 and 5): 1.16–1.79 eV; 1.64–1.75 eV; 2.04–3.30 eV; 0.95–2.03 eV; for o-SnS, c-SnS, SnS_2, and Sn_2S_3, respectively. The band gap energies of these phases depend on various factors such as strain, sulfur impurities, and Sn vacancies [116–122].

2.3. Conduction Type and Electrical Characteristics

In Sn_xS_y, three types of defects, namely, (i) Sn and S vacancies (V_{Sn} and V_S), (ii) Sn and S interstitials (Sn_i and S_i), and (iii) Sn on S antisites (Sn_S) and S on Sn antisites (S_{Sn}) are commonly formed, as shown in Figure 4a. According to the defect energy concepts (Figure 4b), the formation energy of vacancies ($V_{Sn(II)\ or\ Sn(IV)}$, Vs) depends on the coordination number, i.e., it generally increases with increasing coordination number. Thus, $V_{Sn(II)}$ has lower formation energy compared to $V_{Sn(IV)}$ because the coordination number is three for Sn(II) and six for Sn(IV). As a result, $V_{Sn(II)}$ becomes a major defect that acts as an acceptor and contributes to the p-type conducting nature to SnS. In SnS, the primary defects are V_{Sn} and V_S, whereas Sn_i and S_i have higher energies. SnS exhibits the p-type at the Sn-poor condition, whereas the n-type at the Sn-rich condition. The defect-formation energies in SnS_2 differ from those in SnS. The major defects are V_S, S_i, and $S_{Sn(II)}$, which are inert to carrier generation. The defect-formation energies in Sn_2S_3 are typically interpreted as a mixture of those in SnS and SnS_2. On the other hand, the formation energy of interstitials (Sn_i, S_i) associates with the gap of interlayer free spaces, and that gap follows the notation of SnS > Sn_2S_3 > SnS_2. Sn_i always prefers to locate at the center of the gaps, whereas S_i likes to make a covalent bond with neighboring S atoms. The tin interstitial, Sn_i in both SnS_2 and Sn_2S_3, has lower formation energy compared to SnS, and it acts as a deep donor in SnS_2 and Sn_2S_3 and contributes to an n-type conductivity. All the antisites in Sn_xS_y have higher formation energies because they are correlated with chemical bonds. Therefore, these defects do not play a major role in the conduction type of Sn_xS_y phases. The electrical characterization of Sn_xS_y thin films is generally performed by the popular van der Pauw–Hall method.

From the reported electrical parameters of Sn_xS_y films (Table 3), the o-SnS has a hole density in the order of 10^{11}–10^{18} cm^{-3}, hole mobility in the range of 4–500 cm^2 V^{-1} s^{-1}, and electrical resistivity in the range of 13–10^5 Ω cm. In contrast, the c-SnS has a hole density in the order of 10^{11}–10^{18} cm^{-3}, hole mobility in the range of 10^{-2}–78 cm^2 V^{-1} s^{-1}, and electrical resistivity in the range of 70–10^7 Ω cm. In the case of the SnS_2, it has a carrier concentration of the order of 10^{13}–10^{17} cm^{-3}, electron mobility in the range of 15–52 cm^2 V^{-1} s^{-1}, and electrical resistivity in the range of 1.11–10^7 Ω cm, whereas the Sn_2S_3 has a carrier density in the order of 10^{14}–10^{16} cm^{-3} and resistivity in the range of 0.4–10^5 Ω cm, and a very little information related to Sn_2S_3 carrier mobility value is available in the literature. The reported variation in electrical parameters is expected due to the differences in the growth process and chemical composition.

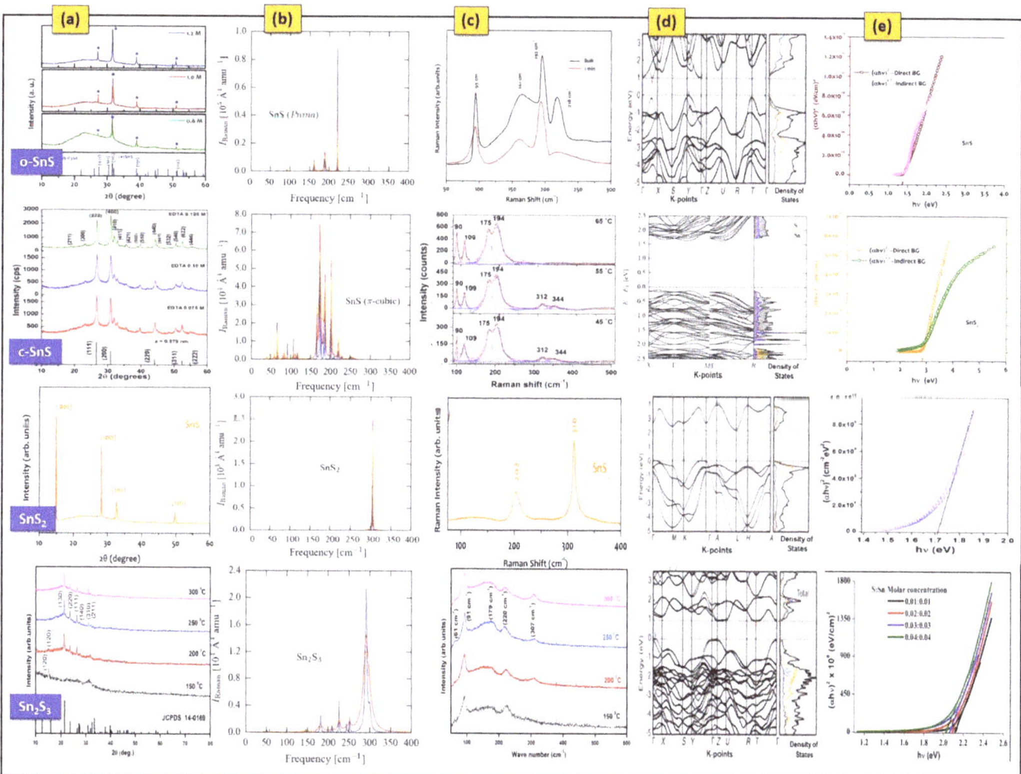

Figure 3. (a) XRD profiles of o-SnS and c-SnS (reprinted with permission [26]. © 2016, Elsevier), hexagonal SnS_2 (reprinted with permission [123]. © 2016, Elsevier), and orthorhombic Sn_2S_3 (reprinted with permission [113]. © 2016, Elsevier). (b) Simulated Raman spectra for Sn_xS_y at different temperatures of 10 K, 150 K, and 300 K (reprinted with permission [51]. © 2017, Royal Society of Chemistry). (c) Experimental Raman spectra of o-SnS (reprinted with permission [110]. © 2017, Elsevier), c-SnS (reprinted with permission [26]. © 2016, Elsevier), hexagonal SnS_2 (reprinted with permission [123]. © 2016, Elsevier), and orthorhombic Sn_2S_3 (reprinted with permission [113]. © 2016, Elsevier). (d) Band structures of Sn_xS_y (reprinted with permission [114]. © 016, American Physical Society), and (e) bandgap estimation of o-SnS (reprinted with permission [123]. © 2016, Elsevier), c-SnS (reprinted with permission [26]. © 2016, Elsevier), hexagonal SnS_2 (reprinted with permission [123]. © 2016, Elsevier), and orthorhombic Sn_2S_3 (reprinted with permission [124]. © 2016, Sciendo).

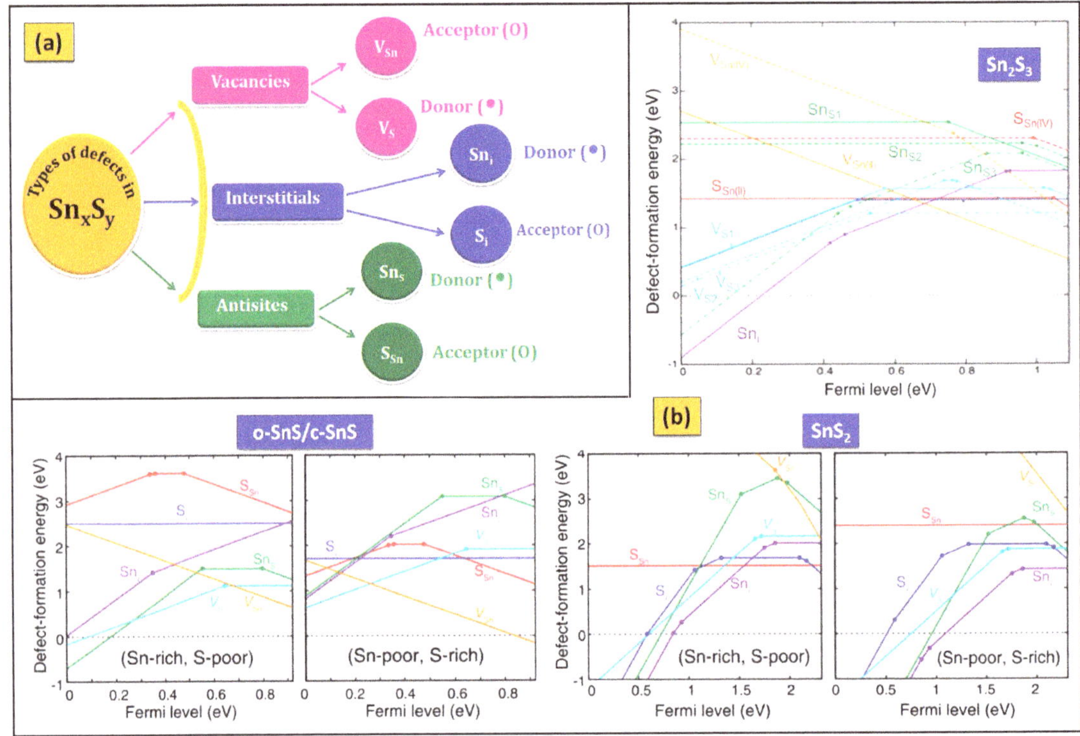

Figure 4. (**a**) Possibility of different defect formation in Sn_xS_y, and (**b**) defect formation energies in Sn_xS_y as a function of Fermi energy under Sn-rich (S-poor) and Sn-poor (S-rich) conditions (reprinted with permission [114]. © 2016, American Physical Society).

Table 3. Optimized theoretical and experimental lattice parameters, reported optical bandgaps, and electrical parameters of Sn_xS_y (Th: theoretical, Exp: experimental).

			Structural Properties			Optical Properties	Electrical Properties		
				Parameters of unit cell					
					Intercepts a (Å), b (Å), c (Å)				
Sn_xS_y Phase	Structure (Space Group)	Oxidation State of Sn	Angles and Rule	Theoretical [51]	Experimental [111,125,126]	Optical Band Gap (eV)	Carrier Concentration (cm^{-3})	Mobility ($cm^2V^{-1}s^{-1}$)	Resistivity (Ω-cm)
o-SnS	Orthorhombic (Pnma)	2+	$\alpha = \beta = \gamma$ $= 90°$ $a \neq b \neq c$	4.251, 11.082, 3.978	4.33, 11.18, 3.98	1.16 [16], 1.30 [17], 1.32 [18], 1.35 [19], 1.42 [20], 1.43 [20], 1.48 [21], 1.70 [22], 1.79 [23].	1×10^{11} [43], 3.6×10^{12} [44], $(1–1.2) \times 10^{15}$ [21,45], 1.5×10^{16} [46], $(1–1.16) \times 10^{17}$ [47,48], $(1–3) \times 10^{18}$ [49].	3.7 [20], 15.3 [46], 90 [43,49], 228 [44], 385 [19], 400–500 [17].	12.98 [17], 14.49 [20], 30 [16], 33.33 [18], 0.63×10^3 [43], 2.1×10^4 [44], $(0.16–0.25) \times 10^5$ [127,128].
c-SnS	Cubic (P2$_1$3)	2+	$\alpha = \beta = \gamma$ $= 90°$ $a = b = c$	11.506	11.603	1.64 [24], 1.66 [25], 1.67 [26], 1.73 [27], 1.74 [28], 1.75 [29].	5.87×10^{11} [29], 7.93×10^{12} [50], 6×10^{18} [51].	1.47×10^{-2} [51], 75 [50], 77.7 [29].	70 [51], 1×10^4 [50], 1.37×10^5 [29], 1×10^6 [25], 1×10^7 [28].
SnS$_2$	Hexagonal (P3m1)	4+	$\alpha = \beta =$ $90°; \gamma =$ $120°$ $a = b \neq c$	3.651, 3.651, 6.015	3.638, 3.638, 5.880	2.04 [36], 2.12 [39], 2.14 [129], 2.18 [37], 2.30 [38], 2.35 [130], 2.40 [131], 2.41 [39], 2.44 [132], 2.45 [133], 2.50 [134], 2.67 [135], 2.75 [136], 2.80 [56], 3.08 [40], 3.30 [41].	1×10^{13} [54], 2×10^{17} [55], 6.8×10^{17} [56].	15 [54], 48 [56], 51.5 [55].	1.11 [55], 11.2 [56], 0.77×10^2 [137], 0.42×10^5 [54], 0.26×10^7 [138].
Sn$_2$S$_3$	Orthorhombic (Pnma)	2+ and 4+	$\alpha = \beta = \gamma$ $= 90°$ $a \neq b \neq c$	8.11, 3.76, 13.83	8.878, 3.751, 14.020	0.95 [30], 1.16 [31] 1.2 [139], 1.65 [32], 1.9 [33], 1.96 [140], 2.0 [141], 2.03 [34].	9.4×10^{14} [52] 1×10^{15} [45], 4.0×10^{16} [53].	20.5 [53]	0.359 [124], 7.57 [53], 0.66×10^2 [45], $(0.22–0.36) \times 10^3$ [52,141], $(0.4–2.5) \times 10^5$ [137].

3. Influence of Deposition Parameters on Sn_xS_y Thin Film Growth and Properties

In CBD, the selection of tin source precursor, sulfur source precursor, complexing agent, and their concentrations is crucial to prepare the high-quality Sn_xS_y thin films using CBD. Moreover, the selection of suitable activation conditions such as solution/bath temperature, solution/bath pH (acidic or basic medium), deposition time, and stirring speed is also important [36,142] because they significantly affect the phase formation, growth, and properties of Sn_xS_y films. In addition to the above parameters, the nature of the substrate and its cleaning procedure also affect the phase formation, growth, and properties of Sn_xS_y films. Therefore, the understanding of the influence of all those parameters on the growth process of Sn_xS_y films and their physical properties is necessary to deposit the quality films for device applications. In Tables 4 and 5, the deposition parameters used for different thin films of tin sulfides made from chemical methods were summarized.

3.1. Overview of CBD Process of Sn_xS_y Thin Films

The CBD refers to "a typical synthesis employing mild conditions [143]". As schematically illustrated in Figure 5a, the experimental setup of CBD consists of the following parts: (i) magnetic stirrer with thermostat (to stir the mixed reactant solution continuously), (ii) oil bath (to maintain the desired temperature), (iii) substrate holder (to keep the substrates stable), (iv) stock chemical solutions to compose the reaction bath (mixture of different reagent solutions and its level always remains below the outer oil level), and (v) cleaned substrates [144]. The deposition of o-SnS, c-SnS, SnS_2, and Sn_2S_3 by CBD was reported in 1987 [145], 2006 [146], 1990 [36,130], and 2012 [34], respectively. Sn_xS_y films were deposited using various Sn precursors such as tin (II) chloride dihydrate ($SnCl_2 \cdot 2H_2O$), tin(IV) chloride pentahydrate ($SnCl_4 \cdot 5H_2O$), and tin ingots; various S precursors such as sodium sulfide (Na_2S), ammonium sulfide ($(NH_4)_2S$), sodium thiosulfate ($Na_2S_2O_3$), thioacetamide (C_2H_5NS), and thiourea (CH_4N_2S); and various complexing agents such as triethanolamine ($C_6H_{15}NO_3$), ammonia (NH_3)/ammonium hydroxide (NH_4OH), ammonium fluoride (NH_4F), ammonium citrate ($C_6H_{17}N_3O_7$), trisodium citrate ($Na_3C_6H_5O_7$), citric acid ($C_6H_8O_7$), tartaric acid ($C_4H_6O_6$), ethylenediaminetetraacetic acid ($C_{10}H_{16}N_2O_8$), and disodium ethylenediaminetetraacetate ($C_{10}H_{14}N_2Na_2O_8$). Among the above-mentioned chemicals, tin (II) chloride, thioacetamide, and triethanolamine, along with ammonia, were widely used as Sn precursor, S precursor, and complexing agents, respectively (Figure 5b). Other types of Sn precursors such as tin ingots [130,147] and tin (IV) chloride [131] were employed to deposit the SnS_2 films. Except for the above reports, tin (II) chloride was used as an Sn source. In the case of S precursors, sodium thiosulfate was used as a second alternative to the regularly used thioacetamide.

The preparation of Sn_xS_y thin films by CBD occurs when a substrate is immersed in the solution mixture of Sn ion (Sn^{2+} or Sn^{4+})-source, S ion (S^{2-})-source, and an appropriate complexing agent. In the deposition process, the Sn^{2+}/Sn^{4+} ions are complexed through the coordinated bond formation by the complexing agent, which controls the rate of reaction [148]. At super saturation condition (Ionic product, Q_{ip} > Solubility product K_{sp}), Sn_xS_y films can be deposited (Figure 5c). However, simply maintaining supersaturation condition in the bath will not provide acceptable quality Sn_xS_y films; managing the solubility product of tin hydroxides is required because when an Sn precursor is dissolved in water, it rapidly binds with hydroxide ions, creating $Sn(OH)_2$ and $Sn(OH)_4$. The differences in K_{sp} values between SnS (1×10^{-25} mol^2 dm^{-6}), irrespective of the polymorphs, and SnS_2 (1×10^{-46} mol^3 dm^{-9}) are very close to those between their hydroxides ($Sn(OH)_2$ (1×10^{-28} mol^3 dm^{-9}) and $Sn(OH)_4$ (1×10^{-56} mol^5 dm^{-15})). Therefore, it is vital to monitor supersaturation with respect to an individual phase as well as the growth kinetics. In addition, the K_{sp} of Sn_xS_y is affected by the concentration of precursor, solvent type, bath temperature, and bath pH [148,149]. Therefore, the optimum condition for the deposition of Sn_xS_y thin film can be achieved by manipulating the above deposition parameters.

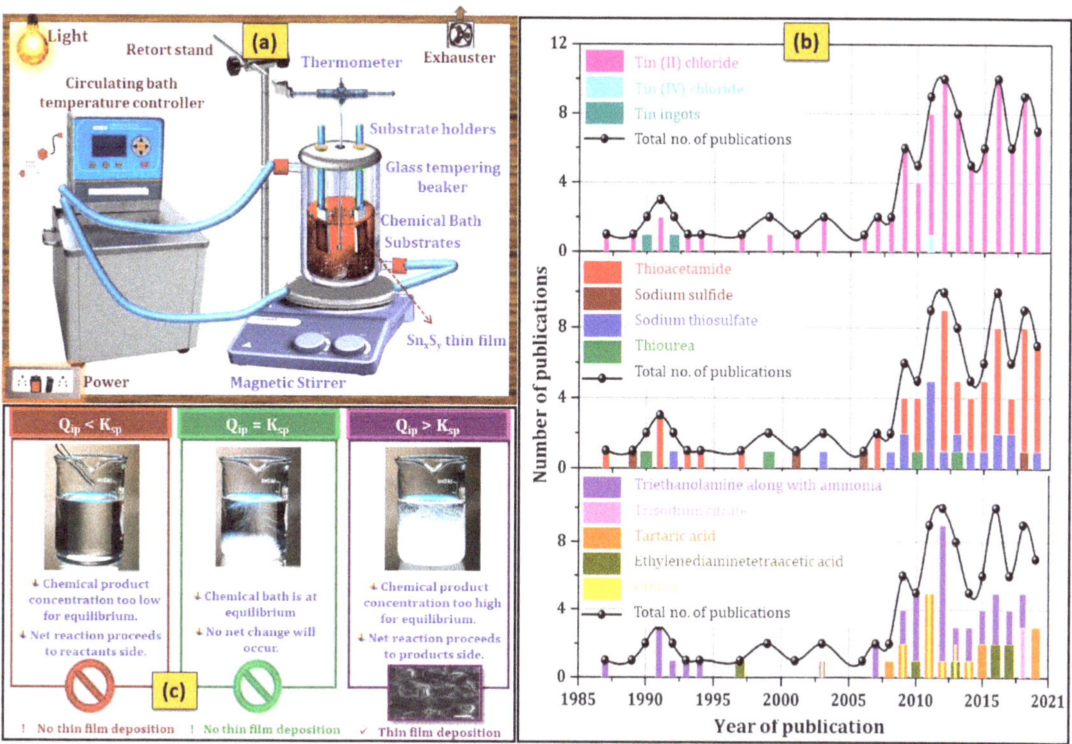

Figure 5. Schematic representation of (**a**) CBD, (**b**) main chemical reagents used for the preparation of Sn_xS_y thin films from 1987 to the present, and (**c**) importance of K_{sp} and Q_{ip} relation on the films by CBD.

According to previous reports, the formation of o-SnS, c-SnS, SnS_2, and Sn_2S_3 thin films is achieved through either an ion-by-ion mechanism (Figure 6a) or a simple cluster (hydroxide) mechanism (Figure 6b) based on the reaction process and parameters maintained in the bath [150]. The formation reaction of Sn_xS_y thin films through the ion-by-ion mechanism and cluster (hydroxide) mechanism is as follows:

Ion-by-ion mechanism:

$$xSn^{p+} + yS^{q-} \rightarrow Sn_xS_y$$

$$[\because p = 2 \text{ or } 4, q = 2; x = 1 \text{ or } 2, y = 1, 2, \text{ or } 3; Sn_xS_y = SnS, SnS_2, \text{ or } Sn_2S_3]$$

Cluster (hydroxide) mechanism:

$$xSn^{p+} + y(OH)^{q-} \rightarrow Sn(OH)_n$$

$$Sn(OH)_n + yS^{q-} \rightarrow Sn_xS_y + n(OH)$$

$$[\because p = 2 \text{ or } 4, q = 2; n = 2 \text{ or } 4; x = 1 \text{ or } 2, y = 1, 2, \text{ or } 3; Sn_xS_y = SnS, SnS_2, \text{ or } Sn_2S_3]$$

The physical properties of the CBD-deposited o-SnS, c-SnS, SnS_2, and Sn_2S_3 thin films can be affected by the growth mechanism, level of supersaturation, and surface energy of the complexing agents [17]. Therefore, it is crucial to understand the actual mechanism undertaken in the solution for tuning the properties of the deposited films. Moreover, in the process of o-SnS, c-SnS, SnS_2, and Sn_2S_3 thin film deposition, controlling the reaction to reduce or remove the spontaneous precipitation is essential, which can only be achieved by complexing the tin ions using an appropriate complexing agent (L). The kinetics of o-SnS, c-SnS, SnS_2, and Sn_2S_3 thin film formation can be comprehended through the following reactions.

The complexing reactions in an aqueous Sn precursor solution are as follows:

$$SnCl_2 \cdot 2H_2O + L \Leftrightarrow Sn(L)^{2+} + 2Cl^- + 2H_2O$$

$$SnCl_4 \cdot 5H_2O + L \Leftrightarrow Sn(L)^{4+} + 4Cl^- + 5H_2O$$

The free Sn^{2+}/Sn^{4+} ions are slowly released by the tin complex in a controlled way. As the tin complex dissociates, then

$$Sn(L)^{2+} \rightleftharpoons Sn^{2+} + L$$

$$Sn(L)^{4+} \rightleftharpoons Sn^{4+} + L$$

Here, the concentration of complex tin ions in the solution, $Sn(L)^{2+}$ or $Sn(L)^{4+}$, can be controlled by adjusting the concentration of the complexing agent and bath temperature [151]. If these ions can be generated, then the deposition of Sn_xS_y thin films can be achieved. On the other hand, controlling the reaction by a slow and uniform generation of sulfur ions in the solution is also a significant factor when thin films are deposited. Thioacetamide (C_2H_5NS) is one of the most frequently employed S precursors. The hydrolysis of the S precursor can produce H_2S and then S^{2-} ions by the following reactions [152]:

$$CH_3CSNH_2 + H_2O \rightleftharpoons CH_3CONH_2 + H_2S$$

When the reaction attains an equilibrium [153], the following reactions are expected at a temperature of 25 °C:

$$H_2S + H_2O \rightleftharpoons H_3O^+ + HS^- \; (K_0 = 10^{-7})$$

$$HS^- \rightleftharpoons H^+ + S^{2-} \; (K_1 = 10^{-17})$$

$$HS^- + OH^- \rightleftharpoons H_2O + S^{2-} \; (K_2 = 10^{-3})$$

At low pH values (<2.5), the reaction is controlled by the rate of hydrolysis of S precursor leading to the formation of hydrogen sulfide (H_2S), whereas at higher pH values (>2.5), the reaction is controlled by the formation and decomposition of the tin–thioacetamide complex. Therefore, the pH of the bath and metal–thioacetamide complexes are also considered as the growth rate- and growth mechanism-determining components in the film formation [154].

The Sn ions react with the S ions and initiate the formation of tin sulfides (o-SnS, c-SnS, SnS_2, and Sn_2S_3). The generation rate of Sn and S ions is controlled primarily by the source concentration, pH, and solution temperature. When the precursor concentration is changed, multiphase or other single-phase films can be formed by the following reactions [155,156]:

$$Sn^{2+} + S^{2-} \rightarrow SnS \; (K_{sp} = 1 \times 10^{-25} \text{ at } 25\,°C)$$

$$Sn^{4+} + 2S^{2-} \rightarrow SnS_2 \; (K_{sp} = 1 \times 10^{-46} \text{ at } 25\,°C)$$

$$Sn^{2+} + Sn^{4+} + 3S^{2-} \rightarrow Sn_2S_3$$

$$Sn(II)S + Sn(I\,V)S_2 \rightarrow Sn(II)(IV)_2S_3 \text{ (or) } SnS + SnS_2 \rightarrow Sn_2S_3$$

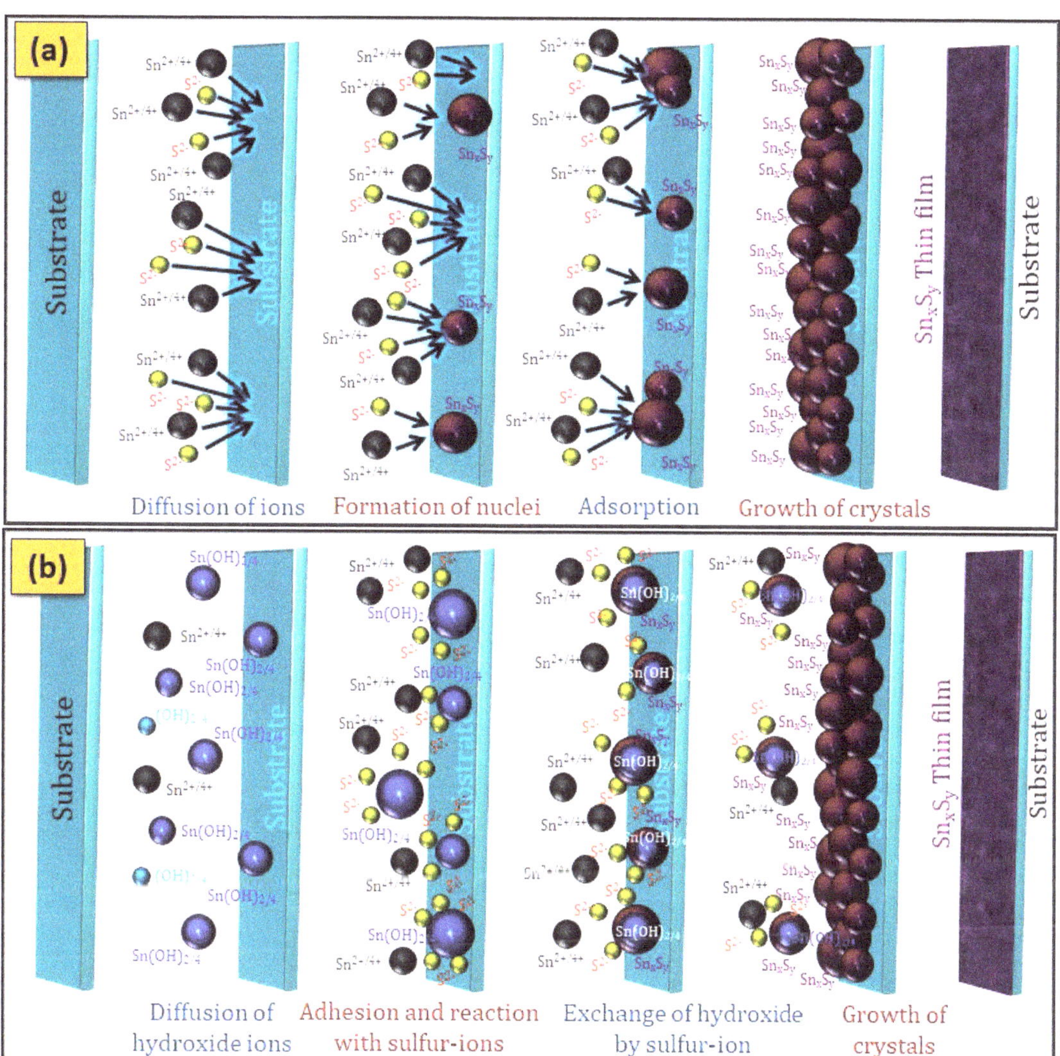

Figure 6. Schematic representation of (**a**) ion-by-ion mechanism and (**b**) cluster-by-cluster (hydroxide) mechanism.

Table 4. Deposition conditions and the physical properties of ORT- and c-SnS films grown by CBD.

Sn_xS_y Phase	Precursors	Complexing Gent	Deposition Parameters	Structure	Band Gap (eV)	Electrical Parameters				Ref
						Type	R (Ωcm)	μ ($cm^2V^{-1}S^{-1}$)	N (cm^{-3})	
				o-SnS						
o-SnS	T(II)C = 0.1 M TA = 0.1 M	TEA = 15 mL NH_3 = 8 mL	T_b = 27 °C t_d = 20 h pH = 10.5 ± 1	Amorphous	1.51 (i)	n	–	–	–	1987 [145]
o-SnS	T(II)C = 0.025 mol SDS/AS = 0.025 mol	–	T_b = – t_d = – pH = 3,10,12	ORT(013)	1.08	p	10^7–10^3	–	–	1989 [36]
o-SnS	T(II)C = – TU = –	–	T_b = – t_d = – pH = –	Polycrystalline	1.3 (i)	–	–	–	–	1990 [157]
o-SnS	T(II)C = 1 g TA = 1 M	TEA = 12 mL NH_3 = 10 mL	T_b = 75 °C, 25 °C t_d = 5 h, 40 h pH = –	Polycrystalline	1.3	p	–	–	–	1991 [128]
o-SnS	T(II)C = 2 mL TA = 1 mol L^{-1}	TEA = 0.5 mL NH_3 = –	T_b = 50 °C, 25 °C t_d = 2–4 h, 5–10 h pH = –	Crystalline	–	–	–	–	–	1991 [158]
o-SnS	T(II)C = 1 g TA = 4 mL, 8 mL	TEA = 12 mL NH_3 = 12 mL	T_b = 75 °C, 25 °C t_d = 5 h, 40 h pH = –	–	–	–	–	–	–	1991 [159]
o-SnS	T(II)C = 1 g TA = 1 M	TEA = 12 mL NH_3 = 10 mL	T_b = 60 °C t_d = 7 h 30 min pH = –	ORT(111)	–	–	–	–	–	1992 [160]
o-SnS	T(II)C = 1 g TA = 1 M	TEA = 12 mL NH3 = 10 mL	T_b = 60 °C t_d = 7 h 30 min pH = 9.5	ORT(111)	–	p	–	–	–	1993 [161]
o-SnS	T(II)C = 1 g TA = 1 M	TEA = 12 mL NH_3 = 13 mL	T_b = 50–75 °C, t_d = 1.5 h, 20 h pH = –	ORT(111)	–	p	–	–	–	1994 [162]

Table 4. *Cont.*

Sn_xS_y Phase	Precursors	Complexing Gent	Deposition Parameters	Structure	Band Gap (eV)	Type	R (Ωcm)	μ ($cm^2V^{-1}S^{-1}$)	N (cm^{-3})	Ref
			o-SnS							
o-SnS	$T(II)C = 15$ g $TU = 5$ g, 10 g	–	$T_b = -$ $t_d = 5$ min $pH = 3$ $S_p = .33$ mm/s	ORT(040)	1.4	–	–	–	–	1999 [163]
o-SnS	$T(II)C = 0.1$ M $SDS = 0.05$ M	–	$T_b = 80\,°C$ $t_d = -$ $pH = 12$	–	–	p	–	–	–	2001 [164]
o-SnS	$T(II)C = 1.125$ g $ST = 2$ M	AF	$T_b = T_r$ $t_d = 18$ h $pH = 7$	ORT(111)	1.38 (d) 0.96–1.14 (i)	–	–	–	–	2003 [165]
o-SnS	$T(II)C = 1$ g $TA = 8$ mL	TEA = 12 mL NH3 = 10 mL	$T_b = 35\,°C$ $t_d = 15$ h $pH = 9.5$	ORT(111)	1.18 (d)	p	10^7–10^4	–	–	2003 [166]
o-SnS	$T(II)C = 0.56$ g $SS = 0.025$ M	–	$T_b = 80\,°C$ $t_c = -$ $pH = 12$	ORT(111)	–	–	–	–	–	2006 [167]
o-SnS	$T(II)C = 1.13$ g $TA = 0.1$	TEA = 30 mL $NH_3 = 16$ mL	$T_b = RT\ 293–298$ K $t_d = 5$–6 h $pH = -$	ORT(111)	1.17 (d) 1.12 (i)	–	10^8–10^6	–	–	2007 [168]
	$T(II)C = 1$ g $TA = 1$ M	TEA = 12 mL $NH_3 = 10$ mL	$T_b = 308$ K $t_d = 20$ h $pH = -$							
o-SnS	$T(II)C = 1$ g $TA = 1$ M	TEA = 10 mL $NH_3 = 5$ mL	$T_b = 45\,°C$ $t_d = -$ $pH = -$	ORT(111)	1.33–1.39 (d)	–	–	–	–	2007 [169]
o-SnS	$T(II)C = 1$ g $TA = 8$ mL	TEA = 12 mL $NH_3 = 10$ mL	$T_b = 55\,°C$ $t_d = 8$ h $pH = -$	ORT(111)	–	p	10^3	90	10^{11}	2008 [43]

Table 4. *Cont.*

Sn_xS_y Phase	Precursors	Complexing Gent	Deposition Parameters	Structure	Band Gap (eV)	Electrical Parameters				Ref
						Type	R (Ωcm)	μ (cm^{2}V^{-1}S^{-1})	N (cm^{-3})	
o-SnS										
o-SnS	T(II)C = 1.12 g ST = 0.5 M	TTA = 10 mL	$T_b = T_r$ $t_d = 24$ h pH = 7	ORT(111)	1.1 (d)	–	10^6	–	–	2008 [170]
o-SnS	T(II)C = 1 g TA = 8 mL	TEA = 12 mL NH_3 = 10 mL	$T_b = 313$ K $t_d = 8$–22 h pH = –	ORT(111)	1.2–1.7 (d)	p	–	–	–	2009 [171]
o-SnS	T(II)C = 0.15 M ST = 2 M	NH_4OH = 6 mL	$T_b = 30$ °C $t_d = 24$ h pH = 7	ORT(040)/(141)	1.31 (d)	–	–	–	–	2009 [172]
o-SnS	T(II)C = 2 × 10^{-2} M TA = 1 × 10^{-2}–8 × 10^{-2} M	–	$T_b = 80$ °C $t_d = 60$ min pH = 1.87	Amorphous	–	–	–	–	–	2009 [142]
o-SnS	T(II)C = 1 g TA = 8 mL	TEA = 12 mL NH_3 = 10 mL	$T_b = 55$ °C $t_d = 8$ h pH = –	ORT(111)	1.12 (i)	–	–	–	–	2009 [173]
o-SnS	T(II)C = 0.15 M ST = 2 M	AH = 6 mL	$T_b = T_r$ $t_d = 24$ h pH = 7	ORT(111)	–	–	–	–	–	2009 [174]
o-SnS	T(II)C = – TA = –	TEA = – NH_3 = –	$T_b = 75$ °C $t_d = $ – pH = –	ORT(111)	0.82–1.22 (i)	–	–	–	–	2009 [175]
o-SnS	T(II)C TA = 0.1 M	TEA = 30 mL NH_4OH = 16 mL	$T_b = $ – $t_d = 5$ h pH = –	ORT(111)/(040)	1.76 (i)	–	–	–	–	2010 [176]
o-SnS	T(II)C = 1 M TA = 1 M	TEA = 10 mL TSS = 5 mL NH_3/NH_4Cl = 5 mL	$T_b = 60$ °C $t_d = 2$–10 h pH = 9.31	ORT(111)/(040)	1.30–1.97 (d) 0.83–1.36 (i)	p	9.9–12.3	–	–	2010 [177]

Table 4. *Cont.*

Sn_xS_y Phase	Precursors	Complexing Gent	Deposition Parameters	Structure	Band Gap (eV)	Electrical Parameters				Ref
						Type	R (Ωcm)	μ (cm^{2}V^{-1}S^{-1})	N (cm^{-3})	
o-SnS										
o-SnS	$T(II)C = 1$ M $TA = 1$ M	$TEA = 10$ mL $TSS = 5$ mL $NH_3/NH_4Cl = 5$ mL	$T_b = 27\ °C$ $t_c = 24$ h $pH = 10.7$	ORT(110)	1.37 (d) 1.05 (i)	p	10^5	9×10^5	–	2010 [178]
o-SnS	–	$TEA = 12.5$ M, 13 M	$T_b = -$ $t_d = -$ $pH = -$	–	1.93–2.16 (d)	–	–	–	–	2010 [179]
o-SnS	$T(II)C = 0.95$ g $TA = 0.1$ M	$TEA = 8$ mL $NH_3 = 6$ mL	$T_b = 75\ °C$ $t_d = 1$ h $pH = -$	ORT(111)/(040)	1.3 (i)	p	–	–	–	2010 [180]
o-SnS	$T(II)C = -$ $TA = -$	TEA, NH_3 TTA	$T_b = T_r, 90\ °C$ $t_d = 24$ h, 3 h $pH = -$	ORT(400)	1.1–1.9 (d)	–	–	–	–	2011 [181]
o-SnS	$T(II)C = 0.2$ M $ST = 0.2$ M	$Na_2EDTA = 25$ mL of 0.2 M	$T_b = 40–80\ °C$ $t_d = 30$ h $pH = 1.5$	–	1.2–1.5 (d)	–	–	–	–	2011 [182]
o-SnS	$T(II)C = 0.15$ M $ST = 0.15$ M	$Na_2EDTA = 25$ mL of 0.2 M	$T_b = 75\ °C$ $t_d = 150$ min $pH = -$	–	1.2–1.6 (d)	–	–	–	–	2011 [183]
o-SnS	$T(II)C = 0.1$ M $ST = 0.25$ M	$AC = 50$ mL of 0.2 M	$T_b = 35\ °C$ $t_d = 10$ h $pH = 5, 6$	ORT(111)	1.75 (d) 1.12 (i)	–	–	–	–	2011 [184]
o-SnS	$T(II)C = 0.1$ M $ST = 0.25$ M	$AC = 50$ mL of 0.2 M	$T_b = 35°C$ $t_d = 10$ h $pH = 5$	ORT(111)	1.75 (d) 1.15 (i)	–	420	–	–	2011 [185]
o-SnS	$T(II)C$ TA	$TEA = -$ $NH_3 = -$	$T_b = 20–50\ °C$ $t_d = -$ $pH = -$	ORT(111)	1.15 (i) 1.35(d)	p	6.3 ± 0.1	11 ± 7	10^{16}–10^{17}	2011 [186]

Table 4. *Cont.*

Sn_xS_y Phase	Precursors	Complexing Gent	Deposition Parameters	Structure	Band Gap (eV)	Electrical Parameters				Ref
						Type	R (Ωcm)	μ (cm^{2}V^{-1}S^{-1})	N (cm^{-3})	
o-SnS										
o-SnS	T(II)C = 0.1 M ST = 0.25–0.75 M	AC = 50 mL of 0.3 M	T_b = 60–80 °C t_d = 3 h pH = 5	ORT(111)/(040))1.01–1.26 (i)		p	10^3	–	–	2012 [187]
o-SnS	T(II)C = – TA = –	TEA = – NH$_4$Cl = –	T_b = 45 °C t_d = 5 h pH = –	ORT(111)	0.7–1.3 (i)	–	–	–	–	2012 [188]
o-SnS	T(II)C = 1 g TA = 1 M	TEA = 12 mL NH$_3$ = 10 mL	T_b = 60 °C t_d = 6 h pH = 6	ORT(111)/(101)	0.9–1.1	–	10^6–10^1	–	–	2012 [189]
o-SnS	T(II)C = 1 M TA = 1 M	TEA = 10 mL NH$_3$ = 2 mL	T_b = RT = 27 °C t_d = 24–72 h pH = 9.7	ORT(111)	1.14–1.18 (i) 1.32–1.44 (d)	–	–	–	–	2012 [190]
o-SnS	T(II)C = 0.06 M–0.12 M TA = 0.1 M	TEA = 1.85 M NH$_3$ = 1.5 M	T_b = 30 °C t_d = 90 min pH = –	ORT(040)/(111) 1.5–1.95 (d)		–	–	–	–	2012 [191]
	T(II)C = 0.1 M TA = 0.1 M	TEA = 1.75–1.90 M NH$_3$ = 1.5 M	T_b = 30 °C t_d = 90 min pH = –							
	T(II)C = 0.1 M TA = 0.1 M	TEA = 1.85 M NH$_3$ = 1.5 M	T_b = 40–60 °C t_d = 90 min pH = –							
o-SnS	T(II)C = 1 g TA = 1 M	TEA = 12 mL NH$_4$OH = 10 mL	T_b = 60 °C t_d = 6 h pH = –	ORT(111)	1.9 (d) 1.1 (i)	–	–	–	–	2013 [192]
o-SnS	T(II)C = 0.1 M TA = 0.6 M	TTA = 1 M	T_b = 50–70 °C t_d = 50 min pH = 1.5	ORT(111)	1.30–1.35 (d)	–	–	–	–	2013 [193]

Table 4. *Cont.*

Sn_xS_y Phase	Precursors	Complexing Gent	Deposition Parameters	Structure	Band Gap (eV)	Electrical Parameters				Ref
						Type	R (Ωcm)	μ ($cm^2V^{-1}S^{-1}$)	N (cm^{-3})	
o-SnS										
o-SnS	T(II)C = 0.05–0.2 M TA = 0.4–0.7 M	Na_2EDTA = 20 mL of 0.1 M	T_b = 50–80 °C t_d = 0.5–3 h pH = 9–12	ORT(200)	–	–	–	–	–	2013 [194]
o-SnS	T(II)C = 0.1 M ST = 0.3 M	Na_2EDTA = 5 mL of 0.1 M TSC = 5 mL of 0.66 M	T_b = T_r t_d = 24 h pH = 10	–	1.50–1.90 (d)	–	–	–	–	2013 [195]
o-SnS	T(II)C = 0.5 M TU = 1 M	NH_3 = 3 M	T_b = T_r t_d = 60–180 min pH = –	–	1.98–2.01(d) 1.82–1.98 (i)	p	–	–	–	2013 [196]
o-SnS	T(II)C TA = 0.1 M	TEA = 30 mL NH_4OH = 16 mL	T_b = – t_d = 5 h pH = –	ORT(111)/(200)	1.64–1.7 (f)	–	–	–	–	2013 [197]
o-SnS	T(II)C = 0.1 M TA = 0.1 M	EDTA = 0.05 M–0.08 M NH_3 = 1.4 M	T_b = – t_d = 3–4 h pH = –	ORT(111)/(101)	1.5–1.60 (d)	p	400	–	–	2013 [198]
o-SnS	T(II)C = 1 g TA = 1 M	TEA = 6 mL NH_3 = 10 mL	T_b = – t_d = – pH = –	ORT(240)	1.78–1.75 (d)	–	10^9–10^8	–	–	2014 [199]
o-SnS	T(II)C = 1 g TA = 1 M	TEA = 12 mL NH_3 = 10 mL	T_b = 20–40 °C t_d = 24 h pH = 11	ORT(111)	ORT 1.1 (i)	p	10^7–10^2	–	–	2014 [146]
o-SnS	T(II)C = 0.5 g TA = 1 M	TEA = 6 mL TSC = 0.006–0.008 M NH_3 = 5 mL	T_b = 30 °C t_d = 24 h pH = –	ORT(111)	1.17–1.40 (d)	–	10^4	148–228	10^{12}	2014 [44]
o-SnS	T(II)C = – TA = –	TSC = –	T_b = 50 °C t_d = 2.5 h pH = 5	ORT(111)	1.25–1.83 (d) 1.1–1.65 (i)	n	10^3	–	–	2014 [200]

Table 4. *Cont.*

Sn_xS_y Phase	Precursors	Complexing Gent	Deposition Parameters	Structure	Band Gap (eV)	Electrical Parameters				Ref
						Type	R (Ωcm)	μ (cm^{2}V^{-1}S^{-1})	N (cm^{-3})	
o-SnS										
o-SnS	T(II)C = 0.03 M ST = 0.03 M	TTA = 0.44 M	$T_b = T_r$ $t_d = 24$ h pH = 7	ORT(400)	1.49–1.39 (i) 1.28–1.5 (i)	–	–	–	–	2014 [201]
o-SnS	T(II)C = 1 g TA = 1 M	TEA = 12 mL NH$_3$ = 10 mL	$T_b = 40$ °C $t_d = 17$ h pH = –	ORT(111)	1.25–1.1 (i)	–	10^3	–	–	2015 [202]
o-SnS	T(II)C = 0.1 M TA = 0.1 M	TEA = 15 mL NH$_3$ = 8 mL	$T_b = 26$ °C $t_d = 22$ h pH = –	ORT(021)	1.76–3.32 (d)	–	–	–	–	2015 [203]
o-SnS	T(II)C = – ST = 0.01–0.09 M	TTA	$T_b = 22$ °C $t_d = 24$ h pH = 7	–	–	–	–	–	–	2015 [204]
o-SnS	T(II)C = 20 mL TA = 20 mL	TTA = 1 M	$T_b = 40$–80 °C $t_d = 50$ min pH = 1.5	–	1.33–1.41 (d)	–	–	–	–	2015 [17]
o-SnS	T(II)C = 0.1 M TA = 0.15 M	TSC = 0.2 M NH$_3$ = –	$T_b = 80$ °C $t_d = 4$ h pH = 7	ORT(040)	1.65 (d)	p	–	–	–	2016 [205]
o-SnS	T(II)C = 1 g TA = 8 mL	TEA = 12 mL NH$_3$ = 10 mL	$T_b = 40$ °C $t_d = 10$ h pH = 11	ORT(111)	ORT = 1.1 (i)	p	10^6	–	–	2016 [25]
o-SnS	T(II)C = 0.1 M TA = 20 mL	TTA = 1 M	$T_b = 70$ °C $t_d = $ – pH = –	ORT(111)	1.31–1.26 (d)	p	6–38	124	10^{15}–10^{16}	2016 [144]
o-SnS	T(II)C = 1 g TA = 0.3 g	TEA = 5.5 mL NH$_3$ = 5 mL	$T_b = 70$ °C $t_d = $ – pH = –	ORT(002)	1.14–1.75 (d)	–	–	–	–	2016 [206]

Table 4. *Cont.*

Sn_xS_y Phase	Precursors	Complexing Gent	Deposition Parameters	Structure	Band Gap (eV)	Electrical Parameters				Ref
						Type	R (Ωcm)	μ (cm^{2}V^{-1}S^{-1})	N (cm^{-3})	
o-SnS										
o-SnS	T(II)C = – TA = –	TTA = 1 M	T_b = 70 °C t_d = – pH = –	ORT(111)	1.3 (d)	p	38–14.2	55–23	10^{15}–10^{19}	2016 [123]
o-SnS	T(II)C = 0.1 M TA = 0.15 M	TSC = 0.15–0.21 M	T_b = 80 °C t_d = 4 h pH = 5.8	ORT(111)	1.64–1.1 (d)	–	–	–	–	2017 [207]
o-SnS	T(II)C = 0.1 M TA = 0.15 M	TSC = 0.2 M	T_b = 80 °C t_d = 4 h pH = 6.5–7.5	ORT(111)	1.51 (d)	–	–	–	–	2018 [27]
o-SnS	T(II)C = 1 g TA = 1 M	TEA = 312 mL NH$_3$ = 10 mL	T_b = 40 °C t_d = 17 h pH = 1.5	ORT(111)	1.1 (i)	–	–	–	–	2018 [208]
o-SnS	T(II)C = 4 mmol TA = 4–8 mmol	TSC = 0.15–0.21 M	T_b = 80 °C t_d = 1–2 h pH = 0.4–1.0	ORT(111)	1.39–1.41 (d)	–	–	–	–	2018 [209]
o-SnS	T(II)C = 0.1 M TA = 0.15 M	TEA = –	T_b = 343 K t_d = 120, 240, 369 min pH = 4	ORT(013)	–	–	–	–	–	2018 [210]
o-SnS	T(II)C = 20 mL TA = 20 mL	TTA = 0.6–1.6 M	T_b = 70 °C t_d = 50 min pH = –	ORT(111)	1.28–1.45 (d)	p	38–62	29–108	1.92×10^{15}–4.12×10^{15}	2019 [211]
o-SnS	T(II)C = 0.1 M TA = 0.6 M	TTA = 1 M	T_b = 40–80 °C t_d = 50 min pH = 1.5	ORT(111)	1.30–1.41 (d)	p	38	55	1.5×10^{15}–3.4×10^{15}	2019 [212]

Table 4. *Cont.*

Sn_xS_y Phase	Precursors	Complexing Gent	Deposition Parameters	Structure	Band Gap (eV)	Electrical Parameters				Ref
						Type	R (Ωcm)	μ (cm^{2}V^{-1}S^{-1})	N (cm^{-3})	
				o-SnS						
o-SnS	$T(II)C = 1$ g $TA = 1$ M	$TEA = 18$ mL $NH_3 = 10$ mL	$T_b = 40–70$ °C $t_d = 3$ H pH = 10	ORT(040)	1.32–2.08 (d)	–	–	–	–	2019 [213]
o-SnS	$T(II)C = 0.2$ M $TA = 0.4$ M	$TTA = 0.5$ M	$T_b = 50–80$ °C $t_d = 90$ min pH = 1.5	ORT(040)	1.55–1.92 (d)	–	–	–	–	2019 [214]
o-SnS	$T(II)C = 0.1$ M $TA = 0.15$ M	$TSC = 0.2$ M	$T_b = 80$ °C $t_d = 4$ h pH = 5.0–6.5	ORT(111)	1.34–1.51 (d)	–	–	–	–	2019 [215]
o-SnS	$T(II)C = 4$ mmol $TA = 6$ mmol	–	$T_b = 80$ °C $t_d = 120$ min pH = 0.7	ORT(111)	1.41–1.49 (d)	–	–	–	–	2019 [216]
o-SnS	$T(II)C = 2$ g $ST = 0.2$ M	$TEA = 70$ mL $CA = 0.4$ M $NH_3 = 10$ mL	$T_b = 55$ °C $t_d = 4$ h pH = 11	ORT(111)	1.33 (i)					2019 [217]
o-SnS	$T(II)C = 20$ mL $TA = 10$ mL $PVA = 2$ g	$TTA = 0.5$ M	$T_b = 80$ °C $t_d = 45–90$ min pH = 10	ORT(040)	1.55–1.79 (d)	–	–	–	–	2019 [218]
o-SnS	$T(II)C = 0.1$ M $TA = 1$ M	$TEA = 10$ mL $TSC = 0.66$ M	$T_b = -$ $t_d = -$ pH = 9.2–9.6	ORT(102)	1.36–1.99 (d)	–	–	–	–	2020 [219]
o-SnS	$T(II)C = 0.1$ mol $TA = 0.4$ mol	$AA = 0.8$ mL	$T_b = 75$ °C $t_d = 70$ min pH = 9.2–9.6	ORT(110)	–	–	–	–	–	2020 [220]
o-SnS	$T(II)C = -$ $TA = 0.1$ M	$TEA = -$ $NH_3 = 15$ mL	$T_b = 25$ °C $t_d = 4$ h pH = - 200–600 °C	–	1.5–1.7 (d)	–	–	–	–	2021 [221]

Table 4. *Cont.*

Sn_xS_y Phase	Precursors	Complexing Gent	Deposition Parameters	Structure	Band Gap (eV)	Electrical Parameters				Ref
						Type	R (Ωcm)	μ (cm^{2}V^{-1}S^{-1})	N (cm^{-3})	
o-SnS										
o-SnS	T(II)C = 1 g TA = 0.6 g	TEA = 12 mL NH$_3$ = 15 mL	T$_b$ = 70 °C t$_c$ = 2 h pH = 10.93	ORT(111)	1.38 (d)	–	–	–	–	2021 [222]
o-SnS	T(II)C = 4 m mol TA = 6 m mol	–	T$_b$ = 80 °C t$_d$ = 120 min pH = 0.7	ORT(111)	0.78–1.13 (d)	–	–	–	–	2021 [223]
o-SnS	–	–	T$_b$ = 65 °C t$_d$ = 3 h pH = 5.5–8.5	ORT(111)	1.41–1.75 (d)	–	–	–	–	2021 [224]
c-SnS										
c-SnS	T(II)C = 2.26 g TA = 10 mL	TEA = 30 mL NH$_3$ = 16 mL	T$_b$ = 25 °C t$_d$ = 6 h pH = –	CUB (111)/(200)	1.64–1.73 (d)	p	10^5	10^4	10^9	2008 [43]
c-SnS	T(II)C = 2.26 g TA = 10 mL	TEA = 30 mL NH$_3$ = 16 mL	T$_b$ = 25 °C t$_d$ = 6 h pH = –	CUB (111)/(200)	1.7 (d)					2009 [171]
c-SnS	T(II)C = 2.26 g TA = 10 mL	TEA = 30 mL NH$_3$ = 16 mL	T$_b$ = 25 °C t$_d$ = 6 h pH = –	CUB (111)/(200)	1.7 (d)	–	–	–	–	2009 [173]
c-SnS	T(II)C = – TA = 0.1 M	TEA = 8964 g NH$_4$OH = 15 M	T$_b$ = 25 °C t$_d$ = 2–4 h 30 min pH = –	CUB (111)/(200)	1. 7 (d)	–	–	–	–	2011 [225]
c-SnS	T(II)C = – TA = 0.1 M	TEA = – NH$_4$OH = 15 M	T$_b$ = 25 °C t$_d$ = – pH = –	–	–	–	–	–	–	2011 [226]
c-SnS	T(II)C = – TA = 0.1 M	TEA = 8964 g NH$_4$OH = 15 M	T$_b$ = 25 °C t$_d$ = – pH = –	CUB (111)/(200)	1.76 (d) 1.44–1.51 (d)	–	–	–	–	2012 [227]

Table 4. *Cont.*

Sn_xS_y Phase	Precursors	Complexing Gent	Deposition Parameters	Structure	Band Gap (eV)	Electrical Parameters				Ref
						Type	R (Ωcm)	μ (cm^{2}V^{-1}S^{-1})	N (cm^{-3})	
c-SnS										
c-SnS	T(II)C = – TA = 0.1 M	TEA = 8964 g NH$_4$OH = 15 M	T$_b$ = 25 °C t$_d$ = – pH = –	CUB (111)/(200)	1.76 (d)	–	–	–	–	2012 [228]
c-SnS	T(II)C = 1 g TA = 1 M	TEA = 12 mL NH$_3$ = 10 mL	T$_b$ = 20–40 °C t$_d$ = 24 h pH = 11	CUB (111)/(200)	1.67 (d)	p	10^7–10^2	–	–	2014 [146]
c-SnS	T(II)C = 2.26 g TA = 0.1 M	TEA = 30 mL NH$_3$ = 16 mL	T$_b$ = 17 °C t$_d$ = 10 h pH = –	CUB (222)/(400)	1.74 (d)	–	–	–	–	2015 [28]
c-SnS	T(II)C TA = 0.1 M	TEA = 0.1 M NH$_4$OH = 15 M	T$_b$ = 25 °C t$_d$ = – pH = –	CUB(111)/(200)	1.70 (d)	–	–	–	–	2015 [229]
c-SnS	T(II)C = 2.26 g T(II)C = 0.1 M	TEA = 30 mL NH$_3$ = 16 mL	T$_b$ = 17 °C t$_d$ = 15 h pH = 11	CUB(222)/(400)	1.73 (d)	–	10^3	–	–	2016 [230]
c-SnS	T(II)C = 2.26 g TA = 10 mL	TEA = 30 mL NH$_3$ = 16 mL	T$_b$ = 17 °C, 10 °C t$_d$ = 4 h, 18 h pH = 11	CUB(222)/(400) 1.66–1.72 (d)		p	10^6	–	–	2016 [25]
c-SnS	T(II)C = 0.1 M TA = 0.1 M	TEA = 30 mL NH$_3$ = 16 mL	T$_b$ = 25 °C t$_d$ = 6 h pH = 11	CUB(222)/(400)	–	–	–	–	–	2016 [231]
c-SnS	T(II)C = 2.26 g TA = 0.1 M	TEA = 30 mL NH$_3$ = 16 mL	T$_b$ = 17 °C t$_d$ = 10 h pH = 11	CUB(222)/(400)	–	–	–	–	–	2016 [232]
c-SnS	T(II)C = 2.26 g ST = 1 M	EDTA = 20 mL of 0.5 M	T$_b$ = 25–65 °C t$_d$ = 6 h pH = 10.5	CUB(222)/(400) 1.74–1.68 (d)		p	10^5–10^4	8.98–28.6	10^{12}–10^{13}	2016 [50]

Table 4. *Cont.*

Sn_xS_y Phase	Precursors	Complexing Gent	Deposition Parameters	Structure	Band Gap (eV)	Electrical Parameters				Ref
						Type	R (Ωcm)	μ ($cm^2V^{-1}S^{-1}$)	N (cm^{-3})	
c-SnS										
c-SnS	$T(II)C = 2.26$ g $ST = 1$ M	EDTA = 15–25 mL of 0.5 M $NH_3 = 5$ mL	$T_b = 45\,°C$ $t_d = 6$ h pH = 10.5 $S_p = -$	CUB(222)/(400) 1.67–1.73 (d)		p	10^5–10^4	0.34–28.6	10^{14}–10^{12}	2016 [26]
c-SnS	$T(II)C = 0.1$ M $TA = 0.15$ M	TSC = 0.2 M	$T_b = 80\,°C$ $t_c = 4$ h pH = 7	CUB(222)/(400) 1.64 (d)		–	–	–	–	2017 [24]
c-SnS	$T(II)C = 0.1$ M $TA = 0.1$ M	TEA = 30 mL $NH_3 = 16$ mL	$T_b = 17\,°C, 80\,°C$ $t_d = 3$ h, 21 h pH = –	–		–	–	–	–	2017 [233]
c-SnS	$T(II)C = 0.1$ M $ST = 0.125$ M	EDTA = 0.1 M	$T_b = 45\,°C$ $t_d = 6$ h pH = –	CUB(222)/(400) 1.67–1.75 (d)		p	10^5–10^4	5.22–77.7	10^{11}–10^{13}	2017 [29]
c-SnS	$T(II)C = 0.1$ M $TA = 0.15$ M	TSC = 0.2 M	$T_b = 80\,°C$ $t_d = 4$ h pH = 6.5–7.5	CUB(222)/(400) 1.64–1.73 (d)		–	–	–	–	2018 [27]
c-SnS	$T(II)C = 0.1$ M $TA = 0.15$ M	TSC = 0.2 M	$T_b = 80\,°C$ $t_d = 4$ h pH = 7	CUB(222)/(400) 1.5 (d)		–	–	–	–	2018 [234]
c-SnS	$T(II)C = 0.5$ M $TA = 0.5$ M	TEA = 30 mL $NH_3 = 16$ mL	$T_b = 35\,°C$ $t_d = 4$ h pH = 9.78	CUB(222)/(400) 1.74 (d)		–	–	–	–	2018 [235]
c-SnS	$T(II)C = 2.26$ g $TA = 0.1$ M	TEA = 30 mL $NH_3 = 16$ mL	$T_b = 17\,°C, 80\,°C$ $t_d = 3$ h, 21 h pH = –	CUB(222)/(400) 1.76 (d)		–	–	–	–	2018 [236]
c-SnS	$T(II)C = 2.26$ g $TA = 10$ mL	TEA = 30 mL $NH_3 = 16$ mL	$T_b = 17–8\,°C$ $t_d = 3–21$ h pH = –	CUB(222)/(400)						2019 [217]

Table 4. *Cont.*

Sn_xS_y Phase	Precursors	Complexing Gent	Deposition Parameters	Structure	Band Gap (eV)	Electrical Parameters				Ref
						Type	R (Ωcm)	μ ($cm^2V^{-1}S^{-1}$)	N (cm^{-3})	
c-SnS										
c-SnS	T(II)C = 0.04 M TA = 0.08 M	TEA = 1.1 M NH_3 = 9.5 mL	T_b = 30 °C t_d = 4 h pH = –	CUB(222)/(400)	1.74 (d)	–	–	–	–	2020 [237]
c-SnS	T(II)C = 0.1 M TA = 0.15 M	TSC = 0.2 M	T_b = – t_d = - pH = –	CUB(222)/(400)	–	–	–	–	–	2020 [238]
c-SnS	T(II)C = 1 g TA = 0.3 g	TEA = 5.5 mL NH_3 = 5 mL	T_b = 24 °C t_d = 4.25 h pH = 9.25	CUB(222)/(400) 1.70–1.74 (d)		–	10^3–10^4	–	–	2020 [239]
c-SnS	T(II)C = 0.2 M TA = 0.1 M	TEA = 5.5 mL NH_3 = 5 mL	T_b = 17–8 °C t_d = 3–21 h pH = 11	CUB(222)/(400)	1.76 (d)	p	10^8	–	–	2020 [240]
c-SnS	T(II)C = 2.25 g ST = 0.1 M	EDTA = 0.5 M NH_3 = 5–7.5 mL	T_b = 50 °C t_d = 6 h pH = 10.3	CUB(222)/(400) 1.75–1.8 (d)		p	10^3–10^4	15–75	10^{12}–10^{13}	2020 [241]
c-SnS	T(II)C = 1 g TA = 0.6 g	TEA = 12 mL NH_3 = 15 mL	T_b = 70 °C t_d = 2 h pH = 8.24	CUB(200)	1.72 (d)	–	–	–	–	2021 [222]
c-SnS	T(II)C = 1.21 g TA = 0.5 M	TTA = 1 M	T_b = 80 °C t_d = 2–6 h pH = 5–8	CUB(222)/(400) 1.72–1.90 (d)		–	10^7–10^8	–	–	2021 [242]
c-SnS	T(II)C = 0.5 g TA = 1 M	NTA = 0.6 M	T_b = 40 °C t_d = 90–182 min pH = 10	CUB(222)/(400) 1.77–1.81 (d)		–	10^6	–	–	2021 [243]
c-SnS	T(II)C = 0.01 mol TA = 0.1 M	TEA = 0.6 M	T_b = 17–8 °C t_d = 3–21 h pH = 10	CUB(222)/(400) 1.70–1.80 (d)		–	–	–	–	2021 [244]

Table 5. Deposition conditions and the physical properties of SnS$_2$ and Sn$_2$S3 films grown by CBD.

Sn$_x$S$_y$ Phase	Precursors	Complexing Agent	Deposition Parameters	Structure	Band Gap (eV)	Electrical Parameters				Ref
						Type	R (Ωcm)	μ (cm^{2}V^{-1}S^{-1})	N (cm^{-3})	
SnS$_2$										
SnS$_2$	T(II)C = 0.025 mol SDS/AS = 0.025 mol	–	T$_b$ = – t$_d$ = – pH = 3, 10, 12	–	2.04	SnS$_2$- n	10^7–10^3	–	–	1989 [36]
SnS$_2$	Tin-ingots (99.9%) ST = 10 mL	–	T$_b$ = T$_r$ t$_d$ = 2 h pH = –	Amorphous	2.35 (d)	n	10^3–10^4	–	–	1990 [130]
SnS$_2$	Tin ingots (99.9%) ST = 10 mL	–	T$_b$ = 27 °C t$_d$ = – pH = 1.4	Amorphous	2.20 (i)	n	10^7–10^8	–	–	1992 [147]
SnS$_2$	T(II)C = 1.13 g TA = 0.1 M	EDTA = 25 mL NH$_3$ = 15 mL	T$_b$ = T$_r$ t$_d$ = 10–120 min pH = 10	–	2.3 (d)	n	4 $\times$ 10^{-1}	–	–	1997 [38]
SnS$_2$	T(II)C = 15 g TU = 5 g, 10 g	–	T$_b$ = – t$_d$ = 5 min pH = 3 S$_p$ = 1.33 mm/s	HEX(001)	2.05 (i)	–	–	–	–	1999 [163]
SnS$_2$	TC(IV) = 0.02 mol TA = 0.5 mol L^{-1}	CA = 0.375, 0.5, 0.625 mol/L	T$_b$ = 35 °C t$_d$ = – pH = 1.3	–	2.40 (d)	–	–	–	–	2011 [131]
SnS$_2$	T(II)C = 1 g TA = 0.5 M	TEA = 24 mL NH$_3$ = 12 mL–20 mL	T$_b$ = 60 °C t$_d$ = 2 h pH = –	HEX(001)	3.3–3.7 (d)	–	–	–	–	2012 [41]
SnS$_2$	T(II)C = 0.8 M TA = 0.5 M	TEA = 3.75 M NH$_3$ = 12 mL	T$_b$ = 60 °C t$_d$ = – pH = –	HEX(001)	2.8–3.0 (d)	–	–	–	–	2013 [245]
SnS$_2$	T(II)C = 2.26 g ST = 1 M	EDTA = 20 mL of 0.5 M NH$_3$ = 5 mL	T$_b$ = 45 °C t$_d$ = 6 h pH = –	HEX(001)	2.58 (d)	–	–	–	–	2017 [246]

Table 5. *Cont.*

Sn_xS_y Phase	Precursors	Complexing Agent	Deposition Parameters	Structure	Band Gap (eV)	Electrical Parameters				Ref
						Type	R (Ωcm)	μ (cm^{2}V^{-1}S^{-1})	N (cm^{-3})	
					SnS_2					
SnS_2	T(II)C TA	TTA = 1 M	T_b = – t_d = 30–120 min pH = –	HEX(001)	2.95–2.80 (d)	n	11.2	48	10^{17}	2017 [56]
SnS_2	T(II)C = 0.84 g TA = 0.5 M	TEA = 24 mL NH_3 = 16 mL	T_b = 60 °C t_d = 2 h pH = –	–	–	–	–	–	–	2018 [247]
SnS_2	T(II)C = 0.1 M TA = 0.1 M	TTA = 0.1 M	T_b = 60 °C t_d = 6 h pH = –	HEX(001)	2.25–2.53 (d)	–	–	–	–	2019 [248]
					Sn_2S_3					
Sn_2S_3	T(II)C = 1 M TA = 1 M	TEA = 10 mL	T_b = 30 °C t_d = 20–24 h pH = 10.7	ORT(131)	2.03–2.12 (d)	–	–	–	–	2012 [34]
Sn_2S_3	T(II)C = 1.4 g TA = 1 M	TEA = 30 mL NH_3 = 50 mL	T_b = RT t_d = 24 h pH = –	ORT(211)	1.2 (d)	–	–	–	–	2012 [139]
Sn_2S_3	T(II)C = 0.05 M SDS = 0.05 M	–	T_b = – t_d = – pH = –	ORT(021)	1.3 (d)	–	–	–	–	2018 [249]
Sn_2S_3	T(II)C = 0.1 M TA = 0.1 M	TEA = 30 mL NH_3 = 16 mL	T_b = 17 °C t_d = 15 h 450 °C (S-powder: 15 mg), 5–75 min	ORT(211)	1.75 (d)	p	10^4	6×10^{-6}	–	2020 [250]

3.2. Sn and S Precursors and Their Concentration Effect

The selection of Sn precursor and its concentration plays a vital role in the growth, phase formation, crystallinity, preferred orientation, morphology, band gap, and other properties of Sn_xS_y thin films [251]. This is because the releasing rate of Sn ions strongly depends on the selection of Sn precursors. As mentioned in Section 3.1, the $SnCl_2 \cdot 2H_2O$ (T(II)C) has been considerably utilized as an Sn precursor for the deposition of Sn_xS_y films. According to the literature (Tables 4 and 5), until recently, there have been no reports related to the study of different types of Sn precursors on the formation of Sn_xS_y films and the Sn precursor concentration effects on the formation of SnS_2 and Sn_2S_3 films and their properties. However, there have been very few quantitative analyses of Sn precursor concentration effect on the formation of o-SnS films and their properties. The primary report related to the effect of Sn precursor concentration ([T(II)C] = 0.06–0.12 M) on the growth of o-SnS films was made in 2012 [191]. A lower T(II)C concentration stimulates the formation of multi phases with a dominant SnS_2 phase, whereas a higher T(II)C concentration reduces the crystallinity. The T(II)C concentration of 0.1 M is beneficial for the deposition of pure, good crystalline o-SnS with (111) preferred orientation (Figure 7a). A small variation in T(II)C concentration (at 0.15 M) changes the preferred orientation of o-SnS from (111) to (200) [194]. Moreover, the change in T(II)C concentration can increase the grain size and decrease the band gap (1.95–1.5 eV) (Figure 7b,c) [191]. Therefore, the manipulation of preferred orientation, crystallinity, and band gap can be achieved by the change in Sn precursor concentration.

In addition to the suitable Sn precursor selection, the choice of S precursor and its concentration are highly desirable to obtain good quality Sn_xS_y films. In CBD, the releasing rate (or reaction rate) of S ions greatly affect the growth kinetics and phase formation, and it can be controlled by the S precursor concentration. According to the previous reports (Tables 4 and 5), TA and ST have been chiefly used as S ion sources (Figure 5b). In those, TA is preferable compared to ST because it works in both acidic and alkaline bath conditions. The influence of TA concentration on o-SnS film growth (thickness) was initially reported in 1987 [145]. An extremely low or high TA concentration produces the o-SnS films of smaller terminal thickness, whereas a moderate TA concentration promotes the growth of maximum thickness (Figure 7d). The reason for the lower film thickness obtained at a lower S precursor concentration is the insufficient number of S ions in the reaction bath that can combine with all the available Sn ions. At a higher S precursor concentration, the releasing rate of S ions is high enough to stimulate the precipitation process, which also results in a lower film thickness [145].

Furthermore, the S precursor concentration can influence the morphology and phase formation of o-SnS films (Figure 7e). A higher TA concentration stimulates the formation of multi phases such as Sn_2S_3 and Sn_3S_4 (Sn_2S_3 + SnS → Sn_3S_4) [209] and a lower TA concentration assists the growth of single-phase o-SnS films, but with lower crystallinity [142,198]. A TA concentration of 0.1 M is preferable for the deposition of a single-phase, polycrystalline o-SnS with (101) preferred orientation [198], and a ST concentration of 0.75 M is advisable for (111)/(040) preferred orientation (Figure 7f). The effect of changes in the S source concentration on the band gap of o-SnS films is controversial until the present. A reduction in band gap from 1.70 eV to 1.25 eV with increasing TA concentrations was reported in [200], although no significant change in band gap was found with TA concentration in [198]. On the other hand, no studies in the literature have focused on the influence of S precursor concentration on the properties of c-SnS, SnS_2, and Sn_2S_3 films.

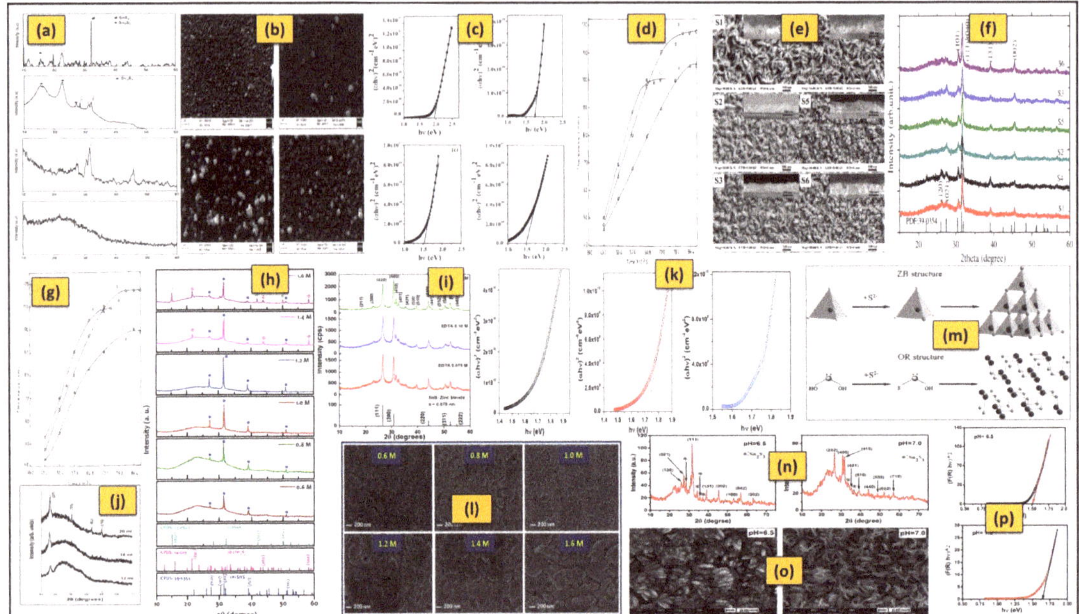

Figure 7. (**a–c**) XRD patterns, SEM images, and $(\alpha h\upsilon)^2$ versus (hυ) graph of o-SnS films grown at various $SnCl_2$ concentrations (reprinted with permission [191]. © 2012, Elsevier), and variation in o-SnS film (**d**) thickness with different TA concentrations (reprinted with permission [145]. © 1987, Elsevier). (**e,f**) Morphology and crystallinity changes of o-SnS films with ST concentrations (reprinted with permission [187]. © 2012, Elsevier). (**g**) Variation in SnS film thickness with different TEA concentrations (reprinted with permission [145]. © 1987, Elsevier). (**h,i**) XRD patterns of the o-SnS films and c-SnS films deposited at different TTA (reprinted with permission [211]. © 2019, Elsevier) and EDTA concentrations, respectively (reprinted with permission [26]. © 2016, Elsevier). (**j**) XRD patterns of SnS_2 films deposited at various volumes of ammonia solution (reprinted with permission [41]. © 2012, Elsevier). (**k**) $(\alpha h\upsilon)^2$ versus (hυ) for c-SnS films prepared using various EDTA amounts reprinted with permission [26]. © 2016, Elsevier). (**l**) SEM images of o-SnS films deposited with various TTA concentrations (reprinted with permission [211]. © 2019, Elsevier). (**m,n**) Scheme of the formation (reprinted with permission [185]. © 2011, Elsevier) and XRD patterns of o-SnS and c-SnS films (reprinted with permission [27]. © 2018, Elsevier). (**o**) Morphologies of o-SnS and c-SnS films (reprinted with permission [184]. © 2011, Elsevier), and (**p**) variation in the band gap o-SnS films at different pH values (reprinted with permission [27]. © 2018, Elsevier).

3.3. Complexing Agents and Their Concentration Effect

As stated in Section 3.1, in order to develop influential Sn_xS_y films, the control of the availability of Sn ions in the reaction bath is essential. It can be successfully attained by the addition of an appropriate concentration of a complexing agent [128,148]. Moreover, adhesion, morphology, crystallinity, and the deposition rate of Sn_xS_y films can be significantly affected by the concentration of the complexing agent [128]. Therefore, knowledge of the behavior of complexing agents in the bath can help to obtain good quality Sn_xS_y films. The behavior of complexing agents is described in terms of their stability constants (K_s), which is the equilibrium constant for the formation of a complex in a solution [252]. It is defined for the equilibrium between an Sn ion ($Sn^{2+/4+}$) and a ligand (L) as [148]

$$K_S = \frac{a_{Sn^{2+/4+}} - L}{a_{Sn^{2+/4+}} + a_L} \tag{1}$$

where a is the activity of subscripted species and can be approximated by its concentration. A large value of K_s implies a strong binding affinity for the metal (Sn) ion, while a small value of K_s implies a weak binding affinity [148]. Generally, complexing agents can prevent the formation of powder/bulk precipitation of tin hydroxides in the reaction bath, and they can easily maintain the supersaturating condition. If a complexing agent has a weak binding affinity, it does not arrest the bulk precipitation of tin hydroxides. On the other hand, if it has an extremely strong binding affinity, it restricts the deposition of the desired film [253]. Therefore, in order to prevent powder/bulk precipitation of tin hydroxides, the complexing binding affinity must be intermediate.

Various complexing agents have been explored to control Sn ions depending on the bath conditions during the deposition of Sn_xS_y films (Tables 4 and 5). However, there are only a few reports on the study of complexing agent concentration. Initially, the influence of TEA complexing agent concentration on the thickness of o-SnS films was made in 1987 [145]. An optimized TEA complexing agent concentration controls the formation of o-SnS films, yielding a thick o-SnS film (Figure 7g). In addition to the growth (thickness), the change in TEA complexing agent concentration can also influence the phase formation and crystallinity of o-SnS, c-SnS, and SnS_2 films. The lower tartaric acid (TTA) complexing agent concentration creates the weak tin complexation, leading to partial homogeneous precipitation, resulting in low-crystalline o-SnS films. As the complexing agent concentration increases, the improved tin complexation controls the reaction, yielding the formation of better crystalline films o-SnS. Over the limit, the availability of free Sn ions is reduced due to strong complexation, resulting in the formation of sulfur-rich tin phases such as Sn_2S_3 and SnS_2 (Figure 7h) [211]. Single-phase, polycrystalline o-SnS films with (111) preferred orientation are produced at 1.85 M of TEA [191] and 1.4 M of TTA [211], while a c-SnS (222)/(400) is formed at 0.125 M of EDTA [26] concentrations (Figure 7i). The crystallinity of o-SnS films can be improved by replacing the lower stability (Sn^{2+}-TEA) complexing agent with the higher stability (Sn^{2+}-EDTA) one [198,254], due to the fact that EDTA (hexaligand) may generate a ligand more quickly than TEA (triligand) [255]. An increase in citric acid and ammonia concentration also improves the crystallinity in the case of SnS_2 films (Figure 7j) [41,131]. The concentration of the complexing agent similarly influences the morphological and optical properties of the o-SnS, c-SnS, and SnS_2 films. The direct optical energy gap for o-SnS films reduces with increasing complexing agent concentration (TSC, 0.06–0.08 M; TEA, 12.5–13 M; TTA, 0.6–1.4 M) from 2.16 eV to 1.17 eV [50,182,214], but rises from 1.67 eV to 1.73 eV [26] for c-SnS films with EDTA (0.075–0.125 M) (Figure 7l). The change in complexing agent concentration (TSC, TTA) improves the compactness and morphology of o-SnS films (Figure 7m). This may improve their electrical properties, such as electrical mobility (~228 $cm^2V^{-1}s^{-1}$) and carrier concentration (~4.1 × 10^{15} cm^{-3}). No previous study has examined the effect of complexing agents on the formation and physical properties of Sn_2S_3 films.

3.4. Solution pH Effect

In CBD, solution pH/bath pH (a measure of the acidity or basicity of a solution) is an important parameter because it directly affects the growth mechanism as well as reaction rate. Therefore, it can influence the formation of phases and physical properties of films [27]. In addition, the bath pH must be at a specific optimum value to maintain supersaturation condition (Q_{ip} > K_{sp}, Figure 5c) for the formation of Sn_xS_y films. The preparation of Sn_xS_y films was reported both in acidic (pH < 7) and alkaline (pH > 7) baths (see Tables 4 and 5). When the bath pH is varied between 1 and 14, the concentration of OH- ions increases, which results in a reduction in the concentration of free Sn^{2+} or Sn^{4+} ions in the solution. Thus, the hydroxide mechanism can predominate during film development, resulting in the creation of $Sn(OH)_{2\,or\,4}$ in addition to Sn_xS_y. A higher bath pH, on the other hand, encourages the hydrolysis of a sulfur source precursor.

A few researchers have investigated the bath pH effect on o-SnS and c-SnS films growth and their physical properties (Table 4). The bath pH effect on the adhesion and growth rate of o-SnS films was first reported in 1989 [36]. According to this report, the good adhesion of o-SnS films on glass can be obtained with a bath pH >3. The growth rate is low at pH~7 and high at pH~10 for o-SnS films due to the formation of $Sn(OH)_{2\,or\,4}$ precipitate from the hydrolysis of an Sn precursor because a part of $Sn(OH)_{2 \text{ or } 4}$ precipitate turns into Na_2SnO_2, which dissolves back in the solution. The change in growth rate by bath pH leads to the variation in grain size of o-SnS films [183]. The bath pH can also influence the growth mechanism, which leads to phase transformation [184,185]. An o-SnS forms at a lower pH of 6.5 via the cluster-by-cluster mechanism, whereas c-SnS forms at a higher pH of 7.0 through the ion-by-ion mechanism (Figure 7n,o) [27].

The phase transition caused by the change in bath pH leads to a change in the morphologies of the film surfaces (Figure 7o) and the energy band gaps (o-SnS: 1.51 eV and c-SnS: 1.64 eV) (Figure 7p) [27,184,185]. The increase in solution pH results in the decrease of free Sn^{2+} ion concentration as well as the concentration of OH^- ions, which are favorable for the hydrolysis of the S ion source [256], leading to the increase in the concentration of S^{2-} ions. Thus, the interaction of Sn^{2+} and S^{2-} ions can form the c-SnS via an ion-by-ion mechanism because the potential barrier of heterogeneous nucleation is lower than that of homogeneous nucleation [25,185]. However, these mechanisms are speculated, and direct in situ measurement evidence such as in situ quartz crystal microbalance and electrochemical impedance is lacking. Therefore, such studies are required for understanding the growth mechanism of Sn_xS_y films [257]. On the other hand, no studies in the literature have examined the effect of bath pH on the properties of SnS_2 and Sn_2S_3 films.

3.5. Solution Temperature (T_b) Effect

In CBD, solution temperature/bath temperature (T_b) also played a crucial role in the preparation of thin films with high quality and desired features. It critically enhances the rate of dissociation of the precursors and thus strongly affects the thickness, growth rate, type of nucleation, crystalline phase, crystallite size, morphology, and optoelectrical properties of thin films. The change in film growth rate as a function of T_b can be determined through the Arrhenius equation [258],

$$k(T) = Ae^{\frac{-E_a}{RT}} \tag{2}$$

where k(T) is the temperature-dependent growth rate for the given deposition conditions, A is a pre-exponential constant related to the initial reagent concentration, E_a is the activation energy (kJ/mol), and R is the gas constant (R = 8.3145 J mol^{-1} K^{-1}).

The deposition of Sn_xS_y films has been reported in the T_b range of room temperature (T_r)—90 °C (Tables 4 and 5). The effect of T_b on the formation of o-SnS, c-SnS films, and their properties (Table 4) was studied extensively. The T_b changes the growth rate due to the variation in the deposition mechanism, i.e., the ion-by-ion mechanism, which is less thermally activated with low activation energy (at lower bath temperatures). In contrast, cluster-by-cluster is believed to occur at relatively higher temperatures [259]. Thus, the thickness of a film has a close relationship with the T_b. First, it increases significantly with the T_b due to the increase in bath supersaturation [145,213] and reaches saturation point very quickly because the hydrolysis of the S precursor is greatly improved by the increase in T_b [182] (Figure 8a). Then, it decreases down to a terminal point because of the ion–ion condensation process and high homogeneous precipitation rate [145,146]. In addition, the T_b significantly affects the microstructures of o-SnS and c-SnS films.

Generally, the films prepared at lower T_b have smaller grains, and those grains increase in size with T_b due to the covering of voids by secondary nucleation (Figure 8b) [50,212]. The change in grain shape may indicate a change in the growth mechanism [146]. The T_b can also influence the composition of c-SnS films. The c-SnS films show a non-stoichiometric composition at higher and lower Tb values due to the relatively faster and slower release of Sn^{2+} ions from the tin complex due to the variation of thermal energy in the solution [50]. Single-phase, polycrystalline o-SnS films with (111) preferred orientation and c-SnS films with (222)/(400) preferred orientation are produced separately at T_b of 70 °C [212] and 65 °C [50] using a different source of materials (Table 4), respectively.

On the other hand, the T_b shows an impact on phase transformations when other bath parameters remained constant. The films predominantly exhibit the o-SnS phase above the T_b range of 30–40 °C, whereas the c-SnS phase is below this range (20–30 °C) for particular deposition conditions [146]. The T_b can directly affect the crystallinity of both o-SnS and c-SnS films. The crystallinity of both films is improved with T_b due to the supply of sufficient thermal energy for further crystallization (Figure 8c) [50,212]. Thus, an average crystallite size is improved with T_b—however, up to a certain extent [212]. As the T_b improves the kinetic energy of the reactants and accelerates the interaction between all ions in the reaction bath, the nuclei formation (crystallite grow) is enhanced on the surface of the substrate [260]. However, at higher T_b, the crystallite size is decreased due to the dissolution of grown film.

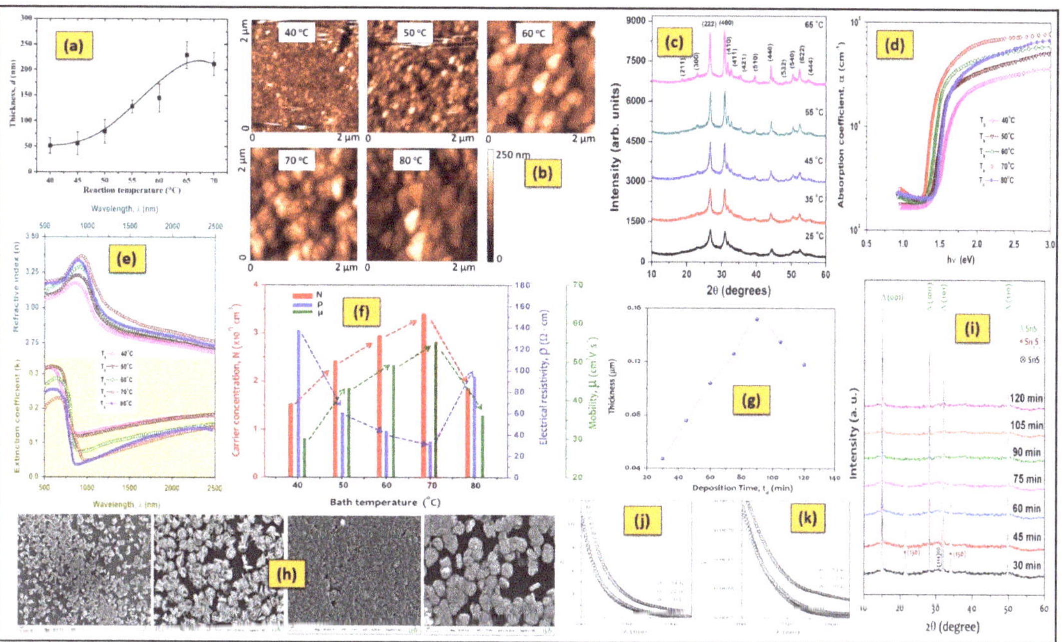

Figure 8. (**a**) Variation in o-SnS film thickness with bath temperature (reprinted with permission [213]. © 2019, Elsevier). (**b,f**) Morphological and electrical properties of o-SnS at different bath temperatures (reprinted with permission [212]. © 2019, Elsevier). (**c**) Change in crystallinity of c-SnS films with bath temperature (reprinted with permission [50]. © 2016, Elsevier). (**d,e**) Plots of absorption coefficient refractive index and extinction coefficient of o-SnS films (reprinted with permission [17]. © 2015, Elsevier). (**g,i**) Variation in thickness and crystallinity of SnS_2 films with deposition time (reprinted with permission [56]. © 2017, Elsevier). (**h**) Morphology of ORT- SnS thin films deposited at various times (reprinted with permission [177]. © 2010, Elsevier). (**j**) Variation in refractive index and (**k**) variation in extinction coefficient of Sn_2S_3 films with deposition time (reprinted with permission [34]. © 2012, Elsevier).

The T_b also influences the optical characteristics of the o-SnS and c-SnS films. In o-SnS films, the T_b improves the sharpness of the absorption edge with a high optical absorption coefficient ($>10^4$ cm^{-1}) [17] (Figure 8d), which is suitable for PV devices. The optical energy gap of o-SnS and c-SnS films decreases from 1.41 eV to 1.30 eV and from 1.74 eV to 1.68 eV, respectively, with the increase in T_b (30–70 °C) [17,50]. As mentioned above, T_b can improve the grain size and simultaneously reduce height (smoothness of the surface) and the number of grain boundaries [261]; this minimizes imperfections in the film and enhances the quality of the film, which can lead to change in density of localized states within the energy gap [262]. Therefore, band gap tuning is easily possible in CBD deposited o-SnS and c-SnS films regarding T_b, which is essential for designing highly efficient solar cells [263]. The optical parameters, namely, refractive index (n), extinction coefficient (k), and real/imaginary dielectric constants of o-SnS films, are in ranges of 2.72–3.24, 0.24–0.13, and 7.34–10.48/0.85–1.32 [17], respectively (Figure 8e). Here, the variation in the optical parameters may be arrived from the change in strain and packing density with T_b [264].

As previously mentioned, T_b improves the crystallinity along with grain size and thickness. Thus, the scattering of charge carriers by grain boundaries decreases with respect to T_b, which makes a significant change in the electrical characteristics of films [265]. These possible reasons may improve the carrier density and a consequent reduction in resistivity in both o-SnS and c-SnS films. The reduction of the dispersing effects of carriers can lead to an increase in the mobility of carriers in those films (~55 cm^2 V^{-1} s^{-1} for o-SnS at 70 °C [212] and 28 cm^2 V^{-1} s^{-1} for c-SnS at 45 °C [50]) (Figure 8f). However, the above-mentioned description confirms the importance of T_b in the CBD process until there are no reports on the T_b influence on both SnS$_2$ and Sn$_2$S$_3$ films.

3.6. Deposition Time Effect

Deposition time (t_d) is the most important bath parameter among the various deposition conditions. It affects the growth rate/thickness and properties of Sn$_x$S$_y$ films. In CBD, the t_d period can be divided into three steps, namely, (i) nucleation or initiation, which requires high activation energy; (ii) linear growth, which includes heterogeneous growth of nuclei; and (iii) termination or saturation, in which chemical reagents become depleted and the reaction begins to slow down and eventually stops [266]. Thus, the rate of formation of nuclei can be in terms of t_d by the following Avrami equation (a conventional diffusion-controlled reaction model) [267]:

$$\alpha = 1 - e^{-(kt_d)^n} \tag{3}$$

where α is the fractional decomposition (or reaction), k is a rate constant, and n is the Avrami exponent.

There are a considerable number of reports on the study of the t_d effect on o-SnS films, but there are only a few reports for c-SnS [210,225], SnS$_2$ [56], and Sn$_2$S$_3$ [34] films (Tables 4 and 5). Typically, a t_d from a few minutes to several hours has been reported to prepare these Sn$_x$S$_y$ films (Tables 4 and 5). The growth of Sn$_x$S$_y$ films with t_d was simply described in terms of thickness. In the initial state, the change in film thickness is insignificant because of the requirement of long incubation time for nucleation [198], and the thickness increases linearly due to the availability of sufficient amounts of Sn^{2+} or Sn^{4+} and S^{2-} ions. Next, the film thickness increases faster, then decreases at a longer deposition time, and attains a maximum value as a terminal/final thickness. Here, the attained terminal thickness is not only t_d-dependent but also T_b-related [145]. Thus, a terminal thickness should be considered when the reaction undergoes at a constant temperature. A terminal thickness in the range of 120–900 nm can be obtained for different t_d varying from 1 h to 24 h at a constant range of T_b (T_r—75 °C) for o-SnS and c-SnS films [128,145,175,198,225], and a thickness of 152 nm can be attained at a t_d of 90 min for SnS$_2$ films (Figure 8g) [56]. The variation in the thickness (growth) of these films with respect to t_d can be explained by considering two competing processes taking place in the deposition bath. One process includes heterogeneous precipitation, which leads to film growth (thickness improves). The other involves the dissolution of the preformed film, which results in the decrease of film thickness.

The t_d has a significant impact on the surface morphology, crystallinity, crystallite size, and phase purity of Sn$_x$S$_y$ films. As the t_d increases, the size and quantity of grains (or aggregations) can be improved to form a more homogeneous film [177] (Figure 8h). This indicates an occurrence of nucleation growth with t_d [56,190]. If t_d exceeds the optimum value, a non-uniform film with porous nature might be formed due to the dissolution of pre-adhered grains in the film [190]. This phenomenon can be experimentally observed for o-SnS, SnS$_2$, and Sn$_2$S$_3$ films when the t_d varies between 2 and 10 h [177], 30–120 min [56], and 20–24 h [34], respectively. The t_d can considerably improve the crystallinity of the Sn$_x$S$_y$ films and simultaneously enhance the crystallite size. However, beyond the limit of t_d, the crystallinity becomes poor, and the crystallite size decreases (Figure 8i) [56,146,177,190,225].

The reduction of crystallite size is due to the lowering of the van der Waals force in between crystallites because the substrate remained in the solution longer than necessary [177]. In addition to the crystallinity of films, the t_d also influences the phase purity of a film. At low t_d, the released Sn^{2+} or Sn^{4+} ions are relatively low in the reaction bath compared to the available S^{2-} ions. These available S^{2-} ions are not balanced by the all released Sn^{2+} or Sn^{4+} ions, leading to the development of other secondary phases, whereas at longer t_d they are counterbalanced, promoting the growth of the pure phase [56].

As mentioned previously, the t_d directly influences the thickness of Sn_xS_y films. Thus, it tremendously shows an impact on their optical transmittance/absorbance. Always, shorter t_d periods generate the thinnest film of high transmittance, which might be affected by abundant porosities [175,177]. Simultaneously, the more extended t_d periods produce thick films of high absorption [34,175,190], essential for solar cell application. The longer t_d period also improves the size of crystallites that can affect the optical absorption and the band gap energy of films [190].

The quantum size effect and changing barrier height (or variation in grain size) are also responsible for the variation in the band gap of films with t_d at other identical growth conditions [22,177]. The increase in t_d period reduces the band gap of o-SnS, SnS_2, and Sn_2S_3 films from 1.83 eV to 1.30 eV [177], 2.95 eV to 2.80 eV [56], and 2.12 to 2.03 eV [34], respectively. In contrast, the longer t_d period generates greater compression impacts with the thickness in o-SnS films, which may enhance the band gap (0.82–1.22 eV) [175]. In addition to the optical gap, t_d shows a significant effect on the optical constants such as refractive index (n, SnS_2:2.57–2.63, Sn_2S_3: 4.89–7.18) and extinction coefficient (k, SnS_2: 0.69–0.61, Sn_2S_3: 0.0015–0.0019) (Figure 8j,k) [34,56]. However, no previous studies had included the variations in optical constants of o-SnS and c-SnS films with t_d. On the other hand, the t_d reduces the electrical resistivity and improves the carrier density and mobility of carriers in the case of both o-SnS and SnS_2 films [56,177] due to the improved crystallinity and suppression of secondary phases with t_d period. In contrast, in the case of c-SnS and Sn_2S_3 films, there are no reports available in the literature.

3.7. Other Parameters

3.7.1. Substrate Nature and Its Cleaning Process Effect

Generally, thin films require proper mechanical support that provides sufficient adhesion. These supports are commonly called substrates. Substrates have a significant effect on the film properties in practice [268]. Therefore, the choice of a suitable substrate with a specific form for a thin film with a particular application is critical since the substrate must be structurally and chemically compatible with the thin film material in terms of thermal and mechanical stability [269,270]. Moreover, the substrate nature strongly affects the preferred orientation of a thin film, which plays a major role in device performance [271]. Therefore, currently, the exploration of feasible substrates has become an active research area. The CBD has the benefit of allowing thin film deposition on unevenly shaped surfaces. However, the substrate nature greatly affects the deposition process and film quality. Usually, substrates with rough surfaces have better anchoring of the initial deposit in the tiny valleys. Substrates such as glass, tin oxide (TO), indium tin oxide (ITO), and silica/quartz are relatively reactive, owing to the presence of hydroxyl surface groups. Furthermore, when the lattice of the deposited material matches well with that of the substrate, the free energy change is smaller; this facilitates fast nucleation with good morphology and structure. Although the substrate nature has more impact on the process of CBD and the deposited thin film characteristics, there are only a few studies on this area in the case of o-SnS and SnS_2 films and no reports for c-SnS and Sn_2S_3 films (Tables 4 and 5). The reports related to the effects of molybdenum (Mo), ITO, and TO and borosilicate glass substrates [204] on the properties of o-SnS films and the glass, TO, and titanium (Ti) substrates [130] on SnS_2 films are available in the literature. At 0.01 M of Sn and S sources concentrations, both Mo and TO substrates generate o-SnS films with a better surface coverage, whereas the borosilicate glass and ITO substrates produce a discontinuous film with separate agglomerated o-SnS particles. When the concentration of sources is 0.03 M, all substrates except the borosilicate glass form a complete and uniform coverage of o-SnS films. At a high concentration of 0.09 M, all substrates produce a complete coverage of o-SnS films but with a lower adhesive nature [204]. In the case of SnS_2, the amorphous and n-type nature films formed on the glass and Ti substrates, respectively.

In addition to the substrate nature, the cleaning process of the substrate also significantly affects the quality of thin films. Improper cleaning of substrates results in the formation of pinholes in the film, which creates major issues on the fabrication of large-area devices and produces short circuits in solar cells [272]. Unfortunately, there is a lack of research on this area for CBD deposited o-SnS, c-SnS, SnS_2, and Sn_2S_3 thin films.

3.7.2. Stirring Speed and Humidity Effect

In the CBD, chemical solutions with a homogeneous distribution of precursors are necessary before starting the process. Continuous mixing of the reaction solution is mandatory for realizing a uniform thin film deposition [273]. This could be achieved by stirring the solutions at appropriate speeds. At the beginning of the deposition, the stirring speed does not have a significant impact on the growth rate of thin films. However, for longer deposition times, it directly affects the growth rate. In addition, stirring with uneven speed may produce a variation in thin film uniformity and improper diffusion of complex ions toward the substrate [274], and stirring provokes precipitation and reduces the final thickness of the film. Therefore, care must be taken in stirring the solution to obtain the desired quality of thin films.

On the other hand, environmental humidity also influences the formation and physical properties of CBD processed films [273] since the CBD can be performed in an open environment where the gas–liquid interface is influenced by moisture. Even after the deposition of films, they considerably degrade because of their colloidal nature [275]. Therefore, the maintenance of environmental humidity is vital for the deposition of defect-free films. Although the control of stirring speed and environmental humidity is essential for producing quality Sn_xS_y films, there is no systematic study on these effects in the literature.

3.8. Summary

Sn_xS_y are binary metal chalcogenides that have attracted considerable attention due to their abundant, low cost, and nontoxic constituent elements. In comparison to other vacuum and chemical approaches, they may be simply synthesized utilizing a simple non-vacuum CBD methodology. Sn precursors, S precursors, and complexing agents are ideally T(II)C, TA, and TEA, respectively. The following lines are made based on the examination of published data (Tables 4 and 5) and the explanation in Sections 3.1–3.7. Changes in Sn precursor concentration, complexing agent concentration, and T_b can be used to manipulate high-intensity plans and crystallinity. Maintaining complexing agent concentration, bath pH, and t_d, may regulate phase transition and growth rate. Controlling S precursor and complexing agent concentrations results in good morphological, optical, and electrical characteristics. As a result, optimizing each deposition parameter is critical for producing high-quality Sn_xS_y thin films for a variety of applications. However, no previous research has looked at the effect of S precursor concentration on c-SnS, SnS_2, and Sn_2S_3 films; complexing agent concentration on Sn_2S_3 films; bath pH on the properties of SnS_2 and Sn_2S_3 films; T_b on both SnS_2 and Sn_2S_3 films; and t_d on c-SnS and Sn_2S_3 films. Furthermore, for all o-SnS, c-SnS, SnS_2, and Sn_2S_3 thin films, there is a dearth of research on the substrate nature-cleaning procedure, stirring speed, and humidity influence.

According to the description in this part, it is confirmed that further research is required to improve the quality of Sn_xS_y films and more studies are necessary related to the optimization of all deposition parameters. Hence, research focusing on this area is essential.

4. Conclusions

Sn_xS_y thin films deposited with CBD are a relatively recent development, and their process–property correlations must be understood for the desired application. Further, the fabrication of single-phase o-SnS, c-SnS, SnS_2, and Sn_2S_3 thin films in CBD is very condition-dependent. Additionally, it is crucial to identify and separate the o-SnS, c-SnS, SnS_2, and Sn_2S_3 phases. However, until recently, there has been a dearth of detailed studies on the optimization of growth parameters. The present review outlined the background and basic properties of Sn_xS_y (o-SnS, c-SnS, SnS_2, and Sn_2S_3) along with the principle, nucleation, growth, and growth mechanism of Sn_xS_y thin films by CBD. Furthermore, the influence of growth parameters such as precursor concentration (tin source, sulfur source, and complexing agent), bath pH, bath temperature (T_b), deposition time (t_d) on the phase formation, and physical properties of Sn_xS_y thin films were comprehensively described. As a result, the reader should be able to prepare single-phase tin sulfide materials with ease after studying the present article. Hence, the present review should motivate readers to conduct extensive investigations on Sn_xS_y films to develop cost-effective, eco-friendly, and earth-abundant tin sulfide materials to meet all future energy requirements. The connection between the physical properties of Sn_xS_y thin films and their photovoltaic application will be discussed in our subsequent article.

Author Contributions: Conceptualization, V.R.M.R.; writing—original draft preparation, S.G.; review, V.R.M.R., T.R.R.K., and C.P.; Writing final version—review and editing, W.K.K.; supervision, W.K.K. All authors have read and agreed to the published version of the manuscript.

Funding: This study was supported by Priority Research Centers Program through the National Research Foundation of Korea (NRF), funded by the Ministry of Education (2014R1A6A1031189) and the 2019 Yeungnam University Research Grant.

Data Availability Statement: Not applicable.

Abbreviations

Abbreviation	Chemical Name
A	Ammonia
AA	Acetic acid
AC	Ammonium citrate
ACE	Acetone
AF	Ammonium fluoride
AH	Ammonium hydroxide
ALD	Atomic layer deposition
AS	Ammonium sulfide
BT	Baking temperature
CA	Citric acid
CALPHAD	CALculation of PHAse diagram
CBD	Chemical bath deposition
CBM	Conduction band minimum
CBO	Conduction band offset
CSS	Close space sublimation
CUB	Cubic
DDT	Dodecanethiol
DIW	Deionized water
DW	Distilled water

EDS	Energy-dispersive X-ray spectroscopy
EDTA	Ethylenediaminetetraacetic acid
EL	Electrolyte
FF	Fill factor
G	Glass
GA	Glacial acetic acid
GIXRD	Grazing incidence X-ray diffraction
HCL	Hydrochloric acid
HEX	Hexagonal
HH	Hydrazine hydrate
HWVD	Hot wall vapor deposition
ITO	Indium tin oxide
JCPDS	Joint committee on powder diffraction standards
Li	Lithium
MeOH	Methanol
Mo	Molybdenum
Na	Sodium
Na_2EDTA	Disodium ethylenediaminetetraacetate
NTA	Nitriloacetic acid
ODE	Ooctadecene
OLA	Oleylamine
ORT	Orthorhombic
PG	Propylene glycol
PL	Photoluminescence
QE	Quantum efficiency
RS	Rock salt
SAED	Selected area electron diffraction
SCR	Space charge region
SDS	Sodium sulfide
Si	Silicon
SILAR	Successive ionic layer adsorption and reaction
SIMS	Secondary ion mass spectrometry
SS	Stainless steel
SnS	Tin monosulfide
SnS_2	Tin disulfide
Sn_2S_3	Tin sesquisulfide
ST	Sodium thiosulfate
TA	Thioacetamide
T(II)C	Tin (II) chloride dehydrate
TC(IV)	Tin(IV) chloride pentahydrate
TEA	Triethanolamine
TEM	Transmission electron microscopy
Ti	Titanium
TO	Tin oxide
TOP	Trioctylphosphine oxide
T_r	Room temperature
TSC	Trisodium citrate
TTA	Tartaric acid
TU	Thiourea
UAED	Ultrasound-assisted electrodeposition
VBM	Valence band maximum
XRD	X-ray diffraction
ZB	Zinc blended

References

1. De Elisa, R. Cadmium telluride solar cells: Selenium diffusion unveiled. *Nat. Energy* **2016**, *1*, 16143–16146.
2. Solar Frontier Achieves World Record Thin-Film Solar Cell Efficiency of 23.35%. Available online: http://www.solar-frontier.com/eng/news/2019/0117_press.html (accessed on 14 July 2021).
3. Chen, X.; Hou, Y.; Zhang, B.; Yang, X.H.; Yang, H.G. Low-cost SnS_x counter electrodes for dye-sensitized solar cells. *Chem. Commun.* **2013**, *49*, 5793–5795. [CrossRef] [PubMed]
4. PV Division | Business Information | KITAGAWA SEIKI Co., Ltd. Available online: http://www.kitagawaseiki.co.jp/en/jigyo_pv.html (accessed on 14 July 2021).
5. Ge, J.; Zhang, Y.; Heo, Y.-J.; Park, S.-J. Advanced Design and Synthesis of Composite Photocatalysts for the Remediation of Wastewater: A Review. *Catalysts* **2019**, *9*, 122. [CrossRef]
6. Liao, C.-H.; Huang, C.-W.; Wu, J.C.S. Hydrogen Production from Semiconductor-based Photocatalysis via Water Splitting. *Catalysts* **2012**, *2*, 490–516. [CrossRef]
7. Zhao, Y.; Yang, Q.; Chang, Y.; Pang, W.; Zhang, H.; Duan, X. Novel Gas Sensor Arrays Based on High-Q SAM-Modified Piezotransduced Single-Crystal Silicon Bulk Acoustic Resonators. *Sensors* **2017**, *17*, 1507. [CrossRef]
8. Campaña, A.; Florez, S.; Noguera, M.; Fuentes, O.; Ruiz Puentes, P.; Cruz, J.; Osma, J. Enzyme-Based Electrochemical Biosensors for Microfluidic Platforms to Detect Pharmaceutical Residues in Wastewater. *Biosensors* **2019**, *9*, 41. [CrossRef] [PubMed]
9. Lee, S.-H.; Park, J.-U.; Kim, G.; Jee, D.-W.; Kim, J.H.; Kim, S. Rigorous Study on Hump Phenomena in Surrounding Channel Nanowire (SCNW) Tunnel Field-Effect Transistor (TFET). *Appl. Sci.* **2020**, *10*, 3596. [CrossRef]
10. Mohammed, M.; Chun, D.-M. Electrochemical Performance of Few-Layer Graphene Nano-Flake Supercapacitors Prepared by the Vacuum Kinetic Spray Method. *Coatings* **2018**, *8*, 302. [CrossRef]
11. Wang, Y.; Mao, Y.; Ji, Q.; Yang, M.; Yang, Z.; Lin, H. Electrostatic Discharge Characteristics of SiGe Source/Drain PNN Tunnel FET. *Electronics* **2021**, *10*, 454. [CrossRef]
12. Li, S.; Lam, K.H.; Cheng, K.W.E. The Thermoelectric Analysis of Different Heat Flux Conduction Materials for Power Generation Board. *Energies* **2017**, *10*, 1781. [CrossRef]
13. Kumar, S.G.; Rao, K.S.R.K. Physics and chemistry of CdTe/CdS thin film heterojunction photovoltaic devices: Fundamental and critical aspects. *Energy Environ. Sci.* **2014**, *7*, 45–102. [CrossRef]
14. Alzoubi, T.; Moustafa, M. Numerical optimization of absorber and CdS buffer layers in CIGS solar cells using SCAPS. *Int. J. Smart Grid Clean Energy* **2019**, *8*, 291–298. [CrossRef]
15. Bag, S.; Gunawan, O.; Gokmen, T.; Zhu, Y.; Todorov, T.K.; Mitzi, D.B. Low band gap liquid-processed CZTSe solar cell with 10.1% efficiency. *Energy Environ. Sci.* **2012**, *5*, 7060–7065. [CrossRef]
16. Jiang, F.; Shen, H.; Gao, C.; Liu, B.; Lin, L.; Shen, Z. Preparation and properties of SnS film grown by two-stage process. *Appl. Surf. Sci.* **2011**, *257*, 4901–4905. [CrossRef]
17. Sreedevi, G.; Vasudeva Reddy, M.; Park, C.; Jeon, C.W.; Ramakrishna Reddy, K.T. Comprehensive optical studies on SnS layers synthesized by chemical bath deposition. *Opt. Mater.* **2015**, *42*, 468–475. [CrossRef]
18. Koteeswara Reddy, N.; Ramakrishna Reddy, K.T. Preparation and characterisation of sprayed tin sulphide films grown at different precursor concentrations. *Mater. Chem. Phys.* **2007**, *102*, 13–18. [CrossRef]
19. Jiang, F.; Shen, H.; Wang, W.; Zhang, L. Preparation of SnS film by sulfurization and SnS/a-Si heterojunction solar cells. *J. Electrochem. Soc.* **2012**, *159*, H235. [CrossRef]
20. Devika, M.; Koteeswara Reddy, N.; Venkatramana Reddy, S.; Ramesh, K.; Gunasekhar, K.R. Influence of rapid thermal annealing (RTA) on the structural and electrical properties of SnS films. *J. Mater. Sci. Mater. Electron.* **2009**, *20*, 1129–1134. [CrossRef]
21. Noguchi, H.; Setiyadi, A.; Tanamura, H.; Nagatomo, T.; Omoto, O. Characterization of vacuum-evaporated tin sulfide film for solar cell materials. *Sol. Energy Mater. Sol. Cells* **1994**, *35*, 325–331. [CrossRef]
22. Ogah, O.E.; Zoppi, G.; Forbes, I.; Miles, R.W. Thin films of tin sulphide for use in thin film solar cell devices. *Thin Solid Films* **2009**, *517*, 2485–2488. [CrossRef]
23. Ragina, A.J.; Murali, K.V.; Preetha, K.C.; Deepa, K.; Remadevi, T.L. A Study of optical parameters of tin sulphide thin films using the swanepoel method. In *AIP Conference Proceedings*; American Institute of Physics: College Park, MD, USA, 2011; Volume 1391, pp. 752–754.
24. Mahdi, M.S.; Ibrahim, K.; Hmood, A.; Ahmed, N.M.; Mustafa, F.I.; Azzez, S.A. High performance near infrared photodetector based on cubic crystal structure SnS thin film on a glass substrate. *Mater. Lett.* **2017**, *200*, 10–13. [CrossRef]
25. Nair, P.K.; Garcia-Angelmo, A.R.; Nair, M.T.S. Cubic and orthorhombic SnS thin-film absorbers for tin sulfide solar cells. *Phys. Status Solidi Appl. Mater. Sci.* **2016**, *213*, 170–177. [CrossRef]
26. Chalapathi, U.; Poornaprakash, B.; Park, S.-H. Growth and properties of cubic SnS films prepared by chemical bath deposition using EDTA as the complexing agent. *J. Alloys Compd.* **2016**, *689*, 938–944. [CrossRef]
27. Mahdi, M.S.; Ibrahim, K.; Ahmed, N.M.; Hmood, A.; Azzez, S.A.; Mustafa, F.I.; Bououdina, M. Influence of pH value on structural, optical and photoresponse properties of SnS films grown via chemical bath deposition. *Mater. Lett.* **2018**, *210*, 279–282. [CrossRef]
28. Garcia-Angelmo, A.R.; Romano-Trujillo, R.; Campos-Álvarez, J.; Gomez-Daza, O.; Nair, M.T.S.; Nair, P.K. Thin film solar cell of SnS absorber with cubic crystalline structure. *Phys. Status Solidi* **2015**, *212*, 2332–2340. [CrossRef]
29. Chalapathi, U.; Poornaprakash, B.; Park, S.H. Effect of post-deposition annealing on the growth and properties of cubic SnS films. *Superlattices Microstruct.* **2017**, *103*, 221–229. [CrossRef]

30. Alpen, U.V.; Fenner, J.; Gmelin, E. Semiconductors of the type MeIIMeIVS3. *Strateg. Surv.* **1975**, *76*, 50–54. [CrossRef]
31. Lopez, S.; Ortiz, A. Spray pyrolysis deposition of Sn_xS_y thin films. *Semicond. Sci. Technol.* **1994**, *9*, 2130–2133. [CrossRef]
32. Godoy-Rosas, R.; Barraza-Felix, S.; Ramirez-Bon, R.; Ochoa-Landin, R.; Pineda-Leon, H.A.; Flores-Acosta, M.; Ruvalcaba-Manzo, S.G.; Acosta-Enriquez, M.C.; Castillo, S.J. Synthesis and characterization of Sn_2S_3 as nanoparticles, powders and thin films, using soft chemistry reactions. *Chalcogenide Lett.* **2017**, *14*, 365–371.
33. Koteeswara Reddy, N.; Ramakrishna Reddy, K.T. Optical behaviour of sprayed tin sulphide thin films. *Mater. Res. Bull.* **2006**, *41*, 414–422. [CrossRef]
34. Guneri, E.; Gode, F.; Boyarbay, B.; Gumus, C. Structural and optical studies of chemically deposited Sn_2S_3 thin films. *Mater. Res. Bull.* **2012**, *47*, 3738–3742. [CrossRef]
35. Kanevce, A.; Reese, M.O.; Barnes, T.M.; Jensen, S.A.; Metzger, W.K. The roles of carrier concentration and interface, bulk, and grain-boundary recombination for 25% efficient CdTe solar cells. *J. Appl. Phys.* **2017**, *121*, 214506. [CrossRef]
36. Ristov, M.; Sinadinovski, G.; Grozdanov, I.; Mitreski, M. Chemical deposition of Tin(II) sulphide thin films. *Thin Solid Films* **1989**, *173*, 53–58. [CrossRef]
37. Acharya, S.; Srivastava, O.N. Electronic behaviour of SnS_2 crystals. *Phys. Status Solidi* **1981**, *65*, 717–723. [CrossRef]
38. Varkey, A.J. Preparation of tin disulphide thin films by solution growth. *Int. J. Mater. Prod. Technol.* **1997**, *12*, 490–495. [CrossRef]
39. Shi, C.; Chen, Z.; Shi, G.; Sun, R.; Zhan, X.; Shen, X. Influence of annealing on characteristics of tin disulfide thin films by vacuum thermal evaporation. *Thin Solid Films* **2012**, *520*, 4898–4901. [CrossRef]
40. Fadavieslam, M.R.; Shahtahmasebi, N.; Rezaee-Roknabadi, M.; Bagheri-Mohagheghi, M.M. Effect of deposition conditions on the physical properties of Sn_xS_y thin films prepared by the spray pyrolysis technique. *J. Semicond.* **2011**, *32*, 113002. [CrossRef]
41. Ramakrishna Reddy, K.T.; Sreedevi, G.; Ramya, K.; Miles, R. Physical properties of nano-crystalline SnS_2 layers grown by chemical bath deposition. *Energy Procedia* **2012**, *15*, 340–346. [CrossRef]
42. Ghorpade, U.; Suryawanshi, M.; Shin, S.W.; Gurav, K.; Patil, P.; Pawar, S.; Hong, C.W.; Kim, J.H.; Kolekar, S. Towards environmentally benign approaches for the synthesis of CZTSSe nanocrystals by a hot injection method: A status review. *Chem. Commun.* **2014**, *50*, 11258. [CrossRef]
43. Avellaneda, D.; Nair, M.T.S.; Nair, P.K. Polymorphic tin sulfide thin films of zinc blende and orthorhombic structures by chemical deposition. *J. Electrochem. Soc.* **2008**, *155*, D517. [CrossRef]
44. Gode, F.; Guneri, E.; Baglayan, O. Effect of tri-sodium citrate concentration on structural, optical and electrical properties of chemically deposited tin sulfide films. *Appl. Surf. Sci.* **2014**, *318*, 227–233. [CrossRef]
45. Koteeswara Reddy, N.; Ramakrishna Reddy, K.T. Electrical properties of spray pyrolytic tin sulfide films. *Solid State Electron.* **2005**, *49*, 902–906. [CrossRef]
46. Sinsermsuksakul, P.; Heo, J.; Noh, W.; Hock, A.S.; Gordon, R.G. Atomic layer deposition of tin monosulfide thin films. *Adv. Energy Mater.* **2011**, *1*, 1116–1125. [CrossRef]
47. Wangperawong, A.; Herron, S.M.; Runser, R.R.; Hagglund, C.; Tanskanen, J.T.; Lee, H.B.R.; Clemens, B.M.; Bent, S.F. Vapor transport deposition and epitaxy of orthorhombic SnS on glass and NaCl substrates. *Appl. Phys. Lett.* **2013**, *103*, 052105. [CrossRef]
48. Djessas, K.; Masse, G. SnS thin films grown by close-spaced vapor transport. *J. Mater. Sci. Lett.* **2000**, *19*, 2135–2137. [CrossRef]
49. Albers, W.; Haas, C.; Vink, H.J.; Wasscher, J.D. Investigations on SnS. *J. Appl. Phys.* **1961**, *32*, 2220–2225. [CrossRef]
50. Chalapathi, U.; Poornaprakash, B.; Park, S.H. Chemically deposited cubic SnS thin films for solar cell applications. *Sol. Energy* **2016**, *139*, 238–248. [CrossRef]
51. Skelton, J.M.; Burton, L.A.; Jackson, A.J.; Oba, F.; Parker, S.C.; Walsh, A. Lattice dynamics of the tin sulphides SnS_2, SnS and Sn_2S_3: Vibrational spectra and thermal transport. *Phys. Chem. Chem. Phys.* **2017**, *19*, 12452–12465. [CrossRef] [PubMed]
52. Kherchachi, I.B.; Attaf, A.; Saidi, H.; Bouhdjar, A.; Bendjdidi, H.; Youcef, B.; Azizi, R. The synthesis, characterization and phase stability of tin sulfides (SnS_2, SnS and Sn_2S_3) films deposited by ultrasonic spray. *Main Group Chem.* **2016**, *15*, 231–242. [CrossRef]
53. Chen, B.; Xu, X.; Wang, F.; Liu, J.; Ji, J. Electrochemical preparation and characterization of three-dimensional nanostructured Sn_2S_3 semiconductor films with nanorod network. *Mater. Lett.* **2011**, *65*, 400–402. [CrossRef]
54. Julien, C.; Eddrief, M.; Samaras, I.; Balkanski, M. Optical and electrical characterizations of SnSe, SnS_2 and $SnSe_2$ single crystals. *Mater. Sci. Eng. B* **1992**, *15*, 70–72. [CrossRef]
55. Madelung, O. IV-VII2 compounds. In *Semiconductors: Data Handbook*; Springer: Berlin/Heidelberg, Germany, 2004; pp. 606–612.
56. Sreedevi, G.; Vasudeva Reddy, M.R.; Babu, P.; Chinho, P.; Chan Wook, J.; Ramakrishna Reddy, K.T. Studies on chemical bath deposited SnS_2 films for Cd-free thin film solar cells. *Ceram. Int.* **2017**, *43*, 3713–3719. [CrossRef]
57. Chen, S.; Gong, X.G.; Walsh, A.; Wei, S.-H. Defect physics of the kesterite thin-film solar cell absorber Cu_2ZnSnS_4. *Appl. Phys. Lett.* **2010**, *96*, 021902. [CrossRef]
58. Ettema, A.R.H.F.; de Groot, R.A.; Haas, C.; Turner, T.S. Electronic structure of SnS deduced from photoelectron spectra and band-structure calculations. *Phys. Rev. B* **1992**, *46*, 7363–7373. [CrossRef]
59. Lippens, P.E.; El Khalifi, M.; Womes, M. Electronic structures of SnS and SnS_2. *Phys. Status Solidi* **2016**, *254*, 1600194. [CrossRef]
60. Skelton, J.M.; Burton, L.A.; Oba, F.; Walsh, A. Chemical and lattice stability of the tin sulfides. *J. Phys. Chem. C* **2017**, *121*, 6446–6454. [CrossRef] [PubMed]
61. Lindwall, G.; Shang, S.; Kelly, N.R.; Anderson, T.; Liu, Z.K. Thermodynamics of the S-Sn system: Implication for synthesis of earth abundant photovoltaic absorber materials. *Sol. Energy* **2016**, *125*, 314–323. [CrossRef]

62. Wang, W.; Winkler, M.T.; Gunawan, O.; Gokmen, T.; Todorov, T.K.; Zhu, Y.; Mitzi, D.B. Device characteristics of CZTSSe thin-film solar cells with 12.6% efficiency. *Adv. Energy Mater.* **2014**, *4*, 1–5. [CrossRef]
63. Loferski, J.J. Theoretical considerations governing the choice of the optimum semiconductor for photovoltaic solar energy conversion. *J. Appl. Phys.* **1956**, *27*, 777–784. [CrossRef]
64. Skelton, J.M.; Burton, L.A.; Oba, F.; Walsh, A. Metastable cubic tin sulfide: A novel phonon-stable chiral semiconductor. *APL Mater.* **2017**, *5*, 036101. [CrossRef]
65. Herzenbergite. Available online: https://www.mindat.org/min-1880.html (accessed on 13 February 2018).
66. Ottemannite: Ottemannite Mineral Information and Data. Available online: https://www.mindat.org/min-3042.html (accessed on 13 February 2018).
67. Berndtite: Berndtite Mineral Information and Data. Available online: https://www.mindat.org/min-637.html (accessed on 13 February 2018).
68. Anthony, J.W.; Bideaux, R.A.; Bladh, K.W.; Nichols, M.C. *Handbook of Mineralogy-Borates, Carbonates, Sulfates*; Mineral Data Publishing: Tucson, AZ, USA, 2004; Volume V, p. 221.
69. Eastaugh, N.; Walsh, V.; Chaplin, T.; Ruth, S. *Pigment Compendium: A Dictionary and Optical Microscopy of Historical Pigments*; Butterworth-Heinemann: Oxfordshire, UK, 2008; ISBN 0750689803.
70. Moh, G.H.; Berndt, F. Two new natural tin sulfides, Sn_2S_3 and SnS_2. *New Yearb. Mineral. Monatshefte* **1964**, *4*, 94–95.
71. Yao, K.; Li, J.; Shan, S.; Jia, Q. One-step synthesis of urchinlike SnS/SnS_2 heterostructures with superior visible-light photocatalytic performance. *Catal. Commun.* **2017**, *101*, 51–56. [CrossRef]
72. Jing, J.; Cao, M.; Wu, C.; Huang, J.; Lai, J.; Sun, Y.; Wang, L.; Shen, Y. Chemical bath deposition of SnS nanosheet thin films for FTO/SnS/CdS/Pt photocathode. *J. Alloys Compd.* **2017**, *726*, 720–728. [CrossRef]
73. Zhang, F.; Zhang, Y.; Zhang, G.; Yang, Z.; Dionysiou, D.D.; Zhu, A. Exceptional synergistic enhancement of the photocatalytic activity of SnS_2 by coupling with polyaniline and N-doped reduced graphene oxide. *Appl. Catal. B Environ.* **2018**, *236*, 53–63. [CrossRef]
74. Sun, Y.; Cheng, H.; Gao, S.; Sun, Z.; Liu, Q.; Leu, Q.; Lei, F.; Yao, T.; He, J.; Wei, S.; et al. Freestanding tin disulfide single-layers realizing efficient visible-light water splitting. *Angew. Chem. Int. Ed.* **2012**, *51*, 8727–8731. [CrossRef]
75. Chauhan, H.; Singh, M.K.; Kumar, P.; Hashmi, S.A.; Deka, S. Development of SnS_2/RGO nanosheet composite for cost-effective aqueous hybrid supercapacitors. *Nanotechnology* **2017**, *28*, 025401. [CrossRef] [PubMed]
76. Wang, Y.; Huang, L.; Wei, Z. Photoresponsive field-effect transistors based on multilayer SnS_2 nanosheets. *J. Semicond.* **2017**, *38*, 034001. [CrossRef]
77. Wu, Q.; Jiao, L.; Du, J.; Yang, J.; Guo, L.; Liu, Y.; Wang, Y. One-pot synthesis of three-dimensional SnS_2 hierarchitectures as anode material for lithium-ion batteries. *J. Power Source* **2013**, *239*, 89–93. [CrossRef]
78. Li, H.; Zhou, M.; Li, W.; Wang, K.; Cheng, S.; Jiang, K. Layered SnS_2 cross-linked by carbon nanotubes as a high performance anode for sodium ion batteries. *RSC Adv.* **2016**, *6*, 35197–35202. [CrossRef]
79. Ou, J.Z.; Ge, W.; Carey, B.; Daeneke, T.; Rotbart, A.; Shan, W.; Wang, Y.; Fu, Z.; Chrimes, A.F.; Wlodarski, W.; et al. Physisorption-based charge transfer in two-dimensional SnS_2 for selective and reversible NO_2 gas sensing. *ACS Nano* **2015**, *9*, 10313–10323. [CrossRef] [PubMed]
80. Sanchez-Juarez, A.; Tiburcio-Silver, A.; Ortiz, A. Fabrication of SnS_2/SnS heterojunction thin film diodes by plasma-enhanced chemical vapor deposition. *Thin Solid Films* **2005**, *480*, 452–456. [CrossRef]
81. Fan, C.; Li, Y.; Lu, F.Y.; Deng, H.X.; Wei, Z.M.; Li, J.B. Wavelength dependent UV-Vis photodetectors from SnS_2 flakes. *RSC Adv.* **2016**, *6*, 422–427. [CrossRef]
82. Singh, D.J. Optical and electronic properties of semiconducting Sn_2S_3. *Appl. Phys. Lett.* **2016**, *109*, 1–5. [CrossRef]
83. Li, Y.; Xie, H.; Tu, J. Nanostructured SnS/carbon composite for supercapacitor. *Mater. Lett.* **2009**, *63*, 1785–1787. [CrossRef]
84. Saito, W.; Hayashi, K.; Nagai, H.; Miyazaki, Y. Preparation and thermoelectric properties of mixed valence compound Sn_2S_3. *Jpn. J. Appl. Phys.* **2017**, *56*, 061201. [CrossRef]
85. Devika, M.; Koteeswara Reddy, N.; Sreekantha Reddy, D.; Ramesh, K.; Gunasekhar, K.; Raja Gopal, E.; Sung Ha, P. Metal–insulator–semiconductor field-effect transistors (MISFETs) using p-type SnS and nanometer-thick Al_2S_3 layers. *RSC Adv.* **2017**, *7*, 11111–11117. [CrossRef]
86. Wang, W.; Shi, L.; Lan, D.; Li, Q. Improving cycle stability of SnS anode for sodium-ion batteries by limiting Sn agglomeration. *J. Power Source* **2018**, *377*, 1–6. [CrossRef]
87. Lian, Q.; Zhou, G.; Liu, J.; Wu, C.; Wei, W.; Chen, L.; Li, C. Extrinsic pseudocapacitve Li-ion storage of SnS anode via lithiation-induced structural optimization on cycling. *J. Power Source* **2017**, *366*, 1–8. [CrossRef]
88. Afsar, M.F.; Rafiq, M.A.; Tok, A.I.Y. Two-dimensional SnS nanoflakes: Synthesis and application to acetone and alcohol sensors. *RSC Adv.* **2017**, *7*, 21556–21566. [CrossRef]
89. Chung, R.J.; Wang, A.N.; Peng, S.Y. An enzymatic glucose sensor composed of carbon-coated nano tin sulfide. *Nanomaterials* **2017**, *7*, 39. [CrossRef]
90. Wang, C.; Chen, Y.; Jiang, J.; Zhang, R.; Niu, Y.; Zhou, T.; Xia, J.; Tian, H.; Hu, J.; Yang, P. Improved thermoelectric properties of SnS synthesized by chemical precipitation. *RSC Adv.* **2017**, *7*, 16795–16800. [CrossRef]
91. Jayalakshmi, M.; Mohan Rao, M.; Choudary, B.M. Identifying nano SnS as a new electrode material for electrochemical capacitors in aqueous solutions. *Electrochem. Commun.* **2004**, *6*, 1119–1122. [CrossRef]

92. Nozaki, H.; Fukano, T.; Ohta, S.; Seno, Y.; Katagiri, H.; Jimbo, K. Crystal structure determination of solar cell materials: Cu_2ZnSnS_4 thin films using X-ray anomalous dispersion. *J. Alloys Compd.* **2012**, *524*, 22–25. [CrossRef]

93. Burton, L.A.; Walsh, A. Phase stability of the earth-abundant tin sulfides SnS, SnS_2, and Sn_2S_3. *J. Phys. Chem. C* **2012**, *116*, 24262–24267. [CrossRef]

94. Ahmet, I.Y.; Hill, M.S.; Johnson, A.L.; Peter, L.M. polymorph-selective deposition of high purity SnS thin films from a single source precursor. *Chem. Mater.* **2015**, *27*, 7680–7688. [CrossRef]

95. Ehm, L.; Knorr, K.; Dera, P.; Krimmel, A.; Bouvier, P.; Mezouar, M. Pressure-induced structural phase transition in the IV–VI semiconductor SnS. *J. Phys. Condens. Matter* **2004**, *16*, 3545–3554. [CrossRef]

96. Brownson, J.R.S.; Georges, C.; Larramona, G.; Jacob, A.; Delatouche, B.; Levy-Clement, C. Chemistry of tin monosulfide (δ-SnS) electrodeposition. *J. Electrochem. Soc.* **2008**, *155*, D40. [CrossRef]

97. Brownson, J.R.S.; Georges, C.; Levy-Clement, C. Synthesis of a δ-SnS polymorph by electrodeposition. *Chem. Mater.* **2006**, *18*, 6397–6402. [CrossRef]

98. Mariano, A.N.; Chopra, K.L. Polymorphism in some IV-VI compounds induced by high pressure and thin-film epitaxial growth. *Appl. Phys. Lett.* **1967**, *10*, 282–284. [CrossRef]

99. Rabkin, A.; Samuha, S.; Abutbul, R.E.; Ezersky, V.; Meshi, L.; Golan, Y. New nanocrystalline materials: A previously unknown simple cubic phase in the SnS binary system. *Nano Lett.* **2015**, *15*, 2174–2179. [CrossRef]

100. Biacchi, A.J.; Vaughn, D.D.; Schaak, R.E. Synthesis and crystallographic analysis of shape-controlled SnS nanocrystal photocatalysts: Evidence for a pseudotetragonal structural modification. *J. Am. Chem. Soc.* **2013**, *135*, 11634–11644. [CrossRef] [PubMed]

101. Wang, X.; Liu, Z.; Zhao, X.-G.; Lv, J.; Biswas, K.; Zhang, L. Computational design of mixed-valence tin sulfides as solar absorbers. *ACS Appl. Mater. Interfaces* **2019**, *11*, 24867–24875. [CrossRef] [PubMed]

102. Sugaki, A.; Kitakaze, A.; Kitazawa, H. Synthesized tin and tin-silver sulfide minerals: Synthetic sulfide minerals (XIII). *Sci. Rep.* **1985**, *3*, 199–211.

103. Guenter, J.R.; Oswald, H. New polytype form of tin (IV) sulfide. *Nat. Sci.* **1968**, *55*, 171–177. [CrossRef]

104. Mosburg, S.; Ross, D.R.; Bethke, P.M.; Toulmin, P. X-ray powder data for herzenbergite, teallite and tin trisulfide. *US Geol. Surv. Prof. Pap. C* **1961**, *424*, 347.

105. Babu, P.; Vasudeva Reddy, M.; Sreedevi, G.; Chinho, P. Review on earth-abundant and environmentally benign Cu–Sn–X(X = S, Se) nanoparticles by chemical synthesis for sustainable solar energy conversion. *J. Ind. Eng. Chem.* **2018**, *60*, 19–52. [CrossRef]

106. Nikolic, P.M.; Mihajlovic, P.; Lavrencic, B. Splitting and coupling of lattice modes in the layer compound SnS. *J. Phys. C Solid State Phys.* **1977**, *10*, L289–L292. [CrossRef]

107. Wiley, J.D.; Buckel, W.J.; Schmidt, R.L. Infrared reflectivity and Raman scattering in GeS. *Phys. Rev. B* **1976**, *13*, 2489–2496. [CrossRef]

108. Smith, A.J.; Meek, P.E.; Liang, W.Y. Raman scattering studies of SnS_2 and $SnSe_2$. *J. Phys. C Solid State Phys.* **1977**, *10*, 1321–1323. [CrossRef]

109. Chandrasekhar, H.R.; Mead, D.G. Long-wavelength phonons in mixed-valence semiconductor SnIISnIVS3. *Phys. Rev. B* **1979**, *19*, 932–937. [CrossRef]

110. Gedi, S.; Reddy, V.R.M.; Kang, J.; Jeon, C.-W. Impact of high temperature and short period annealing on SnS films deposited by E-beam evaporation. *Appl. Surf. Sci.* **2017**, *402*, 463–468. [CrossRef]

111. Abutbul, R.E.; Segev, E.; Zeiri, L.; Ezersky, V.; Makov, G.; Golan, Y. Synthesis and properties of nanocrystalline π-SnS—A new cubic phase of tin sulphide. *RSC Adv.* **2016**, *6*, 5848–5855. [CrossRef]

112. Bharatula, L.D.; Erande, M.B.; Mulla, I.S.; Rout, C.S.; Late, D.J. SnS_2 nanoflakes for efficient humidity and alcohol sensing at room temperature. *RSC Adv.* **2016**, *6*, 105421–105427. [CrossRef]

113. Reddy, T.S.; Kumar, M.C.S. Effect of substrate temperature on the physical properties of co-evaporated Sn_2S_3 thin films. *Ceram. Int.* **2016**, *42*, 12262–12269. [CrossRef]

114. Kumagai, Y.; Burton, L.A.; Walsh, A.; Oba, F. Electronic structure and defect physics of tin sulfides: SnS, Sn_2S_3, and SnS_2. *Phys. Rev. Appl.* **2016**, *6*, 1–14. [CrossRef]

115. Burton, L.A.; Colombara, D.; Abellon, R.D.; Grozema, F.C.; Peter, L.M.; Savenije, T.J.; Dennler, G.; Walsh, A. Synthesis, characterization, and electronic structure of single-crystal SnS, Sn_2S_3, and SnS_2. *Chem. Mater.* **2013**, *25*, 4908–4916. [CrossRef]

116. Vasudeva Reddy, M.R.; Mohan Reddy, P.; Phaneendra Reddy, G.; Sreedevi, G.; Kishore Kumar, Y.B.R.; Babu, P.; Woo Kyoung, K.; Ramakrishna Reddy, K.T.; Chinho, P. Review on Cu_2SnS_3, Cu_3SnS_4, and Cu_4SnS_4 thin films and their photovoltaic performance. *J. Ind. Eng. Chem.* **2019**, *76*, 39–74. [CrossRef]

117. Khoa, D.Q.; Nguyen, C.V.; Phuc, H.V.; Ilyasov, V.V.; Vu, T.V.; Cuong, N.Q.; Hoi, B.D.; Lu, D.V.; Feddi, E.; El-Yadri, M.; et al. Effect of strains on electronic and optical properties of monolayer SnS: Ab-initio study. *Phys. B Condens. Matter* **2018**, *545*, 255–261. [CrossRef]

118. Segev, E.; Abutbul, R.E.; Argaman, U.; Golan, Y.; Makov, G. Surface energies and nanocrystal stability in the orthorhombic and π-phases of tin and germanium monochalcogenides. *CrystEngComm* **2018**, *20*, 4237–4248. [CrossRef]

119. Qiu, G.; Zhang, H.; Liu, Y.; Xia, C. Strain effect on the electronic properties of Ce-doped SnS_2 monolayer. *Phys. B Condens. Matter* **2018**, *547*, 1–5. [CrossRef]

120. Ullah, H.; Noor-A-Alam, M.; Kim, H.J.; Shin, Y.H. Influences of vacancy and doping on electronic and magnetic properties of monolayer SnS. *J. Appl. Phys.* **2018**, *124*. [CrossRef]

121. Zandalazini, C.I.; Navarro Sanchez, J.; Albanesi, E.A.; Gupta, Y.; Arun, P. Contribution of lattice parameter and vacancies on anisotropic optical properties of tin sulphide. *J. Alloys Compd.* **2018**, *746*, 9–18. [CrossRef]

122. Akhoundi, E.; Faghihnasiri, M.; Memarzadeh, S.; Firouzian, A.H. Mechanical and strain-tunable electronic properties of the SnS monolayer. *J. Phys. Chem. Solids* **2019**, *126*, 43–54. [CrossRef]

123. Sreedevi, G.; Vasudeva Reddy, M.; Babu, P.; Chan-Wook, J.; Chinho, P.; Ramakrishna Reddy, K.T. A facile inexpensive route for SnS thin film solar cells with SnS$_2$ buffer. *Appl. Surf. Sci.* **2016**, *372*, 116–124. [CrossRef]

124. Srivind, J.; Nagarethinam, V.S.; Balu, A.R. Optimization of S:Sn precursor molar concentration on the physical properties of spray deposited single phase Sn$_2$S$_3$ thin films. *Mater. Sci. Pol.* **2016**, *34*, 393–398. [CrossRef]

125. Von Schnering, H.G.; Wiedemeier, H. The high temperature structure of ß-SnS and ß-SnSe and the B16-to-B33 type λ-transition path. *J. Crystallogr. Cryst. Mater.* **1981**, *156*, 143–150. [CrossRef]

126. Hazen RM and Finger LW The crystal structres and compressibilities of layer minerals at high pressure. I SnS$_2$, berndite. *Am. Mineral.* **1978**, *63*, 289–292.

127. Park, H.K.; Jo, J.; Hong, H.K.; Song, G.Y.; Heo, J. Structural, optical, and electrical properties of tin sulfide thin films grown with electron-beam evaporation. *Curr. Appl. Phys.* **2015**, *15*, 964–969. [CrossRef]

128. Nair, M.T.S.; Nair, P.K. Simplified chemical deposition technique for good quality SnS thin films. *Semicond. Sci. Technol* **1991**, *6*, 132–134. [CrossRef]

129. Price, L.S.; Parkin, I.P.; Hardy, A.M.E.; Clark, R.J.H.; Hibbert, T.G.; Molloy, K.C. Atmospheric pressure chemical vapor deposition of tin sulfides (SnS, Sn$_2$S$_3$, and SnS$_2$) on glass. *Chem. Mater.* **1999**, *11*, 1792–1799. [CrossRef]

130. Lokhande, C.D. A chemical method for tin disulphide thin film deposition. *J. Phys. D Appl. Phys.* **1990**, *23*, 1703–1705. [CrossRef]

131. Li, J.; Zhang, Y.C.; Zhang, M. Preparation of SnS$_2$ thin films by chemical bath deposition. *Mater. Sci. Forum* **2010**, *663*, 104–107. [CrossRef]

132. Thangaraju, B.; Kaliannan, P. Spray pyrolytic deposition and characterization of SnS and SnS$_2$ thin films. *J. Phys. D Appl. Phys.* **2000**, *33*, 1054–1059. [CrossRef]

133. Gopalakrishnan, P.; Anbazhagan, G.; Vijayarajasekaran, J.; Vijayakumar, K. Effect of substrate temperature on tin disulphide thin films. *Int. J. Thin Film. Sci. Technol.* **2017**, *6*, 73–75. [CrossRef]

134. Tomas, R.; Lukas, S.; Libor, D.; Marek, B.; Milan, V.; Ludvik, B.; Tomas, W.; Roman, J. SnS and SnS$_2$ thin films deposited using a spin-coating technique from intramolecularly coordinated organotin sulfides. *Appl. Organomet. Chem.* **2015**, *29*, 176–180. [CrossRef]

135. Anitha, N.; Anitha, M.; Amalraj, L. Influence of precursor solution volume on the properties of tin disulphide (SnS$_2$) thin films prepared by nebulized spray pyrolysis technique. *Opt. Int. J. Light Electron Opt.* **2017**, *148*, 28–38. [CrossRef]

136. Vijayakumar, K.; Sanjeeviraja, C.; Jayachandran, M.; Amalraj, L. Characterization of tin disulphide thin films prepared at different substrate temperature using spray pyrolysis technique. *J. Mater. Sci. Mater. Electron.* **2011**, *22*, 929–935. [CrossRef]

137. Sanchez-Juarez, A.; Ortz, A. Effects of precursor concentration on the optical and electrical properties of Sn$_x$S$_y$ thin films prepared by plasma-enhanced chemical vapour deposition. *Semicond. Sci. Technol.* **2002**, *17*, 931–937. [CrossRef]

138. Amalraj, L.; Sanjeeviraja, C.; Jayachandran, M. Spray pyrolysised tin disulphide thin film and characterisation. *J. Cryst. Growth* **2002**, *234*, 683–689. [CrossRef]

139. Ragina, A.J.; Murali, K.V.; Preetha, K.C.; Deepa, K.; Remadevi, T.L. UV irradiated wet chemical deposition and characterization of nanostructured tin sulfide thin films. *J. Mater. Sci. Mater. Electron.* **2012**, *23*, 2264–2271. [CrossRef]

140. Joshua Gnanamuthu, S.; Johnson Jeyakumar, S.; Kartharinal Punithavathy, I.; Jobe Prabhakar, P.C.; Suganya, M.; Usharani, K.; Balu, A.R. Properties of spray deposited nano needle structured Cu-doped Sn$_2$S$_3$ thin films towards photovoltaic applications. *Optik* **2016**, *127*, 3999–4003. [CrossRef]

141. Khadraoui, M.; Benramdane, N.; Mathieu, C.; Bouzidi, A.; Miloua, R.; Kebbab, Z.; Sahraoui, K.; Desfeux, R.; Wang, M. Optical and electrical properties of Sn$_2$S$_3$ thin films grown by spray pyrolysis. *Solid State Commun.* **2010**, *150*, 297–300. [CrossRef]

142. Mnari, M.; Kamoun, N.; Bonnet, J.; Dachraoui, M. Chemical bath deposition of tin sulphide thin films in acid solution. *Comptes Rendus Chim.* **2009**, *12*, 824–827. [CrossRef]

143. David Smyth-Boyle's Homepage. Available online: http://www.ch.ic.ac.uk/obrien/barton/dsb/dsbcbd.html (accessed on 14 February 2018).

144. Sreedevi, G.; Vasudeva Reddy, M.; Ramakrishna Reddy, K.T.; Soo Hyun, K.; Chan Wook, J. Chemically synthesized Ag-doped SnS films for PV applications. *Ceram. Int.* **2016**, *42*, 19027–19035. [CrossRef]

145. Pramanik, P.; Basu, P.K.; Biswas, S. Preparation and characterization of chemically deposited tin(II) sulphide thin films. *Thin Solid Films* **1987**, *150*, 269–276. [CrossRef]

146. Garcia-Angelmo, A.R.; Nair, M.T.S.; Nair, P.K. Evolution of crystalline structure in SnS thin films prepared by chemical deposition. *Solid State Sci.* **2014**, *30*, 26–35. [CrossRef]

147. Lokhande, C.D.; Bhad, V.V.; Dhumure, S.S. Conversion of tin disulphide into silver sulphide by a simple chemical method. *J. Phys. D Appl. Phys.* **1992**, *25*, 315–318. [CrossRef]

148. Opasanont, B.; Baxter, J.B. Dynamic speciation modeling to guide selection of complexing agents for chemical bath deposition: Case study for ZnS thin films. *Cryst. Growth Des.* **2015**, *15*, 4893–4900. [CrossRef]

149. Lokhande, C.D. Chemical deposition of metal chalcogenide thin films. *Mater. Chem. Phys.* **1991**, *27*, 1–43. [CrossRef]

150. Chopra, K.L.; Kainthla, R.C.; Pandya, D.K.; Thakoor, A.P. *Physics of Thin Films*; Francombe, M.H., Vossen, J.L., Eds.; Academic Press: New York, NY, USA, 1982; Volume 12, p. 201.

151. Chopra, K.L.; Kainthla, R.C.; Pandya, D.K.; Thakoor, A.P. Chemical solution deposition of inorganic films. In *Physics of Thin Films*; Elsevier: Amsterdam, The Netherlands, 1982; Volume 12, pp. 167–235. ISBN 0079-1970.

152. Savadogo, O.; Mandal, K.C. Studies on new chemically deposited photoconducting antimony trisulphide thin films. *Sol. Energy Mater. Sol. Cells* **1992**, *26*, 117–136. [CrossRef]

153. Licht, S.; Longo, K.; Peramunage, D.; Forouzan, F. Conductometric analysis of the second acid dissociation constant of H_2S in highly concentrated aqueous media. *J. Electroanal. Chem. Interfacial Electrochem.* **1991**, *318*, 111–129. [CrossRef]

154. Pentia, E.; Draghici, V.; Sarau, G.; Mereu, B.; Pintilie, L.; Sava, F.; Popescu, M. Structural, electrical, and photoelectrical properties of $Cd_xPb_{1-x}S$ thin films prepared by chemical bath deposition. *J. Electrochem. Soc.* **2004**, *151*, G729. [CrossRef]

155. Wired Chemist, Solubility Product Constants, Ksp. Available online: http://www.wiredchemist.com/chemistry/data/solubility-product-constants (accessed on 30 April 2018).

156. Vaxasoftware, Solubility Product Constants. Available online: http://www.vaxasoftware.com/doc_eduen/qui/ks.pdf (accessed on 30 April 2018).

157. Engelken, R.D.; Ali, S.; Chang, L.N.; Brinkley, C.; Turner, K.; Hester, C. Study and development of a generic electrochemical ion-exchange process to form M_xS optoelectronic materials from ZnS precursor films formed by chemical-precipitation solution deposition. *Mater. Lett.* **1990**, *10*, 264–274. [CrossRef]

158. Nair, M.T.S.; Nair, P.K. SnS-Cu_xS thin film combination: A desirable solar control coating for architectural and automobile glazings. *J. Phys. D Appl. Phys.* **1991**, *24*, 450–453. [CrossRef]

159. Nair, P.K.; Nair, M.T.S. Chemically deposited SnS-Cu_xS thin films with high solar absorptance: New approach to all-glass tubular solar collectors. *J. Phys. D. Appl. Phys.* **1991**, *24*, 83–87. [CrossRef]

160. Nair, P.K.; Nair, M.T.S.; Campos, J.; Sanchez, A. SnS-SnO_2 conversion of chemically deposited SnS thin films. *Adv. Mater. Opt. Electron.* **1992**, *1*, 117–121. [CrossRef]

161. Nair, P.K. Photoconductive SnO_2 thin films from thermal decomposition of chemically deposited SnS thin films. *J. Electrochem. Soc.* **1993**, *140*, 539. [CrossRef]

162. Nair, P.K.; Nair, M.T.S.; Zingaro, R.A.; Meyers, E.A. XRD, XPS, optical and electrical studies on the conversion of SnS thin films to SnO_2. *Thin Solid Films* **1994**, *239*, 85–92. [CrossRef]

163. Ray, S.C.; Karanjai, M.K.; DasGupta, D. Structure and photoconductive properties of dip-deposited SnS and SnS_2 thin films and their conversion to tin dioxide by annealing in air. *Thin Solid Films* **1999**, *350*, 72–78. [CrossRef]

164. Ristov, M.; Sinadinovski, G.; Mitreski, M.; Ristova, M. Photovoltaic cells based on chemically deposited p-type SnS. *Sol. Energy Mater. Sol. Cells* **2001**, *69*, 17–24. [CrossRef]

165. Tanusevski, A. Optical and photoelectric properties of SnS thin films prepared by chemical bath deposition. *Semicond. Sci. Technol.* **2003**, *18*, 501–505. [CrossRef]

166. Nair, M.T.S.; Lopez-Mata, C.; GomezDaza, O.; Nair, P.K. Copper tin sulfide semiconductor thin films produced by heating SnS–CuS layers deposited from chemical bath. *Semicond. Sci. Technol.* **2003**, *18*, 755–759. [CrossRef]

167. Niinobe, D.; Wada, Y. Controlled deposition of SnS into/onto SnO_2 nano-particle film and application to photoelectrochemical cells. *Bull. Chem. Soc. Jpn.* **2006**, *79*, 495–497. [CrossRef]

168. Avellaneda, D.; Delgado, G.; Nair, M.T.S.; Nair, P.K. Structural and chemical transformations in SnS thin films used in chemically deposited photovoltaic cells. *Thin Solid Films* **2007**, *515*, 5771–5776. [CrossRef]

169. Ge, Y.H.; Guo, Y.Y.; Shi, W.M.; Qiu, Y.H.; Wei, G.P. Influence of In-doping on resistivity of chemical bath deposited SnS films. *J. Shanghai Univ.* **2007**, *11*, 403–406. [CrossRef]

170. Hankare, P.P.; Jadhav, A.V.; Chate, P.A.; Rathod, K.C.; Chavan, P.A.; Ingole, S.A. Synthesis and characterization of tin sulphide thin films grown by chemical bath deposition technique. *J. Alloys Compd.* **2008**, *463*, 581–584. [CrossRef]

171. Avellaneda, D.; Nair, M.T.S.; Nair, P.K. Photovoltaic structures using chemically deposited tin sulfide thin films. *Thin Solid Films* **2009**, *517*, 2500–2502. [CrossRef]

172. Turan, E.; Kul, M.; Aybek, A.S.; Zor, M. Structural and optical properties of SnS semiconductor films produced by chemical bath deposition. *J. Phys. D Appl. Phys.* **2009**, *42*, 245408. [CrossRef]

173. Mathews, N.R.; Avellaneda, D.; Anaya, H.B.M.; Campos, J.; Nair, M.T.S.; Nair, P.K. Chemically and electrochemically deposited thin films of tin sulfide for photovoltaic structures. *Mater. Res. Soc. Symp. Proc.* **2009**, *1165*, 367–373. [CrossRef]

174. Aksay, S.; Ozer, T.; Zor, M. Vibrational and X-ray diffraction spectra of SnS film deposited by chemical bath deposition method. *Eur. Phys. J. Appl. Phys.* **2009**, *47*, 30502. [CrossRef]

175. Wang, Y.; Reddy, Y.B.K.; Gong, H. Large-surface-area nanowall SnS films prepared by chemical bath deposition. *J. Electrochem. Soc.* **2009**, *156*, H157. [CrossRef]

176. Akkari, A.; Guasch, C.; Kamoun, T.N. Chemically deposited tin sulphide. *J. Alloys Compd.* **2010**, *490*, 180–183. [CrossRef]

177. Guneri, E.; Ulutas, C.; Kirmizigul, F.; Altindemir, G.; Gode, F.; Gumus, C. Effect of deposition time on structural, electrical, and optical properties of SnS thin films deposited by chemical bath deposition. *Appl. Surf. Sci.* **2010**, *257*, 1189–1195. [CrossRef]

178. Guneri, E.; Gode, F.; Ulutas, C.; Kirmizigul, F.; Altindemir, G.; Gumus, C. Properties of p-type SnS thin films prepared by chemical bath deposition. *Chalcogenide Lett.* **2010**, *7*, 685–694.

179. Gaied, I.; Akkari, A.; Yacoubi, N.; Kamoun, N. Influence of the triethanolamine concentration on the optical properties of tin sulphide thin films by the photothermal deflection spectroscopy. *J. Phys. Conf. Ser.* **2010**, *214*, 012128. [CrossRef]
180. Wang, Y.; Gong, H.; Fan, B.; Hu, G. Photovoltaic behavior of nanocrystalline SnS/TiO$_2$. *J. Phys. Chem. C* **2010**, *114*, 3256–3259. [CrossRef]
181. Ragina, A.J.; Preetha, K.C.; Murali, K.V.; Deepa, K.; Remadevi, T.L. Wet chemical synthesis and characterization of tin sulphide thin films from different host solutions. *Adv. Appl. Sci. Res.* **2011**, *2*, 438–444.
182. Kassim, A.; Min, H.S.; Sharif, A.; Haron, J.; Nagalingam, S. Chemical bath deposition of SnS thin films: AFM, EDAX and UV-Visible characterization. *Orient. J. Chem.* **2011**, *27*, 1375–1381.
183. Kassim, A.; Min, H.S.; Shariff, A.; Haron, M.J. The effect of the pH value on the growth and properties of chemical bath deposited SnS thin films. *Mater. Chem. Phys.* **2011**, *15*, 45–48. [CrossRef]
184. Gao, C.; Shen, H.; Sun, L.; Shen, Z. Chemical bath deposition of SnS films with different crystal structures. *Mater. Lett.* **2011**, *65*, 1413–1415. [CrossRef]
185. Gao, C.; Shen, H.; Sun, L. Preparation and properties of zinc blende and orthorhombic SnS films by chemical bath deposition. *Appl. Surf. Sci.* **2011**, *257*, 6750–6755. [CrossRef]
186. Herron, S.M.; Wangperawong, A.; Bent, S.F. Chemical bath deposition and microstructuring of tin (II) sulfide films for photovoltaics. In Proceedings of the 37th IEEE Photovoltaic Specialists Conference, Seattle, WA, USA, 19–24 June 2011; pp. 368–371.
187. Gao, C.; Shen, H. Influence of the deposition parameters on the properties of orthorhombic SnS films by chemical bath deposition. *Thin Solid Films* **2012**, *520*, 3523–3527. [CrossRef]
188. Xia, D.L.; Xu, J.; Shi, W.Q.; Lei, P.; Zhao, X.J. Synthesis and properties of SnS thin films by chemical bath deposition. *Eng. Mater.* **2012**, *509*, 333–338. [CrossRef]
189. Martinez, H.; Avellaneda, D. Modifications in SnS thin films by plasma treatments. *Nucl. Instrum. Methods Phys. Res. Sect. B Beam Interact. Mater. Atoms* **2012**, *272*, 351–356. [CrossRef]
190. Patel, T.H. Influence of deposition time on structural and optical properties of chemically deposited SnS thin films. *Open Surf. Sci. J.* **2012**, *4*, 6–13. [CrossRef]
191. Jayasree, Y.; Chalapathi, U.; Uday Bhaskar, P.; Raja, V.S. Effect of precursor concentration and bath temperature on the growth of chemical bath deposited tin sulphide thin films. *Appl. Surf. Sci.* **2012**, *258*, 2732–2740. [CrossRef]
192. Avellaneda, D.; Krishnan, B.; Das Roy, T.K.; Castillo, G.A.; Shaji, S. Modification of structure, morphology and physical properties of tin sulfide thin films by pulsed laser irradiation. *Appl. Phys. A* **2013**, *110*, 667–672. [CrossRef]
193. Sreedevi, G.; Ramakrishna Reddy, K.T. Properties of tin monosulphide films grown by chemical bath deposition. *Conf. Pap. Energy* **2013**, *2013*, 528724. [CrossRef]
194. Soonmin, H.; Kassim, A.; Weetee, T. Thickness dependent characteristics of chemically deposited tin sulfide films. *Univ. J. Chem.* **2013**, *1*, 170–174. [CrossRef]
195. Osuwa, J.C.; Ugochukwu, J. Effects of aluminum and manganese impurity concentrations on optoelectronic properties of thin films of tin sulfide (SnS) using CBD method. *IOSR J. Environ. Sci. Toxicol. Food Technol.* **2013**, *5*, 32–36. [CrossRef]
196. Onwuemeka, J.I.; Ezike, F.M.; Nwulu, N.C. The effect of annealing temperature and time on the optical properties of SnS thin films prepared by chemical bath deposition. *Int. J. Innov. Educ. Res.* **2013**, *1*, 121–130. [CrossRef]
197. Reghima, M.; Akkari, A.; Guasch, C.; Castagné, M.; Kamoun-Turki, N. Synthesis and characterization of Fe-doped SnS thin films by chemical bath deposition technique for solar cells applications. *J. Renew. Sustain. Energy* **2013**, *5*, 063109. [CrossRef]
198. Jayasree, Y.; Chalapathi, U.; Raja, V.S. Growth and characterization of tin sulphide thin films by chemical bath deposition using ethylene diamine tetra-acetic acid as the complexing agent. *Thin Solid Films* **2013**, *537*, 149–155. [CrossRef]
199. Dhanya, A.C.; Deepa, K.; Geetanjali, P.M.; Anupama, M.; Remadevi, T.L. Effect of post deposition by UV irradiation on chemical bath deposited tin sulfide thin films. *Appl. Phys. A* **2014**, *116*, 1467–1472. [CrossRef]
200. He, H.Y.; Fei, J.; Lu, J. Optical and electrical properties of pure and Sn^{4+}-doped n-SnS films deposited by chemical bath deposition. *Mater. Sci. Semicond. Process.* **2014**, *24*, 90–95. [CrossRef]
201. Safonova, M.; Nair, P.K.; Mellikov, E.; Garcia, A.R.; Kerm, K.; Revathi, N.; Romann, T.; Mikli, V.; Volobujeva, O. Chemical bath deposition of SnS thin films on ZnS and CdS substrates. *J. Mater. Sci. Mater. Electron.* **2014**, *25*, 3160–3165. [CrossRef]
202. Avellaneda, D.; Krishnan, B.; Rodriguez, A.C.; Das Roy, T.K.; Shaji, S. Heat treatments in chemically deposited SnS thin films and their influence in CdS/SnS photovoltaic structures. *J. Mater. Sci. Mater. Electron.* **2015**, *26*, 5585–5592. [CrossRef]
203. Joshi, L.P.; Risal, L.; Shrestha, S.P. Effects of concentration of triethanolamine and annealing temperature on band gap of thin film of tin sulphide prepared by chemical bath deposition method. *J. Nepal Phys. Soc.* **2016**, *3*, 1–5. [CrossRef]
204. Safonova, M.; Mellikov, E.; Mikli, V.; Kerm, K.; Revathi, N.; Volobujeva, O. Chemical bath deposition of SnS thin films from the solutions with different concentrations of tin and sulphur. *Adv. Mater. Res.* **2015**, *1117*, 183–186. [CrossRef]
205. Mahdi, M.S.; Ibrahim, K.; Hmood, A.; Ahmed, N.M.; Azzez, S.A.; Mustafa, F.I. A highly sensitive flexible SnS thin film photodetector in the ultraviolet to near infrared prepared by chemical bath deposition. *RSC Adv.* **2016**, *6*, 114980–114988. [CrossRef]
206. Chaki, S.H.; Chaudhary, M.D.; Deshpande, M.P. SnS thin films deposited by chemical bath deposition, dip coating and SILAR techniques. *J. Semicond.* **2016**, *37*, 1–9. [CrossRef]
207. Mahdi, M.S.; Ibrahim, K.; Hmood, A.; Ahmed, N.M.; Mustafa, F.I. Control of phase, structural and optical properties of tin sulfide nanostructured thin films grown via chemical bath deposition. *J. Electron. Mater.* **2017**, *46*, 4227–4235. [CrossRef]

208. Avellaneda, D.; Sánchez-Orozco, I.; Martínez, J.A.A.; Shaji, S.; Krishnan, B. Thin films of tin sulfides: Structure, composition and optoelectronic properties. *Mater. Res. Express* **2018**, *6*, 016409. [CrossRef]
209. Cao, M.; Wu, C.; Yao, K.; Jing, J.; Huang, J.; Cao, M.; Zhang, J.; Lai, J.; Ali, O.; Wang, L.; et al. Chemical bath deposition of single crystal SnS nanobelts on glass substrates. *Mater. Res. Bull.* **2018**, *104*, 244–249. [CrossRef]
210. Patil, A.M.; Lokhande, V.C.; Patil, U.M.; Shinde, P.A.; Lokhande, C.D. High performance all-solid-state asymmetric supercapacitor device based on 3D nanospheres of β-MnO$_2$ and Nanoflowers of O-SnS. *ACS Sustain. Chem. Eng.* **2018**, *6*, 787–802. [CrossRef]
211. Gedi, S.; Minnam Reddy, V.R.; Kotte, T.R.R.; Park, Y.; Kim, W.K. Effect of C$_4$H$_6$O$_6$ concentration on the properties of SnS thin films for solar cell applications. *Appl. Surf. Sci.* **2019**, *465*, 802–815. [CrossRef]
212. Gedi, S.; Minnam Reddy, V.R.; Alhammadi, S.; Reddy Guddeti, P.; Kotte, T.R.R.; Park, C.; Kim, W.K. Influence of deposition temperature on the efficiency of SnS solar cells. *Sol. Energy* **2019**, *184*, 305–314. [CrossRef]
213. Cabrera-German, D.; García-Valenzuela, J.A.; Cota-Leal, M.; Martínez-Gil, M.; Aceves, R.; Sotelo-Lerma, M. Detailed characterization of good-quality SnS thin films obtained by chemical solution deposition at different reaction temperatures. *Mater. Sci. Semicond. Process.* **2019**, *89*, 131–142. [CrossRef]
214. Mallika Bramaramba Devi, P.; Phaneendra Reddy, G.; Ramakrishna Reddy, K.T. Structural and optical studies on PVA capped SnS films grown by chemical bath deposition for solar cell application. *J. Semicond.* **2019**, *40*, 052101. [CrossRef]
215. Mahdi, M.S.; Hmood, A.; Ibrahim, K.; Ahmed, N.M.; Bououdina, M. Dependence of pH on phase stability, optical and photoelectrical properties of SnS thin films. *Superlattices Microstruct.* **2019**, *128*, 170–176. [CrossRef]
216. Huang, J.; Ma, Y.; Yao, K.; Wu, C.; Cao, M.; Lai, J.; Zhang, J.; Sun, Y.; Wang, L.; Shen, Y. Chemical bath deposition of SnS:In thin films for Pt/CdS/SnS:In/Mo photocathode. *Surf. Coat. Technol.* **2019**, *358*, 84–90. [CrossRef]
217. Gonzaez-Flores, V.E.; Mohan, R.N.; Ballinas-Morales, R.; Nair, M.T.S.; Nair, P.K. Thin film solar cells of chemically deposited SnS of cubic and orthorhombic structures. *Thin Solid Films* **2019**, *672*, 62–65. [CrossRef]
218. Mallika Bramaramba Devi, P.; Reddy, G.P.; Reddy, K.T.R. Optical investigations on PVA capped SnS nanocrystalline films deposited by CBD process. *Mater. Res. Express* **2019**, *6*, 115523. [CrossRef]
219. John, S.; Geetha, V.; Francis, M. Effect of pH on the optical and structural properties of SnS prepared by chemical bath deposition method. *IOP Conf. Ser. Mater. Sci. Eng.* **2020**, *872*, 012139. [CrossRef]
220. Muthalif, M.P.A.; Choe, Y. Control of the interfacial charge transfer resistance to improve the performance of quantum dot sensitized solar cells with highly electrocatalytic Cu-doped SnS counter electrodes. *Appl. Surf. Sci.* **2020**, *508*, 145297. [CrossRef]
221. Ben Mbarek, M.; Reghima, M.; Yacoubi, N.; Barradas, N.; Alves, E.; Bundaleski, N.; Teodoro, O.; Kunst, M.; Schwarz, R. Microwave transient reflection in annealed SnS thin films. *Mater. Sci. Semicond. Process.* **2021**, *121*, 105302. [CrossRef]
222. Marquez, I.G.; Romano-Trujillo, R.; Gracia-Jimenez, J.M.; Galeazzi, R.; Silva-González, N.R.; García, G.; Coyopol, A.; Nieto-Caballero, F.G.; Rosendo, E.; Morales, C. Cubic, orthorhombic and amorphous SnS thin films on flexible plastic substrates by CBD. *J. Mater. Sci. Mater. Electron.* **2021**, 1–9. [CrossRef]
223. Cao, M.; Zhang, X.; Ren, J.; Sun, Y.; Cui, Y.; Zhang, J.; Ling, J.; Huang, J.; Shen, Y.; Wang, L.; et al. Chemical bath deposition of SnS:Cu/ZnS for solar hydrogen production and solar cells. *J. Alloys Compd.* **2021**, *863*, 158727. [CrossRef]
224. Higareda-Sanchez, A.; Mis-Fernandez, R.; Rimmaudo, I.; Camacho-Espinosa, E.; Pena, J.L. Evaluation of pH and deposition mechanisms effect on tin sulfide thin films deposited by chemical bath deposition. *Superlattices Microstruct.* **2021**, *151*, 106831. [CrossRef]
225. Akkari, A.; Regima, M.; Guasch, C.; Kamoun Turki, N. Effect of deposition time on physical properties of nanocrystallized SnS zinc blend thin films grown by chemical bath deposition. *Adv. Mater. Res.* **2011**, *324*, 101–104. [CrossRef]
226. Akkari, A.; Guasch, C.; Castagne, M.; Kamoun-Turki, N. Optical study of zinc blend SnS and cubic In$_2$S$_3$:Al thin films prepared by chemical bath deposition. *J. Mater. Sci.* **2011**, *46*, 6285–6292. [CrossRef]
227. Akkari, A.; Reghima, M.; Guasch, C.; Kamoun-Turki, N. Effect of copper doping on physical properties of nanocrystallized SnS zinc blend thin films grown by chemical bath deposition. *J. Mater. Sci.* **2012**, *47*, 1365–1371. [CrossRef]
228. Reghima, M.; Akkari, A.; Guasch, C.; Kamoun-Turki, N. Effect of indium doping on physical properties of nanocrystallized SnS zinc blend thin films grown by chemical bath deposition. *J. Renew. Sustain. Energy* **2012**, *4*, 011602. [CrossRef]
229. Reghima, M.; Akkari, A.; Guasch, C.; Kamoun, N. Structural, optical, and electrical properties of SnS:Ag thin films. *J. Electron. Mater.* **2015**, *44*, 4392–4399. [CrossRef]
230. Barrios-Salgado, E.; Rodríguez-Guadarrama, L.A.; Garcia-Angelmo, A.R.; Campos Álvarez, J.; Nair, M.T.S.; Nair, P.K. Large cubic tin sulfide–tin selenide thin film stacks for energy conversion. *Thin Solid Films* **2016**, *615*, 415–422. [CrossRef]
231. Nair, P.K.; Barrios-Salgado, E.; Nair, M.T.S. Cubic-structured tin selenide thin film as a novel solar cell absorber. *Phys. Status Solidi Appl. Mater. Sci.* **2016**, *213*, 2229–2236. [CrossRef]
232. Abutbul, R.E.; Garcia-Angelmo, A.R.; Burshtein, Z.; Nair, M.T.S.; Nair, P.K.; Golan, Y. Crystal structure of a large cubic tin monosulfide polymorph: An unraveled puzzle. *CrystEngComm* **2016**, *18*, 5188–5194. [CrossRef]
233. Sanal, K.C.; Nair, P.K.; Nair, M.T.S. Band offset in zinc oxy-sulfide/cubic-tin sulfide interface from X-ray photoelectron spectroscopy. *Appl. Surf. Sci.* **2017**, *396*, 1092–1097. [CrossRef]
234. Mahdi, M.S.; Ibrahim, K.; Ahmed, N.M.; Hmood, A.; Mustafa, F.I.; Azzez, S.A.; Bououdina, M. High performance and low-cost UV–Visible–NIR photodetector based on tin sulphide nanostructures. *J. Alloys Compd.* **2018**, *735*, 2256–2262. [CrossRef]
235. Javed, A.; Bashir, M. Controlled growth, structure and optical properties of Fe-doped cubic π-SnS thin films. *J. Alloys Compd.* **2018**, *759*, 14–21. [CrossRef]

236. Flores, V.E.G.; Nair, M.T.S.; Nair, P.K. Thermal stability of "metastable" cubic tin sulfide and its relevance to applications. *Semicond. Sci. Technol.* **2018**, *33*. [CrossRef]

237. Abutbul, R.E.; Golan, Y. Chemical epitaxy of π-phase cubic tin monosulphide. *CrystEngComm* **2020**, *22*, 6170–6181. [CrossRef]

238. Mahdi, M.S.; Al-Arab, H.S.; Al-Salman, H.S.; Ibrahim, K.; Ahmed, N.M.; Hmood, A.; Bououdina, M. A high-performance near-infrared photodetector based on π-SnS phase. *Mater. Lett.* **2020**, *273*, 127910. [CrossRef]

239. Javed, A.; Khan, N.; Bashir, S.; Ahmad, M.; Bashir, M. Thickness dependent structural, electrical and optical properties of cubic SnS thin films. *Mater. Chem. Phys.* **2020**, *246*, 122831. [CrossRef]

240. Barrios Salgado, E.; Lara Llanderal, D.E.; Nair, M.T.S.; Nair, P.K. Thin film thermoelectric elements of p–n tin chalcogenides from chemically deposited SnS–SnSe stacks of cubic crystalline structure. *Semicond. Sci. Technol.* **2020**, *35*, 045006. [CrossRef]

241. Chalapathi, U.; Poornaprakash, B.; Choi, W.J.; Park, S.-H. Ammonia(aq)-enhanced growth of cubic SnS thin films by chemical bath deposition for solar cell applications. *Appl. Phys. A* **2020**, *126*, 583. [CrossRef]

242. Rodriguez-Guadarrama, L.A.; Escorcia-Garcia, J.; Alonso-Lemus, I.L.; Campos-Álvarez, J. Synthesis of π-SnS thin films through chemical bath deposition: Effects of pH, deposition time, and annealing temperature. *J. Mater. Sci. Mater. Electron.* **2021**, *32*, 7464–7480. [CrossRef]

243. Gaitan-Arevalo, J.R.; Gonzalez, L.A.; Escorcia-García, J. Cubic tin sulfide thin films by a Sn-NTA system based chemical bath process. *Mater. Lett.* **2021**, *286*, 129222. [CrossRef]

244. Eswar Neerugatti, K.; Shivaji Pawar, P.; Heo, J. Differential growth and evaluation of band structure of π-SnS for thin-film solar cell applications. *Mater. Lett.* **2021**, *284*, 129026. [CrossRef]

245. Ramakrishna Reddy, K.T.; Sreedevi, G.; Miles, R.W. Thickness effect on the structural and optical properties of SnS_2 films grown by CBD process. *J. Mater. Sci. Eng. A* **2013**, *3*, 182–186. [CrossRef]

246. Chalapathi, U.; Poornaprakash, B.; Purushotham Reddy, B.; Si Hyun, P. Preparation of SnS_2 thin films by conversion of chemically deposited cubic SnS films into SnS2. *Thin Solid Films* **2017**, *640*, 81–87. [CrossRef]

247. Zhou, P.; Huang, Z.; Sun, X.; Ran, G.; Shen, R.; Ouyang, Q. Saturable absorption properties of the SnS_2/FTO thin film. *Optik* **2018**, *171*, 839–844. [CrossRef]

248. Joshi, M.P.; Khot, K.V.; Patil, S.S.; Mali, S.S.; Hong, C.K.; Bhosale, P.N. Investigating the light harvesting capacity of sulfur ion concentration dependent SnS_2 thin films synthesized by self-assembled arrested precipitation technique. *Mater. Res. Express* **2019**, *6*, 086467. [CrossRef]

249. Noppakuadrittidej, P.; Vailikhit, V.; Teesetsopon, P.; Choopun, S.; Tubtimtae, A. Copper incorporation in Mn^{2+} doped Sn_2S_3 nanocrystals and the resultant structural, optical, and electrochemical characteristics. *Ceram. Int.* **2018**, *44*, 13973–13985. [CrossRef]

250. Mohan, R.N.; Nair, M.T.S.; Nair, P.K. Thin film Sn_2S_3 via chemical deposition and controlled heating—Its prospects as a solar cell absorber. *Appl. Surf. Sci.* **2020**, *504*, 144162. [CrossRef]

251. Liu, T.; Ke, H.; Zhang, H.; Duo, S.; Sun, Q.; Fei, X.; Zhou, G.; Liu, H.; Fan, L. Effect of four different zinc salts and annealing treatment on growth, structural, mechanical and optical properties of nanocrystalline ZnS thin films by chemical bath deposition. *Mater. Sci. Semicond. Process.* **2014**, *26*, 301–311. [CrossRef]

252. Salh, A.; Kyoongchan, M.; Hyeonwook, P.; Woo Kyoung, K. Effect of different cadmium salts on the properties of chemical-bath-deposited CdS thin films and Cu(InGa)Se$_2$ solar cells. *Thin Solid Films* **2017**, *625*, 56–61. [CrossRef]

253. Hodes, G. *Chemical Solution Deposition of Semiconductor Films*, 1st ed.; Marcel Dekker: New York, NY, USA, 2002; ISBN 0824708512.

254. Jeffery, G.H.; Bassett, J.; Mendham, J.; Denney, R.C. *Vogel's Textbook of Quantitative Inorganic Analysis*; ELBS Publ.: London, UK, 1989.

255. Flaschka, H.A. *EDTA Titrations. An Introduction to Theory and Practice*; Macmillan: New York, NY, USA, 1959.

256. Kaur, I. Growth kinetics and polymorphism of chemically deposited CdS films. *J. Electrochem. Soc.* **1980**, *127*, 943. [CrossRef]

257. Donna, J.M. Chemical bath deposition of CdS thin films: Electrochemical in situ kinetic studies. *J. Electrochem. Soc.* **1992**, *139*, 2810. [CrossRef]

258. Gonzalez Panzo, I.J.; Martin Varguez, P.E.; Oliva, A.I. Role of thiourea in the kinetic of growth of the chemical bath deposited ZnS films. *J. Electrochem. Soc.* **2014**, *161*, D761–D767. [CrossRef]

259. Kumarage, W.G.C.; Wijesundara, L.B.D.R.P.; Seneviratne, V.A.; Jayalath, C.P.; Dassanayake, B.S. Influence of bath temperature on CBD-CdS thin films. *Procedia Eng.* **2016**, *139*, 64–68. [CrossRef]

260. Bayon, R.; Hernandez-Mayoral, M.; Herrero, J. Growth mechanism of CBD-In(OH)$_x$S$_y$ thin films. *J. Electrochem. Soc.* **2002**, *149*, C59–C67. [CrossRef]

261. Cortes, A.; Gómez, H.; Marotti, R.E.; Riveros, G.; Dalchiele, E.A. Grain size dependence of the bandgap in chemical bath deposited CdS thin films. *Sol. Energy Mater. Sol. Cells* **2004**, *82*, 21–34. [CrossRef]

262. Balasubramanian, V.; Suriyanarayanan, N.; Prabahar, S. Thickness-dependent structural properties of chemically deposited Bi_2S_3 thin films. *Adv. Apllied Sci. Res.* **2012**, *3*, 2369–2373.

263. Patil, S.A.; Mengal, N.; Memon, A.A.; Jeong, S.H.; Kim, H.S. CuS thin film grown using the one pot, solution-process method for dye-sensitized solar cell applications. *J. Alloys Compd.* **2017**, *708*, 568–574. [CrossRef]

264. Gupta, V.; Mansingh, A. Influence of postdeposition annealing on the structural and optical properties of sputtered zinc oxide film. *J. Appl. Phys.* **1996**, *80*, 1063–1073. [CrossRef]

265. Jin Hong, L.; Byung Ok, P. Transparent conducting In_2O_3 thin films prepared by ultrasonic spray pyrolysis. *Surf. Coat. Technol.* **2004**, *184*, 102–107. [CrossRef]

266. Brien, P.O.; Mcaleese, J. bath deposition of ZnS and CdS. *Technology* **1998**, *8*, 2309–2314. [CrossRef]
267. Avrami, M. Granulation, phase change, and microstructure kinetics of phase change. III. *J. Chem. Phys.* **1941**, *9*, 177–184. [CrossRef]
268. Phillips, J.M. Substrate selection for thin-film growth. *MRS Bull.* **1995**, *20*, 35–39. [CrossRef]
269. Vasudeva Reddy, M.; Sreedevi, G.; Chinho, P.; Miles, R.W.; Ramakrishna Reddy, K.T. Development of sulphurized SnS thin film solar cells. *Curr. Appl. Phys.* **2015**, *15*, 588–598. [CrossRef]
270. Devika, M.; Koteeswara Reddy, N.; Ramesh, K.; Ganesan, V.; Gopal, E.S.R.; Ramakrishna Reddy, K.T. Influence of substrate temperature on surface structure and electrical resistivity of the evaporated tin sulphide films. *Appl. Surf. Sci.* **2006**, *253*, 1673–1676. [CrossRef]
271. Vasudeva Reddy, M.; Sreedevi, G.; Babu, P.; Ramakrishna Reddy, K.T.; Guillaume, Z.; Chinho, P. Influence of different substrates on the properties of sulfurized SnS films. *Sci. Adv. Mater.* **2016**, *8*, 247–251. [CrossRef]
272. Camacho Espinosa, E.; Oliva Aviles, A.I.; Oliva, A.I. Effect of the substrate cleaning process on pinhole formation in sputtered CdTe films. *J. Mater. Eng. Perform.* **2017**, *26*, 4020–4028. [CrossRef]
273. Tarkeshwar, S.; Devjyoti, L.; Ayush, K. Effects of various parameters on structural and optical properties of CBD-grown ZnS thin films: A review. *J. Electron. Mater.* **2018**, *47*, 1730–1751. [CrossRef]
274. Hubert, C.; Naghavi, N.; Roussel, O.; Etcheberry, A.; Hariskos, D.; Menner, R.; Powalla, M.; Kerrec, O.; Lincot, D. The Zn(S,O,OH)/ZnMgO buffer in thin film Cu(In,Ga)(S,Se)$_2$-based solar cells part I: Fast chemical bath deposition of Zn(S,O,OH) buffer layers for industrial application on co-evaporated Cu(In,Ga)Se$_2$ and electrodeposited CuIn(S,Se)$_2$ solar cells. *Prog. Photovolt. Res. Appl.* **2009**, *17*, 470–478. [CrossRef]
275. Li, Z.Q.; Shi, J.H.; Liu, Q.Q.; Wang, Z.A.; Sun, Z.; Huang, S.M. Effect of [Zn]/[S] ratios on the properties of chemical bath deposited zinc sulfide thin films. *Appl. Surf. Sci.* **2010**, *257*, 122–126. [CrossRef]

nanomaterials

MDPI

Article

Mass Transfer Study on Improved Chemistry for Electrodeposition of Copper Indium Gallium Selenide (CIGS) Compound for Photovoltaics Applications

Mahfouz Saeed [1,2,*] and Omar Israel González Peña [2,3,*]

[1]   Department of Environmental Engineering, A'Sharqiyah University, P.O. Box 42, Ibra 400, Oman
[2]   Department of Chemical & Biomolecular Engineering, Case Western Reserve University, Cleveland, OH 44106, USA
[3]   Water Center for Latin America and the Caribbean, School of Engineering and Sciences, Tecnologico de Monterrey, Eugenio Garza Sada Sur Avenue 2501, Colonia Tecnológico, Monterrey 64849, NL, Mexico
*   Correspondence: Mahfouz.saeed@asu.edu.om (M.S.); ogonzalez.pena@gmail.com or oig@tec.mx (O.I.G.P.)

Citation: Saeed, M.; González Peña, O.I. Mass Transfer Study on Improved Chemistry for Electrodeposition of Copper Indium Gallium Selenide (CIGS) Compound for Photovoltaics Applications. *Nanomaterials* **2021**, *11*, 1222. https://doi.org/10.3390/nano11051222

Academic Editor: Vlad Andrei Antohe

Received: 26 March 2021
Accepted: 26 April 2021
Published: 6 May 2021

Publisher's Note: MDPI stays neutral with regard to jurisdictional claims in published maps and institutional affiliations.

Abstract: Copper indium gallium selenium (CIGS) films are attractive for photovoltaic applications due to their high optical absorption coefficient. The generation of CIGS films by electrodeposition is particularly appealing due to the relatively low capital cost and high throughput. Numerous publications address the electrodeposition of CIGS; however, very few recognize the critical significance of transport in affecting the composition and properties of the compound. This study introduces a new electrolyte composition, which is far more dilute than systems that had been previously described, which yields much improved CIGS films. The electrodeposition experiments were carried out on a rotating disk electrode, which provides quantitative control of the transport rates. Experiments with the conventional electrolyte, ten times more concentrated than the new electrolyte proposed here, yielded powdery and non-adherent deposit. By contrast, the new, low concentration electrolyte produced in the preferred potential interval of $-0.64 \leq E \leq -0.76$ V vs. NHE, a smooth and adherent uniform deposit with the desired composition across a broad range of rotation speeds. The effects of mass transport on the deposit are discussed. Sample polarization curves at different electrode rotation rates, obtained in deposition experiments from the high and the low concentration electrolytes, are critically compared. Characterization of the overall efficiency, quantum efficiency, open circuit voltage, short circuit current, dark current, band gap, and the fill factor are reported.

Keywords: rotating disk electrode; photovoltaic; electrodeposition; mass transport; thin film; hydrogen evolution; CIGS; PV cells; fill factor; quantum efficiency

1. Introduction

$CuIn_xGa_{(1-x)}Se_2$ (CIGS) is a solid thin film quaternary compound consisting of copper, indium, gallium, and selenium. In addition, CIGS is a highly effective absorber layer for photovoltaic (PV) devices. The value designated by "x" in the formula can range from zero to one and affects mostly the CIGS structure and its band gap energy [1]. It has been reported that the optimal x value is 0.3 [1]. CIGS thin film should be in the order of 1.5 to 2.5 µm; moreover, this film is attractive for photovoltaic applications because of its high optical absorption coefficient [1], about 10^5 cm^{-1}. The PV cells with CIGS as an absorption layer have an efficiency approaching 20% as has been demonstrated in laboratory tests [2]. Vacuum deposition techniques provided the best method to obtain the best CIGS absorber layer and the most efficient CIGS devices between other fabrication methods [3–5]. Germany's Center for Solar Energy claims to have the record efficiency of 22.6% as it is reported in 2016 [6].

Non-vacuum techniques for fabricating CIGS films are gaining interest since they offer significantly less costly production and easier scalability to fabricate them in large areas [6–25]. Among those, the generation of CIGS thin layers by electrodeposition is particularly attractive due to the relatively low capital cost and high throughput [8,9]. The hybrid process, combining electrodeposition with metal addition from the gas phase, provided a 15.4% efficient device [10]. The introduction of Hydrion buffer, which is a mixture of sulfamic acid and potassium biphthalate in the electrolyte bath, has been particularly helpful in reducing indium and gallium oxides and hydroxides species, leading to a more stable absorber layer [10,11]. Different substrates have been reported as adequate back contacts, including molybdenum, copper, nickel, and stainless steel due to their conductivities [12,13]. Likewise, there was a study of the roughness of the electrodeposition of CIGS on different back contacts, such as fluorine tin oxide, fluorine-doped thin oxide, and molybdenum [14]. Moreover, the electrodeposition of CIGS was also studied over a ZnO window layer [15]. The electrodeposition of CIG (copper, indium, and gallium) followed by selenium addition through evaporation has also been reported [16,17]. Even an electrodeposition study of CIGS on patterned molybdenum/glass substrates showed a semi-transparent glazing-based PV cell [18]. Several research groups have focused on characterizing the chemistry of the electrodeposition bath and the growth mechanisms of the CIGS absorber [19–24]. The film produced by low temperature through an electrodeposition method exhibits low crystallinity and requires a post-electrodeposition thermal treatment (annealing) [25]. Annealing under a selenium atmosphere has been recommended in literature to accommodate the reactions of selenium with copper, gallium, and indium to generate the formation reactions, a good recrystallization, and the adjustment of the absorber's final composition [26–30]. In addition, studies of the annealing process on CIGS note that the annealing step takes an important role in the efficiency of the PV cell due to it affecting the grain size and crystallinity on the absorber alloy in collecting more light [31]; in addition, the Cu/In ratio and the type of precursor can affect the morphology of the film [32]. Pulse electrodeposition was suggested to obtain a smoother CIGS layer and it improves the control in the electrodeposition [33–35]. An electrodeposition process followed by physical vapor deposition with an addition of the vapor phase of copper, indium, and gallium to the absorber layer has also been reported [36]. Exhaustive studies of CIGS electrodeposition processes have been provided in literature without considering the hydrogen evolution [36–38].

The CIGS system has been studied mostly using potential sweep methods to characterize the electrochemical reduction reactions of the four components [19,39–41]. Recently, the electrodeposition of CIGS on a rotating disk electrode (RDE) has been studied by two methods: DC electrodeposition and DC electrodeposition plus mechanical perturbations [42]. Furthermore, some studies have been done on a RDE on the copper-indium-selenium (CIS) system [43,44]. This means that most reported CIGS electrodeposition studies were carried out in a beaker under ill-defined transport conditions. Consequently, there is a lack of information on the significance of mass transport and agitation in the CIGS electrodeposition. In other words, despite the electrodeposition of CIGS having been studied for roughly 15 years, very few investigators recognize the critical significance of mass transport and agitation in the CIGS electrodeposition process as a very relevant parameter to obtain an absorption layer with a controlled process to reach a homogeneous atomic composition of the desired alloy in large scale areas. Likewise, it is important to note that in the formation of the CIGS alloy, it is necessary to apply very large overpotentials where there is a significant contribution of the reduction of hydrogen gas that determines the influence of the final composition and the morphology of the film alloy. However, despite the enormous problem of the co-evolution of hydrogen gas, for the moment, no studies have been reported to understand the phenomena associated with mass transfer and/or the kinetics of the reaction that governs each of these precursors separately to form the CIGS alloy. As a result, these enormous issues can be described as a big white elephant in the room that has not been addressed by previous studies. This means that the considerable amount of hydrogen gas produced in the electrodeposition of CIGS and the mass transport effect in the reduction of the alloy are elements that this study is focused on addressing. Therefore, this work provides a systematic study on the application of the RDE for characterizing the significance of transport in CIGS electrodeposition. The RDE provides a quantitative measure of the transport rates as a function of the disk rotating speed, offering well-quantified and uniform transport rates across the entire electrode.

The bath concentration in this study was reduced by a factor of ten as compared to conventional systems, providing much improved deposit properties. An investigation of the ionic mass transport in both the conventional bath and the new dilute system, with the objective of improving the film physical and optical properties, is introduced herein. Moreover, an analysis of the role taken by the hydrogen co-evolution in the electrodeposition of CIGS is done by evaluating the hydrogen current density with respect to the current density of the precursors and the total current density of the alloy reduction.

In addition to the CIGS absorber layer, an entire photovoltaic device, consisting of the layers stainless steel/Mo/Ni/CIGS/CdS/ZnO/ZnO-Al, has been fabricated using only electrodeposition methods. Consequently, quantum efficiency measurements and data collected from a solar simulator were obtained to characterize the integrated photovoltaic device. Therefore, the characterized parameters are overall efficiency, quantum efficiency, open circuit voltage, short circuit current, black current, band gap, and the fill factor.

2. Materials and Methods

2.1. Fabrication Process

Laboratory experiments were conducted using a RDE system. The system consisted of a 0.32 cm^2 stainless-steel (406 SS). This stainless steel electrode was deposited with a layer of molybdenum via electron-beam physical vapor deposition, using 99.95% purity Mo pellets; after that a nickel deposit was also added via electrodeposition process by a rotation rate of 400 rpm at room temperature. The electrolyte composition was: 0.15 M NiSO$_4$, 0.2M Na$_2$SO$_4$, and the pH of the bath was adjusted by sulfuric acid to 2.2; the suitable nickel electrodeposition potential applied was -1.35 V vs. SCE for 25 min to obtain a smooth layer. The nickel layer thickness estimated using a profilometer was between 1.5 to 2 μm.

The process was carried out under a vacuum atmosphere (2×10^{-6} torr). The deposition rate was low (0.3 A/s) to get rid of high temperature effects on the equipment. As a result, the sputtering process was carried out for 60 min to obtain a thickness of about 1 μm. After that, the disk was embedded flush in an insulating Teflon cylinder, the counter electrode was a platinum mesh, and the reference electrode was a saturated calomel electrode (SCE). These electrodes were immersed in a beaker with 50 mL within the electrochemical baths studied. The electrolyte pH was adjusted to 1.9 by the addition of hydrochloric acid. Experiments at the ambient temperature of 20 °C were carried out at rotation speeds ranging from 0 to 700 rpm.

The substrate was first rinsed with acetone, then rinsed with deionized water, and finally dried in air. Prior to electrodeposition, the substrate was electro-activated for two seconds at 1.5 V vs. NHE in 0.1 M sulfuric acid solution. The conventional higher concentration chemical system consisted of 4.2–6.5 mM CuCl$_2$·2H$_2$O (Sigma-Aldrich, St. Louis, MO, USA); 2.9–5 mM InCl$_3$ (Strem-Chemicals, Newburyport, MA, USA); 7–9 mM H$_2$SeO$_3$ (Sigma-Aldrich, St. Louis, MO, USA); 3–7 mM GaCl$_3$ (Strem-Chemicals, Newburyport, MA, USA); pHydrion (pH = 2) (Sigma-Aldrich, St. Louis, MO, USA); and 0.66 M LiCl (Sigma-Aldrich, St. Louis, MO, USA) as supporting electrolyte. The novel bath is at lower concentration system of 0.4–0.6 mM CuCl$_2$·2H$_2$O; 0.28–0.5 mM InCl$_3$; 0.6–0.85 mM H$_2$SeO$_3$; 0.35–0.6 mM GaCl$_3$; pHydrion (pH = 2); and 0.66 M LiCl as supporting electrolyte. The above-listed electrolyte ranges were explored to obtain optimal film composition and performance. The experiments were run potentiostatically, with the electrodeposition potential ranging between -0.64 V to -0.76 V vs. normal hydrogen electrode (NHE). Thus, its equivalence on the scale in the reference electrode was (0 V vs. NHE = 0.242 V vs. SCE).

In experiments where the entire PV device was fabricated, the device was completed by electrodepositing additional layers on the top of the CIGS film. The first layer on top of the CIGS absorber was about 50 nm of CdS. The cadmium sulfide electrodeposition was carried out at 500 rpm by applying -0.8 V vs. SCE for 12 min, at a temperature of 65 °C from an electrolyte consisting of 0.2 M $CdCl_2$, 5 mM $Na_2S_2O_3$, and 0.5 M KCl; the bath was adjusted by HCl to 2. The CdS layer deposition was followed by 200 nm of electrodeposited un-doped zinc oxide layer. This layer was electrodeposited at 300 rpm and 75 °C by applying -0.85 V vs. SCE for 40 min from an electrolyte consisting of 0.1 M $Zn(NO_3)_2$ and 0.4 M KCl; the bath was adjusted by NaOH to 6. Lastly, a 500 nm indium-doped zinc oxide window layer was electrodeposited by applying -1.1 V vs. SCE for 55 min at a rotation speed of 200 rpm at 80 °C from an electrolyte consisting of 0.1 M $Zn(NO_3)_2$, 0.9 mM $InCl_3$, and 0.4 M KCl; the pH bath was adjusted by NaOH to 3.5.

2.2. Instrumental Techniques

A Bio-Logic potentiostat/galvanostat Model VSP A (Seyssinet-Pariset, France) was used as the power source to obtain polarization curves by applying a constant step voltage (chronoamperometry experiments) on the electrochemical system. The final electrodeposit composition at the CIGS surface was determined using a Hitachi S4500 scanning electron microscope (SEM) equipped with a Noran energy dispersive spectrometer (EDS), (Santa Clara, CA, USA). The focused ion beam (FIB) for the atomic composition analysis in the cross section of the CIGS film was determined by the FEI Helios NanoLab 650 Dual-Beam System equipped with an EDS (Milpitas, CA, USA). KLA-Tencor P-6 Stylus Profilometer, (Milpitas, CA, USA) was used to determine the thickness of the CIGS film. X-ray diffraction (XRD) was used to analyze CIGS crystallography (Bruker Discover D8 X-ray diffractometer, Billerica, MA, USA) with a Cu K α (alpha) radiation ($\lambda = 0.15406$ nm) as the source, with a step of 0.01°. The equipment used in the PV measurements was the QEX10 quantum efficiency measurement system (PV Measurements, Boulder, CO, USA), and an Oriel Sol2A solar simulator, (Irvine, CA, USA). As a result, the characterized parameters were overall efficiency, quantum efficiency, open circuit voltage, short circuit current, dark current, band gap, and the fill factor.

2.3. CIGS Electrodeposition—Overriding Considerations

The large difference in the standard deposition potentials of the four co-deposited metals presents a major challenge to CIGS electrodeposition. The cathodic reactions and standard potentials for the four components are:

$$HSeO^{+3} + 4H^+ + 4e^- + OH^- \rightleftharpoons H_2SeO_3 + 4H^+ + 4e^- \rightleftharpoons Se + 3H_2O \ (0.74 \ V \ vs. \ NHE) \tag{1}$$

$$Cu^{2+} + 2e^- \rightleftharpoons Cu \ (0.34 \ V \ vs. \ NHE) \tag{2}$$

$$In^{3+} + 3e^- \rightleftharpoons In \ (-1.34 \ V \ vs. \ NHE) \tag{3}$$

$$Ga^{3+} + 3e^- \rightleftharpoons Ga \ (-G.53 \ V \ vs. \ NHE) \tag{4}$$

The co-deposition must take place at a potential more negative than that of gallium (-0.53 V vs. NHE), leading to the deposition of all other species (Se, Cu, In), in which they reduce at a very cathodic voltage to their standard potential, and therefore, close to, or at their limiting current—a region where they are critically sensitive to transport and convection. The diffusion flux of ionic species j towards the electrode can be written as:

$$N_j = \frac{D_j \left(C_{b,j} - C_{e,j} \right)}{(1 - t_j)\delta_j} \tag{5}$$

where, N_j, D_j, and t_j are the ionic flux, the diffusivity, and the transport number of ionic species j, respectively. $C_{b,j}$ and $C_{e,j}$ are the bulk concentration and the concentration at the electrode of species j, respectively, and δ_j is its equivalent Nernst diffusion layer thickness [47]. The latter depends slightly on the ionic species, however, when the diffusion coefficients of the involved species do not differ greatly, we may assume that δ for each species is independent and depends mainly on the prevailing transport mode and conditions. In a well-supported electrolyte such as the CIGS system, the transport number can be zero, $t_j \sim 0$, and at the limiting current conditions the concentration at the electrode is zero, $C_e = 0$. As a result, Equation (5) yields the transport limited flux condition, $N_{j,L}$:

$$N_j = \frac{D_j C_{b,j}}{\delta_j} \tag{6}$$

Therefore, the corresponding limiting current density can be written as follows:

$$i_{j,L} = \frac{n_j F\, D_j C_{b,j}}{\delta_j} \tag{7}$$

where n_j is the number of electrons transferred in the electrode reaction and F is Faraday's constant.

Accordingly, we expect that the mass transport of selenium, copper, and indium will be dominated in the electrodeposition process; consequently, agitation and convective flow play an important role. It should also be stated that just about all commercial plating processes are carried out under some convective flow mode, typically pumping, air agitation, or translating electrodes. However, because of the challenge in quantifying and scaling those processes, and because the deposit texture produced under mass transport control is typically rough and powdery, the design of most practical processes is under kinetics control.

3. Results

Initially, deposition of the CIGS layer was carried out from the conventional electrolyte composition. However, the deposit was rough and powdery and did not adhere well to the electrode. This was noted particularly at the higher rotation rates (250 to 600 rpm) where the deposit from the conventional electrolyte was shed as powder off the electrode into the solution. This is further discussed, and samples are shown, subsequently.

To improve the quality of the deposit, a dilute electrolyte for CIGS electrodeposition was introduced. This considerably more dilute solution (about ten-fold) yielded significantly improved deposits with the expected composition. The dilute electrolyte produced a smooth deposit with no evidence of powder formation on either the substrate or within the electrolyte, across a broad rotation range up to 600 rpm. In this lower concentration system, the desirable final composition was obtained in the potential interval $-0.66 \leq E \leq -0.76$ V vs. NHE providing uniform atomic composition across the sample.

3.1. Polarization Studies in CIGS

Polarization studies were conducted on the RDE of stainless steel covered with a molybdenum layer obtained with a sputtering process previously described in Section 2.1. These were to perform an electrodeposition process for both electrolyte systems: the conventional concentrated system ($\sim 10^{-3}$ M), and the newly developed dilute system ($\sim 10^{-4}$ M). The quaternary (CIGS) and the binary (CuSe) systems were specifically studied.

The polarization curves obtained for the higher concentration system (Figure 1) demonstrate the difficulties in defining the limiting current, due to a large amount of hydrogen co-evolution associated with the process. We believe, based on visual observations, that the reduction in current past the peak at about -0.75 V vs. NHE is due to a copious amount of co-evolved hydrogen, accumulating as bubbles on the electrode and blocking a portion of its active area. As shown subsequently, the partial current densities were at the limiting currents for a number of the components such as Cu and Se, leading to rough deposits which exhibited pure adhesion and flaking. At the lower current densities and overpotentials, the current density increased rapidly due mainly to roughness; however, at higher potentials, the current density decreased due to bubble blockage.

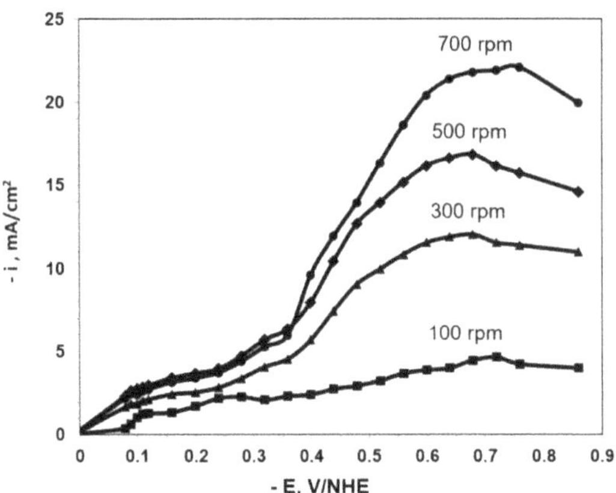

Figure 1. Polarization curves for CIGS electrodeposition on a disk electrode rotated at the indicated rates, from the conventional, higher concentration bath that consisted of 5 mM $CuCl_2 \cdot 2H_2O$; 5.3 mM $InCl_3$; 7.8 mM H_2SeO_3; 6.1 mM $GaCl_3$; pHydrion (pH = 2); and 0.66 M LiCl as supporting electrolyte. The electrodeposition was carried out at ambient temperature = 20 °C.

By contrast, the polarization curves obtained for the lower concentration (by a factor of about 10) electrolyte exhibited better defined plateaus for each rotation rate, possibly because of less roughness in the formation of the film, since the deposit was smooth and adherent. As expected, the deposition current density from the lower concentration electrolyte was significantly lower than the current in the higher concentration system. Under those conditions, the process was below the limiting currents of indium and gallium. Additionally, a shift in the open circuit potential of about 300 mV vs. NHE (in the cathodic direction) was observed at the lower concentration system as compared to the higher concentration systems, most likely due to copper-selenium complexation (Figure 2).

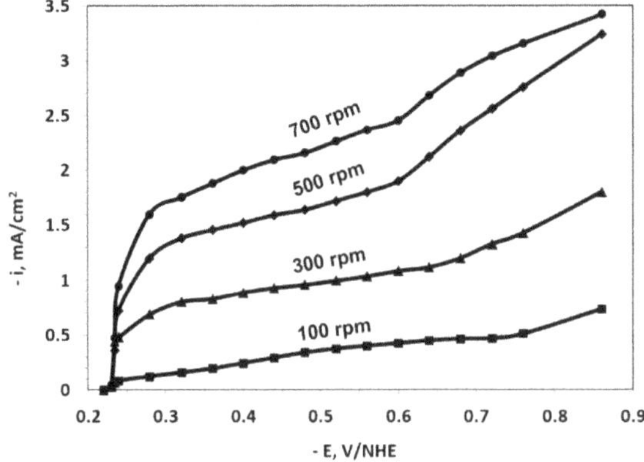

Figure 2. Polarization curves for CIGS electrodeposition on a disk electrode rotated at the indicated rates, from the low concentration bath consisting of 0.45 mM $CuCl_2 \cdot 2H_2O$; 0.44 mM $InCl_3$; 0.85 mM H_2SeO_3; 0.5 mM $GaCl_3$; pHydrion (pH = 2); and 0.66 M LiCl as supporting electrolyte. The electrodeposition was done at ambient temperature = 20 °C.

3.2. Copper-Selenium Co-Deposition

Copper and selenium polarization curves were measured to investigate the cause for the shift of the open circuit potential (E_{ocp}) between the higher and lower concentration systems by about 300 mV vs. NHE. The copper-selenium system showed (Figure 3) a shift of 0.23 V vs. NHE in the open circuit potential between the two systems. This shift is poorly known; thus, it may be due to the formation of secondary phases, in this case the phases could be In_xSe or Cu_ySe, among others [39]; the phases In_xSe and Cu_ySe could be formed, as well as the formation of $CuInSe_2$. As a result, this spontaneous process generates complexity in the reaction mechanism implicated [38]. This shift may be due to another possibility in a different complexion of the pHydrion buffer to both selenium and copper ions, leading to different ion concentration at the electrode not being able to dismiss this shift on the open circuit potential.

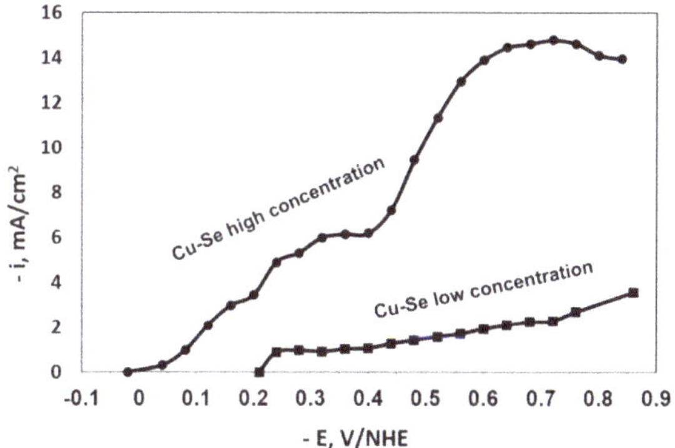

Figure 3. Polarization curves for copper and selenium co-deposition from the higher and lower concentration systems at 300 rpm. The electrolyte compositions were identical to those listed in Figure 1 (high concentration) and Figure 2 (low concentration) of the bath. Temperature = 20 °C.

3.3. Quantifying the Convective Flow Effects on CIGS

In order to determine mass transport effects on the CIGS electrodeposition, the polarization curves for both the conventional electrolyte and the new, low concentration electrolyte were measured on the RDE at three different rotation speeds, as shown in Figures 1 and 2. To avoid a transient response, the polarization curves were generated by holding the potential at a given value until the current stabilized, and after recording this value, the potential was stepped up to the next measurement point. Figures 1 and 2 display quite different current density scales. Both plots indicate that the current density increased with the rotation rate; however, while the conventional electrolyte exhibited a maximum current density at about −0.65 V vs. NHE, the lower concentration of the bath exhibited an increment in the current density across the entire scanned potential range.

In Table 1, the atomic composition of the surface of the film obtained after 40 min at −0.76 V vs. NHE at the three rotation rates (500, 300 and 100) rpm is shown for the lower and higher concentration baths described in Figures 1 and 2. These electrodeposits were obtained at ambient temperature, 20 °C. The electrodeposit composition was analyzed using a Hitachi S4500 scanning electron microscope (SEM) equipped with a Noran energy dispersive spectrometer (EDS).

In the comparison of the deposited films of these two baths, it is possible to observe that the deposit composition was quite sensitive to the mass transport conditions when it was electrodeposited from the conventional electrolyte, with the copper and selenium concentrations increasing with the rotation rates while the indium and gallium content decreased. However, the films deposited from the new bath, low concentration of the precursors, were relatively insensitive to the rotation rate with the copper and selenium content in the deposit varying by less than 10% when the rotation rate was increased from 100 to 500 rpm.

Table 1. Electrodeposit composition obtained after 40 min at -0.76 V vs. NHE at the three rotation rates. The higher concentration bath consisted of 5 mM $CuCl_2 \cdot 2H_2O$; 5.3 mM $InCl_3$; 7.8 mM H_2SeO_3; 6.1 mM $GaCl_3$; pHydrion (pH = 2); and 0.66 M LiCl as supporting electrolyte. The lower concentration bath consisted of 0.45 mM $CuCl_2 \cdot 2H_2O$; 0.44 mM $InCl_3$; 0.85 mM H_2SeO_3; 0.5 mM $GaCl_3$; pHydrion (pH = 2); and 0.66 M LiCl as supporting electrolyte. The electrodeposition process in both baths were carried out at ambient temperature, 20 °C.

Rotation Rate (rpm)		High Concentration Bath				Low Concentration Bath			
		Cu	In	Ga	Se	Cu	In	Ga	Se
500	%	30.1	8	4	57.6	24.3	12	6.8	56.9
	W (mg)	2.6	0.73	0.37	5	0.42	0.24	0.13	0.85
300	%	26.9	13	6.5	52	23.6	13.1	7.4	55.8
	W (mg)	1.6	0.7	0.36	2.9	0.36	0.22	0.11	0.86
100	%	21.2	20	11	44.1	21.8	15.1	9.5	53.1
	W (mg)	0.8	0.74	0.4	1.65	0.23	0.18	0.07	0.36

Comparing experimental data displayed in Figure 1 to the calculated limiting currents of the four species in the higher concentration system, we found that both the selenium and the copper were electrodeposited at their approximate limiting currents (1.49 and 1.22 mA/cm^2, respectively for 100 rpm, as given by Equation (3)) while indium and gallium were kinetically controlled. Therefore, it is proven that mass transport has an important role in CIGS electrodeposition due to metal atomic ratio variation at different rotation speeds. Since the copper and selenium are electrodeposited at their limiting currents, these elements will be more sensitive to rotation-induced mass transport than the kinetically controlled indium and gallium.

By contrast, the lower concentration system exhibited less dependence of the deposit composition on the rotation speed, as indicated in Figure 2. For example, the selenium content in the deposit increased by just about 2% when there was an increasing rotation speed from 300 to 500 rpm. At the low concentration, all metal species were mass transport controlled (Figure 2) and hence they were all affected similarly by the rotation speed; as a result, the composition of the film was not greatly affected by the rotation speed.

3.4. Hydrogen Evolution

In order to characterize the magnitude of hydrogen evolution and gain insight into the controlling deposition regime of the individual components, partial current polarization curves were assembled by analyzing the sample composition (by EDS) and then determining the partial currents from the individual metal components weight using Faraday's law. The partial currents polarization curves are shown in Figures 4 and 5.

The partial currents data were combined to display the total CIGS deposition current density, based on the compositional analysis of the deposit, as displayed in Figures 4 and 5. The difference between the measured total current density and the one based on the compositional deposit analysis is ascribed to hydrogen evolution. Also indicated in Figures 4 and 5 are the theoretical limiting current densities i_L determined using the Levich equation shown below [45]:

$$i_L = i_{\text{Levich}} = 0.62nFAD_j^{2/3}\omega^{1/2}v^{-1/6}C_{b,j} \tag{8}$$

Equation (8) is the expression of limiting current density on a RDE under steady state conditions. Here, ω is the angular rotation rate (rad/s), v is the kinematic viscosity of the electrolyte (≈ 0.01 cm^2/s), and A is the geometric axial area of the working electrode.

The polarization curves associated with the higher concentration system (Figure 4) displayed significantly higher observed current density than expected indicative of a very large amount of hydrogen evolution, accounting for about 75% of the total current. It should be noted that the maximum hydrogen evolution occurred at about the same potential where the maximum total current was observed. At this peak, the current efficiency was about

23%. The hydrogen evolution declined at higher overpotentials, possibly due to bubble blockage of the electrode, as previously discussed.

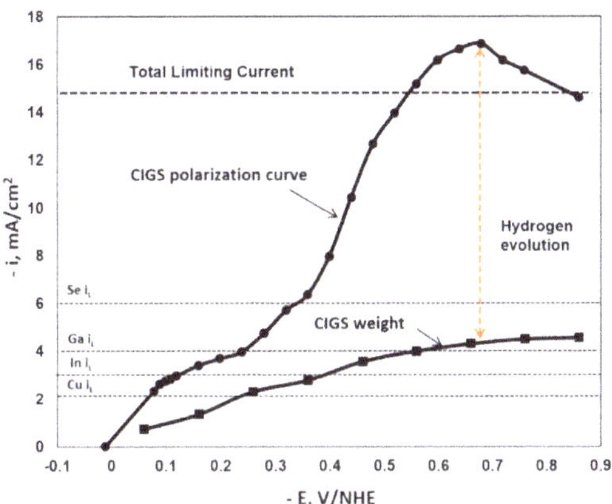

Figure 4. Polarization curves for CIGS deposition from the higher concentration system at 500 rpm. Subtracting the current density of the system by the current density obtained by the weight of the film through Faraday's law provides the current density of hydrogen evolution. The indicated dashed lines are the expected limiting currents based on the non-complexed species concentration obtained with the Levich eq.

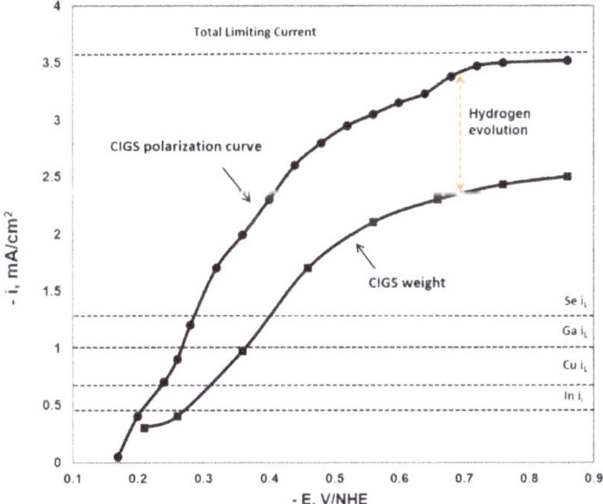

Figure 5. Polarization curves for CIGS deposition from the lower concentration system at 500 rpm. The difference between the overall measured current and the current based on partial currents determined from the deposit weight is due to hydrogen evolution. The electrolyte composition was identical to that listed. The indicated dashed lines are the expected limiting currents based on the non-complexed species concentration obtained with the Levich eq.

The corresponding polarization curves, depicting the hydrogen evolution in the lower concentration system (Figure 5) indicate significantly lower hydrogen evolution as compared to the higher concentration system, both on the absolute scale and as a fraction of the actual deposition current. The amount of evolved hydrogen corresponds in this system to only about 20% of the total current. The smooth deposit and the difference in electrochemical behavior between the two systems may be explained by the amount of hydrogen gas evolved on the surface. This means that the morphology of the surface deposit can be affected by the presence of hydrogen evolution and the concentration of the metal ions since all these species are competing to be reduced at the electrode surface.

A number of common baths that have been described earlier in the literature with composition ranges listed in Table 2 were also investigated. The deposit composition was controlled by adjusting the pH, the overpotential, and the species concentrations in the baths. The deposits were mostly in the desired range for CIGS composition. However, it was difficult to control the process while applying agitation, especially in the 300 to 600 rpm range. The deposit was rough and tended to peel off. By contrast, the new electrolyte introduced herein, where the concentrations of the major components were reduced by a factor of about 10 as compared to the conventional baths, produced high-quality films. The deposit was smooth and adherent across the entire tested rotation range of 0 to 700 rpm and its composition was uniform across the substrate and did not vary much with agitation.

Table 2. Different CIGS electrolyte baths compositions reported in literature and the new chemistry introduced in this study.

Bath	$CuCl_2$ (mM)	H_2SeO_3 (mM)	$InCl_3$ (mM)	$GaCl_3$ (mM)	Supporting Electrolyte	Potential (V vs. SCE)	pH	ω (rpm)
Literature, [14–22]	4–6	6–9	3–6	4–8	LiCl, $Na_3C_6H_5O_7$, pHydrion 2	−0.9 to −1	2–3	None
This work	0.4–0.6	0.7–0.9	0.3–0.5	0.3–0.9	pHydrion 2	−0.8 to −1.1	1–2	0–600

Table 3 indicates the minimum and maximum concentration for each species to provide the final composition. Table 3 lists the deposit weight and its adhesion quality as a function of the disk rotation rate and the current density, for the two systems of the higher and the lower concentrations. Electrodeposition was carried out at a constant potential of −0.76 V vs. NHE at room temperature (20 °C) for 40 min. The metals composition in the electrolyte was adjusted to obtain the desirable CIGS final stoichiometry.

Table 3. Electrodeposited CIGS films at −0.76 V vs. NHE at the listed rotation speeds. (**a**) Left panels: high concentration system; (**b**) right panels: new, low concentration system. Adhesion was tested by a tape peel test. All tests were carried at ambient temperature (20 °C).

Rotation Rate (rpm)	(a) Conventional Composition (~10^{-3} M)			(b) New Composition, Dilute Electrolyte (~10^{-4} M)		
	i (mA/cm^2)	Weight (mg)	Adhesion	i (mA/cm^2)	Weight (mg)	Adhesion
0	1.42	0.7	Good	0.76	0.4	Good
100	3.98	1.1	Good	1.42	0.7	Good
200	6.8	1.2	Good	1.81	0.8	Good
300	9.11	1.1	Poor	2.02	1	Good
400	13.3	0.61	Poor	2.76	1.3	Good
500	17.27	0.32	Poor	2.91	1.5	Good

The stainless steel disk was weighed before and after the deposition in order to determine the deposit weight. As noted in Table 3, less weight was in fact measured at higher rotation rates (>300 rpm) for the conventional, high concentration system as compared to the low concentration system. This unexpected result is due to the non-adherent nature of the deposit obtained from the high concentration electrolyte, which peeled off the substrate at the higher rotation speeds.

3.5. Peel Test

Peel tests were applied to the deposits from both systems (Figure 6). The peel test consisted of applying scotch tape to the deposit (disk with a radius of 0.32 cm) and peeling it rapidly. The deposit was considered to be of a good quality if it remained intact. Deposit fragments observed on the peeled tape were indicative of poor adhesion. Six of the seven samples electrodeposited from the low concentration system passed the peel test successfully, while all seven samples from the high concentration solutions failed the test (Figure 6). The peeled tape on the cylinder (a) (high concentration) shows a significant cylindrical black patch of peeled deposit, while the tape on the cylinder (b) (low concentration) is clear.

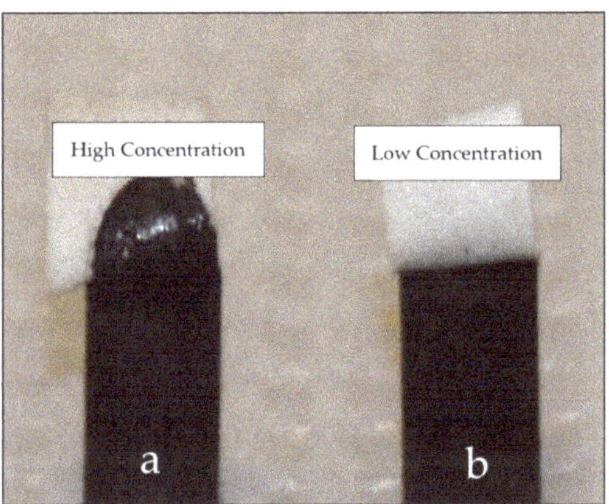

Figure 6. Peel test results, comparing adhesion of CIGS deposits. (**a**) electrodeposit obtained with the bath with high concentration. (**b**) electrodeposit obtained with the bath with low concentration. The substrate was a stainless steel cylinder, 0.32 cm in diameter, electrodeposited on its flat end. A deposit of 1.6 μm CIGS was electrodeposited at −0.76 V vs. NHE for 40 min at 500 rpm. The stainless steel substrate was cleaned in acetone and etched for 2 min in 10% with sulfuric acid prior to deposition.

3.6. Deposit Morphology

SEM inspection of the deposit reveals that the deposit from the lower concentration system had a quite small nuclei size, on the order of 1 μm. Additionally, the deposit provided continuous coverage of the surface and exhibited a fairly narrow (one to two microns) distribution of nuclei size. By contrast, the deposit obtained from the higher concentration system had a much larger nuclei size, on the order of 10 μm. The deposit has a larger variation of nuclei size (5 to 20 μm). The most troublesome characteristic was the larger gaps observed in the deposit, indicating a cracked and rough deposit (Figure 7).

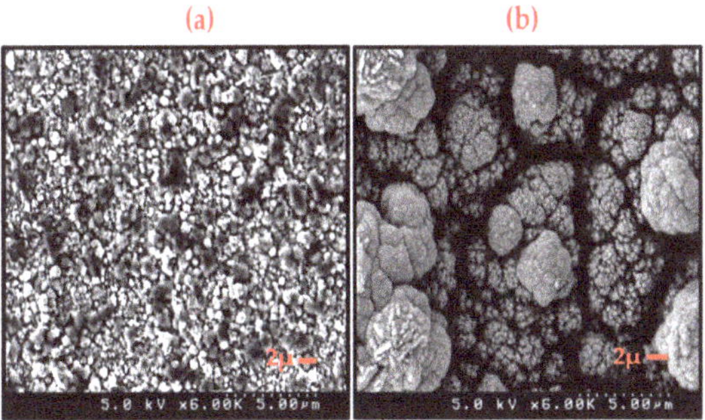

Figure 7. SEM micrographs of CIGS electrodeposited samples. (**a**): deposit from the lower concentration electrolyte. (**b**): deposit from the higher concentration electrolyte. The crystalline size in the deposit from the new, low concentration electrolyte was of the order of 1 μm and exhibited only moderate size distribution. By contrast, the deposit from the conventional, concentrated electrolyte exhibited a much larger crystalline size (10 μm) with a large size distribution and discontinuities. The electrodeposition conditions and substrate preparation were identical to those listed for Figure 6.

3.7. Electrolyte Composition and Potential Range for Obtaining the Desired CIGS Composition

To establish the effect of the applied potential on the deposit composition, electrodeposition studies were carried out in the range between −0.66 and −0.76 V vs. NHE at 500 rpm and ambient temperature. All deposition experiments were carried out for 40 min. The metals ion concentration in the electrolyte was adjusted in the different experiments under the different potentials to obtain the desirable CIGS final stoichiometric atomic composition. The second through the fourth columns in Table 4 list the solution compositions from which the near-optimal deposit compositions were obtained, under the potential listed in the fifth column. The last four columns on the right describe the final alloy atomic compositions as determined by Hitachi S4500 scanning electron microscope (SEM) equipped with a Noran energy dispersive spectrometer (EDS). It was difficult to obtain the desirable gallium amount in the final composition under all the tested conditions at −0.6 V vs. NHE. Good indium gallium distribution and a smooth layer were detected in the range of −0.64 to −0.76 V vs. NHE. The final composition listed was obtained after annealing for one hour at 520 °C under argon atmosphere.

Table 4. CIGS composition after annealing as a function of deposition potential and bath composition. All deposition experiments were carried out at ambient temperature (20 °C) and a rotation rate of 500 rpm for 40 min.

Bath	Potential V/NHE	Electrolyte Concentration (mM)				CIGS Composition %			
		$CuCl_2$	$GaCl_3$	$InCl_3$	H_2SeO_3	Cu	In	Ga	Se
Bath 1	−0.76	0.5	0.37	0.32	0.845	24.5	17.4	7.6	50.3
Bath 2	−0.71	0.45	0.52	0.38	0.78	24.7	17.6	7	50.8
Bath 3	−0.68	0.45	0.69	0.47	0.77	25	17.8	7.1	50.1
Bath 4	−0.66	0.43	0.82	0.54	0.76	25.3	17.8	6.4	50.3
Bath 5	−0.64	0.42	0.9	0.61	0.76	25.4	17.7	6.3	50.2

3.8. Uniformity on the CIGS Electrodeposition

3.8.1. CIGS Composition across Sample

The analysis of the CIGS deposit (by EDS) indicated uniform composition across a 500 μm sample for all four species as illustrated in Figure 8. It is particularly significant to note that the gallium had a uniform composition, at the correct target value, prior to annealing. Typically, due to its high vapor pressure, it is difficult to obtain the optimum gallium composition, and adjusting it during annealing is likely to extend the eventual annealing time.

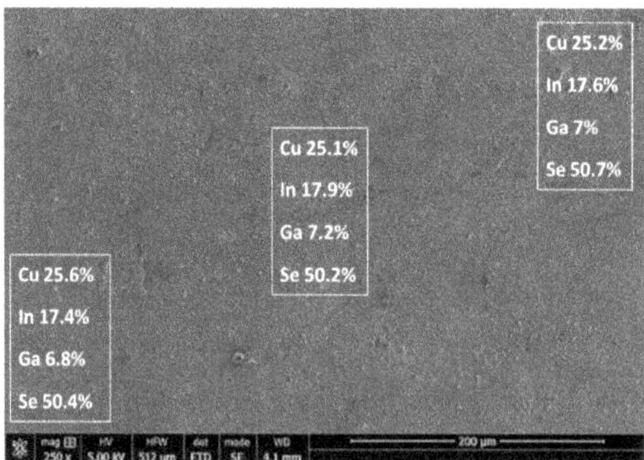

Figure 8. CIGS composition uniformity across a sample electrodeposited from the low concentration, new electrolyte. The listed compositions were measured by EDS at approximately the center of the corresponding rectangles. The sample was electrodeposited for 40 min at −0.76 V vs. NHE, on a disk rotating at 500 rpm, under ambient temperature (20 °C). The electrolyte composition was identical to that listed in Figure 2.

3.8.2. CIGS Thin Film Composition by Depth Profiling

The analysis of the CIGS samples was obtained using FIB. In detail, the atomic composition information in the cross-section was determined by the FEI Helios NanoLab 650 Dual-Beam System equipped with a EDS. As shown in Figure 9, uniform atomic composition throughout the sample depth was obtained for the four components. Gallium again showed uniform composition prior to annealing, thus potentially enabling the reduction of the annealing time. It should also be noted that both gallium and indium exhibited uniform distribution across the sample, thus enabling an increase of the gallium atomic content in the deposit to 9% to increase the energy band.

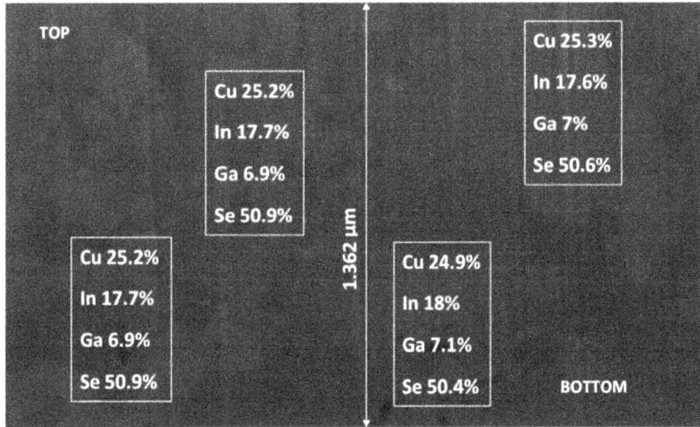

Figure 9. CIGS composition variation with depth as measured by FIB in a sample electrodeposited from the low concentration, new electrolyte. The listed compositions were measured at approximately the center of the corresponding rectangles. The sample was electrodeposited for 40 min at −0.76 V vs. NHE, on a disk rotating at 300 rpm, at ambient temperature (20 °C). The electrolyte composition was identical to that listed in Table 2.

3.9. Effects of the Electrolyte Concentration on the Deposit Morphology and Adhesion

The deposit composition, morphology, and the electrochemical behavior (polarization curves) of the system were determined as a function of the concentration of the major components of the electrodeposited electrolyte in the ranges indicated in Figure 10 at fixed potential −0.76 V vs. NHE, rotation rates 600 rpm, and electrodeposition times 40 min. Deposits obtained from 1/2 and 1/4 of the conventional bath concentration (i.e., bath dilutions by factors of 2 and 4) were compared to deposits obtained from the conventional bath. It was observed that for these dilutions (1/2 and 1/4), the deposit was still rough; the electrochemical behavior (total current plateau, Figures 1 and 3) of this system was still similar to that obtained from the conventional composition. At 1/8 of the conventional bath (eight-fold dilution), smooth deposit was obtained. Moreover, the shape of the polarization curve of this system was still similar to that of the new bath (total current plateau, Figures 2 and 5).

At 1/15 of the conventional bath concentration (15-fold dilution), dark spots were noted on the deposit, indicating voids. Accordingly, it was determined that the best concentration range for CIGS electrodeposition is between 1/8 to the 1/10 of the conventional bath concentration.

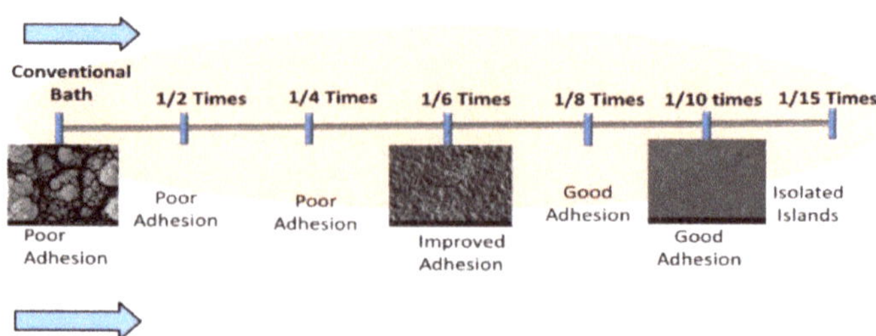

Figure 10. CIGS deposit quality as a function of the bath concentration. The indicated dilutions are with respect to the conventional bath composition. As noted, the best adhesion and deposit morphology was obtained at about 1/10 the concentration of the conventional electrolyte composition.

3.10. Effects of the Electrolyte

Thermal annealing is typically applied after deposition to improve the deposited layer crystal structure, to decrease the recombination of defects, and to obtain a uniform atomic ratio [39–50]. The post treatment parameters depend on the thin film composition, its thickness, the partial vapor pressure, and on the binary stacked layers [51].

The thermal treatment step was carried out in a tube furnace at 520 °C under argon atmosphere for 30 min at a gas flow rate of 8 cm^3/s. The temperature was selected to provide the optimal crystallinity for the CIGS absorber layer [1,37]. Low melting point indium selenides (In_xSe_y) are partially lost during the annealing process as indicated by the data in Table 5.

Table 5. CIGS composition before and after annealing at 520 °C under argon atmosphere for 30 min at a gas flow rate of 8 cm^3/s. Film obtained with the lower concentration bath described in Figure 1.

Metal	Before Annealing	After Annealing
Copper	22.9%	25%
Indium	16.8%	17%
Gallium	6%	7.5%
Selenium	54.3%	50.5%

3.11. Crystallography of CIGS Thin Film

The deposited CIGS thin film was annealed at 500 °C for 30 min under argon atmosphere, and then subjected to XRD analysis to determine the crystallography of the absorber layer. The tested sample was 1.52 µm thick (KLA-Tencor P-6 Stylus Profilometer) (Milpitas, California, USA), and had a composition of 24.9% Cu, 17.3% In, 7.7% Ga, and 50.1% Se. The sample was electrodeposited at −0.76 V vs. NHE for 40 min, under ambient temperature (20 °C) on a stainless steel substrate disk rotated at 500 rpm. The electrolyte composition was 4.4 mM $CuCl_2 \cdot 2H_2O$; 4.85 mM $InCl_3$; 8.1 mM H_2SeO_3; 6.7 mM $GaCl_3$; pHydrion (pH = 3); and 0.66 M LiCl as supporting electrolyte.

Two main diffraction peaks associated with <112> and <220> planes that correspond to the CIGS crystal structure are observed in Figure 11. The XRD analysis showed that the deposited absorber thin film had a $Cu(GaIn)Se_2$ structure, with no other peaks detected. Likewise, the most intense diffraction peak <112> was located at 26.81°, which is indicative of the desirable CIGS crystallography due to the crystal phase corresponding to chalcopyrite material (JCPDS card 35–1102). The thickness of the CIGS layer was about 1.4 µm, which should be sufficient to absorb 97% of the incident light.

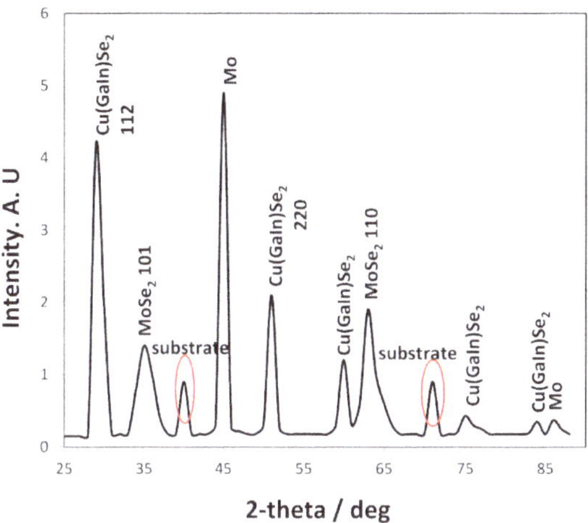

Figure 11. Diffraction peaks of the CIGS layer deposited from the low concentration electrolyte, after annealing for 30 min at 500 °C. The tabulated insets indicate the deposit composition obtained after 40 min at −0.76 V vs. NHE at 500 rpm. The electrolyte consisted of 0.45 mM $CuCl_2 \cdot 2H_2O$; 0.44 mM $InCl_3$; 0.85 mM H_2SeO_3; 0.5 mM $GaCl_3$; pHydrion (pH = 2); and 0.66 M LiCl as supporting electrolyte. Temperature = 20 °C.

3.12. Mechanistic Discussion of Surface Coverage and Nucleation Density

A model proposed by Scharifker and Hills [52], providing a relationship between the nucleation density and the electrolyte concentration, was adapted to analyze the process and results of the present research. Figure 12 is a schematic representation of probable steps in the deposit nucleation process. The two panels provide a suggested comparison for the effect of the electrolyte concentration on the deposit morphology.

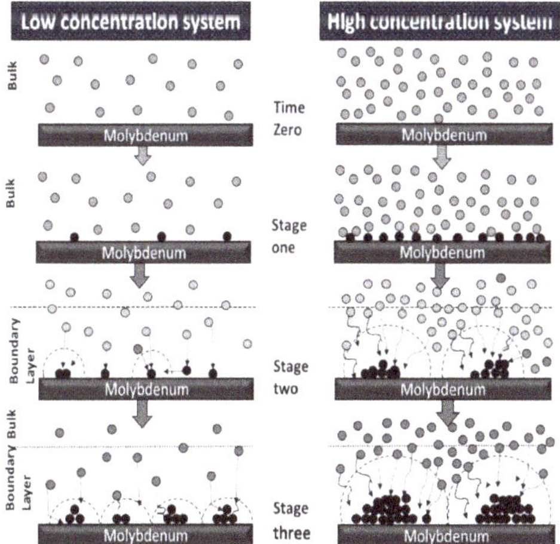

Figure 12. Schematic illustrating the three steps of CIGS electronucleation at the molecular level following the Scharifker/Hills model. Stage one shows more ions near the substrate surface at the high concentration system as compared to the lower concertation system. Stage two indicates that adatoms start to accumulate at the higher concentration system while the adatoms still travel longer in the low concentration system. In stage three, the higher concentration system displays larger nuclei as compared to the lower concentration system.

At the initial stage of the electrodeposition process, the number of adatoms deposited on the substrate surface is a function of the electrolyte bulk concentration [52], which is larger for the higher concentration electrolyte. Consequently, the distance between metal adatoms is larger in the low concentration system as compared to the higher concentration system. The adatoms migrate towards one another to reach a minimal energy state. In order to form nuclei in the lower concentration system, adatoms need to travel a longer distance as compared to the high concentration system. Therefore, in the low concentration system, the adatoms may reach the minimal energy state (stop moving) before meeting the closest adatoms, resulting in a large number of small nuclei [45]. In contrast, in the higher concentration system, since the density of electrodeposited adatoms on the surface is high, the closest adatoms are likely to group together to create larger nuclei, resulting in a lower nuclei population density [52].

The Scharifker/Hills model uses the peak of transient current (I_{max}) and corresponding peak time (t_{max}) at different metals concentrations to determine the nuclei population density for the system [52]. Accordingly, the nuclei population density is given by:

$$N_0 = 0.065 \left(\frac{1}{8\pi C_0 V_m} \right)^{1/2} \left(\frac{nFC_0}{i_{max} \cdot t_{max}} \right)^2 \tag{9}$$

Here, C_0 is the concentration of species in the bulk; V_m is the molar volume; t_{max} is the peak time; and i_{max} is the peak current density (A/cm^2).

When substituting the corresponding numerical values into the Scharifker/Hills model, we find that the expected nucleation density for the conventional bath, higher concentration system was 2×10^6 nuclei/cm^2, while the lower concentration system (1/10 dilution) was expected to yield a far lower 1.5×10^8 nuclei/cm^2. This was in close agreement to measurements of deposits from the low concentration electrolyte, which indicated nuclei density of about 3×10^8 nuclei/cm^2. Clearly, the higher nuclei density is expected to provide better surface coverage and improved adhesion. A description to explain why it is better to perform electrodeposition experiments at lower concentrations to inhibit roughness and enhance adhesion on the film can be explained in Figure 12. Here, it is possible to notice that in a higher concentration system, the presence of large clusters in scattered areas of the surface is favored, thereby proving heterogeneity in the composition of the alloy since not all precursors follow the same dynamic; that is, some of the precursors are governed by mass transfer while others are controlled by a kinetics reaction to the potential applied to form the CIGS quaternary alloy.

3.13. Optical Characterization Test

This thin film semiconductor of CIGS study was done in an electrodeposition process of a single bath (one step) without the need to add a gas phase from an additional precursor, which is not reported in the literature. It is therefore relevant to evaluate if the alloy has a photovoltaic response. Thus, the characterizations tests were done on the CIGS device after its completion process. The solar cell equivalent is shown in Figure 13. The equivalent circuit is helpful for characterizing the electrical response of the device and identifying the important factors that may affect its behavior. The Illumination current (I_L) is possibly the most important measured parameter in the solar cell and it is also called the photogenerated current. Another measured current is the dark current (I_D), which is the reverse saturation current, and it is measured under dark conditions. Dark current represents a standard diode curve of the photovoltaic equivalent circle. The difference between the illuminating current and the dark current is the solar cell generated current [53].

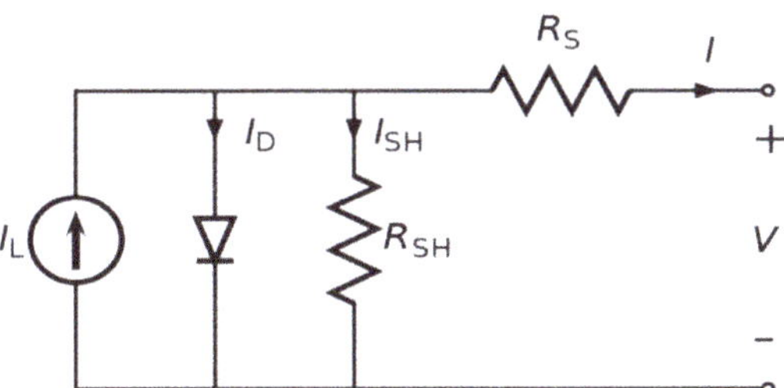

Figure 13. A typical equivalent circuit of a PV cell.

The shunt resistance (R_{SH}) is due to electron-hole recombination; in the shunt, resistance passes the shunt current (I_{SH}); the series resistance (R_S) is due to conductivity and connection wire imperfection. Consequently, output current (I) of the photovoltaic cell's can be described as follow [54]:

$$I = I_L - I_D \left\{ exp \left[\frac{q(V + I \cdot R_S)}{Q_D kT} \right] - 1 \right\} - \frac{V + I \cdot R_S}{R_{SH}} \tag{10}$$

Hence, Q_D is the diode ideality factor, q is the elementary charge of an electron, V is the voltage across terminals, T is the absolute temperature of the PV cell, and k is the Boltzmann's constant.

Furthermore, the connection device was designed in the lab in order to have a complete electric circuit to characterize the solar device. The characterized parameters were overall efficiency, quantum efficiency, open circuit voltage, short circuit current, dark current, band gap, and the fill factor. As a result, the equipment used was a PV Measurements QEX10 quantum efficiency (PV Measurements, Boulder, CO, USA) measurement system and an Oriel Sol2A solar simulator (Irvine, CA, USA). The solar simulator provides light and radiation similar to that provided from the Sun. The solar simulator tests the solar device efficiency and determines additional outcome parameters, including fill factor (*FF*), open circuit voltage (V_{OC}), and the short circuit current (I_{SC}). Fill factor is a parameter used to determine the maximum operating power point. It is given by:

$$FF = \frac{P_m}{V_{OC} \cdot I_{SC}} = \frac{\eta \cdot A_C \cdot E}{V_{OC} \cdot I_{SC}} \tag{11}$$

where (η) is the solar cell's energy conversion efficiency, (E) is the input power, and (P_m) is the maximum power.

In Figure 14, it is shown the "current density-voltage" (i vs. V) plot obtained with the set up described in Figure 13 and for the SS/Ni/Mo/CIGS/CdS/i-ZnO/Al:ZnO device; this device was prepared with the electrodeposition of CIGS of the lower concentration electrolyte.

The open short circuit current and open-circuit voltage were about 23.5 mA/cm^2 and 0.41 V, respectively. The fill factor of the CIGS device was about 49 %.

The external quantum efficiency is the ratio of exited electrons to incident photons. The current produced per incident photon was measured as a function of the corresponding wavelength. The quantum efficiency for CIGS solar device under illumination was done at AM1.5, (1000 W/m^2), and it is presented in Figure 15. The quantum efficiency measurements collected is about 6% efficient as it is shown in Figure 15.

This characterization test proves that the CIGS absorbs a higher number of photons in the visible light spectrum than in the longer wavelength spectrum. The quantum efficiency was high because the CIGS thin film obtained from the low electrolyte concentration had good quality. The band gap calculations match the experimental one, which is about 1.29 eV. The quantum efficiency was about 0.62 at the optimal band gap as it is shown in Figure 15. Finally, in this study there was not an optical characterization of the CIGS thin film obtained from the higher concentration bath because this layer did not show a smooth morphology or adhesion with the back contact barrier.

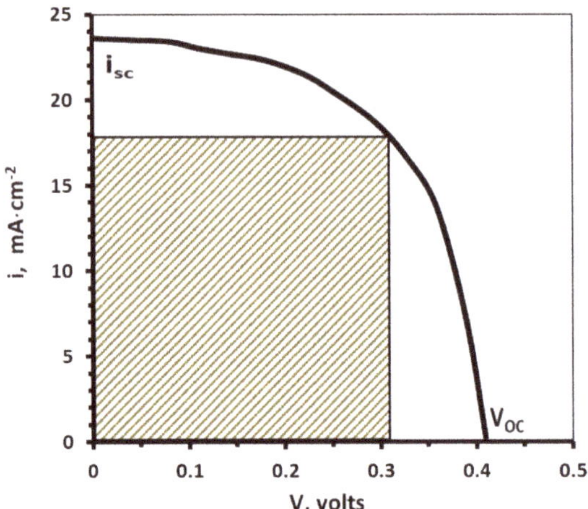

Figure 14. "Current density-voltage" (i vs. V) characteristics of SS/Mo/Ni/CIGS/CdS/i-ZnO/Al:ZnO structure.

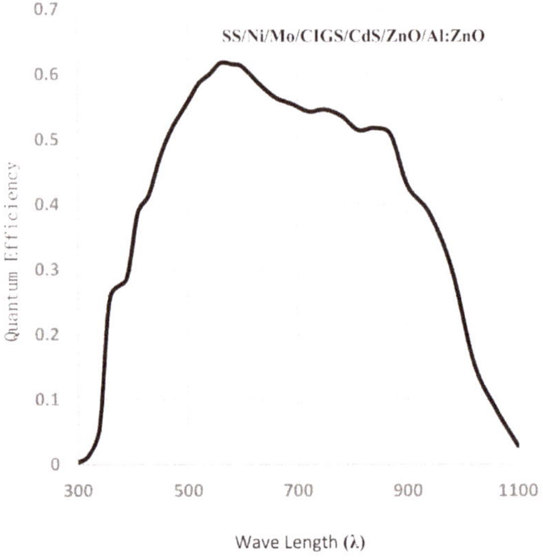

Figure 15. Quantum efficiency characteristics of the SS/Ni/Mo/CIGS/CdS/i-ZnO/Al:ZnO structure.

4. Conclusions

A single step electroplating process for the deposition of high-quality quaternary CIGS absorber layer from a new, low concentration electrolyte composition has been demonstrated. The deposition process and the deposit composition as well as its properties have been characterized. The improved low concentration electrolyte, consisting of deposited ions in the 10^{-4} M range (corresponding to a dilution factor of about ten-fold in comparison to previously reported systems), provided an adherent and smooth deposit. Smaller nuclei sizes (~1 µm) at a density of about 3×10^8 nuclei/cm^2 with a uniform atomic composition at the optimal CIGS composition were observed across the samples, as well as in the deposit bulk.

The flow effects on the deposit composition and its morphology were characterized for both the conventional and the new dilute electrolyte. It was determined that the transport conditions have a major effect on the deposit properties in both systems. The effect of the electrolyte composition and the applied voltage on the deposit composition has been characterized. The process parameters for the deposition of the CIGS absorber layer with near optimal composition have been identified. It was determined that unlike previously reported systems, there is no need for metal addition from the gas phase during the annealing process, and the annealing time under argon atmosphere could be significantly reduced (to about 30 min as compared to two hours in the conventional process).

Moreover, a weight analysis was done to describe the main role of the co-evolution of hydrogen gas in the electrodeposition process on the CIGS alloy. As a result, the presence of the hydrogen evolution in the quality and morphology of the electrodeposit film in which the electrochemical behavior changes is strongly associated with the presence of different concentrations of the precursors on the bath. Moreover, the optical characterization yielded results of an open short circuit current and open-circuit voltage about 23.5 mA/cm^2 and 0.41 V, respectively. Likewise, the fill factor of the CIGS device was about 49%. In the quantum efficiency experiments, the device showed a value of 0.62 at the optical band gap.

Finally, this study and novel electrolyte system showed interesting results when obtaining a thin film electrodeposited with photovoltaic response in a single bath. Therefore, this study successfully provides the basis for designing an improved CIGS electrodeposition system on large areas required in commercial scales because the process has been conveniently governed by mass transfer.

Author Contributions: Both authors participated equally throughout the entire research and writing process of the article. Both authors have read and agreed to the published version of the manuscript.

Funding: This research received no external funding.

Institutional Review Board Statement: Not applicable.

Informed Consent Statement: Not applicable.

Data Availability Statement: Data are contained within the article.

Acknowledgments: Authors acknowledge the department of chemical engineering and biomolecular at Case Western Reserve University for the facilities provided to carry out the experiments.

Conflicts of Interest: The authors declare no conflict of interest. The funders had no role in the design of the study, in the collection, analyses or interpretation of data; in the writing of the manuscript or in the decision to publish the results.

References

1. Hamakawa, Y. Thin-Film Solar Cells, Next Generation Photovoltaics and Its Applications. In *Springer Series in Photonics*; Springer Science & Business Media: Amsterdam, The Netherlands, 2004; Volume 13.
2. Repins, I.; Contreras, M.; Romero, M.; Yan, Y.; Metzger, W.; Li, J.; Johnston, S.; Egaas, B.; DeHart, C.; Scharf, J.; et al. Characterization of 19.9% Efficient CIGS Absorbers. In *33rd IEEE Photovoltaic Specialists Conference*; NREL/PR, 520-43247; IEEE: San Diego, CA, USA, 2008.
3. Ramanathan, K.; Contreras, M.A.; Perkins, C.L.; Asher, S.; Hasoon, F.S.; Keane, J.; Young, D.; Romero, M.; Metzger, W.; Noufi, R.; et al. Properties of 19.2% efficiency ZnO/CdS/CuInGaSe$_2$ thin-film solar cells. *Prog. Photovolt. Res. Appl.* **2003**, *11*, 225–230. [CrossRef]
4. Jo, Y.H.; Mohanty, B.C.; Cho, Y.S. Crystallization and surface segregation in CuIn$_{0.7}$Ga$_{0.3}$Se$_2$ thin films on Cu foils grown by pulsed laser deposition. *Appl. Surf. Sci.* **2010**, *256*, 6819–6823. [CrossRef]
5. Caballero, R.; Kaufmann, C.A.; Efimova, V.; Rissom, T.; Hoffmann, V.; Schock, H.W. Investigation of Cu(In,Ga)Se$_2$ thin-film formation during the multi-stage co-evaporation process. *Prog. Photovolt. Res. Appl.* **2013**, *21*, 30–46. [CrossRef]
6. Jackson, P.; Wuerz, R.; Hariskos, D.; Lotter, E.; Witte, W.; Powalla, M. Effects of heavy alkali elements in Cu(In,Ga)Se$_2$ solar cells with efficiencies up to 22.6%. *Phys. Status Solidi R* **2016**, *10*, 583–586. [CrossRef]

7. Marudachalam, M.; Hichri, H.; Klenk, R.; Birkmire, R.W.; Shafarman, W.N.; Schultz, J.M. Preparation of homogeneous Cu(InGa)Se$_2$ films by selenization of metal precursors in H$_2$Se atmosphere. *Appl. Phys. Lett.* **1995**, *67*, 3978–3980. [CrossRef]

8. Yen, F.; Wei, W.; Kit, L. Electrochemical properties of solid-liquid interface of CuIn$_{1-x}$Ga$_x$Se$_2$ prepared by electrodeposition with various gallium concentrations. *J. Electrochem. Soc.* **2009**, *156*, E133–E138. [CrossRef]

9. Chaure, N.; Samantilleke, A.P.; Burton, R.; Young, J.; Dharmadasa, I. Electrodeposition of p$^+$, p, i, n and n$^+$-type copper indium gallium diselenide for development of multilayer thin film solar cells. *Thin Solid Films* **2005**, *472*, 212–216. [CrossRef]

10. Bhattacharya, R.; Hiltner, J.; Batchelor, W.; Contreras, M.; Noufi, R.; Sites, J. 15.4% CuIn$_{1-x}$Ga$_x$Se$_2$-Based photovoltaic cells from solution-based precursor films. *Thin Solid Films* **2000**, *361–362*, 396–399. [CrossRef]

11. Bhattacharya, N.; Fernandez, M. CuIn$_{1-x}$Ga$_x$Se$_2$-based photovoltaic cells from electrodeposited precursor films. *Sol. Energy Mater. Sol. Cells* **2003**, *76*, 331–337. [CrossRef]

12. Kampmann, A.; Rechid, J.; Raitzig, A.; Wulff, S.; Mihhailova, M.; Thyen, R.; Kalberlah, K. Electrodeposition of CIGS on metal substrates, compound semiconductors photovoltaics. In Proceedings of the Materials Research Society Symposium, Osaka, Japan, 11–18 May 2003; Volume 763, pp. 323–328.

13. Rechid, J.; Thyen, R.; Raitzig, A.; Wulff, S.; Mihhailova, M.; Kalberlah, K.; Kampmann, A. Electrodeposition of CIGS on Metal Substrates. In Proceedings of the Third World Conference on Photovoltaic Solar Energy Conversion, Osaka, Japan, 11–18 May 2003; pp. 559–561.

14. Bouich, A.; Ullah, S.; Ullah, H.; Mari, B.; Hartiti, B.; Touhami, M.E.; Santos, D.M.F. Deposit on different back contacts: To high-quality CuInGaS$_2$ thin films for photovoltaic application. *J. Mater. Sci. Mater. Electron.* **2019**, *30*, 20832–20839. [CrossRef]

15. Ao, J.; Fu, R.; Jeng, M.-J.; Bi, J.; Yao, L.; Gao, S.; Sun, G.; He, Q.; Zhou, Z.; Sun, Y.; et al. Formation of Cl-doped ZnO thin films by a cathodic electrodeposition for use as a window layer in CIGS solar cells. *Materials* **2018**, *11*, 953. [CrossRef]

16. Tinoco, T.; Rincón, C.; Quintero, M.; Pérez, G.S. Phase diagram and optical energy gaps for CuIn$_y$Ga$_{1-y}$Se$_2$ Alloys. *Phys. Status Solidi (a)* **1991**, *124*, 427. [CrossRef]

17. Kapur, J.; Basol, M.; Tseng, S. Preparation of thin Films of Chalcopyrites for Photovoltaics. In *Ternary and Multinary Compounds*; Material Research Society: Pittsburgh, PA, USA, 1987; pp. 219–224.

18. Sidali, T.; Bou, A.; Coutancier, D.; Chassaing, E.; Theys, B.; Barakel, D.; Garuz, R.; Thoulon, P.-Y.; Lincot, D. Semi-transparent photovoltaic glazing based on electrodeposited CIGS solar cells on patterned molybdenum/glass substrates. *EPJ Photovolt.* **2018**, *9*, 2. [CrossRef]

19. Su, C.-Y.; Ho, W.-H.; Lin, H.-C.; Nieh, C.-Y.; Liang, S.-C. The effects of the morphology on the CIGS thin films prepared by CuInGa single precursor. *Sol. Energy Mater. Sol. Cells* **2011**, *95*, 261–263. [CrossRef]

20. Guillemoles, J.-F. Stability of Cu(In,Ga)Se$_2$ solar cells: A thermodynamic approach. *Thin Solid Films* **2000**, *361–362*, 338–345. [CrossRef]

21. Kampmann, A.; Sittinger, V.; Rechid, J.; Reineke-Koch, R. Large area electrodeposition of Cu(In,Ga)Se$_2$. *Thin Solid Films* **2000**, *361*, 309–313. [CrossRef]

22. Fahoume, M.; Chraibi, F.; Aggour, M.; Delplancke, L.; Ennaoui, A.; Lux-Steiner, M. Electrodeposition of CuInGaSe$_2$ Thin Films. In Proceedings of the 17th European Photovoltaic Solar Energy Conference, Munich, Germany, 22–26 October 2001; pp. 1247–1250.

23. Londhe, P.U.; Rohom, A.B.; Fernandes, R.; Kothari, D.C.; Chaure, N.B. Development of superstrate CuInGaSe$_2$ thin film solar cells with low-cost electrochemical route from nonaqueous Bath. *ACS Sustain. Chem. Eng.* **2018**, *6*, 4987–4995. [CrossRef]

24. Qu, J.-Y.; Guo, Z.-F.; Pan, K.; Zhang, W.-W.; Wang, X.-J. Influence of annealing conditions on the properties of Cu(In,Ga)Se$_2$ thin films fabricated by electrodeposition. *J. Zhejiang Univ. A* **2018**, *19*, 399–408. [CrossRef]

25. Calixto, M.E.; Sebastian, P.; Bhattacharya, R.; Noufi, R. Compositional and optoelectronic properties of CIS and CIGS thin films formed by electrodeposition. *Sol. Energy Mater. Sol. Cells* **1999**, *59*, 75–84. [CrossRef]

26. Calixto, M.E.; Dobson, K.D.; McCandless, B.E.; Birkmire, R.W. Controlling growth chemistry and morphology of single-bath electrodeposited Cu(In,Ga)Se$_2$ Thin Films for Photovoltaic Application. *J. Electrochem. Soc.* **2006**, *153*, G521–G528. [CrossRef]

27. Fernández, A.; Bhattacharya, R. Electrodeposition of CuIn$_{1-x}$Ga$_x$Se$_2$ precursor films: Optimization of film composition and morphology. *Thin Solid Films* **2005**, *474*, 10–13. [CrossRef]

28. Zhang, L.; Jiang, F.; Feng, J. Formation of CuInSe$_2$ and Cu(In,Ga)Se$_2$ films by electrodeposition and vacuum annealing treatment. *Sol. Energy Mater. Sol. Cells* **2003**, *80*, 483–490. [CrossRef]

29. Lai, Y.; Liu, F.; Kuang, S.; Liu, J.; Zhang, Z.; Li, J.; Liu, Y. Electrodeposition-based preparation of Cu(In,Ga)(Se,S)$_2$ thin films. *Electrochem. Solid State Lett.* **2009**, *12*, D65–D67. [CrossRef]

30. Hermann, A.; Gonzalez, C.; Ramakrishnan, P.; Balzar, D.; Popa, N.; Rice, P.; Marshall, C.; Hilfiker, J.; Tiwald, T.; Sebastian, P.; et al. Fundamental studies on large area Cu(In,Ga)Se$_2$ films for high efficiency solar cells. *Sol. Energy Mater. Sol. Cells* **2001**, *70*, 345–361. [CrossRef]

31. Bouich, A.; Ullah, S.; Ullah, H.; Mollar, M.; Marí, B.; Touhami, M.E. Electrodeposited CdZnS/CdS/CIGS/Mo: Characterization and solar cell performance. *JOM* **2020**, *72*, 615–620. [CrossRef]

32. Esmaeili-Zare, M.; Behpour, M. Influence of deposition parameters on surface morphology and application of CuInS$_2$ thin films in solar cell and photocatalysis. *Int. J. Hydrog. Energy* **2020**, *45*, 16169–16182. [CrossRef]

33. Fu, Y.-P.; You, R.-W.; Lew, K.K. CuIn$_{1-x}$Ga$_x$Se$_2$ Absorber Layer Fabricated by Pulse-Reverse Electrodeposition Technique for Thin Films Solar Cell. *J. Electrochem. Soc.* **2009**, *156*, D553–D557. [CrossRef]

34. Mandati, S.; Dey, S.R.; Joshi, S.V.; Bulusu, S.V. Cu(In,Ga)Se$_2$ Films with branched nanorod architectures fabricated by economic and environmentally friendly pulse-reverse electrodeposition route. *ACS Sustain. Chem. Eng.* **2018**, *6*, 13787–13796. [CrossRef]

35. Mandati, S.; Dey, S.R.; Joshi, S.V.; Sarada, B.V. Two-dimensional CuIn$_{1-x}$Ga$_x$Se$_2$ nano-flakes by pulse electrodeposition for photovoltaic applications. *Sol. Energy* **2019**, *181*, 396–404. [CrossRef]

36. Lincot, D.; Guillemoles, J.; Taunier, S.; Guimard, D.; Sicx-Kurdi, J.; Chaumont, A.; Roussel, O.; Ramdani, O.; Hubert, C.; Fauvarque, J.; et al. Chalcopyrite thin film solar cells by electrodeposition. *Sol. Energy* **2004**, *77*, 725–737. [CrossRef]

37. Powalla, M.; Paetel, S.; Hariskos, D.; Wuerz, R.; Kessler, F.; Lechner, P.; Wischmann, W.; Friedlmeier, T.M. Advances in Cost-Efficient Thin-Film Photovoltaics Based on Cu(In,Ga)Se$_2$. *Engineering* **2017**, *3*, 445–451. [CrossRef]

38. Bermudez, V. An overview on electrodeposited Cu(In,Ga)(Se,S)$_2$ thin films for photovoltaic devices. *Sol. Energy* **2018**, *175*, 2–8. [CrossRef]

39. Bhattacharya, R.N.; Wiesner, H.; Berens, T.A.; Matson, R.J.; Keane, J.; Ramanathan, K.; Swartzlander, A.; Mason, A.; Noufi, R.N. 12.3% Efficient CuIn$_{1-x}$Ga$_x$Se$_2$-Based Device from Electrodeposited Precursor. *J. Electrochem. Soc.* **1997**, *144*, 1376–1379. [CrossRef]

40. Vasekar, P.S.; Dhere, N.G. Effect of sodium addition on Cu-deficient CuIn$_{1-x}$Ga$_x$Se$_2$ thin film solar cells. *Sol. Energy Mater. Sol. Cells* **2009**, *93*, 69–73. [CrossRef]

41. Romeo, A.; Terheggen, M.; Abou-Ras, D.; Bätzner, D.L.; Haug, F.-J.; Kälin, M.; Rudmann, D.; Tiwari, A.N. Development of thin-film Cu(In,Ga)Se$_2$ and CdTe solar cells. *Prog. Photovolt. Res. Appl.* **2004**, *12*, 93–111. [CrossRef]

42. Lara-Lara, B.; Fernández, A.M. Growth improved of CIGS thin films by applying mechanical perturbations to the working electrode during the electrodeposition process. *Superlattices Microstruct.* **2019**, *128*, 144–150. [CrossRef]

43. Molin, A.; Dikusar, A.; Kiosse, G.; Petrenko, P.; Sokolovsky, A.; Saltanovsky, Y. Electrodeposition of CuInSe$_2$ thin films from aqueous citric solutions I. Influence of potential and mass transfer rate on composition and structure of films. *Thin Solid Films* **1994**, *237*, 66–71. [CrossRef]

44. Molin, A.; Dikusar, A. Electrodeposition of CuInSe$_2$ thin films from aqueous citric solutions II. Investigation of macrokinetics of electrodeposition by means of rotating disk electrode. *Thin Solid Films* **1994**, *237*, 72–77. [CrossRef]

45. Levich, B. *Physicochemical Hydrodynamics*, 1st ed.; Prentice Hall: Hoboken, NJ, USA, 1962.

46. Lin, Y.; Ke, J.; Yen, W.; Liang, S.; Wu, C.; Chiang, C. Preparation and characterization of Cu(In,Ga)(Se,S)$_2$ films without selenization by co-sputtering from Cu(In,Ga)Se$_2$ quaternary and In$_2$S$_3$ targets. *Appl. Surf. Sci.* **2011**, *257*, 4278–4284. [CrossRef]

47. Lincot, D. Electrodeposition of semiconductors. *Thin Solid Films* **2005**, *487*, 40–48. [CrossRef]

48. Kang, F.; Ao, J.; Sun, G.; He, Q.; Sun, Y. Properties of CuIn$_{1-x}$Ga$_x$Se$_2$ thin films grown from electrodeposited precursors with different levels of selenium content. *Curr. Appl. Phys.* **2010**, *10*, 886–888. [CrossRef]

49. Bhattacharya, R.N. Electrodeposited Two-Layer Cu-In-Ga-Se/In-Se Thin Films. *J. Electrochem. Soc.* **2010**, *157*, D406–D410. [CrossRef]

50. Lai, Y.-Q.; Kuang, S.-S.; Liu, F.-Y.; Zhang, Z.-A.; Liu, J.; Li, J.; Liu, Y.-X. Preparation and characterization of Cu(In,Ga)(Se,S)$_2$ thin film by sulfurization of electrodeposited Cu(In,Ga)Se$_2$ Precursor. *App. Phys.* **2009**, *36*, L1394.

51. Ishizuka, S.; Yamada, A.; Matsubara, K.; Fons, P.; Sakurai, K.; Niki, S. Development of high-efficiency flexible Cu(In,Ga)Se$_2$ solar cells: A study of alkali doping effects on CIS, CIGS, and CGS using alkali-silicate glass thin layers. *Curr. Appl. Phys.* **2010**, *10*, S154–S156. [CrossRef]

52. Grujicic, D.; Pesic, B. Electrodeposition of copper: The nucleation mechanisms. *Electrochim. Acta* **2002**, *47*, 2901–2912. [CrossRef]

53. Nelson, J.A. *The Physics of Solar Cells*; World Scientific Publishing Company: Singapore, 9 May 2003.

54. Rosa-Clot, M.; Tina, G.M. Introduction to PV Plants. In *Submerged and Floating Photovoltaic Systems*; Elsevier: London, UK, 2018; pp. 33–64. [CrossRef]

MDPI

nanomaterials

Communication

Efficient Nanocrystal Photovoltaics via Blade Coating Active Layer

Kening Xiao [1], Qichuan Huang [1], Jia Luo [1], Huansong Tang [1], Ao Xu [1], Pu Wang [1], Hao Ren [1], Donghuan Qin [1,2,*], Wei Xu [1,2,*] and Dan Wang [1,2,*]

[1] School of Materials Science and Engineering, South China University of Technology, Guangzhou 510640, China; ddkndyx@163.com (K.X.); hqc666666nb@163.com (Q.H.); l5865421014@126.com (J.L.); 201865320283@mail.scut.edu.cn (H.T.); m13856403915@163.com (A.X.); 18145717332@163.com (P.W.); renhns@foxmail.com (H.R.)

[2] State Key Laboratory of Luminescent Materials & Devices, Institute of Polymer Optoelectronic Materials & Devices, South China University of Technology, Guangzhou 510640, China

* Correspondence: qindh@scut.edu.cn (D.Q.); xuwei@scut.edu.cn (W.X.); wangdan@scut.edu.cn (D.W.)

Abstract: CdTe semiconductor nanocrystal (NC) solar cells have attracted much attention in recent year due to their low-cost solution fabrication process. However, there are still few reports about the fabrication of large area NC solar cells under ambient conditions. Aiming to push CdTe NC solar cells one step forward to the industry, this study used a novel blade coating technique to fabricate CdTe NC solar cells with different areas (0.16, 0.3, 0.5 cm^2) under ambient conditions. By optimizing the deposition parameters of the CdTe NC's active layer, the power conversion efficiency (PCE) of NC solar cells showed a large improvement. Compared to the conventional spin-coated device, a lower post-treatment temperature is required by blade coated NC solar cells. Under the optimal deposition conditions, the NC solar cells with 0.16, 0.3, and 0.5 cm^2 areas exhibited PCEs of 3.58, 2.82, and 1.93%, respectively. More importantly, the NC solar cells fabricated via the blading technique showed high stability where almost no efficiency degradation appeared after keeping the devices under ambient conditions for over 18 days. This is promising for low-cost, roll-by-roll, and large area industrial fabrication.

Keywords: CdTe nanocrystals; blade coating; large area fabrication

Citation: Xiao, K.; Huang, Q.; Luo, J.; Tang, H.; Xu, A.; Wang, P.; Ren, H.; Qin, D.; Xu, W.; Wang, D. Efficient Nanocrystal Photovoltaics via Blade Coating Active Layer. *Nanomaterials* **2021**, *11*, 1522. https://doi.org/10.3390/nano11061522

Academic Editor: Vlad Andrei Antohe

Received: 10 May 2021
Accepted: 5 June 2021
Published: 9 June 2021

Publisher's Note: MDPI stays neutral with regard to jurisdictional claims in published maps and institutional affiliations.

1. Introduction

Recently, semiconductor nanocrystals (NCs), such as CdTe or PbS NCs, have attracted much attention due to their potential application in photovoltaics [1–9], light-emitting diodes [10,11], photodetectors [12,13] or biosensors [14,15] and so on. Compared to other semiconductor NCs, CdTe NCs possess ideal bandgap (~1.5 eV), a high optical absorption coefficient (>10^4/cm in the visible light region), and high stability, which makes them suitable for photovoltaic application [16]. In addition, CdTe NC solar cells can be fabricated through a simple solution process under ambient conditions. Since 2005, CdTe NC solar cells have been developed rapidly, and a power conversion efficiency (PCE) of up to 12% has been realized by optimizing device architecture and fabricating techniques [17–19]. The stability of CdTe NC solar cells with conventional structure, however, remains a challenge because of the misalignment of the work function between ITO (Indium Tin Oxide ~4.7 eV) and the CdTe NC (~5.2 eV) thin film [20]. CdTe NC solar cells with inverted structure have therefore received intensive attention as the holes transfer layer can be easily applied in this case for efficient carrier collecting. In previous reports, back-contact layers like MoOx, Cr, etc., had been used to improve carrier collection efficiency in CdTe NC solar cells [21–23]. Although the open-circuit voltage (V_{oc}) had been improved, the PCE of devices was limited due to the non-ideal device designed. Later on, solution-processed organic hole transfer materials (HTMs) such as Spiro-OMeTAD

(2,2,7,7-tetrakis (*N*,*N*-di-4-methoxyphenylamino)-9,9-spirobitluorene) [24] and P3KT (poly-[3-(potassium-6-hexanoate) thiophene-2, 5-diy]) [25] were applied to CdTe NC solar cells, and such devices exhibited a PCE exceeding 6%. Researchers also found that a dipole layer was formed between the CdTe NC and the holes transfer layer (HTL), which reinforced the built-in electric field and improved carrier collection efficiency. Following this, cross-linkable conjugated polymer poly (diphenylsilane-co-4-vinyl-triphenylamine) (Si-TPA) and poly(phenylphosphine-co-4-vinyl-triphenylamine) (P-TPA) were recently presented as HTLs for CdTe NC solar cells [26,27]. Compared to other organic HTLs, cross-linkable conjugated polymer possesses many merits such as high stability, easily adjustable highest occupied molecular orbital (HOMO) and lowest unoccupied molecular orbital (LUMO) levels, and excellent adhesion to the NC thin film surface. By optimizing the thickness and annealing temperature of these HTLs, a notable PCE of ~9% was obtained, which was the highest PCE ever reported for CdTe NC solar cells with inverted structure. However, the NC solar cells listed above are fabricated by the spin-coating method, which is limited to lab-scale, and exhibit high material consumption as spin-coated thin film has poor uniformity over a large area and most of the NCs have been removed during the spin-casting process. Therefore, a new scaled-up fabricating method for large areas must be developed to meet the requirements of industrial mass production of NC solar cells.

Spray coating, slot-die coating, offset coating, gravure coating, and inkjet printing methods have been well developed in recent years for scaled-up manufacturing of thin film solar cells [28–31]. In 2011, Gianpaolo et al. [32] developed airbrush coating technicques for polymer solar cells. A PCE of 4.1% was obtained in poly (3-hexylthiophene): [6,6]-Phenyl C61 butyric acid methyl-ester (P3HT:PCBM) blend solar cells after optimizing the parameters of the spray system. Later on, Lim et al. [33] presented a new way for polymer solar cells with large areas (~10 cm^2) to be processed with non-halogenated solvents in air via a blade coating method. Modular devices with a high PCE of ~5% were obtained in this case via optimizing the fabricating parameters. In 2017, Hou and coworkers [34] reported a high PCE of 10.6% on fullerene-free polymer solar cells with a 1 cm^2 active area using the blade coating method. Poly(2,6′-4,8-di(5-ethylhexylthienyl)benzo1,2-b, 3,3-bdithiophene{3-fluoro-2(2-ethylhexyl)carbonylthieno3,4-bthiophenediyl})(PTB7-th), Poly[(5,6-difluoro-2,1,3-benzothiadiazole-4,7-diyl)[3,3‴-bis(2-octyldodecyl)[2,2′:5′,2″:5″,2‴-quaterthiophene]-5,5‴-diyl]] (PffBT4T-2OD), and other donor polymers with different properties were also developed successfully for large area organic solar cells [35,36]. Most of them delivered a higher PCE of up to 10% and exhibited low efficiency losses, which is important for their industrial applications. In addition to organic solar cells, perovskite solar cells (PSCs), a new type of thin film solar cell, have attracted much attention in recent year due to their cheap ingredients, simple fabricating process, and ultrahigh efficiency (>24%) [37]. In the case of large area devices, Microquanta Semiconductor (China) presented a large area PSC on a rigid substrate with a PCE of 17.4% [38]. Recently, WonderSolar launched a 110 m^2 photovoltaic (PV) system using large area screen-printed triple mesoscopic PSCs [39]. For PSC large area modules, uniformity of the thin film, pinholes in the active layer, toxicity of the solvents, and stability are all drawbacks that limit their further application [40,41]. In the case of NC solar cells, Matthew et al. [42] first proposed PbS quantum dot solar cells using the solar paint approach and obtained an efficiency of >1%. Although the value is obviously lower than that prepared by the successive ionic layer adsorption and reaction (SILAR)-based method, this solar paint method offers the opportunity to prepare the active layer in a short time, which is preferable for large area fabrication. Later, Kramer et al. [43] developed a room temperature spray coating technique for PbS colloidal quantum dot (CQD) solar cells. In this case, the PbS CQD solution was automatically sprayed on the substrate under computer control and a high PCE of 8.1% was obtained in the final device they made, which implies that there is no compromise between large area manufacturability and lab level. Up to now, printed PbS CQD solar cells with a PCE >10% have been fabricated by modifying PbS colloid ink composition and optimizing deposition methods, which paved the way to commercial application of CQD solar cells [44].

Although the scaled-up fabrication method has been successfully applied in PbS CQD solar cells, there are still few reports on these methods for the fabrication of CdTe NC solar cells. In this work, we develop a novel blade coating method for fabricating CdTe NC solar cells with a device structure of ITO/ZnO/CdSe/CdTe/Au, which allows depositing NC thin films with different areas (0.16, 0.3, and 0.5 cm^2) fabricated under ambient conditions with low material consumption. The effects of NC ink concentration, active layer thickness, annealing temperature, and blading parameter on the solar cell performance is investigated and discussed in detail. By optimizing the device fabrication parameters, PCEs of 3.58, 2.83, and 1.93% can be obtained for NC solar cells with areas of 0.16, 0.3, and 0.5 cm^2, respectively. Stability characterization shows that all the NC solar cells with different areas are extremely stable when stored under ambient conditions, with less than 5% degradation for 18 days storage. As all the active layers can be prepared by this simple blade coating method with low material consumption, this work demonstrates a potential way to realize industrial mass production of low-cost NC solar cells.

2. Device Fabrication

The Zn^{2+} precursor, CdSe and CdTe NCs were prepared based on the methods reported in [26]. The CdSe NCs were dissolved in pyridine with a concentration of 30 mg/mL, while CdTe NCs were dispersed into a pyridine/n-propanol (with volume ratio of 1:1) solvent with a concentration of 40 mg/mL. In the first step, the ZnO thin film with a thickness of ~40 nm was prepared by spin casting Zn^{2+} precursor on ITO substrate followed by an annealing process of 400 °C for 10 min, as described in [27]. After that, CdSe and CdTe NC films were deposited on the ITO/ZnO substrate in sequence by the blade coating method, as shown in Figure 1. During the blade coating process, several drops of CdSe NC solution were put onto the ITO/ZnO substrate and blade coated evenly using a 10-micron wire rod at a speed of 150 mm/min. Then the substrate was transferred to a hot plate and annealed at 150 °C for 10 min and 350 °C for 40 s to eliminate any organic solvent and impurities. Two or more layers of CdSe NC thin film were fabricated by repeating the above process. The thickness of each layer was about 45 nm. The ITO/ZnO/CdSe substrates were then soaked in methanol and washed ultrasonically. After drying in a N$_2$ flow, the CdTe NC thin film was then blade coated on the ITO/ZnO/CdSe substrate at a speed of 150 mm/min. The sample ITO/ZnO/CdSe/CdTe was then placed on a hotplate at 150 °C for 3 min, followed by immediate dipping into saturated CdCl$_2$/methanol solution for 10 s and rinsed in 1-propanol for 3 s to remove excess CdCl$_2$. The sample was then placed on a hot plate at 350 °C for 40 s and cooled down to room temperature. This process was repeated several times to obtain CdTe NC thin films with different thicknesses (Figure S1 shows the blade coating device). The thickness of each single layer of the CdTe NC was around 110 nm. Finally, the samples were dipped into a saturated CdCl$_2$/methanol solution for 3 s and placed on a hot plate at 230~380 °C for 25 min to increase the crystallinity of the CdTe NC thin film. After cleaning in methanol solution, 80 nm Au was then deposited on the ITO/ZnO/CdSe/CdTe through a shadow mask by vacuum thermal evaporation. The final solar cells with areas of 0.16, 0.3, and 0.5 cm^2 were achieved by using different shadow masks (Figure S2).

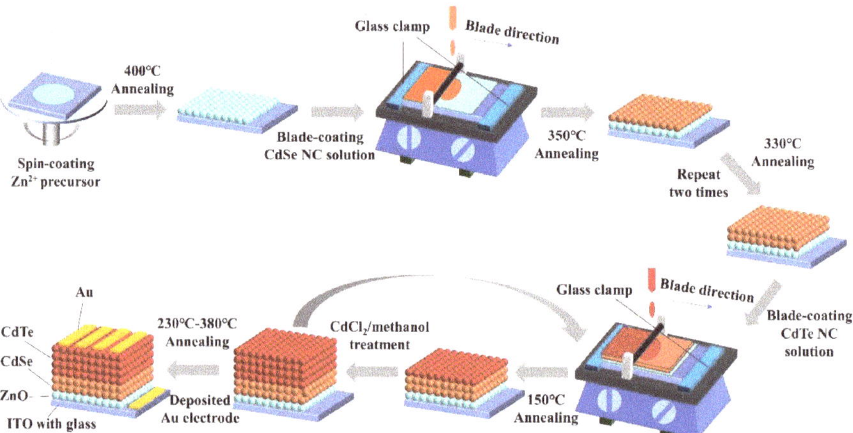

Figure 1. A schematic of the nanocrystal (NC) solar cells fabricated via blade coating.

3. Results and Discussion

To study the optical properties of the CdTe NC thin films, different thicknesses of CdTe NC thin films were deposited by blade coating on the glass substrate. The transmission spectra for glass/ZnO, glass/CdSe, and glass/CdTe with different thicknesses are showed in Figure 2a. It should be noted that ZnO thin films were transparent from 400 to 800 nm, while significant absorption from 400 to 800 nm was found for the CdTe film with the thickness increase from 110 to 440 nm. From the $(\alpha h v)^2$ (α is the absorption coefficient obtained in Figure 2a) versus photon energy (hv) plots (Figure 2b), it is obvious that all samples had a linear onset, indicating a direct bandgap for CdSe and CdTe NC thin films. The bandgaps of CdSe and CdTe NC thin films were 1.78 eV and 1.51 eV, respectively, which was calculated by extrapolating the linear region of this plot to the x-axis.

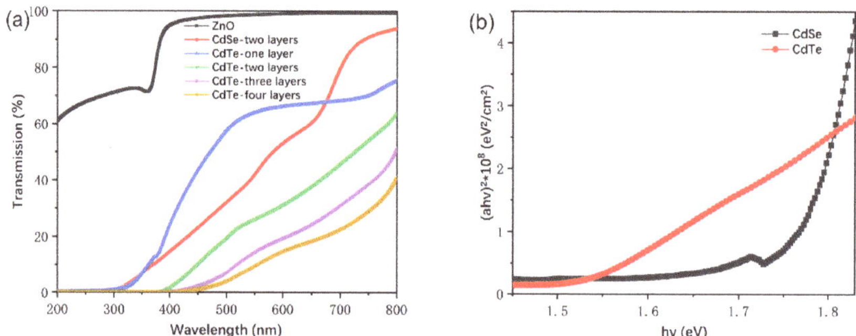

Figure 2. (**a**) Transmission spectra of different ZnO, CdSe NC thin films and CdTe NC thin films with different thicknesses; (**b**) the plots of $(\alpha h v)^2$ versus photon energy (hv) of the CdSe and CdTe NC thin films.

The morphologies of ITO/ZnO, ITO/ZnO/CdSe, and ITO/ZnO/CdSe/CdTe with different annealing temperatures are presented in Figure 3. It is obvious that the ZnO and CdSe NC thin films were very smooth and compact with root mean square (RMS) roughness values of 1.03 and 5.46 nm, respectively (Figure 3a,b). The morphology of the CdTe NC thin film was affected by the annealing temperature. The RMS roughness value of the CdTe NC films with annealing temperatures of 250, 290, 310, and 380 °C were 6.74, 9.39, 10.5, and 13.1 nm, respectively (Figure 3c–f). This suggests that the blade coated CdTe NC thin films are homogeneous and annealing temperature does not have a significant influence on the roughness of the thin film. A smooth CdTe NC thin film is essential to eliminate interface recombination and increase carrier collection efficiency. Besides, it should be noted that there are some pinholes for the NC samples treated under high annealing temperature (Figure 3f), which may result in large leakage current. From this point of view, a high annealing temperature is not recommended for the post-treatment of our NC solar cells.

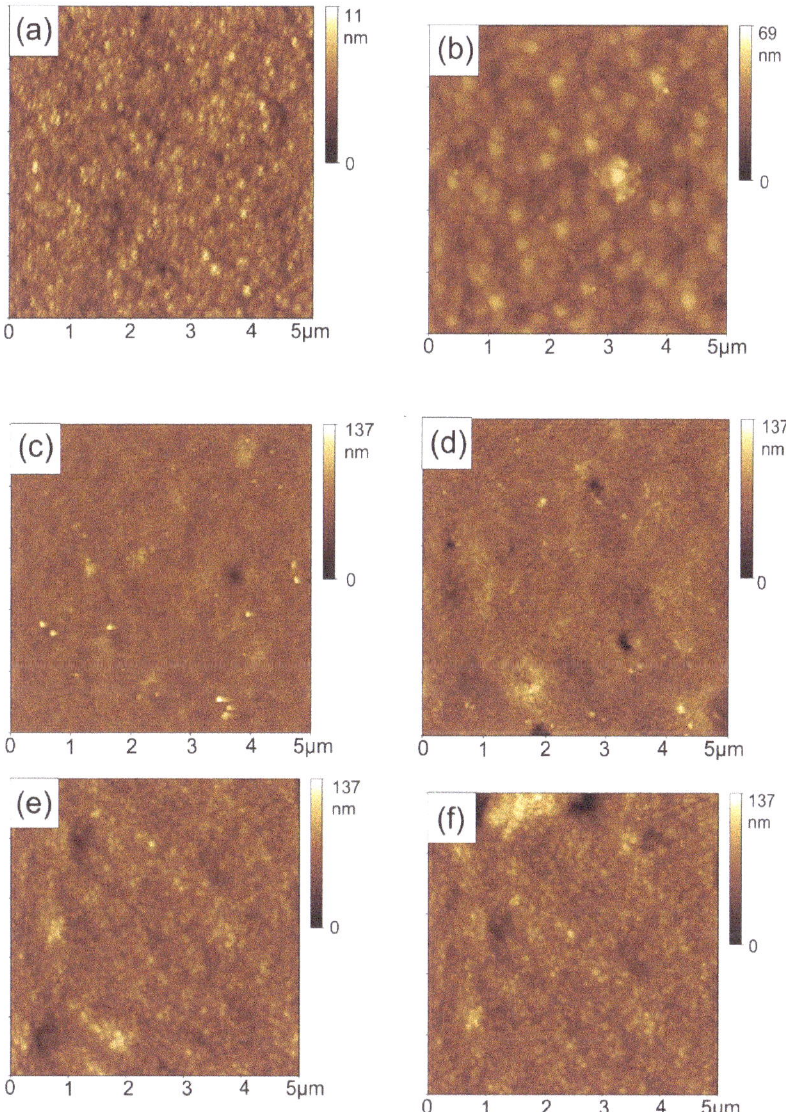

Figure 3. Atomic force microscopy (AFM) images of (**a**) ITO/ZnO thin film; (**b**) ITO/ZnO/CdSe thin film; and ITO/ZnO/CdSe/CdTe with annealing temperatures of (**c**) 250; (**d**) 290; (**e**) 310; and (**f**) 380 °C.

A sintering process could help promote the growth of CdTe crystal grains, eliminate defects inside the thin film or at the interface, and increase the carrier concentration effectively. Therefore, the annealing temperature is crucial to improve the device performance of solution-processed CdTe NC solar cells, which was confirmed in a previous work [8]. In those reports, a moderate temperature of 350 °C was chosen to be the best annealing temperature for CdTe NC solar cells. Devices fabricated via blade coating, however, are quite different from the spin-coated ones, and the best annealing temperature might be different in this case. To investigate the effects of annealing temperature on device performance, the thickness of the CdTe NC was fixed at ~400 nm, prepared by depositing four layers of CdTe NC on the ITO/ZnO/CdSe substrate. The ITO/ZnO/CdSe/CdTe was treated with $CdCl_2$ in methanol and then annealed at 230~380 °C under ambient conditions. Figure 4a shows the cross-section SEM images of an NC device with a configuration of ITO/ZnO/CdSe/CdTe/Au. The CdTe/CdSe NC layer was compact with thickness of ~600 nm, which consisted of two layers of CdSe NC and four layers of CdTe NC. Figure 4b presents the J–V curves of CdTe NC solar cells with different annealing temperatures under 1000 W m^{-2} (AM 1.5 G), while the device performance parameters are summarized in Table 1. It is noted that although the short-circuit current density (J_{sc} = 10.09 mA/cm^2) was high when annealed under a low temperature of 250 °C, the device had a low V_{oc} (0.39 V) and fill factor (FF = 27.49%), which resulted in a low PCE of 1.08%. When the annealing temperature increased to the range of 290~330 °C, a PCE up to 3% was obtained, which was mainly attributed to the increase of V_{oc}. The champion device was obtained at an annealing temperature of 310 °C, the J_{sc}, V_{oc}, and FF were 12.33 mA/cm^2, 0.60 V, and 48.38% respectively, delivering a relatively high PCE of 3.58%. On the other hand, when the annealing temperature was further increased from 330 to 380 °C, the PCE dropped down linearly and only a 0.54% PCE was obtained at 380 °C. All the device parameters (J_{sc}, V_{oc} and FF) deteriorated at a relatively high annealing temperature (>350 °C). The changes in devices performance with different annealing temperatures can be explained from multiple perspectives. First of all, under low annealing temperature, the diffusion of Cl$^-$ is inadequate, leading to low hole concentration in the CdTe NC thin film. Such low hole concentration might affect the Fermi level of the *p*-type region, resulting in a low built-in electric field in the space charge region. In this case, a low V_{oc} and low device performance are expected. On the other hand, if the annealing temperature is too high (>360 °C), the quantity of pinholes or metal oxide may increase inside the CdTe NC thin film or at the interface, which will dramatically deteriorate the device performance. From Table 1, one can see that the device shunt resistance (R_{sh}) is below 140 Ω, implying large leakage current and low *p–n* junction quality at high temperature. Similar results can also be found in our previous report on CdTe NC solar cells with inverted structure [8]. When the annealing temperature was in the range of 290 to 330 °C, all the devices showed high V_{oc} (>0.6 V) and FF (>40%), leading to a PCE up to 3%. In these cases, the series resistance (R_s) was in the range of 10~20 Ω·cm^2 and R_{sh} was higher than 200 Ω·cm^2, indicating high *p–n* junction quality. The typical dark J–V curves of devices annealing at 250, 330, and 380 °C are presented in Figure 4c. Obviously, for device treatment at 330 °C, the dark current at reverse bias voltage was significantly lower than the device at 250 or 380 °C. Therefore, the optimized annealing temperature for blade coated CdTe NC solar cells is ~300 °C. Figure 4d presents the typical external quantum efficiency (EQE) spectra of devices, the EQE value was around 40 between 450 to 900 nm. When EQEs are integrated, J_{sc} of 8.22, 10.83, and 5.10 mA/cm^2 are predicted for devices annealed at 250, 330, and 380 °C respectively, which are consistent with the J–V curves presented in Figure 4b. The J_{sc} (~10 mA/cm^2) value of NC devices prepared by blade coating was lower than that (~20 mA/cm^2) for devices prepared by the spin-coating method with similar device architecture [8], which may attributed to the difference in CdTe and CdSe NC film quality. The NC solar cells with different thicknesses of CdTe NC films treated under optimized sintering conditions were also investigated, and a low PCE was found for CdTe NCs with three or five layers (as shown in Figure S3).

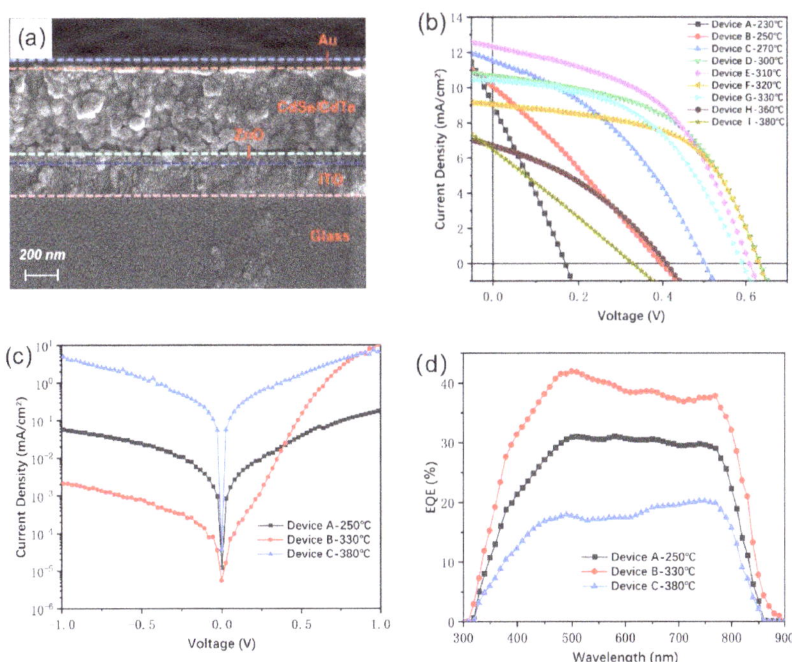

Figure 4. (**a**) Cross-section SEM images of ITO/ZnO/CdSe/CdTe/Au; J–V curves of ITO/ZnO/CdSe/CdTe/Au with different annealing temperatures (**b**) under light and (**c**) under dark; (**d**) the corresponding external quantum efficiency (EQE) spectra.

Table 1. Summarized performance of NC solar cells with different annealing temperatures (Figure 4b).

Device	Temperature (°C)	V_{oc} (V)	J_{sc} (mA/cm^2)	FF (%)	PCE (%)	R_s ($\Omega \cdot$cm^2)	R_{sh} ($\Omega \cdot$cm^2)
A	230	0.16	9.06	27.98	0.41	16.62	20.60
B	250	0.39	10.09	27.49	1.08	38.57	50.37
C	270	0.50	11.55	37.97	2.19	21.29	111.19
D	300	0.63	10.70	51.34	3.46	15.89	351.93
E	310	0.60	12.33	48.38	3.58	15.97	201.59
F	320	0.63	9.05	55.66	3.17	15.34	357.88
G	330	0.59	10.41	48.06	2.95	18.72	694.60
H	360	0.40	6.70	35.77	0.96	32.99	137.42
I	380	0.32	6.47	26.26	0.54	49.24	54.70

The CdTe NC thin film was proved to be extremely stable and could be fabricated under ambient conditions. To ascertain the performance of NC solar cells with different areas, devices with an inverted structure of ITO/ZnO (40 nm)/CdSe (90 nm)/CdTe (400 nm)/Au (80 nm) were fabricated by blade coating four CdTe NC layers on CdSe film. Devices with 0.3 and 0.5 cm^2 were fabricated by depositing a contact metal electrode through shadow masks with active areas of 0.3 and 0.5 cm^2 (as shown in Figure S2). For the device with an active area of 0.3 cm^2, a short-circuit current density (J_{sc}) of 11.86 mA/cm^2, an open-circuit voltage (V_{oc}) of 0.53 V, a fill factor (FF) of 44.77%, and a PCE of 2.82% were obtained, while these value for 0.5 cm^2 were 8.23 mA/cm^2, 0.51 V, 45.38%, and 1.93%, respectively (Figure 5a and Table 2). The device performance degraded as the active area increased, which is primarily ascribed to the reduced in V_{oc}. For the 0.16 cm^2 device the V_{oc} was 0.6 V, while this value was 0.53 and 0.51 V for 0.3 and 0.5 cm^2 devices, respectively. We speculate that the defects in the CdSe and CdTe NC thin films increased when the active area increased. Moreover, the junction quality may have dropped down due to the homogeneity decrease in a large area. Finally, the series resistance of ITO and interfacial contact resistance would increase with increasing active area. All these factors listed above resulted in low device performance for large area devices, and similar results were also reported in the case of PSC [39]. It is noted that the pattern ITO substrate with an active area of 1.5 cm × 1 cm (Length × width) was used here. However, the 0.5 cm^2 device showed length × width = 0.63 cm × 0.8 cm (Figure S2), and the width of the NC devices was close to that of ITO substrate. In this case, the boundary effects may increase defects and decrease device performance. To further investigate the drying process on the large area NC devices (0.5 cm^2) performance, devices with different annealing temperature are presented. The J–V curves are presented in Figure S4 and the solar cell parameters are summarized in Table S1. NC solar cells showed an improved PCE of 2.11% in the case of 320 °C annealing temperature. The annealing temperature affects the drying process of NC thin film, high quality NC thin film can be formed at optimized annealing temperature. The selected solvent may also affect the drying process and the device performance; more work and design experiment will be carried out in our subsequent research. We believe that the PCE of NC solar cells with large areas can certainly be improved via designing and using a suitable pattern ITO substrate. Figure 5b shows the EQE curves of devices with different areas. It should be noted that all the devices show similar shape at the whole short wavelength. The device with a large area prepared by blade coating was not further tuned in this paper, and we believe that there may be room for further improvement.

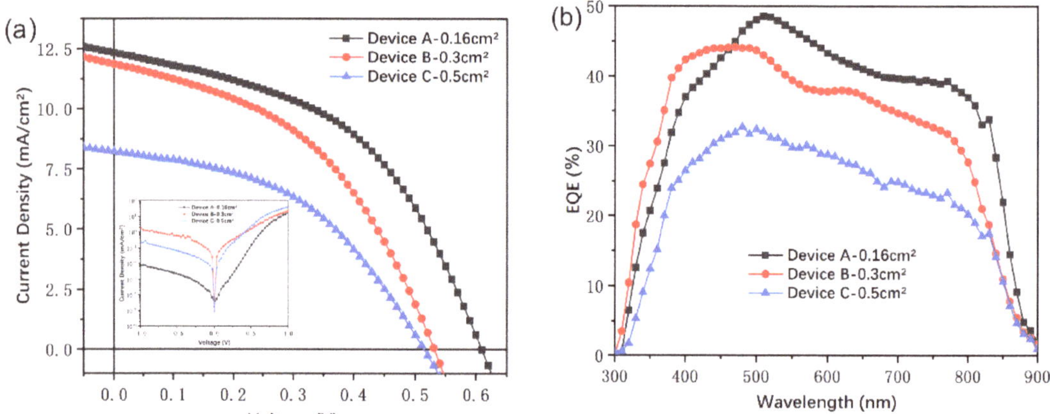

Figure 5. (**a**) J–V curves of ITO/ZnO/CdSe/CdTe/Au with different active areas (Device A—0.16 cm^2, Device B—0.3 cm^2, Device C—0.5 cm^2) under light (inset is J–V under dark); (**b**) the corresponding external quantum efficiency (EQE) spectrum.

Table 2. Summarized performance of NC solar cells with different active areas (Figure 5).

Device	Active Area (cm^2)	V_{oc} (V)	J_{sc} (mA/cm^2)	FF (%)	PCE (%)	R_s ($\Omega \cdot$cm^2)	R_{sh} ($\Omega \cdot$cm^2)
A	0.16	0.60	12.33	48.38	3.58	15.97	201.59
B	0.3	0.53	11.86	44.77	2.82	15.42	175.56
C	0.5	0.51	8.23	45.38	1.93	23.82	313.39

One challenge for commercial application is the stability of NC solar cells. For CdTe NC solar cells, the interfacial layer, the grain boundaries, the contact electrodes, and the diffusion of dopant elements are the main sources of instability. In the previous report, we found that CdTe NC solar cells with inverted structure (ITO/ZnO/CdSe/CdTe/Au) were very stable, with negligible change in PCE when stored under ambient conditions [45]. As the post-treatment process and the device structure in this research are similar to those devices previously reported, blade coated NC solar cells prepared the same way should have similar stability. Here, the long-term operation stability of CdTe NC solar cells with different areas is presented in Figure 6. The V_{oc}, J_{sc}, FF, and PCE of NC solar cells increased in the first several days. As shown in Figure 6a, the PCE for original devices with areas of 0.16 and 0.5 cm^2 were 2.34 and 1.48%, respectively, while these values were 2.95 and 1.93%, respectively on Day 6. From the J_{sc}, V_{oc}, and FF curves of NC devices (Figure 6b–d), one can see that the improvement in device performance was mainly attributed to the increase in V_{oc} and FF. The performance enhancement after storage was found in the case of PbS quantum dot solar cells and CdTe NC solar cells [46,47]. We speculate that the diffusion between CdTe and CdSe NC thin films and the Au on the CdTe NC interface with storage time will eliminate some interface defects, which will increase the carrier collecting efficiency and improve device performance. It should be noted that the NC devices prepared by blade coating showed outstanding stability and less than 5% degradation in PCE after storage at ambient conditions for 18 days.

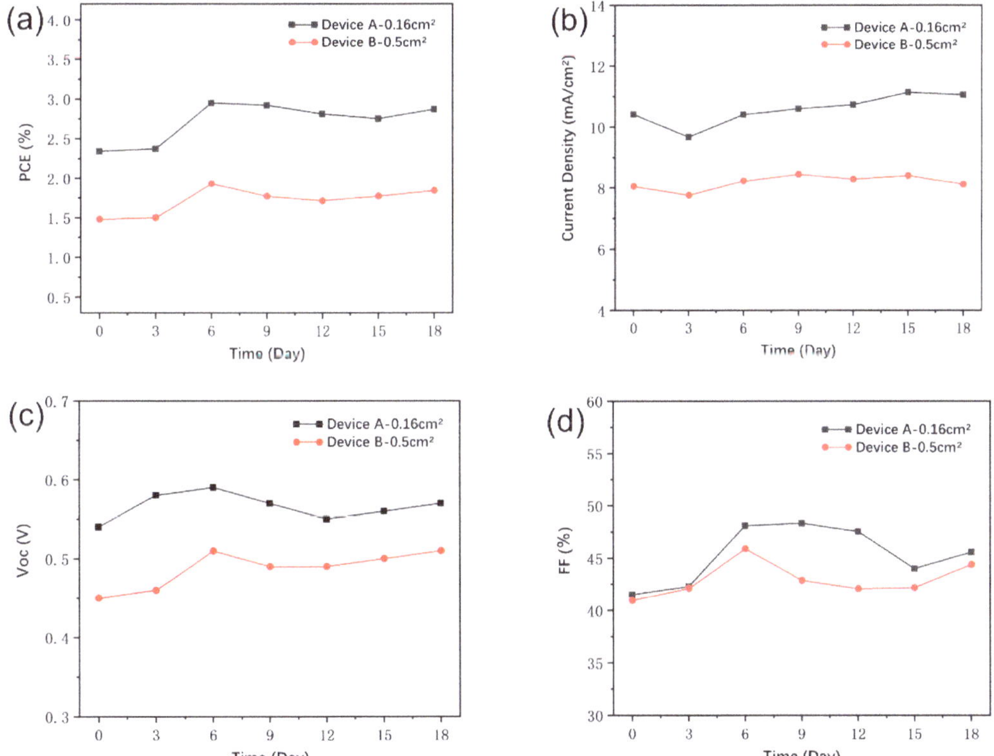

Figure 6. Evolution of parameters (**a**) PCE (power conversion efficiency), (**b**) J_{sc}, (**c**) V_{oc}, and (**d**) FF for devices with different active areas stored under ambient conditions (Device A—0.16 cm^2, Device B—0.5 cm^2).

In conclusion, we developed a blade coating method for fabricating CdTe NC solar cells with large areas. We found that the annealing temperature has significant influence on the NC solar cells. The optimized annealing temperature for CdTe NC solar cells is in the range of 300 to 330 °C, which is different from those NC solar cells prepared by the spin-coating method. At an annealing temperature of 310 °C, a champion device shows a J_{sc} of 12.33 mA/cm^2, a V_{oc} of 0.60 V, an FF of 48.38%, delivering a PCE of 3.58%. It should be noted that NC solar cells with large areas of 0.3 and 0.5 cm^2 can also be easily prepared through the blade coating process and only a small decrease in PCE is found, implying the scalability of this method. The NC solar cells show good stability and less than 5% degradation is found when the device is stored under ambient condition for 18 days. The exploration in this work provides valuable suggestions and solutions for low-cost and large-scale fabricating of NC solar cells.

Supplementary Materials: The following are available online at https://www.mdpi.com/article/10.3390/nano11061522/s1, Figure S1. Equipment for doctor-blading preparation devices, Figure S2. Devices with different active areas (**a**) 0.16 cm^2; (**b**) 0.3 cm^2; (**c**) 0.5 cm^2, Figure S3. J–V curves of ITO/ZnO/CdSe/CdTe/Au with different CdTe layers at 310 °C (**a**) three layers (**b**) five layers, Figure S4. J–V curves of ITO/ZnO/CdSe/CdTe/Au at different annealing temperatures (Device A: 320 °C, Device B: 340 °C, Device C: 360 °C) under light with 0.5 cm^2 active area.

Author Contributions: D.Q. and K.X. conceived and designed the experiments; K.X., Q.H., J.L., H.T., A.X., P.W. and H.R. conducted the experiments; K.X. and J.L. analyzed the data; D.Q., W.X. and D.W. contributed reagents/materials/analysis tools; and D.Q. and K.X. compiled the paper. All authors have read and agreed to the published version of the manuscript.

Funding: We thank the financial support of the Key-Area Research and Development Program of Guangdong Province (No. 2019B010924001), the National Natural Science Foundation of China (No. 21875075, 61774077, 61274062, 61775061), Guangdong Natural Science Fund (No. 2018A0303130041, 2018A0303130211), Guangzhou Science and Technology Plan Project (No. 201804010295, 2018A0303130211), National Undergraduate Innovative and Entrepreneurial Training Program (No. 202010561010), and the Fundamental Research Funds for the Central Universities for financial support.

Data Availability Statement: Data is contained within the article or Supplementary Material.

Acknowledgments: We would like to give our thanks to Songwei Liu, from the Department of Electronic Engineering of the Chinese University of Hong Kong, who helped to revise this manuscript.

Conflicts of Interest: The authors declare no conflict of interest.

References

1. Kirmani, A.R.; García de Arquer, F.P.; Fan, J.Z.; Khan, J.I.; Walters, G.; Hoogland, S.; Wehbe, N.; Said, M.M.; Barlow, S.; Laquai, F.; et al. Molecular Doping of the Hole-Transporting Layer for Efficient, Single-Step-Deposited Colloidal Quantum Dot Photovoltaics. *ACS Energy Lett.* **2017**, *2*, 1952–1959. [CrossRef]
2. Kim, T.; Firdaus, Y.; Kirmani, A.R.; Liang, R.-Z.; Hu, H.; Liu, M.; El Labban, A.; Hoogland, S.; Beaujuge, P.M.; Sargent, E.H.; et al. Hybrid Tandem Quantum Dot/Organic Solar Cells with Enhanced Photocurrent and Efficiency via Ink and Interlayer Engineering. *ACS Energy Lett.* **2018**, *3*, 1307–1314. [CrossRef]
3. Baek, S.-W.; Ouellette, O.; Jo, J.W.; Choi, J.; Seo, K.-W.; Kim, J.; Sun, B.; Lee, S.-H.; Choi, M.-J.; Nam, D.-H.; et al. Infrared Cavity-Enhanced Colloidal Quantum Dot Photovoltaics Employing Asymmetric Multilayer Electrodes. *ACS Energy Lett.* **2018**, *3*, 2908–2913. [CrossRef]
4. Liu, M.; Che, F.; Sun, B.; Voznyy, O.; Proppe, A.; Munir, R.; Wei, M.; Quintero-Bermudez, R.; Hu, L.; Hoogland, S.; et al. Controlled Steric Hindrance Enables Efficient Ligand Exchange for Stable, Infrared-Bandgap Quantum Dot Inks. *ACS Energy Lett.* **2019**, *4*, 1225–1230. [CrossRef]
5. Kim, J.; Ouellette, O.; Voznyy, O.; Wei, M.; Choi, J.; Choi, M.J.; Jo, J.W.; Baek, S.W.; Fan, J.; Saidaminov, M.I.; et al. Butylamine-Catalyzed Synthesis of Nanocrystal Inks Enables Efficient Infrared CQD Solar Cells. *Adv. Mater.* **2018**, *30*, e1803830. [CrossRef] [PubMed]
6. MacDonald, B.I.; Martucci, A.; Rubanov, S.; Watkins, S.E.; Mulvaney, P.; Jasieniak, J.J. Layer-by-Layer Assembly of Sintered CdSexTe1-x Nanocrystal Solar Cells. *ACS Nano* **2012**, *6*, 5995–6004. [CrossRef]
7. Xie, S.; Li, X.; Jiang, Y.; Yang, R.; Fu, M.; Li, W.; Pan, Y.; Qin, D.; Xu, W.; Hou, L. Recent Progress in Hybrid Solar Cells Based on Solution-Processed Organic and Semiconductor Nanocrystal: Perspectives on Device Design. *Appl. Sci.* **2020**, *10*, 17. [CrossRef]
8. Jiang, Y.; Pan, Y.; Wu, W.; Luo, K.; Rong, Z.; Xie, S.; Zuo, W.; Yu, J.; Zhang, R.; Qin, D.; et al. Hole Transfer Layer Engineering for CdTe Nanocrystal Photovoltaics with Improved Efficiency. *Nanomaterials* **2020**, *10*, 11. [CrossRef] [PubMed]
9. Guo, X.; Rong, Z.; Wang, L.; Liu, S.; Liu, Z.; Luo, K.; Chen, B.; Qin, D.; Ma, Y.; Wu, H.; et al. Surface passivation via acid vapor etching enables efficient and stable solution-processed CdTe nanocrystal solar cells. *Sustain. Energy Fuels* **2020**, *4*, 399–406. [CrossRef]

10. Pu, C.; Ma, J.; Qin, H.; Yan, M.; Fu, T.; Niu, Y.; Yang, X.; Huang, Y.; Zhao, F.; Peng, X. Doped Semiconductor-Nanocrystal Emitters with Optimal Photoluminescence Decay Dynamics in Microsecond to Millisecond Range: Synthesis and Applications. *ACS Cent. Sci.* **2016**, *2*, 32–39. [CrossRef] [PubMed]

11. Dai, X.; Zhang, Z.; Jin, Y.; Niu, Y.; Cao, H.; Liang, X.; Chen, L.; Wang, J.; Peng, X. Solution-processed, high-performance light-emitting diodes based on quantum dots. *Nature* **2014**, *515*, 96–99. [CrossRef]

12. Konstantatos, G.; Sargent, E.H. Colloidal quantum dot photodetectors. *Infrared Phys. Technol.* **2011**, *54*, 278–282. [CrossRef]

13. Konstantatos, G.; Sargent, E.H. Nanostructured materials for photon detection. *Nat. Nanotechnol.* **2010**, *5*, 391–400. [CrossRef] [PubMed]

14. Liu, H.; Li, M.; Voznyy, O.; Hu, L.; Fu, Q.; Zhou, D.; Xia, Z.; Sargent, E.H.; Tang, J. Physically flexible, rapid-response gas sensor based on colloidal quantum dot solids. *Adv. Mater.* **2014**, *26*, 2718–2724. [CrossRef]

15. Wan, Y.; Zhou, Y.G.; Poudineh, M.; Safaei, T.S.; Mohamadi, R.M.; Sargent, E.H.; Kelley, S.O. Highly specific electrochemical analysis of cancer cells using multi-nanoparticle labeling. *Angew. Chem. Int. Ed. Engl.* **2014**, *53*, 13145–13149. [CrossRef] [PubMed]

16. Chen, H.C.; Lai, C.W.; Wu, I.C.; Pan, H.R.; Chen, I.W.; Peng, Y.K.; Liu, C.L.; Chen, C.H.; Chou, P.T. Enhanced performance and air stability of 3.2% hybrid solar cells: How the functional polymer and CdTe nanostructure boost the solar cell efficiency. *Adv. Mater.* **2011**, *23*, 5451–5455. [CrossRef] [PubMed]

17. MacDonald, B.I.; Della Gaspera, E.; Watkins, S.E.; Mulvaney, P.; Jasieniak, J.J. Enhanced photovoltaic performance of nanocrystalline CdTe/ZnO solar cells using sol-gel ZnO and positive bias treatment. *J. Appl. Phys.* **2014**, *115*, 6. [CrossRef]

18. Panthani, M.G.; Kurley, J.M.; Crisp, R.W.; Dietz, T.C.; Ezzyat, T.; Luther, J.M.; Talapin, D.V. High efficiency solution processed sintered CdTe nanocrystal solar cells: The role of interfaces. *Nano Lett.* **2014**, *14*, 670–675. [CrossRef]

19. Crisp, R.W.; Panthani, M.G.; Rance, W.L.; Duenow, J.N.; Parilla, P.A.; Callahan, R.; Dabney, M.S.; Berry, J.J.; Talapin, D.V.; Luther, J.M. Nanocrystal Grain Growth and Device Architectures for High-Efficiency CdTe Ink-Based Photovoltaics. *ACS Nano* **2014**, *8*, 9063–9072. [CrossRef]

20. Han, J.-f.; Krishnakumar, V.; Schimper, H.J.; Cha, L.-m.; Liao, C. Investigation of Structural, Chemical, and Electrical Properties of CdTe/Back Contact Interface by TEM and XPS. *J. Electron. Mater.* **2015**, *44*, 3327–3333. [CrossRef]

21. Yoon, W.; Townsend, T.K.; Lumb, M.P.; Tischler, J.G.; Foos, E.E. Sintered CdTe Nanocrystal Thin Films: Determination of Optical Constants and Application in Novel Inverted Heterojunction Solar Cells. *IEEE Trans. Nanotechnol.* **2014**, *13*, 551–556. [CrossRef]

22. Zeng, Q.; Chen, Z.; Zhao, Y.; Du, X.; Liu, F.; Jin, G.; Dong, F.; Zhang, H.; Yang, B. Aqueous-Processed Inorganic Thin-Film Solar Cells Based on CdSe(x)Te(1-x) Nanocrystals: The Impact of Composition on Photovoltaic Performance. *ACS Appl. Mater. Interfaces* **2015**, *7*, 23223–23230. [CrossRef]

23. Liu, F.; Chen, Z.; Du, X.; Zeng, Q.; Ji, T.; Cheng, Z.; Jin, G.; Yang, B. High efficiency aqueous-processed MEH-PPV/CdTe hybrid solar cells with a PCE of 4.20%. *J. Mater. Chem. A* **2016**, *4*, 1105–1111. [CrossRef]

24. Du, X.; Chen, Z.; Liu, F.; Zeng, Q.; Jin, G.; Li, F.; Yao, D.; Yang, B. Improvement in Open-Circuit Voltage of Thin Film Solar Cells from Aqueous Nanocrystals by Interface Engineering. *ACS Appl. Mater. Interfaces* **2016**, *8*, 900–907. [CrossRef]

25. Zeng, Q.; Hu, L.; Cui, J.; Feng, T.; Du, X.; Jin, G.; Liu, F.; Ji, T.; Li, F.; Zhang, H.; et al. High-Efficiency Aqueous-Processed Polymer/CdTe Nanocrystals Planar Heterojunction Solar Cells with Optimized Band Alignment and Reduced Interfacial Charge Recombination. *ACS Appl. Mater. Interfaces* **2017**, *9*, 31345–31351. [CrossRef]

26. Rong, Z.; Guo, X.; Lian, S.; Liu, S.; Qin, D.; Mo, Y.; Xu, W.; Wu, H.; Zhao, H.; Hou, L. Interface Engineering for Both Cathode and Anode Enables Low-Cost Highly Efficient Solution-Processed CdTe Nanocrystal Solar Cells. *Adv. Funct. Mater.* **2019**, *29*, 9. [CrossRef]

27. Guo, X.; Tan, Q.; Liu, S.; Qin, D.; Mo, Y.; Hou, L.; Liu, A.; Wu, H.; Ma, Y. High-efficiency solution-processed CdTe nanocrystal solar cells incorporating a novel crosslinkable conjugated polymer as the hole transport layer. *Nano Energy* **2018**, *46*, 150–157. [CrossRef]

28. Krebs, F.C. Fabrication and processing of polymer solar cells: A review of printing and coating techniques. *Sol. Energy Mater. Sol. Cells* **2009**, *93*, 394–412. [CrossRef]

29. Schilinsky, P.; Waldauf, C.; Brabec, C.J. Performance analysis of printed bulk heterojunction solar cells. *Adv. Funct. Mater.* **2006**, *16*, 1669–1672. [CrossRef]

30. Krebs, F.C.; Jorgensen, M.; Norrman, K.; Hagemann, O.; Alstrup, J.; Nielsen, T.D.; Fyenbo, J.; Larsen, K.; Kristensen, J. A complete process for production of flexible large area polymer solar cells entirely using screen printing-First public demonstration. *Sol. Energy Mater. Sol. Cells* **2009**, *93*, 422–441. [CrossRef]

31. Hoth, C.N.; Choulis, S.A.; Schilinsky, P.; Brabec, C.J. High photovoltaic performance of inkjet printed polymer: Fullerene blends. *Adv. Mater.* **2007**, *19*, 3973. [CrossRef]

32. Susanna, G.; Salamandra, L.; Brown, T.M.; Di Carlo, A.; Brunetti, F.; Reale, A. Airbrush spray-coating of polymer bulk-heterojunction solar cells. *Sol. Energy Mater. Sol. Cells* **2011**, *95*, 1775–1778. [CrossRef]

33. Lim, S.L.; Ong, K.H.; Li, J.; Yang, L.; Chang, Y.F.; Meng, H.F.; Wang, X.Z.; Chen, Z.K. Efficient, large area organic photovoltaic modules with active layers processed with non-halogenated solvents in air. *Org. Electron.* **2017**, *43*, 55–63. [CrossRef]

34. Zhao, W.; Zhang, S.; Zhang, Y.; Li, S.; Liu, X.; He, C.; Zheng, Z.; Hou, J. Environmentally Friendly Solvent-Processed Organic Solar Cells that are Highly Efficient and Adaptable for the Blade-Coating Method. *Adv. Mater.* **2018**, *30*, 7. [CrossRef]

35. Wang, G.; Adil, M.A.; Zhang, J.; Wei, Z. Large-Area Organic Solar Cells: Material Requirements, Modular Designs, and Printing Methods. *Adv. Mater.* **2019**, *31*, e1805089. [CrossRef]

36. Shin, I.; Ahn, H.; Yun, J.H.; Jo, J.W.; Park, S.; Joe, S.-y.; Bang, J.; Son, H.J. High-Performance and Uniform 1 cm^2 Polymer Solar Cells with D-1-A-D-2-A-Type Random Terpolymers. *Adv. Energy Mater.* **2018**, *8*, 13. [CrossRef]
37. Jiang, Q.; Zhao, Y.; Zhang, X.W.; Yang, X.L.; Chen, Y.; Chu, Z.M.; Ye, Q.F.; Li, X.X.; Yin, Z.G.; You, J.B. Surface passivation of perovskite film for efficient solar cells. *Nat. Photonics* **2019**, *13*, 460–466. [CrossRef]
38. Di Giacomo, F.; Shanmugam, S.; Fledderus, H.; Bruijnaers, B.J.; Verhees, W.J.H.; Dorenkamper, M.S.; Veenstra, S.C.; Qiu, W.; Gehlhaar, R.; Merckx, T.; et al. Up-scalable sheet-to-sheet production of high efficiency perovskite module and solar cells on 6-in. substrate using slot die coating. *Sol. Energy Mater. Sol. Cells* **2018**, *181*, 53–59. [CrossRef]
39. Rong, Y.; Hu, Y.; Mei, A.; Tan, H.; Saidaminov, M.I.; Seok, S.I.; McGehee, M.D.; Sargent, E.H.; Han, H. Challenges for commercializing perovskite solar cells. *Science* **2018**, *361*, 7. [CrossRef] [PubMed]
40. Lunardi, M.M.; Ho-Baillie, A.W.Y.; Alvarez-Gaitan, J.P.; Moore, S.; Corkish, R. A life cycle assessment of perovskite/silicon tandem solar cells. *Prog. Photovolt.* **2017**, *25*, 679–695. [CrossRef]
41. Binek, A.; Petrus, M.L.; Huber, N.; Bristow, H.; Hu, Y.; Bein, T.; Docampo, P. Recycling Perovskite Solar Cells to Avoid Lead Waste. *ACS Appl. Mater. Interfaces* **2016**, *8*, 12881–12886. [CrossRef]
42. Genovese, M.P.; Lightcap, I.V.; Kamat, P.V. Sun-Believable Solar Paint. A Transformative One-Step Approach for Designing Nanocrystalline Solar Cells. *ACS Nano* **2012**, *6*, 865–872. [CrossRef]
43. Kramer, I.J.; Minor, J.C.; Moreno-Bautista, G.; Rollny, L.; Kanjanaboos, P.; Kopilovic, D.; Thon, S.M.; Carey, G.H.; Chou, K.W.; Zhitomirsky, D.; et al. Efficient spray-coated colloidal quantum dot solar cells. *Adv. Mater.* **2015**, *27*, 116–121. [CrossRef]
44. Kirmani, A.R.; Sheikh, A.D.; Niazi, M.R.; Haque, M.A.; Liu, M.; de Arquer, F.P.G.; Xu, J.; Sun, B.; Voznyy, O.; Gasparini, N.; et al. Overcoming the Ambient Manufacturability-Scalability-Performance Bottleneck in Colloidal Quantum Dot Photovoltaics. *Adv. Mater.* **2018**, *30*, e1801661. [CrossRef] [PubMed]
45. Liu, H.; Tian, Y.; Zhang, Y.; Gao, K.; Lu, K.; Wu, R.; Qin, D.; Wu, H.; Peng, Z.; Hou, L.; et al. Solution processed CdTe/CdSe nanocrystal solar cells with more than 5.5% efficiency by using an inverted device structure. *J. Mater. Chem. C* **2015**, *3*, 4227–4234. [CrossRef]
46. Chuang, C.H.; Brown, P.R.; Bulovic, V.; Bawendi, M.G. Improved performance and stability in quantum dot solar cells through band alignment engineering. *Nat. Mater.* **2014**, *13*, 796–801. [CrossRef] [PubMed]
47. Xie, Y.; Tan, Q.; Zhang, Z.; Lu, K.; Li, M.; Xu, W.; Qin, D.; Zhang, Y.; Hou, L.; Wu, H. Improving performance in CdTe/CdSe nanocrystals solar cells by using bulk nano-heterojunctions. *J. Mater. Chem. C* **2016**, *4*, 6483–6491. [CrossRef]

nanomaterials

MDPI

Article

One-Step Solution Deposition of Antimony Selenoiodide Films via Precursor Engineering for Lead-Free Solar Cell Applications

Yong Chan Choi * and Kang-Won Jung

Division of Energy Technology, Daegu Gyeongbuk Institute of Science & Technology (DGIST), Daegu 42988, Korea; kw.jung@dgist.ac.kr
* Correspondence: ycchoi@dgist.ac.kr

Abstract: Ternary chalcohalides are promising lead-free photovoltaic materials with excellent optoelectronic properties. We propose a simple one-step solution-phase precursor-engineering method for antimony selenoiodide (SbSeI) film fabrication. SbSeI films were fabricated by spin-coating the precursor solution, and heating. Various precursor solutions were synthesized by adjusting the molar ratio of two solutions based on $SbCl_3$-selenourea and SbI_3. The results suggest that both the molar ratio and the heating temperature play key roles in film phase and morphology. Nanostructured SbSeI films with a high crystallinity were obtained at a molar ratio of 1:1.5 and a temperature of 150 °C. The proposed method could be also used to fabricate (Bi,Sb)SeI.

Keywords: antimony selenoiodide; SbSeI; solution process; solar cells; one-step method

Citation: Choi, Y.C.; Jung, K.-W. One-Step Solution Deposition of Antimony Selenoiodide Films via Precursor Engineering for Lead-Free Solar Cell Applications. *Nanomaterials* **2021**, *11*, 3206. https://doi.org/10.3390/nano11123206

Academic Editor: Vlad Andrei Antohe

Received: 29 October 2021
Accepted: 24 November 2021
Published: 26 November 2021

Publisher's Note: MDPI stays neutral with regard to jurisdictional claims in published maps and institutional affiliations.

1. Introduction

Ternary chalcohalides of antimony and bismuth (MChX, where M = Sb, Bi; Ch = S, Se; X = I, Br, Cl) have recently emerged as potential candidates for lead-free solar cell applications because of their promising optoelectronic properties, high stability, low toxicity, and earth-abundant constituents [1–4]. Solar cells based on these materials are expected to exhibit high device performance because of the ns^2 electronic configuration of Sb^{3+}/Bi^{3+}, such as Pb^{2+} in Pb perovskites, which enables defect-tolerant features. These features make them attractive alternatives to Pb perovskites, which are being widely studied in terms of their use in next-generation solar cells. However, the highest device efficiency of solar cells using ternary chalcohalides that has been reported so far is less than 5% [5], which is far below that of Pb perovskites solar cells (>25%) [6,7], and little work has been conducted to date on improving their efficiency. Further work is therefore required to improve device performance. However, fabrication methods for both obtaining and controlling the properties of high-purity chalcohalides suitable for solar cell applications are lacking.

Sb/Bi chalcohalides have been fabricated for solar cell applications using various methods [2]. Among these, the two-step solution-phase method has been demonstrated to be effective in the fabrication of various materials [5,8–12]. In this approach, chalcohalides are obtained via the conversion of chalcogenides formed in the first step of the two-step method, and so the compositions of the final products can be controlled depending on the chalcogenide and halide species used in the first and second steps, respectively. To date, various materials, such as SbSI [8,10], (Sb,Bi)SI [9], BiSI [11], SbSeI [5], and Sb(S,Se)I [12], have been fabricated using this method. In addition, an efficiency of ~4.1% was obtained using this method from solar cells based on antimony selenoiodide SbSeI [5]. Despite remarkable progress, however, this method has limitations when it comes to obtaining a pure-phase film. To form a pure phase, all chalcogenides formed in the first step must react with halides during the second step. However, chalcogenides often cannot react with halides because of their undesirable morphology, leaving a portion of the chalcogenides

---

unconverted in the final product. For example, we have previously found that the intertwined Bi_2S_3 morphology formed in the first step prevents the BiI_3 solution from reaching a deeper region near the bottom, leaving unreacted nanostructures [11]. This problem may be addressed by applying a one-step method based on the precursor solution. To this end, the precursor solution must be designed to form the desired single phase. In order the achieve this, first we selected the chalcohalide SbSeI, and then we began to prepare a precursor solution for it.

Due to the fact that SbSeI belongs to the SbSI family [13], SbSeI formation may be expressed by the following chemical reaction, similar to the case of SbSI [8,10,12,13]: Sb_2Se_3 + $SbI_3 \rightarrow 3SbSeI$. In addition, according to the Sb-Se-I phase diagram [14,15], the SbSeI phase is formed through a competing process of two phases, Sb_2Se_3 and SbI_3, under controlled molar ratios and temperature conditions. These results suggest that the control of Sb_2Se_3 and SbI_3 is a key factor in the development of a method for SbSeI fabrication. As a first step, we explored different solutions that could produce Sb_2Se_3 and SbI_3. As a result, we found that two solutions based on $SbCl_3$-SeU (Sol A, where SeU is selenourea) and SbI_3 (Sol B) may be used to form Sb_2Se_3 and SbI_3 phases, respectively, at a low temperature of 150 °C (as shown in Figure S1 of the Supplementary Material). Based on these findings, we developed a solution-processing method for the fabrication of SbSeI thin films. Specifically, we designed a precursor solution that may be used to produce a pure-phase film in a single step by mixing the two solutions.

In this work, we report a facile one-step solution-processing method based on precursor engineering using Sol A and Sol B solutions. The precursor solutions were synthesized by mixing Sol A and Sol B at different molar ratios. This controlled molar ratio allowed for the manipulation of Sb_2Se_3 and SbSeI phases, leading to the formation of a pure SbSeI film under specific conditions. Moreover, the pure phase was obtained at a low temperature of 150 °C. We also applied this approach to the fabrication of other selenoiodides, namely (Bi,Sb)SeI, to prove the versatility of the proposed method in terms of the preparation of various chalcoiodides for solar cell applications.

2. Materials and Methods

2.1. Chemical and Materials

Antimony (III) chloride ($SbCl_3$, 99+ %), antimony (III) iodide (SbI_3, 99.999%), SeU (NH_2CSeNH_2, 99.97%), *N*-methyl-2-pyrrolidinone (NMP, C_5H_9NO, anhydrous, 99.5%), and *N,N*-dimethylformamide (DMF, $HCON(CH_3)_2$, anhydrous, 99.8%) were purchased from Alfa Aesar (Seoul, Korea). Cadmium sulfate hydrate ($CdSO_4 \cdot 8/3H_2O$, $\geq$99.0%), thiourea (TU, NH_2CSNH_2, $\geq$99.0%), and bismuth (III) iodide (BiI_3, 99%) were purchased from Sigma-Aldrich (Seoul, Korea). Ammonium hydroxide solution (NH_4OH, 28% NH_3 in H_2O) was purchased from Junsei (Tokyo, Japan). All chemicals were used as received without further purification. FTO glass with a sheet resistance of 15 Ω sq^{-1} was purchased from Pilkington (AMG, Yongin-si, Korea).

2.2. Preparation of the CdS/FTO Substrate

A 50 nm-thick CdS layer was deposited on the FTO glass using a chemical bath deposition method. CdS deposition was performed according to a previously reported procedure [16], in which FTO glass was dipped in an aqueous solution containing $CdSO_4 \cdot 8/3H_2O$, NH_4OH, and TU. During immersion, the temperature and pH of the solution were maintained at 65 °C and 11–11.5, respectively. After being dipped for 12 min 30 s, the glass was removed from the solution and washed with deionized water several times, before being dried. The sample was then immediately transferred into an N_2-filled glove box with a moisture control system, in which the H_2O level was maintained below 1 ppm, to anneal it in an inert gas. Finally, the CdS/FTO substrate was obtained after heating at 400 °C for 1 h in the glove box.

2.3. Synthesis of Precursor Solution and Deposition of SbSeI Thin Films

The precursor solution was synthesized by mixing two stock solutions, Sol A and Sol B, as shown in Figure 1a. In order to synthesize Sol A, 0.5 mmol $SbCl_3$ and 1.25 mmol SeU were dissolved in 1 mL of DMF. Sol B was prepared by dissolving 0.5 mmol SbI_3 in 1 mL of NMP. After stirring the two solutions for 1 h, they were mixed at different molar ratios and stirred for a further 1 h to prepare the precursor solution. After this step, 180 µL of the final solution was spin-coated at 5000 rpm on the pre-cleaned CdS/FTO substrate, followed by heating at 150 °C for 5 min (Figure 1b). This process was repeated five times, and all procedures were performed in a glove box. During solution synthesis, the CdS/FTO substrate was cleaned by UV/O_3 treatment for 20 min outside the glove box, before being immediately returned to the glove box prior to spin-coating.

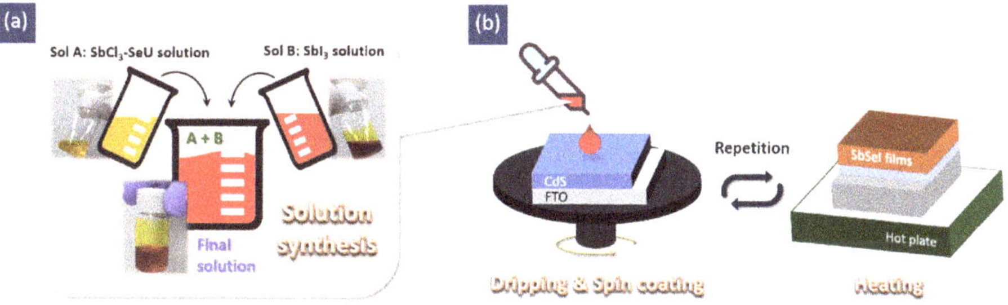

Figure 1. Schematic illustrations of (**a**) the synthesis of the one-step solution, and (**b**) the deposition process for SbSeI thin films.

2.4. Characterization

Optical absorption was measured using a UV-VIS absorption spectrophotometer (Shimadzu UV-2600) in the wavelength range of 400–1200 nm. A sample crystal structure was measured using an X-ray diffractometer (Malvern Panalytical Empyrean, Malvern, UK) in the θ/2θ scan mode. The phase quantification was performed by the Rietveld method using X'Pert HighScore Plus (version 3.0.0) software. A field emission scanning electron microscope (Hitachi S-4800, Tokyo, Japan) was used to investigate sample morphology. Electronic structure was investigated by ultraviolet photoelectron spectroscopy (UPS) using an X-ray photoelectron spectrometer (Thermo Scientific ESCALAB 250Xi, Lexington, MA, USA).

3. Results and Discussion

The precursor solution was synthesized by mixing Sol A and Sol B, and so it was expected that the molar ratio of Sol A and Sol B would affect film formation. To verify this hypothesis, we investigated the absorption properties, crystalline structures, and morphologies of films fabricated using precursor solutions with different Sol A:Sol B molar ratios. At a ratio of 1:0.75 (denoted as a black line in Figure 2a), an absorption edge of ~1050 nm, which is consistent with the value of Sb_2Se_3 [17,18], was observed. The sample also exhibited a dominant Sb_2Se_3 phase (ICDD # 98-065-1518), as shown in the X-ray diffraction (XRD) pattern (Figure 2b, Tables S1 and S2). Concerning the morphology shown in the field emission scanning electron microscopy (FESEM) image of Figure 2c, nanorods with a diameter of ~50 nm were found to grow randomly on the substrate. These results indicate that Sb_2Se_3 nanorods were mainly formed under these conditions. As the SbI_3 content increased to 1:1.5, the absorption edge shifted toward a wavelength of 740 nm (bandgap E_G of 1.68 eV), corresponding to the value for SbSeI [5], as indicated by the yellow arrow in Figure 2a. The absorption intensity in the short-wavelength region below 740 nm also gradually increased (denoted by a red arrow). In addition, at a molar ratio of 1:1.5, the SbSeI phase (ICDD # 98-003-1292) became dominant, whereas the Sb_2Se_3 phase decreased and then disappeared (Figure 2b and Table S1). As the SbI$_3$ content increased, the nanorods aggregated to form nanostructures (Figure 2c). Furthermore, the increase in SbI$_3$ induced a decrease in absorption intensity, although this did not affect the XRD patterns and morphology. These results imply that a nanostructured SbSeI film with high crystallinity may be fabricated at a specific molar ratio of 1:1.5.

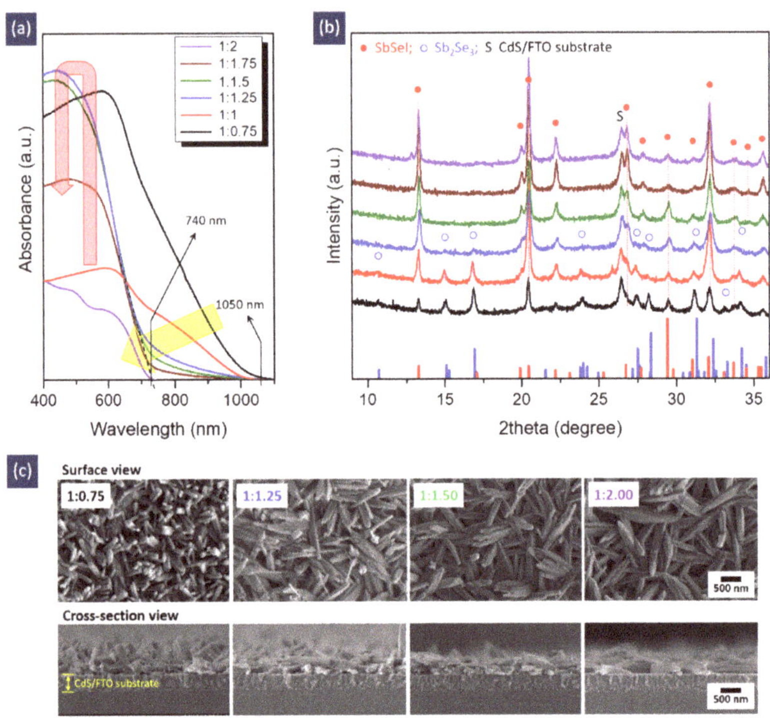

Figure 2. (**a**) Absorption spectra, (**b**) XRD patterns, and (**c**) FESEM images of samples prepared using precursor solutions with different Sol A:Sol B molar ratios. The peak positions of Sb$_2$Se$_3$ and SbSeI are shown in (**b**), based on the reference data of Sb$_2$Se$_3$ (ICDD # 98-065-1518, blue column) and SbSeI (ICDD # 98-003-1292, red column).

In addition to the molar ratio in the precursor solution, we found that annealing temperature played a key role in the formation of pure-phased SbSeI films, as shown in Figure 3. At a temperature of 200 °C, the absorption spectrum was almost equal to that at 150 °C (Figure 3a). However, an unknown peak (the green arrow pointing downwards in Figure 3b) appeared with a decreased SbSeI phase (Figure 3b and Table S3). Further increasing the temperature to 250 °C caused a shift in the absorption edge from 740 nm to 1050 nm (denoted as a blue arrow), revealing a phase change from SbSeI to Sb$_2$Se$_3$. This change was confirmed by the XRD results (Figure 3b and Table S3), in which the Sb$_2$Se$_3$ phase predominantly appeared when the temperature was increased to 250 °C. Note that the unknown phase may be considered to be an intermediate Sb-Se-I phase because it is formed in a temperature region where Sb$_2$Se$_3$ and SbSeI phases can coexist. When the temperature reached 300 °C, only XRD peaks corresponding to the Sb$_2$Se$_3$ phase were observed. The morphology was very similar to that at 150 °C, as shown in Figure 3c, although several voids were observed in the nanostructures, as indicated by the green arrows in the magnified image (Figure 3d). This similarity in morphology suggests that SbSeI was formed at an early stage during formation at 300 °C. However, because the SbSeI phase is unstable at a higher temperature and is prone to decomposition [14,15], SbI$_3$ may evaporate from the initially formed SbSeI as the reaction proceeds, creating voids in the nanostructures. As a result, Sb_2Se_3, which is similar in morphology to SbSeI despite containing many pores, was formed. A low temperature of 150 °C was therefore required to obtain pure SbSeI films. This temperature was also confirmed by an investigation of the XRD pattern in the low-temperature region of 120–180 °C (Figure S2 of the Supplementary Material).

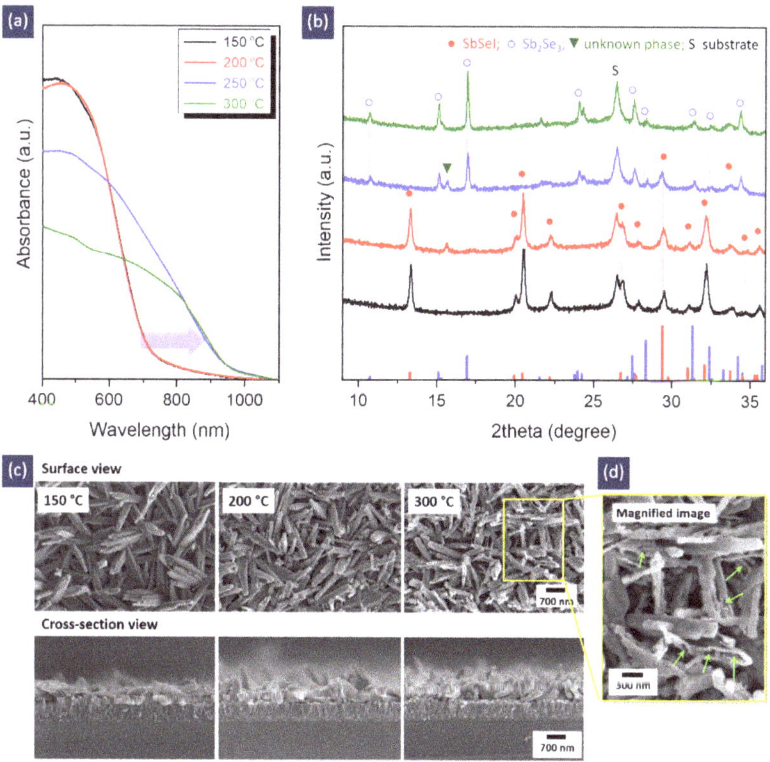

Figure 3. (a) Absorption spectra, (b) XRD patterns, and (c) FESEM images of samples prepared at different annealing temperatures. (d) Magnified image, marked by a yellow box in (c). All samples were fabricated using the precursor solution with a Sol A:Sol B molar ratio of 1:1.5. In (b), two reference patterns of Sb_2Se_3 ((ICDD # 98-065-1518) and SbSeI (ICDD # 98-003-1292) are shown as blue columns and red columns, respectively.

Given that the phase of the fabricated films was determined by the type of solution used in the precursor solution, it was possible to fabricate various chalcohalides by changing the starting solutions. To verify this possibility, we modified the precursor solution by introducing BiI_3 instead of SbI_3 and deposited it on a CdS/FTO substrate following the optimized procedures. For convenience, the sample fabricated using the optimized solution shown in Figure 2 is denoted as 'control' in Figure 4. The sample fabricated using the BiI_3-modified solution is also denoted as 'Bi-SbSeI'. As shown in Figure 4a, the Bi-SbSeI sample exhibits an absorption edge of ~880 nm, which corresponds to an E_G of ~1.41 eV. This E_G value is lower than that of SbSeI (~1.68 eV) but higher than that of BiSeI (~1.32 eV) [19,20]. Figure 4b and Table S4 show that the XRD peaks of Bi-SbSeI are located between the two references for BiSeI (ID: mp-23020, The Materials Project) [21] and SbSeI (ICDD # 98-003-1292). The detected peaks were symmetrical, indicating that a single phase was formed. Based on these results, it can be concluded that a single-phase material composed of (Bi,Sb)SeI was successfully formed by modifying the precursor solution. This suggests that the proposed method can be used to fabricate various chalcohalides, such as SbSI, BiSI, and related alloys.

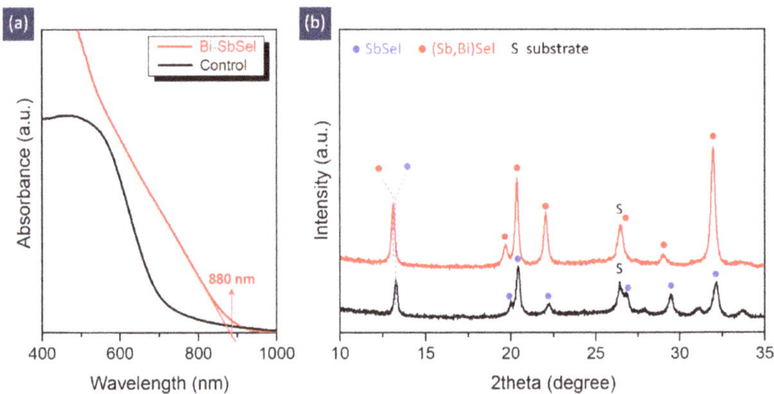

Figure 4. (**a**) Absorption spectra and (**b**) XRD patterns of control and Bi-SbSeI samples.

We investigated the electronic structures of the samples (shown in Figure 4) further by analyzing their UPS spectra (Figure 5a). We obtained two values from the spectra, one for the cut-off energy, E_{cutoff}, and another for the valence band edge energy, E_{VE}. From the two equations $E_V = E_F + E_{VE}$ and $E_V = h\upsilon - (E_{cutoff} - E_{VE})$, we also calculated the following three values: conduction band minimum (E_C), valence band maximum (E_V), and Fermi level energy (E_F) [22]. These values are listed in Table S5 of the Supplementary Material and are shown in Figure 5b. The (Bi,Sb)SeI sample presented a similar E_V to SbSeI, but a lower E_C value. This result indicates that the incorporation of Bi into SbSeI induces a downshift in E_C. This result also suggests that electronic structure may be controlled via compositional engineering. The proposed method could therefore be applied to optimizing electronic properties in order to render materials suitable for solar cell applications via compositional engineering. However, the current method is limited when it comes to forming a dense film, as shown in Figure 2c. It is generally accepted that morphology plays a key role in obtaining high efficiency [7]. Typically, dense films with a large grain size contribute significantly to achieving high performance. Thus, it can be deduced that it is very difficult to obtain a high level of efficiency from our films. A preliminary result confirmed that the device exhibited a very poor efficiency of 0.23% (Table S6 of the Supplementary Materials). Therefore, we are currently adapting the proposed method in order to both improve and optimize the morphology of SbSeI films. The morphology may be improved by various approaches, such as annealing optimization, post-treatment, the use of additive effects, and solvent annealing [7].

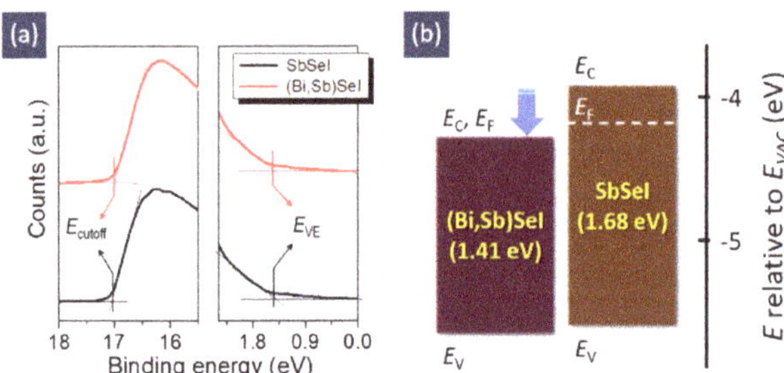

Figure 5. (**a**) UPS spectra of SbSI and (Bi,Sb)SeI, and (**b**) the derived energy level diagram.

Nanomaterials **2021**, *11*, 3206

4. Conclusions

A simple one-step solution-processing method for the fabrication of SbSeI films is presented. The pure-phase SbSeI film (E_G = 1.68 eV) was obtained at a low temperature of 150 °C using a precursor solution with a molar ratio of Sol A:Sol B = 1:1.5. Modifying the precursor solution resulted in the formation of (Bi,Sb)SeI. These results suggest that the proposed method can be readily applied to the fabrication of various chalcohalides, as well as to optimize their electronic structures in order to render them suitable for solar cell applications. However, further improvements are necessary to optimize key factors, such as morphology, crystalline orientation, and defects, so as to make them suitable for high performance Sb/Bi chalcohalide solar cells.

Supplementary Materials: The following are available online at https://www.mdpi.com/article/10.3390/nano11123206/s1: Figure S1: absorption spectra and XRD patterns of samples fabricated with Sol A and sol B; Table S1: Quantitative ratio of Sb_2Se_3 and SbSeI phases; Table S2: Lattice spacings for each phase; Table S3: Quantitative ratio of Sb_2Se_3, SbSeI, and unknown phases; Figure S2: XRD patterns of samples fabricated at 120–180 °C; Table S4: Comparison of XRD peaks of (Bi,Sb)SeI with references; Table S5: E_{cutoff}, E_{VE}, E_C, E_V, and E_F values of SbSI and (Bi,Sb)SeI samples; Table S6: Device parameters of the SbSeI solar cells.

Author Contributions: Conceptualization, Y.C.C.; methodology, Y.C.C.; formal analysis, Y.C.C. and K.-W.J.; investigation, Y.C.C.; writing—original draft preparation, Y.C.C.; writing—review and editing, Y.C.C. and K.-W.J.; supervision, Y.C.C.; project administration, Y.C.C. All authors have read and agreed to the published version of the manuscript.

Funding: This work was supported by the National Research Foundation of Korea (NRF) grant funded by the Korea government (MSIT) (No. 2019R1F1A1049014). This work was also supported by the DGIST R&D program of the Ministry of Science and ICT, Korea (No. 21-ET-08).

Data Availability Statement: All relevant data supporting the findings of this study are available within the article and its Supplementary Material.

Conflicts of Interest: The authors declare no conflict of interest.

References

1. Nie, R.; Sumukam, R.R.; Reddy, S.H.; Banavoth, M.; Seok, S.I. Lead-free perovskite solar cells enabled by hetero-valent substitutes. *Energy Environ. Sci.* **2020**, *13*, 2363–2385. [CrossRef]
2. Choi, Y.C.; Jung, K.-W. Recent Progress in Fabrication of Antimony/Bismuth Chalcohalides for Lead-Free Solar Cell Applications. *Nanomaterials* **2020**, *10*, 2284. [CrossRef]
3. Huang, Y.T.; Kavanagh, S.R.; Scanlon, D.O.; Walsh, A.; Hoye, R.L.Z. Perovskite-inspired materials for photovoltaics and beyond-from design to devices. *Nanotechnology* **2021**, *32*, 132004. [CrossRef] [PubMed]
4. Li, T.; Luo, S.; Wang, X.; Zhang, L. Alternative Lone-Pair ns^2-Cation-Based Semiconductors beyond Lead Halide Perovskites for Optoelectronic Applications. *Adv. Mater.* **2021**, *33*, 2008574. [CrossRef]
5. Nie, R.; Hu, M.; Risqi, A.M.; Li, Z.; Seok, S.I. Efficient and Stable Antimony Selenoiodide Solar Cells. *Adv. Sci.* **2021**, *8*, 2003172. [CrossRef]
6. Li, D.; Zhang, D.; Lim, K.S.; Hu, Y.; Rong, Y.; Mei, A.; Park, N.G.; Han, H. A Review on Scaling Up Perovskite Solar Cells. *Adv. Funct. Mater.* **2021**, *31*, 2008621. [CrossRef]
7. Tailor, N.K.; Abdi-Jalebi, M.; Gupta, V.; Hu, H.; Dar, M.I.; Li, G.; Satapathi, S. Recent progress in morphology optimization in perovskite solar cell. *J. Mater. Chem. A* **2020**, *8*, 21356–21386. [CrossRef]
8. Nie, R.; Yun, H.-s.; Paik, M.-J.; Mehta, A.; Park, B.-w.; Choi, Y.C.; Seok, S.I. Efficient Solar Cells Based on Light-Harvesting Antimony Sulfoiodide. *Adv. Energy Mater.* **2018**, *8*, 1701901. [CrossRef]
9. Nie, R.; Im, J.; Seok, S.I. Efficient Solar Cells Employing Light-Harvesting $Sb_{0.67}Bi_{0.33}SI$. *Adv. Mater.* **2019**, *31*, 1808344. [CrossRef]
10. Choi, Y.C.; Hwang, E.; Kim, D.-H. Controlled growth of SbSI thin films from amorphous Sb_2S_3 for low-temperature solution processed chalcohalide solar cells. *APL Mater.* **2018**, *6*, 121108. [CrossRef]
11. Choi, Y.C.; Hwang, E. Controlled Growth of BiSI Nanorod-Based Films Through a Two-Step Solution Process for Solar Cell Applications. *Nanomaterials* **2019**, *9*, 1650. [CrossRef] [PubMed]
12. Jung, K.-W.; Choi, Y.C. Compositional Engineering of Antimony Chalcoiodides via a Two-Step Solution Process for Solar Cell Applications. *ACS Appl. Energy Mater.* **2021**. [CrossRef]
13. Shiozaki, Y.; Nakamura, E.; Mitsui, T. *Ferroelectrics and Related Substances: Inorganic Substances other than Oxides. Part 1: SbSI Family ... TAAP*, 1st ed.; Springer: Berlin/Heidelberg, Germany, 2004.

14. Aliev, Z.; Musaeva, S.; Babanly, D.; Shevelkov, A.; Babanly, M. Phase diagram of the Sb–Se–I system and thermodynamic properties of SbSeI. *J. Alloys Compd.* **2010**, *505*, 450–455. [CrossRef]
15. Aliev, Z.S. The A^V–B^{VI}–I Ternary Systems: A Brief Review on the Phase Equilibria Review. *Condens. Matter Interphases* **2019**, *21*, 338–349. [CrossRef]
16. Yang, K.J.; Kim, S.; Kim, S.Y.; Ahn, K.; Son, D.H.; Kim, S.H.; Lee, S.J.; Kim, Y.I.; Park, S.N.; Sung, S.J.; et al. Flexible $Cu_2ZnSn(S,Se)_4$ solar cells with over 10% efficiency and methods of enlarging the cell area. *Nat. Commun.* **2019**, *10*, 2959. [CrossRef]
17. Choi, Y.C.; Mandal, T.N.; Yang, W.S.; Lee, Y.H.; Im, S.H.; Noh, J.H.; Seok, S.I. Sb_2Se_3-Sensitized Inorganic-Organic Heterojunction Solar Cells Fabricated Using a Single-Source Precursor. *Angew. Chem. Int. Ed.* **2014**, *53*, 1329–1333. [CrossRef]
18. Choi, Y.C.; Lee, Y.H.; Im, S.H.; Noh, J.H.; Mandal, T.N.; Yang, W.S.; Seok, S.I. Efficient Inorganic-Organic Heterojunction Solar Cells Employing $Sb_2(S_x/Se_{1-x})_3$ Graded-Composition Sensitizers. *Adv. Energy Mater.* **2014**, *4*, 1301680. [CrossRef]
19. Chepur, D.V.; Bercha, D.M.; Tubyanitsa, I.D.; Slivka, V.Y. Peculiarities of the Energy Spectrum and Edge Absorption in the Chain Commpounds $A^V B^{VI} C^{VII}$. *Phys. Stat. Sol.* **1968**, *30*, 461–468. [CrossRef]
20. Kunioku, H.; Higashi, M.; Abe, R. Low-Temperature Synthesis of Bismuth Chalcohalides: Candidate Photovoltaic Materials with Easily, Continuously Controllable Band gap. *Sci. Rep.* **2016**, *6*, 32664. [CrossRef]
21. Jain, A.; Ong, S.P.; Hautier, G.; Chen, W.; Richards, W.D.; Dacek, S.; Cholia, S.; Gunter, D.; Skinner, D.; Ceder, G.; et al. Commentary: The Materials Project: A materials genome approach to accelerating materials innovation. *APL Mater.* **2013**, *1*, 011002. [CrossRef]
22. Shao, G. Work Function and Electron Affinity of Semiconductors: Doping Effect and Complication due to Fermi Level Pinning. *Energy Environ. Mater.* **2021**, *4*, 273–276. [CrossRef]

nanomaterials

Article

Suppressing the Effect of the Wetting Layer through AlAs Capping in InAs/GaAs QD Structures for Solar Cells Applications

Nazaret Ruiz [1,*], Daniel Fernández [1,*], Lazar Stanojević [2], Teresa Ben [1], Sara Flores [1], Verónica Braza [1], Alejandro Gallego Carro [2], Esperanza Luna [3], José María Ulloa [2] and David González [1]

[1]   University Research Institute on Electron Microscopy & Materials, (IMEYMAT), Universidad de Cádiz, 11510 Puerto Real, Spain; teresa.ben@uca.es (T.B.); sara.flores@uca.es (S.F.); veronica.braza@uca.es (V.B.); david.gonzalez@uca.es (D.G.)

[2]   Institute for Systems Based on Optoelectronics and Microtechnology (ISOM), Universidad Politécnica de Madrid, Avda. Complutense 30, 28040 Madrid, Spain; lazar@isom.upm.es (L.S.); alejandro.gallego@alumnos.upm.es (A.G.C.); jmulloa@isom.upm.es (J.M.U.)

[3]   Paul-Drude-Institut für Festkörperelektronik, Leibniz-Institut im Forschungsverbund Berlin e.V., Hausvogteiplatz 5-7, D-10117 Berlin, Germany; luna@pdi-berlin.de

*   Correspondence: nazaret.ruiz@uca.es (N.R.); daniel.fernandez@uca.es (D.F.)

Abstract: Recently, thin AlAs capping layers (CLs) on InAs quantum dot solar cells (QDSCs) have been shown to yield better photovoltaic efficiency compared to traditional QDSCs. Although it has been proposed that this improvement is due to the suppression of the capture of photogenerated carriers through the wetting layer (WL) states by a de-wetting process, the mechanisms that operate during this process are not clear. In this work, a structural analysis of the WL characteristics in the AlAs/InAs QD system with different CL-thickness has been made by scanning transmission electron microscopy techniques. First, an exponential decline of the amount of InAs in the WL with the CL thickness increase has been found, far from a complete elimination of the WL. Instead, this reduction is linked to a higher shield effect against QD decomposition. Second, there is no compositional separation between the WL and CL, but rather single layer with a variable content of InAlGaAs. Both effects, the high intermixing and WL reduction cause a drastic change in electronic levels, with the CL making up of 1–2 monolayers being the most effective configuration to reduce the radiative-recombination and minimize the potential barriers for carrier transport.

Keywords: InAs quantum dots solar cells; AlAs capping; (S)TEM

Citation: Ruiz, N.; Fernández, D.; Stanojević, L.; Ben, T.; Flores, S.; Braza, V.; Carro, A.G.; Luna, E.; Ulloa, J.M.; González, D. Suppressing the Effect of the Wetting Layer through AlAs Capping in InAs/GaAs QD Structures for Solar Cells Applications. *Nanomaterials* **2022**, *12*, 1368. https://doi.org/10.3390/nano12081368

Academic Editor: Efrat Lifshitz

Received: 10 March 2022
Accepted: 11 April 2022
Published: 15 April 2022

Publisher's Note: MDPI stays neutral with regard to jurisdictional claims in published maps and institutional affiliations.

1. Introduction

The cost of electricity produced from a photovoltaic system depends directly on the efficiency of the solar cells (SC), so the main driver of innovation in this field is focused on increasing it. Although theory predicts for a single-junction solar cell a maximum efficiency of up to 85%, the Shockley–Queisser limit lowers this energy conversion efficiency to below 31% [1]. The two main physical processes involved are thermal dissipation losses and non-absorption of photons below the bandgap energy of the material of the cell. Therefore, the next generation of SC with high conversion efficiency needs to develop methods to reduce these losses. Among them, intermediate bandgap solar cells (IBSCs) have attracted a great deal of attention since the concept was first put forward in 1997 [2]. This concept is based on the absorption of energy photons below the bandgap by an "intermediate" electronic band (IB) suitably situated between the conduction and valence bands (CB and VB) that preserves the cell output voltage. Theoretical calculation of the conversion efficiency based on the ideal detailed balance model predicts 47% under one sun and 63% under the maximum concentration [3].

Among the different strategies to introduce intermediate levels, a promising one is to take advantage of nanostructures such as quantum dots (QDs) or quantum wells (QWs) [3–5]. In this framework, self-assembled (Al)GaAs/InAs/GaAs QDs arrays grown via epitaxial Stranski–Krastanov (SK) mode within the intrinsic region of a GaAs pin diode have been widely used as a means of implementing the concept of IBSCs due to their discrete density of states and well-established technology, where the IB arises from the energy states associated to the electrons confined in the CB of the QDs [6,7]. Although several of the principles of IBSC operation have been demonstrated [8,9], the reported efficiencies are nowadays well below expectations [10,11]. Several issues need to be overcome in QDSCs, such as to avoid an important reduction of the open-circuit voltage (V_{OC}) that leads to a drop in the efficiency in comparison to their single junction GaAs SC counterparts [3,12]. It is said that the key prerequisite for improvement of the output voltage is the suppression of photoelectron capture from the conduction band into QD states which mainly occurs through the extended wetting layer (WL) state [13,14]. Certainly, it is demonstrated that the large phase volume of WL plays a major role in this V_{OC} reduction [15], so QD properties can be radically altered if WL states are non-existent. It has been proposed that conventional SK QDs should be advantageously replaced with their no-WL counterparts [16,17].

Recently, the use of thin AlAs capping layers (CLs) on InAs QDs has attracted special attention since important improvements [18,19] or even a complete recovery [14] of the V_{OC} in these QDSCs have been reported. It has been proposed that the cause of this improvement is the elimination of the In(Ga)As WL due to phase separation [20,21]. This hypothesis comes from earlier studies by Tsatsul'nikov et al. [22] using cross-sectional transmission electron microscopy (TEM) imaging. The dark contrast under g002 dark field (DF) conditions of the WL in InAs/GaAs QDs layers is replaced by a white contrast in case of AlAs/InAs/GaAs QDs [14,18]. They suggested that the interpretation of these images is that the In in the WL in the regions between the QDs is strongly depleted or almost completely gone. For them, the only possible explanation for this behavior is that there is a replacement of the In atoms in the WL by Al atoms of the CL, followed by a surface migration of In into the uncapped QDs like what occurs during the formation of QDs. This reasoning has been used by several authors to explain the performance improvements observed in QDSC with this system [14,23].

However, the mechanism proposed in this model is controversial. First, it is true that phenomena of atomic surface exchange could occur during epitaxial growth, where certain atoms in the subsurface are promoted to the surface being substituted by other newcomers [24–27]. Although the strong surface segregation of In that happens during the GaAs capping of InAs QDs is well-known [28,29], there is no consensus on the role that AlAs capping could play in this system. Indeed, some authors proposed that AlAs CL strongly suppresses both In segregation and In–Ga intermixing [30,31], while others pointed to a similar behavior compared to GaAs capping [32,33]. Second, the proposed surface migration of In toward the QDs that removes the WL, like what occurs during the QD formation, is opposite to the process observed during QD capping. In fact, it is broadly admitted that a massive transport of In atoms occur from the dots to the WL during capping due to a decomposition of QD [34,35].

The aim of this work is to shed light on the structural and compositional changes at the nanometer level that occur in the WL during AlAs capping of InAs QD for SC applications. For this purpose, analyses have been performed using different CL thicknesses of AlAs combining different state-of-the-art scanning (S)TEM techniques. As will be shown in the manuscript, capping with AlAs does not lead to a complete removal of the WL, but to a large intermixing between the CL and the WL that substantially alters the energetic states associated with it. These results further clarify the controversy over the interpretation of the DCTEM g002 DF images that existed.

2. Materials and Methods

Materials growth: One sample was grown by solid source molecular beam epitaxy on Si-doped (100) n+ GaAs substrate under As4-stabilized conditions. In the bottom block, 4 thin layers of AlAs were grown at 580 °C for calibration with 1, 2, 3, and 5 monolayers (MLs), respectively and separated by 30 ML of GaAs. In the upper block, over a 100-nm-thick GaAs buffer deposited at 580 °C, 5 QD layers with 2.8 MLs of InAs were deposited at 460 °C and 0.045 ML/s. Each QD layer was covered with 0, 1, 2, 3, or 5 ML of AlAs, respectively, plus 30 ML of GaAs, all grown at 480 °C. Each one is named in the following as CL0, CL1, CL2, CL3, and CL5. These capped QD layers are separated by 50 nm GaAs grown at 580 °C to avoid strain coupling. The growth rate of the AlAs and GaAs in the CLs was 0.5 and 1 ML/s, respectively. The active region was finally capped with a 200 nm GaAs layer. A scheme of the sample is showed in Figure 1.

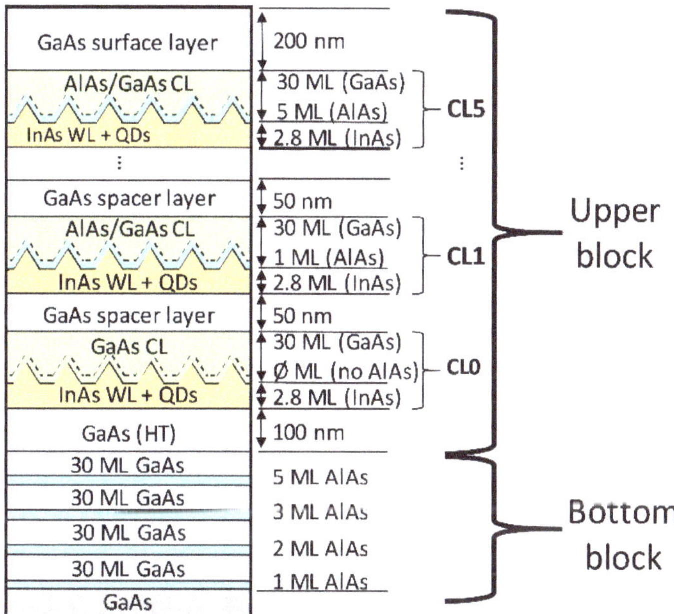

Figure 1. Scheme of the sample. In the first block, AlAs thin layers were grown for calibration. In the second, five layers of InAs QDs were covered by AlAs CLs with different thicknesses.

Characterization methods: The structural and compositional analysis was performed by transmission electron microscopy techniques on conventionally prepared cross-sectional samples. Diffraction contrast (DC) TEM imaging was carried out in a JEM-2100 LaB6 (JEOL, Tokyo, Japan) operating at 200 kV. Energy dispersive X-ray spectroscopy (EDX, (FEI Europe B. V., Eindhoven, The Netherlands)) was simultaneously performed in scanning (S)TEM mode with a double aberration corrected FEI Titan Cubed Themis operated at 200 kV. EDX mapping was carried out with four embedded Bruker bd-4 sx detectors (Bruker, Billerica, Massachusetts) using ChemiSTEM technology. EDX maps were processed using Velox® software (Version: 2.12.1.37).

3. Results

3.1. DCTEM Analyses Using g002 DF Conditions of the CL/WL Regions

As was mentioned in the introduction, it has been suggested that the InAs WL can be reduced or even removed if a thin AlAs layer is deposited on the InAs QD layer. The experimental evidence for this claim is based on the contrast change in DCTEM imaging under g002 DF conditions. For a long time, g002 DF imaging has been widely used to visualize composition changes in III-V semiconductors [36,37]. Roughly, the diffracted intensity leaving the specimen depends on the atomic scattering factors, f, as $A\left(\overline{f_{III}} - \overline{f_V}\right)^2$, where A is essentially constant for similar alloys and TEM conditions (volume of the cell, Bragg angle, electron wavelength, sample thickness, etc.) [38–40]. In the case of ternary alloys, normalizing the intensities with respect to a material with a fixed composition (i.e., GaAs of the substrate) could eliminate the term A, so that parabolic-like curves of normalized intensities could be obtained as a function of the alloy composition. Thus, pure $In_xGa_{1-x}As$ regions appear darker if x < 0.4 as compared to the GaAs reference while the AlGaAs ones appear brighter [39].

Figure 2(left) shows g002 DF images of WL regions with CLs with 0, 1, 2, 3, and 5 ML of AlAs taken close to the edge of the TEM sample to be near the kinematical conditions (sample thickness below 50 nm). The dark streak of layer CL0 that corresponds to the In(Ga)As WL seems to disappear progressively being substituted by the white streak that increases as the CL thickness rises. The initial dark streak is barely observable for the layer CL5. In addition, QDs become larger in size as the CL thickness increases (see ref [41]) which is in accordance with the appearance of the shield effect that blocks the QD decomposition [30,42]. The intensity profiles normalized to GaAs for the same layers are shown in Figure 2(right). Since the profiles were normalized with respect to the GaAs substrate, Al presence gives rise to intensity ratios greater than 1 (peaks of brighter contrast) while In yields to intensity ratios were less than 1 (valleys of darker contrast) for In contents below 40% [36,39]. Valleys are only observed for the first two QD layers, but with a huge shortening for CL1. From CL2 onwards, the peak enlarges with the AlAs CL thickness. Certainly, having only these results and thinking on the nominal design of two consecutive thin layers, it seems that In atoms of the WL are being progressively substituted by Al atoms from the CL. However, as we will see below, care must be taken with this hypothesis based on the nominal design of sequential ternary layers without intermixing, since in the case of a quaternary InAlGaAs alloy, the normalized intensity regarding the GaAs depends on the relative percentage of In and Al contents and can vary from bright to dark.

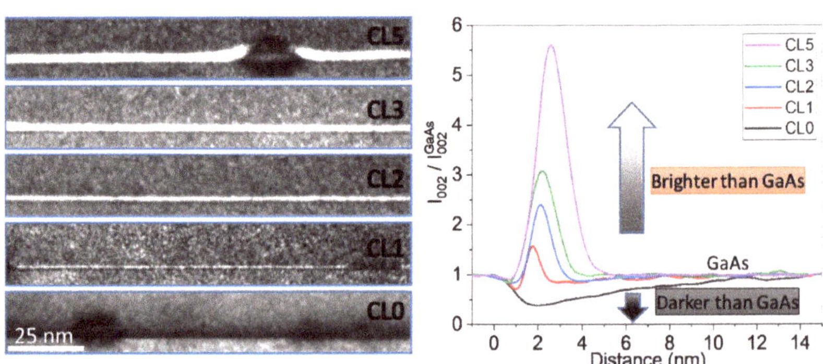

Figure 2. (Left) DCTEM g002 DF images of the QD layers with 0, 1, 2, 3, and 5 ML of AlAs CL taken close to the edge of the TEM sample to be near the kinematical conditions (sample thickness below 50 nm). **(Right)** Average intensity profiles along the growth direction of the WL regions normalized to the GaAs intensity.

Small lateral differences are observed for the same WL depending on which part is examined, and this is especially true for the more buried layers. During the thinning of the sample for electron transparency, a small thickness gradient can occur, so that the thickness of the sample in the more buried layers is thicker than that of the upper layers. Thus, the upper layers (TEM sample thickness < 80 nm) appear more homogeneous than the lower layers (TEM sample thickness greater than 120 nm) due to these thickness differences. As the distribution of QDs is random, the probability of finding a WL region between QDs where we are sure that there is no interference from neighboring QDs increases as the TEM sample becomes thinner. We had chosen this region of the sample with the intention of showing an image with all the layers. As a result, the TEM sample thickness varies along the growth direction which has the above-mentioned noticeable effect on the image. Taking all into account, DCTEM images for each layer were taken individually in regions with thicknesses below 50 nm that are near the kinematical conditions of electron diffraction.

3.2. EDX Analyses of the CL/WL Regions

EDX analyses could supply directly interpretable elemental mappings for Al and In distribution in the structure. Figure 3(left) shows a low magnification elemental EDX map of the five QD layers where green and red hue intensities correspond to Al and In contents, respectively. As shown earlier in the g002 DF images, the nominal InAs WL regions between QD seem to fade as the thickness of the AlAs CL increases, going from a deep red hue in the CL0 layer to a deep green hue on the CL5 top layer. However, multi-channel images can be misinterpreted if both elements are in the same column position. To separate both signals, average compositional profiles along the growth direction for both elements are plotted in Figure 3(right). On the one hand, Al peaks linearly increase with AlAs thickness and considering the typical segregation in these systems, the AlAs is incorporated in the regions between QDs in a quantity proportional to the intended AlAs deposition. On the other hand, In peaks are always present and show similar maximum contents (around 13–15%) for all five layers although they become narrower as AlAs deposition increases. These results point out that the total amount of InAs in the WL regions may decrease in the different QD layers but not until its complete deletion.

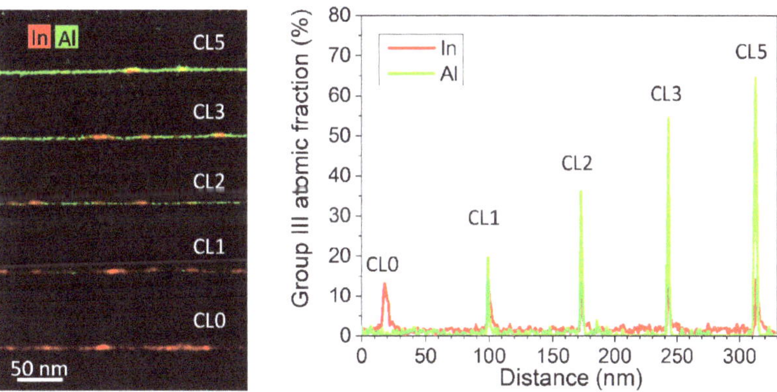

Figure 3. (**left**) Low magnification elemental EDX map for In and Al elements together with (**right**) In and Al atomic fraction profiles along the growth direction.

EDX mappings at low magnification have proved that the WL is not eliminated, but rather a strong superposition between In and Al profiles is present. However, at this magnification, small shifts during EDX map acquisition could lead to mistaken conclusions with respect to spatial distribution and the maximum content detected. With the aim of investigating the overlapping between WL and CL regions, EDX analyses with atomic-column-resolution of the CL/WL region were carried out for the different QD layers using drift correction software during acquisition (Figure 4), where red, yellow, green, and blue colors correspond to Ga, In, As, and Al atoms, respectively. Channeling, probe scattering, and fluorescence could influence the atomic detection and quantification in high spatial resolution EDX maps. A compromise between all parameters (thickness sample, converge angle, etc.) is necessary to ensure practical confidence in the results [43]. At the expense of a certain hindrance in quantitative interpretation, a small convergence angle (16 mrad) was used in areas of the TEM sample with thicknesses between 20 and 40 nm. These column resolved EDX elemental maps allow resolving the dumbbell structure in which As anions are at the lower position and cations (Ga, In, or Al) in the upper. Ga atoms are predominant in the areas at the top and bottom of the image while the In and Al atoms are in the center. Note that In atoms presence seems to appear a little bit before blue fringes linked to Al, which expand as the AlAs thickness increases.

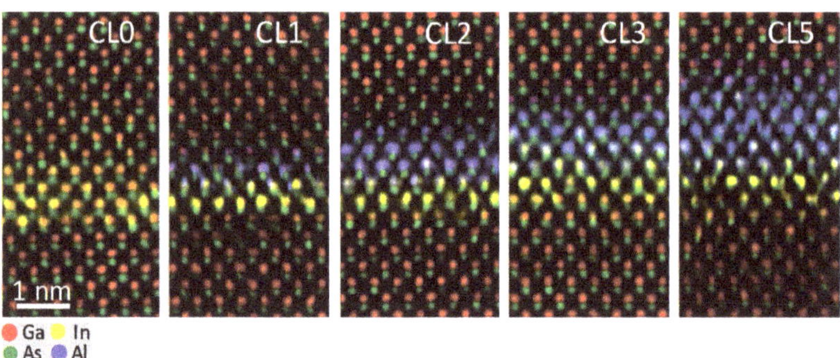

Figure 4. Atomic-level-resolution EDX elemental maps of WL/CL regions for the different InAs QD layers with 0, 1, 2, 3, and 5 ML of AlAs acquired along the [110] zone axis.

To know the exact positions where In (of the WL) and Al atoms (of the CL) begin to be incorporated, a careful layer by layer study of the EDX spectra has been performed for each QD layer. Thus, for the case of the QD layer CL2, Figure 5 shows (a) an EDX map and (b) the average elemental profiles along the growth direction for In and Al using net counts. In these profiles, each peak corresponds to the position of every atomic layer in the growth direction. Considering this, we defined a rectangular region of interest (ROI) with the height of 1 cation width in (a) as reference position 1. Figure 5c shows different EDX spectra obtained by shifting this ROI to distinct positions along the growth direction. First, the position number #-1 corresponds to the GaAs buffer without In and Al, just before the beginning of the InAs WL. Spectra taken on inferior positions (not shown here) present a high similarity, which discard the possibility of a significant channeling effect of the upper layers with Al and In on the spectrum of position number #-1. Since the presence of spurious peaks at 1.487 keV (Kα1 for Al) and 3.279 keV (Lα2 for In) in the GaAs spacer also appear on the GaAs substrate spectra, they are considered as artefacts (fluorescence and/or backscattered electrons from support grid, objective pole-pieces, surface contamination, etc.,) which do not correspond to the X-rays generated by the specimen [44]. Thus, the signal of In starts in position number #1, where there is no evidence of Al atoms. From position #2 onwards, the Al signal appears coexisting with the In one. It should be noted that there is a delay of only 1 ML in the beginning of the incorporation of Al with respect to In. Al atoms hardly reach position #12 but In atoms do, being present until position #15.

Several composition profiles, from regions with sample thicknesses below 50 nm, along the growth direction were averaged using different EDX maps for each QD layer. The counts were discretized considering the sequential monolayer structure and the In, Ga, and Al contents were calculated for each ML. The compositional profiles along the growth direction in different regions are remarkably similar, so we deduce that these column resolved EDX maps are representative of the overall behavior for each QD layer. Figure 6 shows the results of these profiles in WL regions without QDs for each QD layers. Focusing on the In profiles, all In curves show the effects of surface segregation with a typical shark-fin shape. For all layers, the In-content increases during the first 3–4 MLs reaching up to 15–17% and then slowly diminishes in an exponential way. All the In profiles are remarkably similar at the beginning, almost independent of the amount of the AlAs deposited, the differences are found during the decline after the maximum. The number of MLs with the presence of In is reduced almost four times, falling from more than 40 ML for the case without AlAs CL (CL0) down to ~10 MLs for the QD layer CL5. According to these results, there is a reduction of the In amount in WL but only in the decay part.

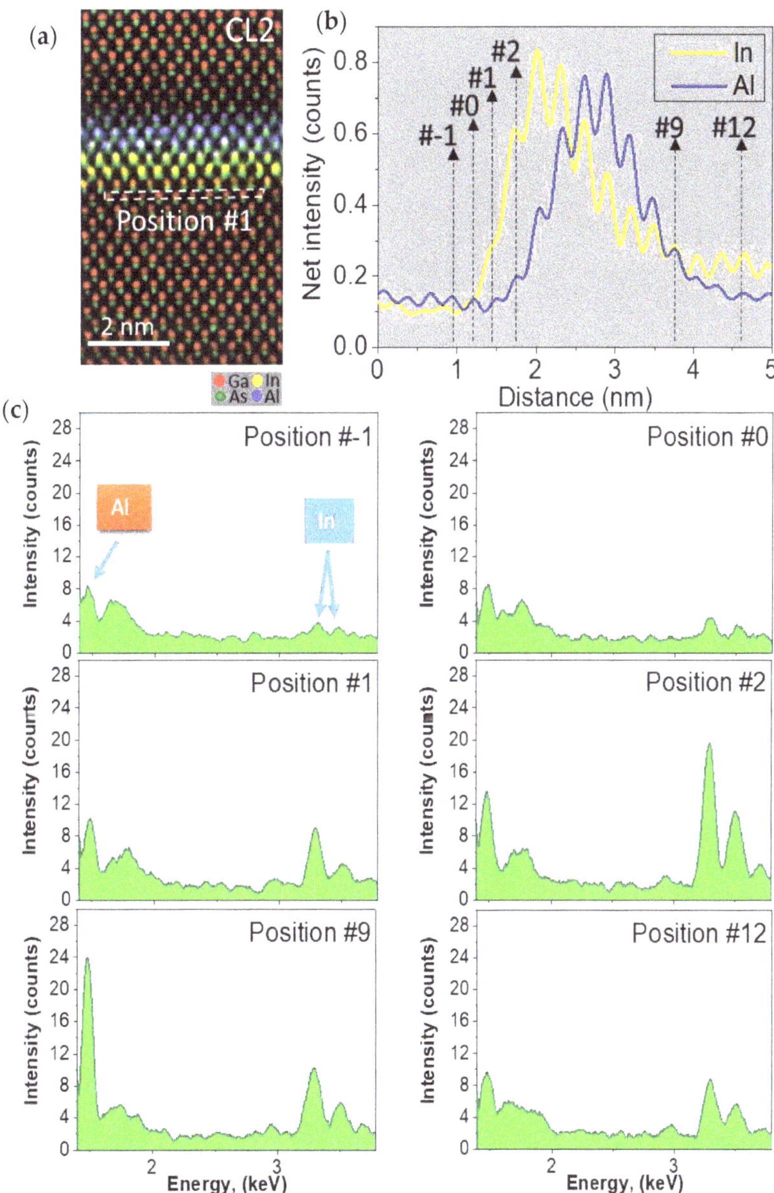

Figure 5. For the WL of CL2, (**a**) EDX map showing the distribution of Ga, In, As, and Al. A rectangular ROI is showed at position #1 (**b**) Net counts profiles for In and Al along the growth direction. (**a**,**c**) EDX spectra for ROI displacements along the growth direction in position #−1, #0, #1, #2, #11, and #12. Spectra of 4096 points in length were smoothed on the graph using Savitzky-Golay algorithm using a 30-point window to remove some background noise.

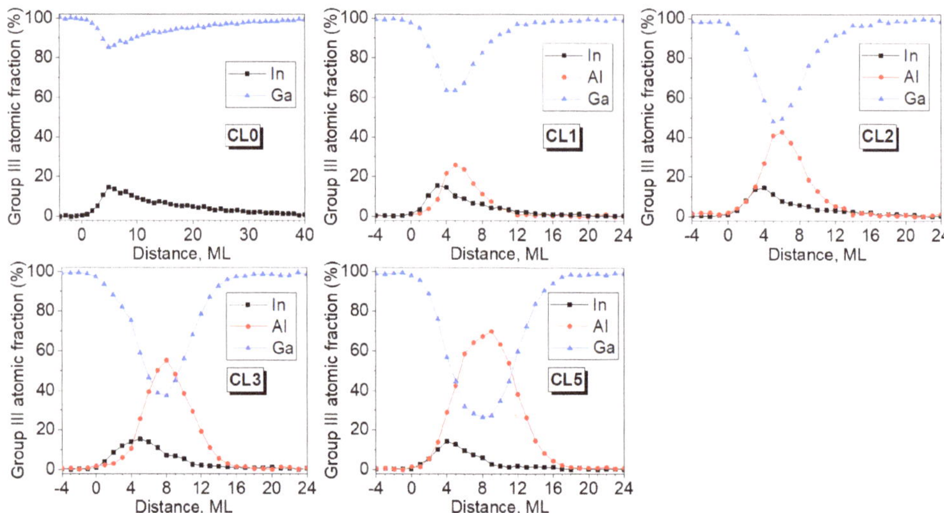

Figure 6. Compositional profiles at ML resolution level for elements of the III group along the growth direction for the WL region in the different AlAs/InAs QD layers.

On the other side, Al profiles are almost symmetrical displacing the peak maxima to taller positions when Al deposition is higher, from the position #5 up to position #10 for CL1 and CL5 layers, respectively. Comparing both In and Al profiles, it can be seen in every case that the Al peak maximum is displaced with respect to the In peak a distance proportional to the CL thickness. The Al contents are always higher than the In ones when the In decay begins and even in the growing part of the profiles from 2 ML of AlAs onwards. The great overlapping between both profiles together with the reduction of the In tail would explain the contrast changes in the g002 DF images as the AlAs CL thickness increases. Of note is the significant continuous presence of Ga atoms throughout the CL/WL region, always exceeding 30% of the III-sites. The regions between QDs must be considered as a quaternary InAlGaAs alloy with a variable composition where In is the minority component.

3.3. Simulation of g002 DF Images

As we have commented above, the intensities of g002 DF images for semiconductors with zinc blende structure are strongly sensitive to local composition changes and are poorly coupled to the local strain [45]. These images are quite simple to simulate using only atomic scattering factors in kinematical conditions (sample thicknesses below ~50 nm), albeit at semi-quantitative level [38]. Figure 7 shows the simulation of the corresponding diffracted intensity under two-beam kinematic g002 DF conditions normalized to GaAs of WL/CL region using the compositional profiles from Figure 6 together with the experimental profiles extracted from g002 DF images (Figure 2(Left)).

For the CL0 layer, the WL in g002 DF image conditions appear as a wide dark band expanding up to 12 nm with intensities less than 1. The addition of only 1 ML of AlAs changes the profile intensely. The thickness of the contrast of the WL/CL is reduced to 3 nm with a first thin dark valley followed by a thin bright peak. With the AlAs incorporation, the first dark valley progressively vanishes, increasing the area of the peak associated with the bright contrast. Simulations even predict the presence of a second dark stripe after the white band (see the inset of Figure 7). This second stripe occurs because the segregation of the In atoms surpasses the position of the AlAs layer. The simulation is in remarkably good agreement with the experimental profiles of Figure 2. Certainly, experimental measurements of the g002 DF intensity even for ternary layers do not exactly match the simulations, probably because it is sensitive to other parameters such as the deviation parameter from the Bragg condition, contributions from 004 reflection, amorphous layers on the sample surface, strain fields, and so on. All attempts to advance the quantification of these materials using DCTEM images have never provided reliable results [39]. In any case, the high intermixing between the CL and the WL from the very beginning of deposition and the increasing amount of AlAs with respect to InAs explain the "apparent" dissolution of the WL.

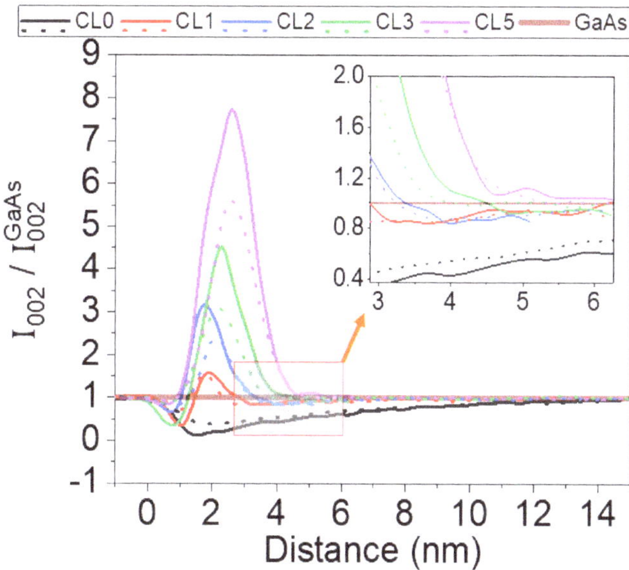

Figure 7. Normalized profiles of the intensities in g002 DF images of WL/CL region along the growth direction simulated using the compositional profiles of Figure 6 (solid line) and experimental g002 DF images (dot line) for the different AlAs/InAs/GaAs QD layers. The inset shows the presence of a second darker streak after the brighter AlAs contrast due to In surface segregation.

3.4. Global Elemental Quantification of the WL and CL

In a thin layer, the area under the EDX compositional profiles along the growth direction has been confirmed as the best parameter for assessing the total amount of material deposited. Previous calibration with different InAs thin layers without QDs permitted us to quantify the total amount of In that remained in the WL regions of QDs layers expressed in MLs of pure InAs [46]. For the case of AlAs layers, similar method was applied in this work, using the first layer block of the sample for calibration. Figure 8 displays the global In and Al amounts of the CL/WL expressed as deposited MLs using this procedure for every QD layer. On the one hand, In amounts in the WL show an exponential decay with an important reduction in comparison with the InAs deposited in CL0 (1.8 ML). The global In content seems to reach a plateau in layer CL5, so we should not expect further reductions for InAs amounts in the WL with increasing thickness of the AlAs CL. This decrease in In content in the WL implies that a higher percentage of the In content is kept in the QDs. This result agrees with previous data [41], which showed a progressive increase in the heights and average content of the QDs as the thickness of the CL increases. It is proposed that AlAs capping has a large shielding effect against the decomposition of QDs, since it presents a significant percentage of pyramidal QDs, which is their stable form before capping. Moreover, in the case of very thick CLs (>5 MLs), the excessive shielding of AlAs against the erosion of QDs has detrimental effects because it preserves the giant QDs existing in uncapped surfaces.

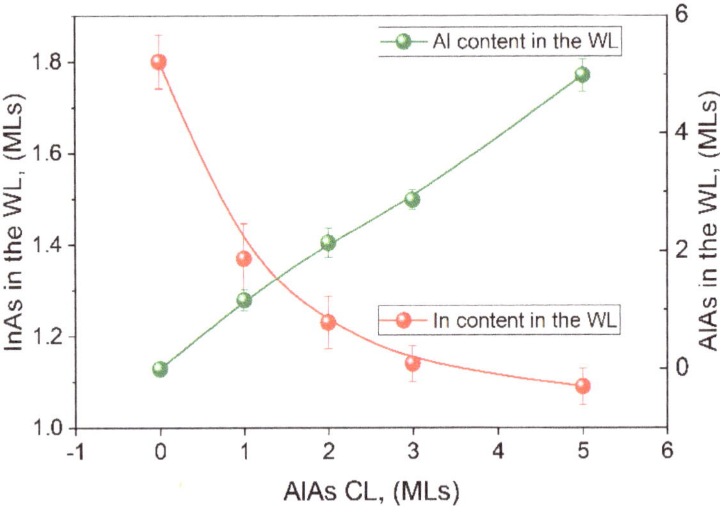

Figure 8. In and Al amount in the CL/WL regions expressed as deposited MLs. The results are obtained from the calibrations of the area in averaged profiles along the growth direction from EDX measurements in thin layers. Error bars shows the upper and lower 95% confidence limits using the standard error for the sample mean, multiplied by 1.96.

Contrary to the large variations of In content in the WL regions between QDs during the capping process, the amount of Al in these regions follows a linear trend that agrees with the nominal number of MLs introduced during growth. That means that Al is homogeneously spread on the surface without significant accumulation in the QDs and regardless of the changes that occur in the QDs. This is consistent with the higher energy barrier for surface diffusion of Al (0.8 eV) [47] compared to Ga (0.62 eV) [48] or In (0.4 eV) [49]. At these growth temperatures, surface diffusion lengths ($L = \sqrt{D_s \tau}$, where Ds is the surface diffusion coefficient and τ the transient time), are of the order of a few, dozens, or hundreds of nm for Al, Ga, and In, respectively. In atoms are very mobile, while Al atoms are almost immobile, staying close to the first surface position where adsorption occurs. Since Al atoms show the shortest surface displacements once adsorbed on the surface at this temperature, the degree of coating of QDs from the first stages of CL growth is remarkably high. The nearly immobile Al atoms rapidly cover the surface of the QDs, freezing the surface redistribution of In and slowing down the loss of In by desorption from the QDs during overgrowth. Consequently, the mass transfer of In to the WL from the QDs is reduced, resulting in larger QDs and thinner WLs [41].

4. Discussion

4.1. WL Reduction

The deposition of a thin layer of AlAs on InAs QDs layers has been found to work as a way to improve the V_{OC} in InAs/GaAs QDSCs [14,18], where it has been suggested that the elimination of the WL during the capping process of AlAs is its cause [22]. Our work has shown that a very thin layer of AlAs induces a significant reduction of the WL thickness but not its complete removal. Thus, the WL of the CL5 layer only keeps an amount of In equivalent to 1.1 MLs of InAs compared to the initial 2.8 MLs deposited. Remarkably, this reduction of the WL is stronger than in other capping strategies such as the use of strain-reducing layers (SRL) or increased GaAs CL growth rate [34,50]. On the one hand, the amount of InAs in the WL after accelerating the growth rate of the GaAs CL to 2 ML/s is 1.7 MLs of InAs, but no substantial improvement is expected for faster growth [46,51]. On the other hand, using a SRL of GaAs$_{0.8}$Sb$_{0.2}$, which is almost the high Sb content limit for GaAsSb capping without plastic relaxation, results in a WL with 1.4 ML of InAs [34]. Certainly, capping with AlAs entails a further reduction of the WL but in line with what has been observed with other materials.

In addition, the analysis of the profiles allows us to assess the Tsatsul'nikov's hypothesis of migration of In toward the dots by substitution of In for Al in the WL [22]. As can be seen by comparing the In profiles in Figure 6, the upward sections of the curves are almost identical for all layers, reaching the same maximum, while the downward sections differ, corresponding to the segregation tail of the In accumulated in the floating layer. Therefore, only this decay section takes part in the comparison of the WL mass reduction. First, the fact that the ascending sections of the In profiles are the same is not compatible with a mechanism of substitution of In for Al in the WL. Al is present from the beginning and in different amounts, so the incorporation of Al in the CL does not affect the already incorporated In profile. Second, the decrease of the downward profile cannot be explained by the supposed migration of In from the WL to the QDs either. All segregation models to describe the decay profiles due to surface segregation in InGaAs QWs [25,27] cannot be applied for the case of WLs in InAs QDs systems. The reason is that the segregation profiles in the case of QWs are always much smaller, being necessary to add an extra amount of In to correctly fit the segregation tails of the WL profiles in QD systems [35]. This added amount of In can only come from the decomposition of the QDs, where it has been found that there is a relationship between the size of the QDs and the In content in the WL. The larger the average QD size, the thinner the WL between the QDs [34,46]. Therefore, the longer segregation tail of the WL without AlAs capping (CL0) is explained by the accumulation of In from the greater decomposition of its QDs, which showed the smallest size. In contrast, shorter tails of In profiles in the case of thicker AlAs layers are linked to the presence of larger QDs, with shapes and sizes similar to uncapped surface QDs [41].

As we have seen, capping with AlAs occurs with processes like those seen with other alloys. First, individually, QDs undergo a decomposition, changing their structure by shortening their heights and reducing their In content and deformation state [52–54]. One of the main driving forces is the intermixing process due to the lattice mismatch between the QDs and the WL, leading to mass transport away from the QDs during capping [55]. Second, collectively, there could be a significant decrease in the QD density after capping [34,56]. Therefore, the region with the WL away from the QDs becomes enriched with In due to lateral migration from the decomposition of the small QDs [35,50,57]. Our results suggest that capping with AlAs does not remove In from the WL during the capping as it has been proposed but acts against the QD decomposition hindering the surface diffusion of In from the QDs and reducing the amount of In available for surface segregation in the WL region.

4.2. CL/WL Mixing

Though the WL reduction is considerable, it is not the only reason for the V_{OC} recovery. A key factor could be the high intermixing showed in our work that happens between the AlAs CL and the InAs WL. EDX profiles at atomic resolution of the In and Al content along the growth direction showed a high degree of overlapping from the first MLs, that is, there is almost no physical separation between these layers with a defined interface but only one graded layer with variable content of these elements. The first 3–4 MLs of this InAlGaAs layer exhibit almost superposed profiles with similar contents of In and Al, but in the next layers, the peak of the Al profile always surpasses the peak of In, even for the sample with 1 ML of AlAs (see Figure 6).

This result is consistent with the small thickness of the WL before capping. The deposition of 2.8 ML of InAs is in fact distributed between the QDs and the WL regions. As can be seen in Figure 8, the amount of InAs in the WL in the case of the layer CL5 is only a little higher than 1 ML. In the case of the layer CL0, the In amount is higher (about 1.8 ML) but this is due to the added incorporation of In due to the QD decomposition process during capping that yields a long tail in the profile. Our results point to a thickness of InAs in the WL before capping of approximately 1 ML or less. During AlAs capping,

Al incorporation occurs over this WL of 1 ML thickness. From here, Al and In from the floating segregated layer are incorporated simultaneously. This explains the huge overlap of both profiles with only a small lag of 1 ML (see Figure 6) at the onset of InAs growth without the need for any substitutional mechanism.

In addition, the high simultaneity of both profiles could be explained by the surface segregation energies. Surface atomic exchanges require overcoming energetic barriers for bulk-to-surface switches that depend on the cohesive energies in the case of III-V compounds [24,58], being 178.9, 154.7, and 144.3 kcal/mol for AlAs, GaAs and InAs, respectively [59]. Then, Al/Ga exchange favors surface segregation of Ga instead of Al [60], while In is always strongly segregated to the surface in the presence of both Ga and Al [61]. As a result, Al atoms, which are incorporated on the first ML of the WL, are in comparison sessile while In always segregates, crossing the AlAs layer and overcoming it.

However, the situation is somewhat different for other CLs of a different nature, such as GaAsSb, where there is a large shift of the CL profile with respect to the WL profile [62,63]. The reasons are multiple. First, the lower cohesive energy of GaSb (138.6 kcal/mol) compared to InAs [59], which means a higher surface segregation of Sb with respect to In. Second, the lower nominal contents of $GaAs_{1-x}Sb_x$ CLs ($x < 0.2$) implies that the amount of Sb incorporated under the surface at the beginning of the CL growth is low, since the surface can accommodate significantly more Sb before saturating [27]. Consequently, the two profiles have little overlap, and there is even an intermediate zone rich in Ga between both.

4.3. Bandgap Approximation of the CL/WL Regions

Finally, the high intermixing and the reduction of the In content in the regions between QDs have a noticeable impact on the electronic states. Due to the difficulty in preparing perfectly controlled InAlGaAs QW structures, little experimental data are available to obtain a set of material parameters for estimating the bandgap in the InAlGaAs system. Because of this limitation, only two equations for the room temperature dependence of the bandgap have been obtained experimentally [64,65], which have a realistic agreement with the published expression for AlGaAs, InGaAs, and AlInAs ternary alloys. Due to the lack of experimental studies at low temperatures, we have used the low-temperature bandgap relation for the InAlGaAs system proposed by E. H. Li [66], based on an interpolation method using the material parameters of the ternaries [67]. In general, a good agreement between the calculated and experimental results over a wide range of QW width and alloy concentration is achieved in the QW structures of III-V alloys using these approaches.

Figure 9 shows the evolution of the bandgap energy (E_g) at 4.2 K along the growth direction for the different layers in the regions between QDs using the EDX compositional profiles and considering the interpolated bowing factor for an InAlGaAs alloy [66]. For the CL0 layer, the WL works as a quantum well with an energy minimum around 1.3 eV.

For comparison, the experimental bandgap of regular InAs QDs ranges from 1 to 1.15 eV [14,16,19]. The energy gap between the WL and the QDs protects the electrons and holes of QDs from coupling, but only to a certain extent, since a low-energy tail of the WL continuum can extend to the confined states of QDs. This energy gap can be bridged by many phenomena [16], which has negative consequences for quantum applications. However, the deposition of very thin layers of AlAs changes the behavior radically. With the introduction of AlAs, the energy well of the WL/CL shrinks largely until it disappears and a potential barrier for electron and holes is additionally formed. As can be appreciated in Figure 9, the formation of an InAlGaAs layer shifts the bandgap to higher energies, limiting the undesirable hybridization of states between the WL and the QDs. This barrier grows with increasing CL thickness, which could be detrimental for carrier transport in the solar cell [16] but is far from the values expected for pure AlAs layer (3.13 eV). In this sense, the large intermixing reduces not only the depth of the potential well due to In, but also limits the height of the barrier due to Al. Remarkably, the layers with 1–2 mL of AlAs show the most compensated configurations, avoiding, on the one hand, big barriers achieved with large CL thicknesses and, on the other hand, deep potential wells which can be observed without AlAs.

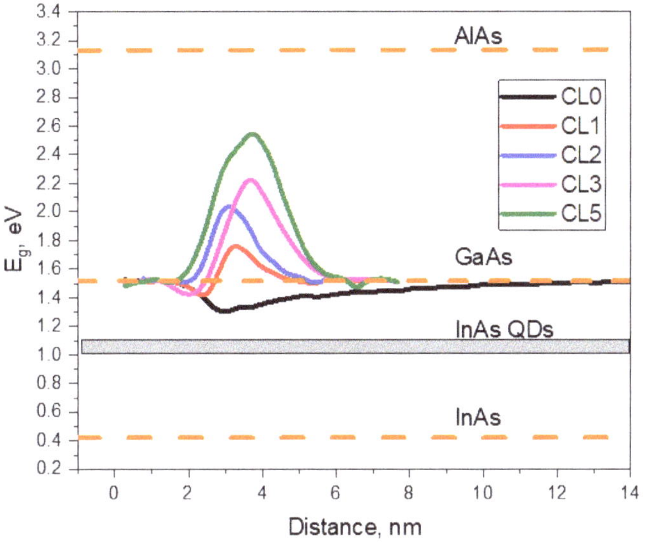

Figure 9. Bandgap approximation at 4.2 K of the WL/CL system along the growth direction, for the different layers in the regions between QDs.

5. Conclusions

The use of thin layers of AlAs in InAs QDSC devices has been proposed to improve the open circuit voltage by intending to make the WL vanish. In this work, the structure of the CL/WL regions between QDs has been characterized at the atomic column level applying the CL thickness variation. First, our results have shown a significant reduction of the WL due to a QDs decomposition reduction, but far from its complete elimination such as those suggested in earlier TEM works. Second, both CL and WL show a high intermixing from the first layers resulting in an InAlGaAs layer with gradual composition. The latter explains the mistaken dissolution of the WL proposed in the CDTEM studies using g002 DF conditions. Both factors cause a drastic change in electronic levels, from a QW for the case without AlAs to a huge barrier for the case of thickest CL, the CL with 1–2 ML being the most adequate configuration for SC applications.

Author Contributions: Conceptualization, D.F., methodology, D.F. and T.B.; formal analysis, N.R. and D.F.; investigation, N.R., V.B. and S.F.; writing—original draft preparation, N.R., D.G., D.F. and E.L.; writing—review and editing, D.F., V.B., A.G.C., E.L., T.B., S.F., L.S., J.M.U. and D.G.; project administration, J.M.U. and D.G. All authors have read and agreed to the published version of the manuscript.

Funding: The work has been co-financed by the Spanish National Research Agency (AEI projects MAT2016-77491-C2-2-R and PID2019-106088RB-C33), Regional Government of Andalusia (project FEDER-UCA18-108319) and the European Regional Development Fund (ERDF).

Data Availability Statement: Not applicable.

Conflicts of Interest: The authors declare no conflict of interest.

References

1. Shockley, W.; Queisser, H.J. Detailed Balance Limit of Efficiency of P-n Junction Solar Cells. *J. Appl. Phys.* **1961**, *32*, 510–519. [CrossRef]
2. Luque, A.; Martí, A. Increasing the Efficiency of Ideal Solar Cells by Photon Induced Transitions at Intermediate Levels. *Phys. Rev. Lett.* **1997**, *78*, 5014. [CrossRef]
3. Okada, Y.; Ekins-Daukes, N.J.; Kita, T.; Tamaki, R.; Yoshida, M.; Pusch, A.; Hess, O.; Phillips, C.C.; Farrell, D.J.; Yoshida, K.; et al. Intermediate Band Solar Cells: Recent Progress and Future Directions. *Appl. Phys. Rev.* **2015**, *2*, 021302. [CrossRef]
4. Ramiro, I.; Marti, A.; Antolin, E.; Luque, A. Review of Experimental Results Related to the Operation of Intermediate Band Solar Cells. *IEEE J. Photovolt.* **2014**, *4*, 736–748. [CrossRef]
5. Delamarre, A.; Suchet, D.; Cavassilas, N.; Okada, Y.; Sugiyama, M.; Guillemoles, J.F. An Electronic Ratchet Is Required in Nanostructured Intermediate-Band Solar Cells. *IEEE J. Photovolt.* **2018**, *8*, 1553–1559. [CrossRef]
6. Okada, Y.; Yoshida, K.; Shoji, Y.; Tamaki, R. *Semiconductor Quantum Dot Solar Cells*; Elsevier: Amsterdam, The Netherlands, 2021; Volume 21, ISBN 9780128220832.
7. Beattie, N.S.; See, P.; Zoppi, G.; Ushasree, P.M.; Duchamp, M.; Farrer, I.; Donchev, V.; Ritchie, D.A.; Tomić, S. Design and Fabrication of InAs/GaAs QD Based Intermediate Band Solar Cells by Quantum Engineering. In Proceedings of the 2018 IEEE 7th World Conference on Photovoltaic Energy Conversion, WCPEC 2018—A Joint Conference 45th IEEE PVSC, 28th PVSEC 34th EU PVSEC, Waikoloa, HI, USA, 10–15 June 2018; pp. 2747–2751. [CrossRef]
8. Villa, J.; Ramiro, I.; Ripalda, J.M.; Tobias, I.; Garcia-Linares, P.; Antolin, E.; Marti, A. Contribution to the Study of Sub-Bandgap Photon Absorption in Quantum Dot InAs/AlGaAs Intermediate Band Solar Cells. *IEEE J. Photovolt.* **2021**, *11*, 420–428. [CrossRef]
9. Ramiro, I.; Villa, J.; Hwang, J.; Martin, A.J.; Millunchick, J.; Phillips, J.; Martí, A. Demonstration of a GaSb/GaAs Quantum Dot Intermediate Band Solar Cell Operating at Maximum Power Point. *Phys. Rev. Lett.* **2020**, *125*, 247703. [CrossRef]
10. Blokhin, S.A.; Sakharov, A.V.; Nadtochy, A.M.; Pauysov, A.S.; Maximov, M.V.; Ledentsov, N.N.; Kovsh, A.R.; Mikhrin, S.S.; Lantratov, V.M.; Mintairov, S.A.; et al. AlGaAs/GaAs Photovoltaic Cells with an Array of InGaAs QDs. *Semiconductors* **2009**, *43*, 514–518. [CrossRef]
11. Sablon, K.; Li, Y.; Vagidov, N.; Mitin, V.; Little, J.W.; Hier, H.; Sergeev, A. GaAs Quantum Dot Solar Cell under Concentrated Radiation. *Appl. Phys. Lett.* **2015**, *107*, 073901. [CrossRef]
12. Sellers, D.G.; Polly, S.; Hubbard, S.M.; Doty, M.F. Analyzing Carrier Escape Mechanisms in InAs/GaAs Quantum Dot p-i-n Junction Photovoltaic Cells. *Appl. Phys. Lett.* **2014**, *104*, 223903. [CrossRef]
13. Cappelluti, F.; Khalili, A.; Gioannini, M. Open Circuit Voltage Recovery in Quantum Dot Solar Cells: A Numerical Study on the Impact of Wetting Layer and Doping. *IET Optoelectron.* **2017**, *11*, 44–48. [CrossRef]
14. Varghese, A.; Yakimov, M.; Tokranov, V.; Mitin, V.; Sablon, K.; Sergeev, A.; Oktyabrsky, S. Complete Voltage Recovery in Quantum Dot Solar Cells Due to Suppression of Electron Capture. *Nanoscale* **2016**, *8*, 7248–7256. [CrossRef] [PubMed]
15. Guimard, D.; Morihara, R.; Bordel, D.; Tanabe, K.; Wakayama, Y.; Nishioka, M.; Arakawa, Y. Fabrication of InAs/GaAs Quantum Dot Solar Cells with Enhanced Photocurrent and without Degradation of Open Circuit Voltage. *Appl. Phys. Lett.* **2010**, *96*, 203507. [CrossRef]
16. Löbl, M.C.; Scholz, S.; Söllner, I.; Ritzmann, J.; Denneulin, T.; Kovács, A.; Kardynał, B.E.; Wieck, A.D.; Ludwig, A.; Warburton, R.J. Excitons in InGaAs Quantum Dots without Electron Wetting Layer States. *Commun. Phys.* **2019**, *2*, 93. [CrossRef]
17. Zieliński, M. Vanishing Fine Structure Splitting in Highly Asymmetric InAs/InP Quantum Dots without Wetting Layer. *Sci. Rep.* **2020**, *10*, 13542. [CrossRef]
18. Tutu, F.K.; Lam, P.; Wu, J.; Miyashita, N.; Okada, Y.; Lee, K.-H.; Ekins-Daukes, N.J.; Wilson, J.; Liu, H. InAs/GaAs Quantum Dot Solar Cell with an AlAs Cap Layer. *Appl. Phys. Lett.* **2013**, *102*, 163907. [CrossRef]
19. Kim, D.; Tang, M.; Wu, J.; Hatch, S.; Maidaniuk, Y.; Dorogan, V.; Mazur, Y.I.; Salamo, G.J.; Liu, H. Si-Doped InAs/GaAs Quantum-Dot Solar Cell with AlAs Cap Layers. *IEEE J. Photovolt.* **2016**, *6*, 906–911. [CrossRef]
20. Shchukin, V.; Ledentsov, N.; Rouvimov, S. Formation of Three-Dimensional Islands in Subcritical Layer Deposition in Stranski-Krastanow Growth. *Phys. Rev. Lett.* **2013**, *110*, 1–5. [CrossRef]
21. Lu, X.M.M.; Matsubara, S.; Nakagawa, Y.; Kitada, T.; Isu, T. Suppression of Photoluminescence from Wetting Layer of InAs Quantum Dots Grown on (311)B GaAs with AlAs Cap. *J. Cryst. Growth* **2015**, *425*, 106–109. [CrossRef]
22. Tsatsul'nikov, A.F.; Kovsh, A.R.; Zhukov, A.E.; Shernyakov, Y.M.; Musikhin, Y.G.; Ustinov, V.M.; Bert, N.A.; Kop'ev, P.S.; Alferov, Z.I.; Mintairov, A.M.; et al. Volmer–Weber and Stranski–Krastanov InAs-(Al,Ga)As Quantum Dots Emitting at 1.3 Mm. *J. Appl. Phys.* **2000**, *88*, 6272–6275. [CrossRef]
23. Tutu, F.K.; Sellers, I.R.; Peinado, M.G.; Pastore, C.E.; Willis, S.M.; Watt, A.R.; Wang, T.; Liu, H.Y. Improved Performance of Multilayer InAs/GaAs Quantum-Dot Solar Cells Using a High-Growth-Temperature GaAs Spacer Layer. *J. Appl. Phys.* **2012**, *111*, 3–6. [CrossRef]
24. Moison, J.M.; Guille, C.; Houzay, F.; Barthe, F.; Van Rompay, M. Surface Segregation of Third-Column Atoms in Group III-V Arsenide Compounds: Ternary Alloys and Heterostructures. *Phys. Rev. B* **1989**, *40*, 6149–6162. [CrossRef] [PubMed]
25. Muraki, K.; Fukatsu, S.; Shiraki, Y.; Ito, R. Surface Segregation of In Atoms during Molecular Beam Epitaxy and Its Influence on the Energy Levels in InGaAs/GaAs Quantum Wells. *Appl. Phys. Lett.* **1992**, *61*, 557–559. [CrossRef]
26. Dehaese, O.; Wallart, X.; Mollot, F. Kinetic Model of Element III Segregation during Molecular Beam Epitaxy of III-III'-V Semiconductor Compounds. *Appl. Phys. Lett.* **1995**, *66*, 52–54. [CrossRef]
27. Godbey, D.J.; Ancona, M.G. Modeling of Ge Segregation in the Limits of Zero and Infinite Surface Diffusion. *J. Vac. Sci. Technol. A Vac. Surf. Film.* **1997**, *15*, 976–980. [CrossRef]

28. Cullis, A.G.; Norris, D.J.; Walther, T.; Migliorato, M.A.; Hopkinson, M. Stranski-Krastanow Transition and Epitaxial Island Growth. *Phys. Rev. B* **2002**, *66*, 081305. [CrossRef]
29. Litvinov, D.; Gerthsen, D.; Rosenauer, A.; Schowalter, M.; Passow, T.; Feinäugle, P.; Hetterich, M. Transmission Electron Microscopy Investigation of Segregation and Critical Floating-Layer Content of Indium for Island Formation in InGaAs. *Phys. Rev. B* **2006**, *74*, 165306. [CrossRef]
30. Arzberger, M.; Käsberger, U.; Böhm, G.; Abstreiter, G. Influence of a Thin AlAs Cap Layer on Optical Properties of Self-Assembled InAs/GaAs Quantum Dots. *Appl. Phys. Lett.* **1999**, *75*, 3968–3970. [CrossRef]
31. Dorogan, V.G.; Mazur, Y.I.; Lee, J.H.; Wang, Z.M.; Ware, M.E.; Salamo, G.J. Thermal Peculiarity of AlAs-Capped InAs Quantum Dots in a GaAs Matrix. *J. Appl. Phys.* **2008**, *104*, 104303. [CrossRef]
32. Oktyabrsky, S.; Tokranov, V.; Agnello, G.; Van Eisden, J.; Yakimov, M. Nano-Engineering Approaches to Self-Assembled InAs Quantum Dot Laser Medium. *J. Electron. Mater.* **2006**, *35*, 822–833. [CrossRef]
33. Schowalter, M.; Rosenauer, A.; Gerthsen, D.; Arzberger, M.; Bichler, M.; Abstreiter, G. Investigation of In Segregation in InAs/AlAs Quantum-Well Structures. *Appl. Phys. Lett.* **2001**, *79*, 4426–4428. [CrossRef]
34. González, D.; Flores, S.; Ruiz-Marín, N.; Reyes, D.F.F.; Stanojević, L.; Utrilla, A.D.D.; Gonzalo, A.; Gallego Carro, A.; Ulloa, J.M.M.; Ben, T. Evaluation of Different Capping Strategies in the InAs/GaAs QD System: Composition, Size and QD Density Features. *Appl. Surf. Sci.* **2021**, *537*, 148062. [CrossRef]
35. Eisele, H.; Ebert, P.; Liu, N.; Holmes, A.L.; Shih, C.-K. Reverse Mass Transport during Capping of In 0.5 Ga 0.5 As/GaAs Quantum Dots. *Appl. Phys. Lett.* **2012**, *101*, 233107. [CrossRef]
36. Cerva, H. Transmission Electron Microscopy of Heteroepitaxial Layer Structures. *Appl. Surf. Sci.* **1991**, *50*, 19–27. [CrossRef]
37. Bithell, E.G.; Stobbs, W.M. III-V Ternary Semiconductor Heterostructures: The Choice of an Appropriate Compositional Analysis Technique. *J. Appl. Phys.* **1991**, *69*, 2149–2155. [CrossRef]
38. Rosenauer, A.; Schowalter, M.; Glas, F.; Lamoen, D. First-Principles Calculations of 002 Structure Factors for Electron Scattering in Strained InxGa1-XAs. *Phys. Rev. B—Condens. Matter Mater. Phys.* **2005**, *72*, 1–10. [CrossRef]
39. Beanland, R. Dark Field Transmission Electron Microscope Images of III-V Quantum Dot Structures. *Ultramicroscopy* **2005**, *102*, 115–125. [CrossRef]
40. Grillo, V.; Albrecht, M.; Remmele, T.; Strunk, H.P.; Egorov, A.Y.; Riechert, H. Simultaneous Experimental Evaluation of In and N Concentrations in InGaAsN Quantum Wells. *J. Appl. Phys.* **2001**, *90*, 3792–3798. [CrossRef]
41. Ruiz-Marín, N.; Reyes, D.F.; Stanojević, L.; Ben, T.; Braza, V.; Gallego-Carro, A.; Bárcena-González, G.; Ulloa, J.M.; González, D. Effect of the AlAs Capping Layer Thickness on the Structure of InAs/GaAs QD. *Appl. Surf. Sci.* **2022**, *573*, 151572. [CrossRef]
42. Song, H.Z.Z.; Tanaka, Y.; Yamamoto, T.; Yokoyama, N.; Sugawara, M.; Arakawa, Y. AlGaAs Capping Effect on InAs Quantum Dots Self-Assembled on GaAs. *Phys. Lett. A* **2011**, *375*, 3517–3520. [CrossRef]
43. Chen, Z.; Weyland, M.; Sang, X.; Xu, W.; Dycus, J.H.; LeBeau, J.M.; D'Alfonso, A.J.; Allen, L.J.; Findlay, S.D. Quantitative Atomic Resolution Elemental Mapping via Absolute-Scale Energy Dispersive X-Ray Spectroscopy. *Ultramicroscopy* **2016**, *168*, 7–16. [CrossRef] [PubMed]
44. Liao, Y. Practical Electron Microscopy and Database. 2013. Available online: www.globalsino.com/EM/ (accessed on 10 April 2022).
45. Twesten, R.D.; Follstaedt, D.M.; Lee, S.R.; Jones, E.D.; Reno, J.L.; Millunchick, J.M.; Norman, A.G.; Ahrenkiel, S.P.; Mascarenhas, A. Characterizing Composition Modulations in InAs/AlAs Short-Period Superlattices. *Phys. Rev. B* **1999**, *60*, 13619–13635. [CrossRef]
46. González, D.; Braza, V.; Utrilla, A.D.; Gonzalo, A.; Reyes, D.F.; Ben, T.; Guzman, A.; Hierro, A.; Ulloa, J.M. Quantitative Analysis of the Interplay between InAs Quantum Dots and Wetting Layer during the GaAs Capping Process. *Nanotechnology* **2017**, *28*, 425702. [CrossRef] [PubMed]
47. Kasu, M.; Kobayashi, N. Surface-Diffusion and Step-Bunching Mechanisms of Metalorganic Vapor-Phase Epitaxy Studied by High-Vacuum Scanning Tunneling Microscopy. *J. Appl. Phys.* **1995**, *78*, 3026–3035. [CrossRef]
48. Kasu, M.; Kobayashi, N. Surface Diffusion of AlAs on GaAs in Metalorganic Vapor Phase Epitaxy Studied by High-vacuum Scanning Tunneling Microscopy. *Appl. Phys. Lett.* **1995**, *67*, 2842–2844. [CrossRef]
49. Fujiwara, K.; Ishii, A.; Aisaka, T. First Principles Calculation of Indium Migration Barrier Energy on an InAs(001) Surface. *Thin Solid Films* **2004**, *464–465*, 35–37. [CrossRef]
50. González, D.; Reyes, D.F.; Utrilla, A.D.; Ben, T.; Braza, V.; Guzman, A.; Hierro, A.; Ulloa, J.M. General Route for the Decomposition of InAs Quantum Dots during the Capping Process. *Nanotechnology* **2016**, *27*, 125703. [CrossRef]
51. Utrilla, A.D.; Grossi, D.F.; Reyes, D.F.; Gonzalo, A.; Braza, V.; Ben, T.; González, D.; Guzman, A.; Hierro, A.; Koenraad, P.M.; et al. Size and Shape Tunability of Self-Assembled InAs/GaAs Nanostructures through the Capping Rate. *Appl. Surf. Sci.* **2018**, *444*, 260–266. [CrossRef]
52. Costantini, G.; Rastelli, A.; Manzano, C.; Acosta-Diaz, P.; Songmuang, R.; Katsaros, G.; Schmidt, O.G.; Kern, K. Interplay between Thermodynamics and Kinetics in the Capping of InAs/GaAs(001) Quantum Dots. *Phys. Rev. Lett.* **2006**, *96*, 226106. [CrossRef]
53. Eisele, H.; Lenz, A.; Heitz, R.; Timm, R.; Dähne, M.; Temko, Y.; Suzuki, T.; Jacobi, K. Change of InAs/GaAs Quantum Dot Shape and Composition during Capping. *J. Appl. Phys.* **2008**, *104*, 124301. [CrossRef]
54. Gong, Q.; Offermans, P.; Nötzel, R.; Koenraad, P.M.; Wolter, J.H. Capping Process of InAs/GaAs Quantum Dots Studied by Cross-Sectional Scanning Tunneling Microscopy. *Appl. Phys. Lett.* **2004**, *85*, 5697–5699. [CrossRef]
55. Xie, Q.H.; Madhukar, A.; Chen, P.; Kobayashi, N.P. Vertically Self-Organized InAs Quantum Box Island on GaAs (100). *Phys. Rev. Lett.* **1995**, *75*, 2542–2545. [CrossRef] [PubMed]

56. Ferdos, F.; Wang, S.; Wei, Y.; Larsson, A.; Sadeghi, M.; Zhao, Q. Influence of a Thin GaAs Cap Layer on Structural and Optical Properties of InAs Quantum Dots. *Appl. Phys. Lett.* **2002**, *81*, 1195–1197. [CrossRef]
57. Ulloa, J.M.; Gargallo-Caballero, R.; Bozkurt, M.; del Moral, M.; Guzmán, A.; Koenraad, P.M.; Hierro, A. GaAsSb-Capped InAs Quantum Dots: From Enlarged Quantum Dot Height to Alloy Fluctuations. *Phys. Rev. B* **2010**, *81*, 165305. [CrossRef]
58. Flores, S.; Reyes, D.F.; Braza, V.; Richards, R.D.; Bastiman, F.; Ben, T.; Ruiz-Marín, N.; David, J.P.R.; González, D. Modelling of Bismuth Segregation in InAsBi/InAs Superlattices: Determination of the Exchange Energies. *Appl. Surf. Sci.* **2019**, *485*, 29–34. [CrossRef]
59. Aresti, A.; Garbato, L.; Rucci, A. Some Cohesive Energy Features of Tetrahedral Semiconductors. *J. Phys. Chem. Solids* **1984**, *45*, 361–365. [CrossRef]
60. Spencer, G.S.; Menéndez, J.; Pfeiffer, L.N.; West, K.W. Optical-Phonon Raman-Scattering Study of Short-Period GaAs-AlAs Superlattices: An Examination of Interface Disorder. *Phys. Rev. B* **1995**, *52*, 8205–8218. [CrossRef]
61. Dorin, C.; Mirecki Millunchick, J. Lateral Composition Modulation in AlAs/InAs and GaAs/InAs Short Period Superlattices Structures: The Role of Surface Segregation. *J. Appl. Phys.* **2002**, *91*, 237–244. [CrossRef]
62. González, D.; Reyes, D.F.; Ben, T.; Utrilla, A.D.; Guzman, A.; Hierro, A.; Ulloa, J.M. Influence of Sb/N Contents during the Capping Process on the Morphology of InAs/GaAs Quantum Dots. *Sol. Energy Mater. Sol. Cells* **2016**, *145*, 154–161. [CrossRef]
63. Haxha, V.; Drouzas, I.; Ulloa, J.M.; Bozkurt, M.; Koenraad, P.M.; Mowbray, D.J.; Liu, H.Y.; Steer, M.J.; Hopkinson, M.; Migliorato, M.A. Role of Segregation in InAs/GaAs Quantum Dot Structures Capped with a GaAsSb Strain-Reduction Layer. *Phys. Rev. B* **2009**, *80*, 165334. [CrossRef]
64. Ishikawa, T.; Bowers, J.E. Band Lineup and In-Plane Effective Mass of InGaAsP or InGaAlAs on InP Strained-Layer Quantum Well. *IEEE J. Quantum Electron.* **1994**, *30*, 562–570. [CrossRef]
65. Olego, D.; Chang, T.Y.; Silberg, E.; Caridi, E.A.; Pinczuk, A. Compositional Dependence of Band-Gap Energy and Conduction-Band Effective Mass of In1-x-YGaxAlyAs Lattice Matched to InP. *Appl. Phys. Lett.* **1982**, *41*, 476–478. [CrossRef]
66. Li, E.H.H. Material Parameters of InGaAsP and InAlGaAs Systems for Use in Quantum Well Structures at Low and Room Temperatures. *Phys. E Low-Dimens. Syst. Nanostruct.* **2000**, *5*, 215–273. [CrossRef]
67. Hirayama, Y.; Choi, W.Y.; Peng, L.H.; Fonstad, C.G. Absorption Spectroscopy on Room Temperature Excitonic Transitions in Strained Layer InGaAs/InGaAlAs Multiquantum-Well Structures. *J. Appl. Phys.* **1993**, *74*, 570–578. [CrossRef]

Article

Atomic Layer Deposition of Ultrathin ZnO Films for Hybrid Window Layers for Cu(In$_x$,Ga$_{1-x}$)Se$_2$ Solar Cells

Jaebaek Lee [1,2,3], Dong-Hwan Jeon [1,2], Dae-Kue Hwang [1,2], Kee-Jeong Yang [1,2], Jin-Kyu Kang [1,2], Shi-Joon Sung [1,2,*], Hyunwoong Park [3,*] and Dae-Hwan Kim [1,2,*]

[1] Research Center for Thin Film Solar Cells, Daegu-Gyeongbuk Institute of Science and Technology (DGIST), Daegu 42988, Korea; tglee@dgist.ac.kr (J.L.); dhjeon@dgist.ac.kr (D.-H.J.); dkhwang@dgist.ac.kr (D.-K.H.); kjyang@dgist.ac.kr (K.-J.Y.); apollon@dgist.ac.kr (J.-K.K.)

[2] Division of Energy Technology, Daegu-Gyeongbuk Institute of Science and Technology (DGIST), Daegu 42988, Korea

[3] School of Energy Engineering, Kyungpook National University, Daegu 41566, Korea

* Correspondence: sjsung@dgist.ac.kr (S.-J.S.); hwp@knu.ac.kr (H.P.); monolith@dgist.ac.kr (D.-H.K.)

Citation: Lee, J.; Jeon, D.-H.; Hwang, D.-K.; Yang, K.-J.; Kang, J.-K.; Sung, S.-J.; Park, H.; Kim, D.-H. Atomic Layer Deposition of Ultrathin ZnO Films for Hybrid Window Layers for Cu(In$_x$,Ga$_{1-x}$)Se$_2$ Solar Cells. *Nanomaterials* **2021**, *11*, 2779. https://doi.org/10.3390/nano11112779

Academic Editor: Vlad Andrei Antohe

Received: 24 September 2021
Accepted: 18 October 2021
Published: 20 October 2021

Publisher's Note: MDPI stays neutral with regard to jurisdictional claims in published maps and institutional affiliations.

Abstract: The efficiency of thin-film chalcogenide solar cells is dependent on their window layer thickness. However, the application of an ultrathin window layer is difficult because of the limited capability of the deposition process. This paper reports the use of atomic layer deposition (ALD) processes for fabrication of thin window layers for Cu(In$_x$,Ga$_{1-x}$)Se$_2$ (CIGS) thin-film solar cells, replacing conventional sputtering techniques. We fabricated a viable ultrathin 12 nm window layer on a CdS buffer layer from the uniform conformal coating provided by ALD. CIGS solar cells with an ALD ZnO window layer exhibited superior photovoltaic performances to those of cells with a sputtered intrinsic ZnO (i-ZnO) window layer. The short-circuit current of the former solar cells improved with the reduction in light loss caused by using a thinner ZnO window layer with a wider band gap. Ultrathin uniform A-ZnO window layers also proved more effective than sputtered i-ZnO layers at improving the open-circuit voltage of the CIGS solar cells, because of the additional buffering effect caused by their semiconducting nature. In addition, because of the precise control of the material structure provided by ALD, CIGS solar cells with A-ZnO window layers exhibited a narrow deviation of photovoltaic properties, advantageous for large-scale mass production purposes.

Keywords: ZnO; atomic layer deposition; ultrathin; window layer; CIGS; solar cells

1. Introduction

Thin-film chalcogenide solar cells such as Cu(In$_x$,Ga$_{1-x}$)Se$_2$ (CIGS), Cu$_2$ZnSnSe$_4$ (CZTSe), and SnS typically require an intrinsic ZnO (i-ZnO) window layer, deposited between their CdS buffer layer and transparent conducting oxide (TCO) layers, to improve device performance without increasing light absorption loss [1–4]. This i-ZnO window layer effectively blocks the short-circuit pathways through the voids in the CdS buffer layer, and consequently enhances the shunt resistance (R_{sh}), fill factor (FF), and open-circuit voltage (V_{OC}) of the solar cells. In addition, the i-ZnO window layer protects the CdS buffer layer during the subsequent deposition of a TCO layer [5–8].

Sputtering processes are typically used for deposition of the window layers of CIGS thin-film solar cells because of their fast deposition rate, and the strong adhesion of i-ZnO to CdS [9,10]. However, as such techniques are incapable of depositing ultrathin uniform ZnO films, only a narrow range of process parameters are suitable for the sputtering process of i-ZnO window layers. In general, the thickness of a ZnO window layer must be greater than 50 nm to cover the rough CdS/CIGS surface. The thickness of the ZnO window layer is closely related to its parasitic absorption loss, which induces a decrease in the short-circuit current (J_{SC}) of CIGS solar cells. Therefore, a reduction in the thickness of the ZnO window layer is required to achieve a higher value of J_{SC}. However, the surface coverage of a thin sputtered i-ZnO layer on a CdS buffer layer is insufficient to prevent current leakage from CIGS solar cells. In addition, the severe sputter process conditions required for deposition of i-ZnO can cause critical damage to the underlying CdS buffer layer [11,12].

To overcome the drawbacks of sputter coating, atomic layer deposition (ALD) processes have been proposed for fabrication of the ZnO window layers in CIGS thin-film solar cells. Although the ALD technique is a well-established deposition method for precise fabrication of ultrathin films [13–15], few studies have reported on ALD of ZnO window layers for thin-film chalcogenide solar cells [5,16,17]. Such reports that exist have focused on the material properties of atomic-layer-deposited ZnO (A-ZnO). In these studies, the thickness of the A-ZnO window layer remained over 50 nm, and the efficiencies of the corresponding CIGS thin-film solar cells were below 13%. Hence, the strengths of ALD for fabrication of ZnO window layers in chalcogenide solar cells were not practically utilized.

In this work, we discuss the fabrication of ultrathin uniform A-ZnO window layers for high-efficiency CIGS solar cells. To maximize the strengths of the ALD technique, we focus on deposition of thin films with thicknesses between 12 and 23 nm. ALD enables conformal and uniform coating of ZnO on the rough surface of a CdS buffer layer. Hence, ultrathin A-ZnO window layers provide sufficient passivation and protection to the CdS buffer layer, enabling lower thicknesses of ZnO to be used in CIGS solar cells, compared to those required with conventional sputtering. In addition, reducing the thickness of the ZnO window layer is advantageous for the formation of a built-in field over the ZnO window layer, as the spreading of the space charge region can be depressed [18]. As well as reducing its thickness, using ALD instead of sputtering can increase the band gap of the window layer [19], which is advantageous for reducing the light absorption loss of CIGS solar cells in the short wavelength range. This reduced loss can subsequently induce an improvement in the J_{SC} of CIGS solar cells.

Unlike sputtered i-ZnO, which is highly resistive, A-ZnO has electrical properties characteristic of a weak n-type semiconductor [20], making it beneficial for improving the V_{OC} of CIGS solar cells. Carrier transport through the void regions of the buffer layer is easier with an A-ZnO window layer than with a sputtered i-ZnO window layer, as the former material infiltrates the CdS, which increases the V_{OC} of CIGS solar cells. Weak n-type semiconducting A-ZnO window layers can thus act as secondary buffer layers in CIGS solar cells. In contrast, ultrathin ZnO layers deposited by sputtering cannot play this role, as the material does not fill the voids in the CdS buffer layer.

The discussion above demonstrates the potential of A-ZnO window layers in high-efficiency CIGS solar cells, by explaining how the ALD technique can improve both the V_{OC} and J_{SC} of CIGS solar cells. In addition, because of the uniform and conformal coating of ALD, CIGS solar cells with an A-ZnO window layer can be expected to exhibit a narrow statistical deviation of photovoltaic parameters, useful for large-scale mass production of CIGS solar cells. The proposed A-ZnO window layer can also be applied to other chalcogenide solar cells, regardless of their interfacial structures.

2. Materials and Methods

2.1. Device Fabrication

The solar cells investigated in this study consisted of a soda lime glass (SLG) substrate, a 500 nm thick Mo back-contact layer, an approximately 2.5 µm thick CIGS absorber layer, a 60 nm thick CdS buffer layer, i-ZnO or ZnO window layers, a 300 nm thick Al-doped ZnO (AZO) transparent conducting oxide layer, and a 1 µm thick Al collection grid. The Mo layer was deposited on the SLG substrate via DC magnetron sputtering using a Mo target with a purity of 99.99%. Following this, the CIGS absorber was deposited on the back-contact layer by thermal co-evaporation from elemental sources. Here, evaporation was completed in a multi-stage process. First, In, Ga, and Se were deposited on the substrate at a temperature of 430 °C. Following this, Cu and Se were evaporated on the substrate at a temperature of 650 °C, until the absorber became copper-rich. The process was completed by evaporation of In, Ga, and Se, to make the absorber copper-poor again. Next, the CdS layer was deposited onto the device using a wet-chemical process, followed by window layer deposition using sputtering processes or ALD (see Section 2.2), as appropriate. Finally, the 300 nm thick AZO front-contact layer was deposited by radio frequency (RF) sputtering, followed by thermal evaporation of 1 μm Al grids. Solar cells with a total area of 0.5535 cm^2 were defined by mechanical scribing.

2.2. ALD Process

A-ZnO window layers were obtained through ALD of diethyl zinc ((C$_2$H$_5$)$_2$Zn; UP Chemical Co., Ltd., Gyeonggi-do, Korea) and deionized water on the surface of the CdS buffer layer. Deposition was completed using a showerhead-type ALD system (CN1 Co., Ltd. Gyeonggi-do, Korea). ATOMIC PREMIUM), with each ALD cycle performed at a temperature of 120 °C. The thickness of the thin films was modified by varying the number of ALD cycles, and 100 ALD cycles performed at 120 °C obtained a final ZnO thickness of 23 nm. The sub-ALD cycle consisted of 0.5 s of dosing with diethyl zinc precursor, 15 s of purging with N$_2$, 0.5 s of dosing with H$_2$O, and a further 15 s of purging with N$_2$.

2.3. Device Characterization

The solar cell samples were characterized using a combination of field-emission transmission electron microscopy (FE-TEM), atomic force microscopy (AFM), solar simulator, and external quantum efficiency (EQE) measurement. The cross-sectional morphologies of the CIGS solar cells were investigated using an FE-TEM system (HF-3300, Hitachi, Saitama, Japan) with a focused ion beam (FIB) instrument (NB5000, Hitachi, Saitama, Japan). The surface morphologies of A-ZnO and sputtered i-ZnO were investigated using a Park NX10 AFM instrument (Park systems, Gyeonggi-do, Korea). The current-voltage characteristics of the solar cells under a simulated air mass 1.5 global (AM 1.5 G) spectrum were measured at an illumination of 100 mW cm^{-2} (1 sun) using a solar simulator (model 94022A, Newport Co., Irvine, CA, USA). The EQE spectra were measured using a QuantX-300 measuring kit (Newport Co., Irvine, CA, USA).

3. Results and Discussion

3.1. Formation of A-ZnO Window Layers on a CdS/CIGS Interface

Exploiting the advantages of ALD with respect to conventional sputtering depends on the precise conformal deposition of the ultrathin material. Hence, cross-sectional STEM-HADDF and EDS images of the interface of the CdS buffer layer and the CIGS absorber layer were acquired to confirm the formation of a window layer (Figure 1). The differing nature of ALD and sputtering processes was analyzed based on the morphology of ZnO thin films with a nominal thickness of 12 nm. In the case of ALD, a uniform ZnO thin film was deposited on the rough surface of the CdS buffer layer. In contrast, there were large variations in the thickness of the sputtered i-ZnO thin film, including the presence of regions on the CdS layer where no material was deposited. This non-uniform window layer confirms the inability of the sputtering process to deposit ultrathin conformal coatings, explaining why the minimum window layer thickness previously reported has been 50 nm.

We conducted AFM analysis of 46 nm thick ZnO thin films on SLG, for further comparison of the difference between ALD and sputtering processes (Figure 2). Again, the A-ZnO thin film exhibited a uniform morphology, with a regular distribution of small protrusions. In contrast, the sputtered i-ZnO thin film exhibited a non-uniform surface morphology, with large protrusions occupying the spaces between the small protrusions. The different surface morphologies of the atomic-layer-deposited and sputtered ZnO thin films are also reflected in their root-mean-square roughness (R$_q$), with values of 6.367 nm obtained for the A-ZnO film, and 16.992 nm for the sputtered i-ZnO film. This result confirms that, by exploiting the unique characteristics of ALD, an ultrathin uniform ZnO window layer can be successfully deposited on a rough CdS buffer layer.

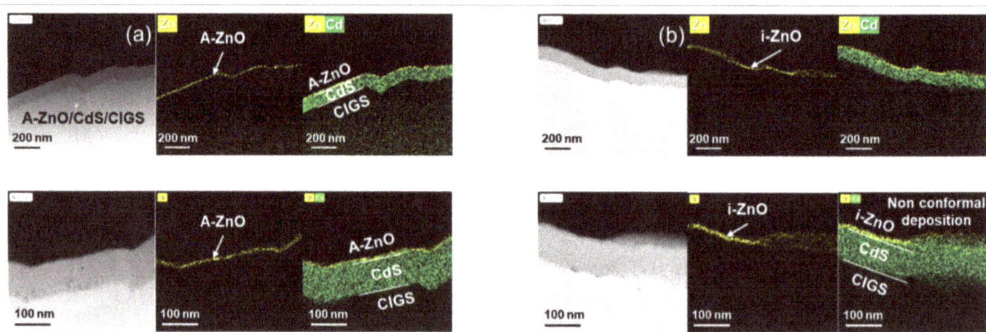

Figure 1. Cross-sectional scanning transmission electron microscopy (STEM) high-angle annular dark-field imaging (HAADF) and energy-dispersive X-ray spectroscopy (EDS) images of (**a**) atomic-layer-deposited ZnO (A-ZnO) on a CdS/CIGS surface and (**b**) sputtered intrinsic ZnO (i-ZnO) on a CdS/CIGS surface.

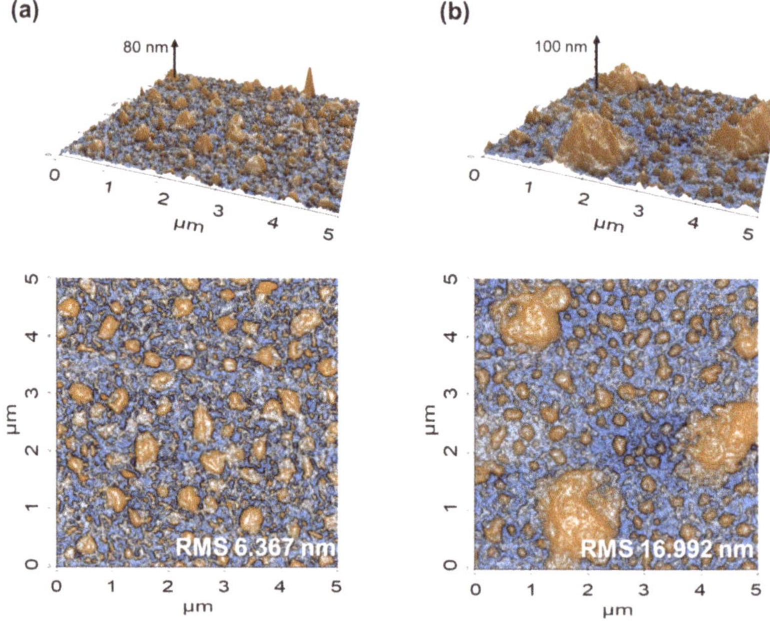

Figure 2. Atomic force microscopy (AFM) images of (**a**) A-ZnO on SLG, and (**b**) sputtered i-ZnO on SLG.

3.2. Performance of A-ZnO Window Layers in CIGS Solar Cells

We acquired cross-sectional TEM images of complete CIGS solar cell devices (Figure 3) to confirm that the window layer remained uniform following deposition of the front contact layers, as this uniformity is essential for improved device performance. The energy dispersive spectroscopy (EDS) mapping images of the CIGS solar cells (Figure 3a) depict a uniform A-ZnO window layer between the CdS buffer layer and the AZO TCO layer. However, it is difficult to distinguish the A-ZnO window layer from the AZO TCO layer because of signal noise from elemental Al. Hence, we performed an EDS line scan to verify the formation of the A-ZnO window layer. These measurements highlighted a region without elemental Al between the CdS buffer layer (120–170 nm) and the AZO TCO layer (0–80 nm) (Figure 3b), indicating the formation of a separated A-ZnO window layer inside the CIGS solar cell device. No signals were observed in this region during EDS mapping of elemental Al at different locations in the CIGS solar cell device (Supplementary Figure S1), confirming our hypothesis.

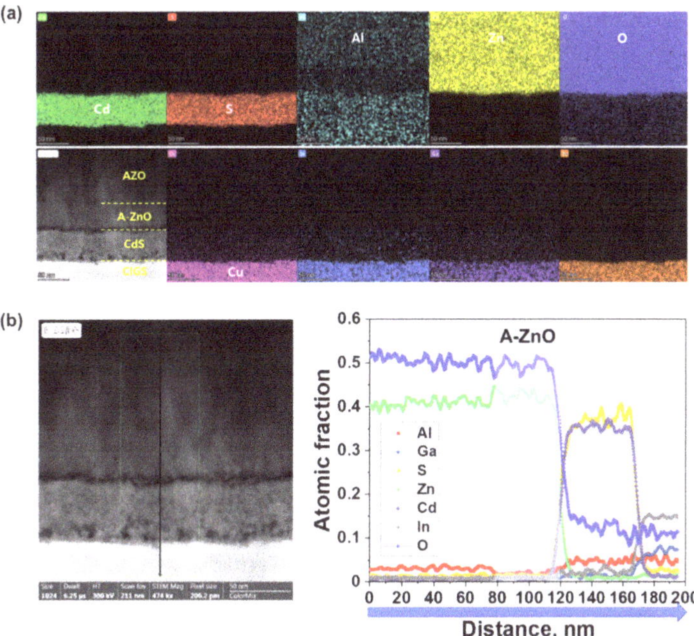

Figure 3. (**a**) Cross-sectional TEM energy dispersive spectroscopy (EDS) mapping images and (**b**) EDS line scan of atomic fraction of elements in the AZO/A-ZnO/CdS/CIGS/Mo/SLG structure of a complete solar cell.

The photovoltaic properties of CIGS solar cells with A-ZnO (with thicknesses of 12–23 nm) were investigated to evaluate the performances of the thin films as window layers. CIGS solar cell devices with sputtered i-ZnO window layers with thicknesses of 12 and 46 nm were also prepared, for comparison. Figure 4 and Table 1 summarize the photovoltaic characteristics of both sets of CIGS solar cell devices.

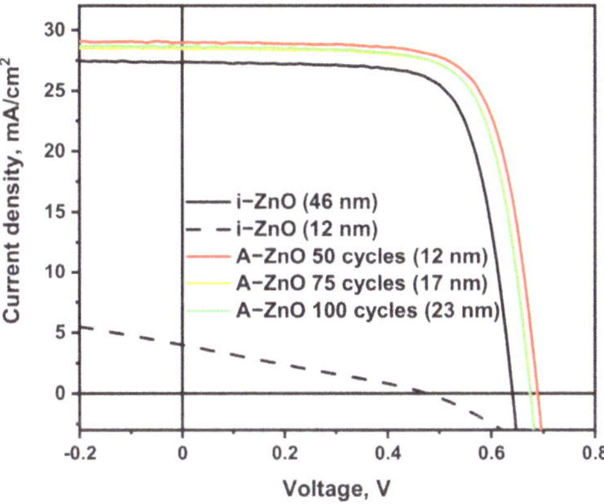

Figure 4. Characteristic current-voltage curves of CIGS solar cells with sputtered i-ZnO and A-ZnO window layers with different thicknesses.

Table 1. Photovoltaic parameters of CIGS solar cells with sputtered i-ZnO and A-ZnO window layers of different thickness.

	Thickness (nm)	V_{OC} (V)	J_{SC} (mA/cm^2)	FF (%)	Efficiency (%)	R_s (Ω)	R_{sh} (Ω)
i-ZnO (Ref.)	46	0.64068	27.3669	73.0578	12.809	2.4	723.6
i-ZnO	12	0.47154	14.0397	25.9519	1.736	16.4	66.8
A-ZnO (50 cycles)	12	0.68862	28.9264	73.1868	14.578	2.0	539.3
A-ZnO (75 cycles)	17	0.67545	28.4352	73.2842	14.075	2.6	365.8
A-ZnO (100 cycles)	23	0.67429	28.5909	72.9267	14.059	2.6	490.5

Our characterization indicates that CIGS solar cells with A-ZnO window layers exhibited higher efficiency than cells with sputtered i-ZnO window layers, regardless of the thickness of the A-ZnO window layer. This improved efficiency originates from the higher V_{OC} and J_{SC} of the former devices, indicating that ultrathin A-ZnO films are suitable for use as the window layer of CIGS solar cells. Although the device with the 12 nm A-ZnO window layer exhibited the best photovoltaic properties, overall, modifying the thickness of the A-ZnO window layers had little effect on the photovoltaic properties of the CIGS solar cells. In contrast, the device with a 12 nm thick sputtered i-ZnO window layer exhibited a poor efficiency of under 2%. This inferior photovoltaic performance can be attributed to the non-uniform formation of the sputtered ultrathin i-ZnO window layer on the CdS buffer layer, which was depicted in Figure 1. The non-uniform coverage of the CdS buffer layer by the sputtered i-ZnO window layer is insufficient to protect the CdS buffer layer from damage during the AZO TCO sputtering process, which deteriorates the photovoltaic performance of the CIGS solar cells. This result indicates that ALD is necessary for the application of ultrathin ZnO window layers in CIGS solar cells.

Figures 1 and 2 highlighted that a distinguishing characteristic of ALD is uniform formation of ultrathin conformal ZnO window layers. For additional evaluation of the merit of this characteristic, we conducted a statistical analysis of the photovoltaic parameters of CIGS solar cells with both A-ZnO and sputtered i-ZnO window layers (Figure 5). With the exception of the short-circuit current (J_{SC}), CIGS solar cells with a sputtered ZnO window layer exhibited large deviations in photovoltaic parameters compared to devices with A-ZnO window layers, with the FF, series resistance (R_s), and efficiency exhibiting particularly large variances. These large deviations can be attributed to the non-uniformity of the sputtered i-ZnO window layer. The large variation in the thickness of the 12 nm sputtered i-ZnO film hinders the ability of the window layer to prevent current leakage. The large deviation in FF is thus closely related to the large deviation in the efficiency of CIGS solar cells. In contrast, CIGS solar cells with A-ZnO thin films showed very narrow deviations in V_{OC}, FF, and R_s, indicating that, because of their uniform and conformal deposition on the CdS buffer layer, they are more effective as window layers than sputtered i-ZnO films. The small deviation in the photovoltaic parameters of CIGS solar cells with A-ZnO window layers is also meaningful for the purposes of scalable manufacture, as it suggests the potential for rapid fabrication of devices with repeatable performance.

3.3. Effect of A-ZnO Window Layers on Photovoltaic Parameters

The analysis of the photovoltaic properties of CIGS solar cells above demonstrates that ultrathin A-ZnO films can act as the window layer in such devices, without the deterioration of their performance. The unique structural properties of the A-ZnO thin films enable them to provide sufficient protection to the CdS buffer layer during deposition of the front-contact layer, despite their reduced thickness. In addition, both CIGS solar cells with A-ZnO window layers and the device with the 46 nm sputtered i-ZnO window layer exhibited similar values of R_{sh}, indicating that an ultrathin A-ZnO window layer suitably prevents current leakage. Since ALD creates a uniform conformal ZnO window layer using the ALD process, the deviation of R_{sh} with devices with such coatings is smaller than that of devices with a sputtered i-ZnO window layer. This explains why the photovoltaic properties of CIGS solar cells with A-ZnO films are similar, regardless of the thickness of the window layer. The narrow distribution of the V_{OC} of the CIGS solar cells with A-ZnO window layers can thus be attributed to the narrow deviation of R_{sh}.

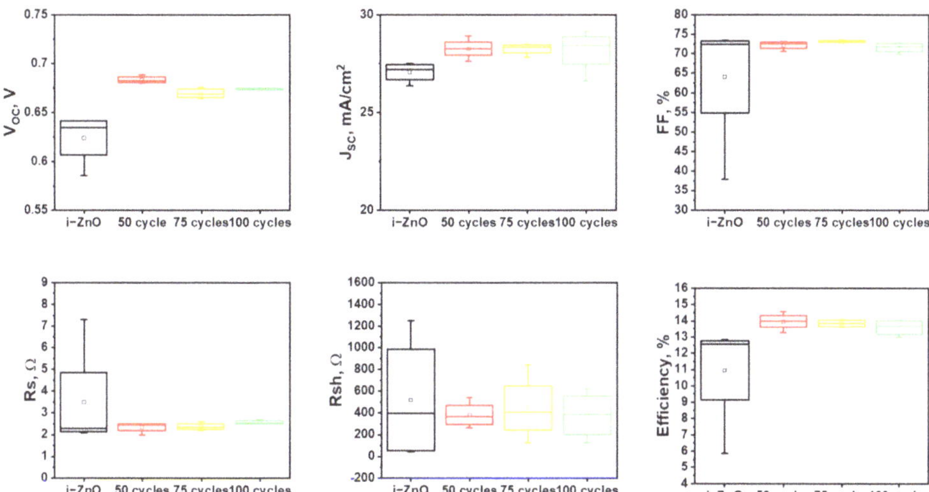

Figure 5. Photovoltaic parameters of CIGS solar cells with i-ZnO and A-ZnO window layers of different thickness.

The improved photovoltaic performance of CIGS solar cells with A-ZnO window layers results from the improvement of J_{SC} and V_{OC}. The origin of these improved photovoltaic characteristics can be explained by considering how the change in the physical characteristics of A-ZnO films modify their optical and electrical properties with respect to those of sputtered i-ZnO films. It has previously been demonstrated that the optical band gap of ZnO thin films is closely related to the conditions used in chemical vapor deposition (CVD) [20]. Accordingly, the differing processes used in deposition of the window layers affected their band gaps, with A-ZnO films exhibiting a wider band gap (3.33 eV) than the sputtered i-ZnO film (3.21 eV) (Supplementary Figure S2). In addition, in this study, the thickness of the A-ZnO window layer was less than half that of the sputtered i-ZnO window layer. The combination of the wider band gap and smaller thickness of the A-ZnO window layer are advantageous for minimizing its light absorption loss, and improving the optical performance of CIGS solar cells. To verify this assertion, we obtained the EQE spectra of both CIGS solar cells with A-ZnO window layers and the CIGS solar cells with a 50 nm sputtered i-ZnO window layer (Figure 6). In the short wavelength range under 550 nm, CIGS solar cells with A-ZnO window layers exhibited a higher EQE, confirming the effect of their wider band gap. In addition, the EQE decreased with increasing ALD cycles, confirming that layer thickness affects light absorption loss. Although the spectra of A-ZnO and sputtered i-ZnO films varied differently at wavelengths over 550 nm, the average EQE of both types of thin film was similar in this range, indicating that ALD is most effective in improving the light absorption loss of CIGS solar cells in the short-wavelength region. The higher values of EQE for wavelengths under 550 nm thus contribute to the increased J_{SC} of CIGS solar cells with A-ZnO window layers.

Similarly, the modified electrical properties of the A-ZnO thin films can explain the improvement in the V_{OC} of CIGS solar cells with such coatings. A comparison of the electrical properties of the A-ZnO and sputtered i-ZnO thin films from the Hall measurement is included in Table 2. The A-ZnO films exhibited higher conductivities, lower resistivities, and higher carrier concentrations than the sputtered i-ZnO film, attributes that are consistent with their well-known similarity to weak n-type semiconductors [20]. The higher V_{OC} of CIGS solar cells with A-ZnO window layers can thus be explained by the additional buffering effect provided by the weak n-type semiconductor layer. By infiltrating the void region of the CdS buffer layer, the A-ZnO thin film can function as a secondary buffer layer, facilitating smooth carrier transport. Sputtered i-ZnO window layers are unable to function in this way, explaining the improved V_{OC} observed with A-ZnO thin films.

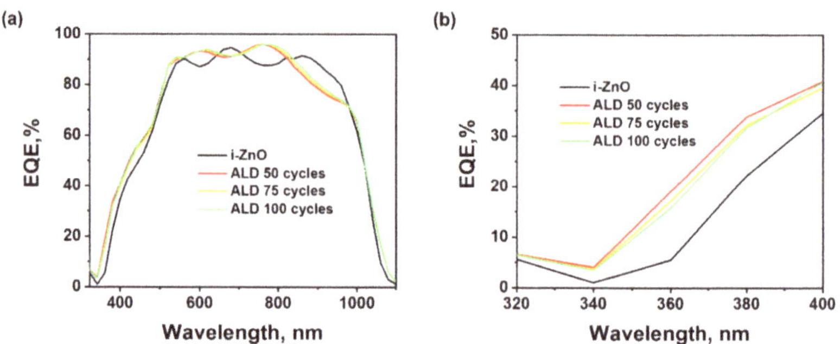

Figure 6. External quantum efficiency (EQE) of CIGS solar cells with i-ZnO and A-ZnO window layers with different thickness in (**a**) the full wavelength range (300–1100 nm), and the (**b**) short-wavelength region (~320–400 nm).

Table 2. Electrical properties of sputtered i-ZnO and A-ZnO thin films.

	Mobility (cm^2/V·s)	Conductivity (1/Ω·cm)	Resistivity (Ω·cm)	Carrier Concentration (cm^{-3})
Sputtered i-ZnO	14.77	7.684×10^{-6}	1.301×10^{5}	-3.247×10^{12}
A-ZnO	12.99	14.59	6.852×10^{-2}	-7.015×10^{18}

4. Conclusions

In this study, we substituted the conventionally sputtered i-ZnO window layers in CIGS solar cells with A-ZnO thin films. Our characterization highlighted that devices using the latter material exhibited superior photovoltaic performance to those using the former. We also demonstrated that ultrathin A-ZnO films (~12 nm) acted suitably well as the window layer of CIGS solar cells, because the ALD process enabled uniform conformal coating of the CdS buffer layer. The CIGS solar cell using an ultrathin A-ZnO window layer showed higher values of efficiency (14.578%), V_{OC} (0.68862 V), J_{SC} (28.9264 mAcm^{-2}), and FF (73.1868%) than the CIGS solar cell using a sputtered i-ZnO. The improved performance of CIGS solar cells with A-ZnO window layers can be explained by enhancements to their J_{SC} (from 27.3669 to 28.9264 mAcm^{-2}) and V_{OC} (from 0.64068 to 0.68862 V). These enhancements result from the modified optical and electrical properties of the window layer, caused by the ALD process' creation of a ZnO film with a more uniform structure. The enhancement of the J_{SC} of these CIGS solar cells was caused by their improved light absorption loss, resulting from the fabrication of a thin film with a wider band gap, and the ability to employ thinner ZnO films in the window layer using ALD. In addition, we observed that the electrical properties of A-ZnO thin films are characteristic of weak n-type semiconductors, enabling the window layers to facilitate smooth carrier transport through the void region of the CdS buffer layer. By infiltrating the CdS layer, the A-ZnO thin film functioned as a secondary buffer layer in CIGS solar cells, enabling the improvement in V_{OC}. In addition to the improvement in the photovoltaic performance of the CIGS solar cells with A-ZnO window layers, we observed smaller deviations in the photovoltaic parameters of individual solar cells because of the precise control of window layer structure enabled by our use of ALD. Hence, as well as providing a promising substitute to conventionally sputtered ZnO window layers, the proposed ALD process can also facilitate the large-scale mass production of CIGS solar cells.

Supplementary Materials: The following are available online at https://www.mdpi.com/article/10.3390/nano11112779/s1, Figure S1: EDS data of an aluminum (Al) element in different positions of a CIGS solar cell device, Figure S2: Transmittance and optical band gap properties of the i-ZnO and A-ZnO: (a) transmittance spectra of i-ZnO and A-ZnO, (b) optical band gap of i-ZnO, and (c) optical band gap of A-ZnO.

Author Contributions: Conceptualization, J.L., S.-J.S. and D.-H.K.; device fabrication, D.-H.J. and D.-K.H.; ALD process, J.L.; device characterization, J.L., K.-J.Y. and J.-K.K.; writing—original draft preparation, J.L and S.-J.S.; writing—review and editing, S.-J.S., H.P. and D.-H.K.; supervision, H.P. and D.-H.K.; project administration and funding acquisition, D.-H.K. All authors have read and agreed to the published version of the manuscript.

Funding: This work was supported by the Technology Development Program to Solve Climate Change of the National Research Foundation (NRF), funded by the Ministry of Science and ICT, Republic of Korea (2016M1A2A2936781), and the DGIST R&D Programs of the Ministry of Science and ICT, Republic of Korea (21-ET-08, 21-CoE-ET-01).

Institutional Review Board Statement: Not applicable.

Informed Consent Statement: Not applicable.

Data Availability Statement: The data presented in this study are available on request from the corresponding author.

Conflicts of Interest: The authors declare no conflict of interest.

References

1. Chantana, J.; Kato, T.; Sugimoto, H.; Minemoto, T. Thin-film Cu(In,Ga)(Se,S)$_2$ -based solar cell with (Cd,Zn)S buffer layer and Zn1−x Mg x O window layer. *Prog. Photovolt. Res. Appl.* **2017**, *25*, 431–440. [CrossRef]
2. Kartopu, G.; Williams, B.; Zardetto, V.; Gürlek, A.; Clayton, A.; Jones, S.; Kessels, W.; Creatore, M.; Irvine, S. Enhancement of the photocurrent and efficiency of CdTe solar cells suppressing the front contact reflection using a highly-resistive ZnO buffer layer. *Sol. Energy Mater. Sol. Cells* **2018**, *191*, 78–82. [CrossRef]
3. Jo, E.; Gil Gang, M.; Shim, H.; Suryawanshi, M.P.; Ghorpade, U.V.; Kim, J.H. 8% Efficiency Cu$_2$ZnSn(S,Se)$_4$ (CZTSSe) Thin Film Solar Cells on Flexible and Lightweight Molybdenum Foil Substrates. *ACS Appl. Mater. Interfaces* **2019**, *11*, 23118–23124. [CrossRef] [PubMed]
4. Yago, A.; Sasagawa, S.; Akaki, Y.; Nakamura, S.; Oomae, H.; Katagiri, H.; Araki, H. Comparison of buffer layers on SnS thin-film solar cells prepared by co-evaporation. *Phys. Status Solidi C* **2017**, *14*, 1600194. [CrossRef]
5. Williams, B.L.; Zardetto, V.; Kniknie, B.; Verheijen, M.A.; Kessels, W.M.; Creatore, M. The competing roles of i-ZnO in Cu(In,Ga)Se$_2$ solar cells. *Sol. Energy Mater. Sol. Cells* **2016**, *157*, 798–807. [CrossRef]
6. Nagoya, Y.; Sang, B.; Fujiwara, Y.; Kushiya, K.; Yamase, O. Improved performance of Cu(In,Ga)Se$_2$-based submodules with a stacked structure of ZnO window prepared by sputtering. *Sol. Energy Mater. Sol. Cells* **2003**, *75*, 163–169. [CrossRef]
7. Ishizuka, S.; Sakurai, K.; Yamada, A.; Matsubara, K.; Fons, P.; Iwata, K.; Nakamura, S.; Kimura, Y.; Baba, T.; Nakanishi, H.; et al. Fabrication of wide-gap Cu(In$_{1−x}$Ga$_x$)Se$_2$ thin film solar cells: A study on the correlation of cell performance with highly resistive i-ZnO layer thickness. *Sol. Energy Mater. Sol. Cells* **2005**, *87*, 541–548. [CrossRef]
8. Misic, B.; Pieters, B.E.; Theisen, J.P.; Gerber, A.; Rau, U. Shunt mitigation in ZnO:Al/i-ZnO/CdS/Cu(In,Ga)Se$_2$ solar modules by the i-ZnO/CdS buffer combination. *Phys. Status Solidi* **2014**, *212*, 541–546. [CrossRef]
9. Liu, W.-S.; Hsieh, W.-T.; Chen, S.-Y.; Huang, C.-S. Improvement of CIGS solar cells with high performance transparent conducting Ti-doped GaZnO thin films. *Sol. Energy* **2018**, *174*, 83–96. [CrossRef]
10. Yu, X.; Ma, J.; Ji, F.; Wang, Y.; Zhang, X.; Cheng, C.; Ma, H. Preparation and properties of ZnO:Ga films prepared by RF magnetron sputtering at low temperature. *Appl. Surf. Sci.* **2005**, *239*, 222–226. [CrossRef]
11. Macco, B.; Deligiannis, D.; Smit, S.; van Swaaij, R.A.C.M.M.; Zeman, M.; Kessels, W.M.M. Influence of transparent conductive oxides on passivation of a-Si:H/c-Si heterojunctions as studied by atomic layer deposited Al-doped ZnO. *Semicond. Sci. Technol.* **2014**, *29*, 122001. [CrossRef]
12. Nandakumar, N.; Dielissen, S.B.; Garcia, D.D.G.-A.; Liu, Z.Z.; Gortzen, R.; Kessels, W.M.M.; Aberle, A.G.; Hoex, B. Resistive Intrinsic ZnO Films Deposited by Ultrafast Spatial ALD for PV Applications. *IEEE J. Photovolt.* **2015**, *5*, 1462–1469. [CrossRef]
13. George, S.M. Atomic Layer Deposition: An Overview. *Chem. Rev.* **2009**, *110*, 111–131. [CrossRef] [PubMed]
14. Knez, M.; Nielsch, K.; Niinistö, L. Synthesis and Surface Engineering of Complex Nanostructures by Atomic Layer Deposition. *Adv. Mater.* **2007**, *19*, 3425–3438. [CrossRef]
15. Puurunen, R.L. Surface chemistry of atomic layer deposition: A case study for the trimethylaluminum/water process. *J. Appl. Phys.* **2005**, *97*, 121301. [CrossRef]
16. Williams, B.L.; Smit, S.; Kniknie, B.J.; Bakker, K.J.; Keuning, W.; Kessels, W.M.M.; Schropp, R.E.I.; Creatore, M. Identifying parasitic current pathways in CIGS solar cells by modelling dark J-V response. *Prog. Photovolt.* **2015**, *23*, 1516–1525. [CrossRef]

17. Arepalli, V.K.; Lee, W.-J.; Chung, Y.-D.; Kim, J. Growth and device properties of ALD deposited ZnO films for CIGS solar cells. *Mater. Sci. Semicond. Process.* **2021**, *121*, 105406. [CrossRef]
18. Jahagirdar, A.H.; Kadam, A.A.; Dhere, N.G. Role of i-ZnO in Optimizing Open Circuit Voltage of CIGS$_2$ and CIGS Thin Film Solar Cells. In Proceedings of the 2006 IEEE 4th World Conference on Photovoltaic Energy Conference, Waikoloa, HI, USA, 7–12 May 2006. [CrossRef]
19. Tan, S.T.; Chen, B.J.; Sun, X.; Fan, W.J.; Kwok, H.S.; Zhang, X.H.; Chua, S.J. Blueshift of optical band gap in ZnO thin films grown by metal-organic chemical-vapor deposition. *J. Appl. Phys.* **2005**, *98*, 13505. [CrossRef]
20. Guziewicz, E.; Godlewski, M.; Krajewski, T.; Wachnicki, L.; Szczepanik, A.; Kopalko, K.; Wójcik-Głodowska, A.; Przeździecka, E.; Paszkowicz, W.; Łusakowska, E.; et al. ZnO grown by atomic layer deposition: A material for transparent electronics and organic heterojunctions. *J. Appl. Phys.* **2009**, *105*, 122413. [CrossRef]

nanomaterials

MDPI

Article

Effect of RF Power on the Physical Properties of Sputtered ZnSe Nanostructured Thin Films for Photovoltaic Applications

Ovidiu Toma [1,†], Vlad-Andrei Antohe [1,2,†], Ana-Maria Panaitescu [1], Sorina Iftimie [1], Ana-Maria Răduță [1], Adrian Radu [1], Lucian Ion [1] and Ştefan Antohe [1,3,*]

[1] Faculty of Physics, R&D Center for Materials and Electronic & Optoelectronic Devices (MDEO), University of Bucharest, Atomiştilor Street 405, 077125 Măgurele, Romania; thtoma72@yahoo.com (O.T.); vlad.antohe@fizica.unibuc.ro (V.-A.A.); anamaria.pam11@yahoo.ro (A.-M.P.); sorina.iftimie@fizica.unibuc.ro (S.I.); ana.raduta@fizica.unibuc.ro (A.-M.R.); adrian.radu@fizica.unibuc.ro (A.R.); lucian@solid.fizica.unibuc.ro (L.I.)
[2] Institute of Condensed Matter and Nanosciences (IMCN), Université catholique de Louvain (UCLouvain), Place Croix du Sud 1, B-1348 Louvain-la-Neuve, Belgium
[3] Academy of Romanian Scientists, Splaiul Independenţei 54, 050094 Bucharest, Romania
* Correspondence: santohe@solid.fizica.unibuc.ro
† These authors contributed equally to this work.

Abstract: Zinc selenide (ZnSe) thin films were deposited by RF magnetron sputtering in specific conditions, onto optical glass substrates, at different RF plasma power. The prepared ZnSe layers were afterwards subjected to a series of structural, morphological, optical and electrical characterizations. The obtained results pointed out the optimal sputtering conditions to obtain ZnSe films of excellent quality, especially in terms of better optical properties, lower superficial roughness, reduced micro-strain and a band gap value closer to the one reported for the ZnSe bulk semiconducting material. Electrical characterization were afterwards carried out by measuring the current–voltage (I-V) characteristics at room temperature, of prepared "sandwich"-like Au/ZnSe/Au structures. The analysis of I-V characteristics have shown that at low injection levels there is an Ohmic conduction, followed at high injection levels, after a well-defined transition voltage, by a Space Charge Limited Current (SCLC) in the presence of an exponential trap distribution in the band gap of the ZnSe thin films. The results obtained from all the characterization techniques presented, demonstrated thus the potential of ZnSe thin films sputtered under optimized RF plasma conditions, to be used as alternative environmentally-friendly Cd-free window layers within photovoltaic cells manufacturing.

Keywords: zinc selenide (ZnSe); thin films; radio frequency (RF) magnetron sputtering; physical properties; spectroscopic ellipsometry; electrical measurements

Citation: Toma, O.; Antohe, V.-A.; Panaitescu, A.-M.; Iftimie, S.; Răduță, A.-M.; Radu, A.; Ion, L.; Antohe, Ş. Effect of RF Power on the Physical Properties of Sputtered ZnSe Nanostructured Thin Films for Photovoltaic Applications. *Nanomaterials* **2021**, *11*, 2841. https://dx.doi.org/10.3390/nano11112841

Academic Editor: Aurora Rizzo

Received: 2 July 2021
Accepted: 22 October 2021
Published: 25 October 2021

Publisher's Note: MDPI stays neutral with regard to jurisdictional claims in published maps and institutional affiliations.

1. Introduction

There is currently a high drive to develop semiconducting materials with easily-tunable properties that allow improved light-matter interactions in order to expand the performance and functionality of various optoelectronic devices, such as: infrared-sensitive elements [1,2], light emitting diodes (LEDs) [3,4], or photovoltaic (PV) cells [5]. In this context, zinc selenide (ZnSe) is a very attractive material from A^{II}-B^{VI} binary semiconducting compounds with unique physical properties, such as, large direct band gap of 2.67 eV (at room temperature), low optical absorption in visible and infrared regions, high refractive index, high electrical conductivity, very good photosensitivity, and it is environmentally friendly [6]. Consequently, there are already numerous reported applications based on ZnSe thin films, such as LEDs [7,8], infrared devices [9,10], diodes [11], photodetectors [12], lasers [13], sensors [14], and solar cells [15], to mention only a few. Various techniques have been used to prepare ZnSe thin films including vacuum thermal evaporation [16–18], chemical bath deposition [19–21], chemical vapor deposition [22], sintering [23], close-spaced sublimation [24], electrodeposition [25], molecular beam epitaxy [26], laser deposition [27], or RF magnetron sputtering [9,28,29]. Regarding the use of ZnSe thin films in solar cells technology, among the Cd-free buffer layers, ZnSe is one of the most important candidates to replace cadmium sulfide (CdS) as a window material for solar cells. Besides environment considerations, ZnSe wider band gap energy (around 2.7 eV) as compared to the commonly-employed CdS (around 2.4 eV) is leading to the possibility of improving the transmission of blue light radiation, and enhancing the photocurrent. Notably, ZnSe material also has a good lattice match with $Cu(In,Ga)(S,Se)_2$, therefore there are numerous reports showing the use of ZnSe thin films as buffer material in different solar cell configurations, such as CIGS-based architectures (copper indium gallium selenide), DSSCs (dye-sensitized solar cells), or CdTe-based photovoltaic devices [29–34]. This paper presents the preparation of ZnSe thin films by radio frequency (RF) magnetron sputtering, since this deposition technique proved to ensure high quality chalcogen compounds-based films with excellent thickness uniformity, as demonstrated in our previous studies focused on the synthesis of CdS [35] and ZnSe [29] thin films, respectively. In this paper, we particularly investigated the influence of the RF sputtering power on the structural, morphological, optical and electrical properties of the prepared ZnSe thin films. To the best of our knowledge, there are yet limited reports focusing on the correlation between the RF plasma power and the physical properties of the obtained ZnSe thin films, which are all important to get ZnSe-based solar cells with an expectedly better performance. Moreover, our results are promising, as they pointed out the optimal preparation conditions for ZnSe thin films, to be further used as a more environmentally-friendly, Cd-free buffer or window layer within various optoelectronic devices.

2. Materials and Methods

The ZnSe thin films were deposited by RF magnetron sputtering onto optical glass substrates, using a commercial ZnSe target (PI-KEM). The frequency of RF generator was fixed at 13.56 MHz. The deposition chamber was first evacuated at 10^{-3} Pa, before admission of argon (Ar) gas up to the working pressure of 0.86 Pa. Prior to the deposition, the substrates were ultrasonically cleaned in acetone, isopropyl alcohol and deionized water for 15 minutes in each bath, and dried over gaseous nitrogen flow. Other deposition parameters used in the sputtering process of the ZnSe films were target-to-substrate distance of 8 cm, the substrate temperature of 220°C and the sputtering time of 30 min, while the RF power was varied as 60 W, 80 W, 100 W and 120 W. Films with different thicknesses were thus obtained and they will be indicated as follows: ZnSe1 (60 W), ZnSe2 (80 W), ZnSe3 (100 W) and ZnSe4 (120 W). The structural features of the fabricated samples were determined by X-ray diffraction (XRD), with a Bruker D8 Discover equipment (using $CuK_{\alpha1}$ radiation at $\lambda = 1.5406$ Å). The topography of the fabricated ZnSe films' surface was analyzed by atomic force microscopy (AFM) in tapping mode, using an A.P.E. Research A100-SGS instrument. The acquired AFM micrographs have been afterwards post-processed in Gwyddion software package for extracting the roughness average (R_A), root mean square roughness (RMS), as well as Skewness (Ssk) and Excess Kurtosis (Sku) specific statistical parameters. Subsequently, cross-sectional morphological observations of the prepared ZnSe thin films was carried out by scanning electron microscopy (SEM) using a Tescan Vega XMU-II equipment, mainly to allow an overall direct evaluation of the films thickness. The optical properties of ZnSe thin films deposited on glass substrates were next investigated by optical spectroscopy (OS), i.e., transmission and absorption, using a spectrophotometer Perkin-Elmer Lambda 750, within the spectral range between 200 nm and 1200 nm, in air and at room temperature. Going further, optical constants (refractive indices and extinction coefficients) were also investigated using spectroscopic ellipsometry (SE) with a phase modulated spectro-ellipsometer (PME) from Horiba. The angle of incidence was 75° for all ZnSe samples while the modulation frequency of the photoelastic modulator was set at 50 kHz. Ultimately, the dark current–voltage (I-V) characteristics of the sandwich structures, i.e., Au/ZnSe/Au, were measured at room temperature using a computer-controlled experimental setup including a Keithley 6517a electrometer, a Keithley 2400 SourceMeter and a Lakeshore 332 temperature controller.

3. Results and Discussion

3.1. Structural Characterizations

The structural analysis was performed first in grazing incidence (GIXRD) geometry at an angle of 1°. This type of measurement provides a better signal from thin films and minimizes the signal from the substrate. In Figure 1, the results from GIXRD analysis are shown for ZnSe samples and one can observe that generally the crystalline structure is improved with increasing the RF power and thickness. In particular, when comparing the GIXRD patterns of ZnSe samples deposited at different RF powers (i.e., 60–120 W), it was found that the 60 W RF power was not sufficient to form a ZnSe crystalline layer (Figure 1a). This behavior can be understood in terms of the low energy atoms which are easily adsorbed on the glass substrate, moving over and interacting to form clusters, but without posing enough energy to overcome the nucleation barrier, hence to reach thermodynamic stability that favors the formation of ZnSe crystallites. In contrast, by increasing the RF power towards 120 W (Figure 1b–d), a reducing background noise and an increasing intensity peak reflected from the (111) planes of the ZnSe cubic structure could be observed, together with low intensity peaks emanating from the (220) and (311) plane orientations, respectively. Moreover, as described further, the analysis of the most intense peak (111)—Figure 2, also reveals improvements of the crystalline quality of the samples with increasing the RF power, as the narrowness of the reflected (111) peaks increases with RF power. This behavior could be explained by the fact that an increased RF power improves the electrons mobility, thus further the Ar gas atoms ionization efficiency. Consequently, the highly energized inert Ar ions provide translational kinetic energy to the adatoms sputtered on the growing surface, enhancing their surface diffusion that finally leads to obtaining films with high-quality crystalline structure. This approach was also used to explain a similar behavior in the case of other zinc blend-like crystalline structure materials [36,37].

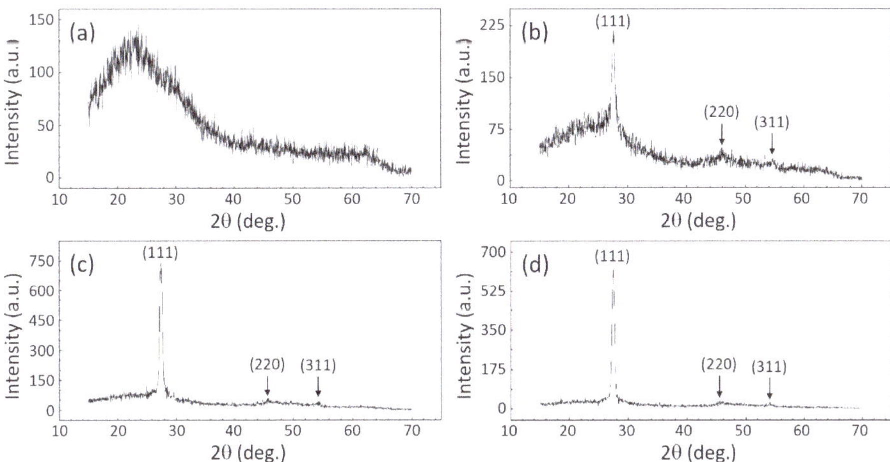

Figure 1. GIXRD patterns of the ZnSe thin films deposited by RF magnetron sputtering onto optical glass substrates at an RF power of: (**a**) 60 W, (**b**) 80 W, (**c**) 100 W, and (**d**) 120 W.

Crystalline structural parameters of the samples were determined by analyzing in Bragg–Brentano theta–theta geometry the most intense reflected plane (111). This characterization is suitable for finding parameters, such as grain size (D_{ef}), micro-strain ($\langle \varepsilon^2 \rangle^{1/2}$) and lattice constant ($a$) that can be calculated using the information obtained after a proper evaluation of the data (see Table 1) [38]. Figure 2 presents the recorded profiles and the fitting curves with Voigt profiles, as well as the residuals of the experimental data after the processing using the theoretical model (the lower plots). It can be observed from Table 1 that when the RF power is increased, bigger crystallite size is obtained, with a small mean-square strain and a value for lattice constant which is closer to that found in bulk crystal [39], the ideal ZnSe lattice constant being $a_0 = 5.669$ Å (according to PDF2 37-1463 card). In particular, the crystallite size (D_{ef}) increased from 51.8 nm to 107.4 nm when raising the RF power, which is a predictable result taking into account the sharpening of the (111) peak as a result of a higher crystallinity, as can be observed within XRD patterns presented in Figure 2. It can be thus noticed from Table 1 that at higher thickness, with the relating sputtering parameters, the structural texture is improved. As explained above, when the nucleation barrier is overcame as a consequence of RF power increase, thermo-dynamically stabilized crystallites are formed, causing their size to become larger, until a saturation of the nucleation density occurs, causing the ZnSe crystalline layers formation, and determining further a natural increase of the films thickness and so of the crystallites size with RF power, as observed in Table 1.

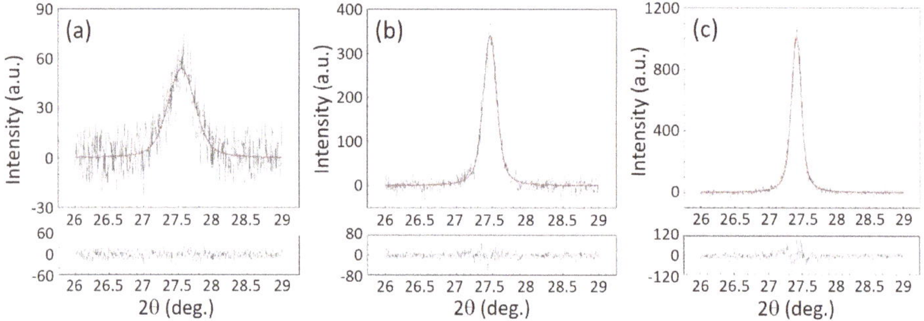

Figure 2. XRD profiles recorded for (111) peak in Bragg–Brentano theta–theta geometry for samples: (**a**) ZnSe2, (**b**) ZnSe3, and (**c**) ZnSe4.

Table 1. Structural parameters of the ZnSe thin films deposited by RF magnetron sputtering at an RF power of: 80 W (ZnSe2), 100 W (ZnSe3), and 120 W (ZnSe4).

Sample	RF Power (W)	D_{ef} (nm)	$\langle \varepsilon^2 \rangle^{1/2}$	a (Å)
ZnSe2	80	51.8	$5.39 \cdot 10^{-3}$	5.603
ZnSe3	100	79.7	$2.31 \cdot 10^{-3}$	5.619
ZnSe4	120	107.4	$2.19 \cdot 10^{-3}$	5.628

3.2. Morphological Characterizations

The surface topography of the fabricated ZnSe samples was analyzed by AFM, in tapping mode. The 2D AFM images of the surface topography of the grown ZnSe thin films are shown in Figure 3, while the evaluated morphological parameters are summarized in Table 2. For all samples, the scanned area was 5×5 μm². As can be noticed from Figure 3, all samples present good uniformity of the ZnSe thin films and a granular surface, with homogenously arranged grains, that slightly changes with RF power. Notably, the overall Z-range reduces with increasing RF power towards 100 W (see Figure 3a–c), then it raises a little at 120 W (see Figure 3d). The latter remark is confirmed by the calculated average roughness (R_A) and root mean square roughness (RMS) parameters (see Tabel 2), which are decreasing, i.e., from 1.9 nm and 2.4 nm, respectively, at 60 W, to 0.8 nm and 1.0 nm, respectively, at 100 W, proving that for this RF power domain the sticking coefficient is increased by the increase of the kinetic energy when the ionized atoms hit the substrate [37,40]. Notably, this observation is also strengthened by the increase of the crystallite size with increasing the RF power (see Table 1), confirming that in given conditions, for a certain range of RF power, the crystallite size raises, while the surface roughness decreases with the sputtering power. This trend is consistent with previous studies, mentioning that at an exceedingly high RF power, the surface roughness could increase due to excessive scattering effects close to the substrate' surface that typically reduce the overall sputtering rate and favor the formation of defects and coarse grains at the surface of the sputtered films [41], effect observed in our studies, too, by an increase of R_A and RMS values, i.e., to 1.3 nm and 1.6 nm, respectively, for the ZnSe film deposited at 120 W (see Table 2). Notably, the ZnSe film grown at 100 W (ZnSe3) features the smallest R_A and RMS values, i.e., 0.8 nm and 1.0 nm, respectively, in respect to the rest of the fabricated samples, eventually demonstrating that the latter sputtering conditions could allow the fabrication of ZnSe thin films with lowest defects density and excellent flatness, suitable to be used as "window" or buffer layers within various solar cells architectures.

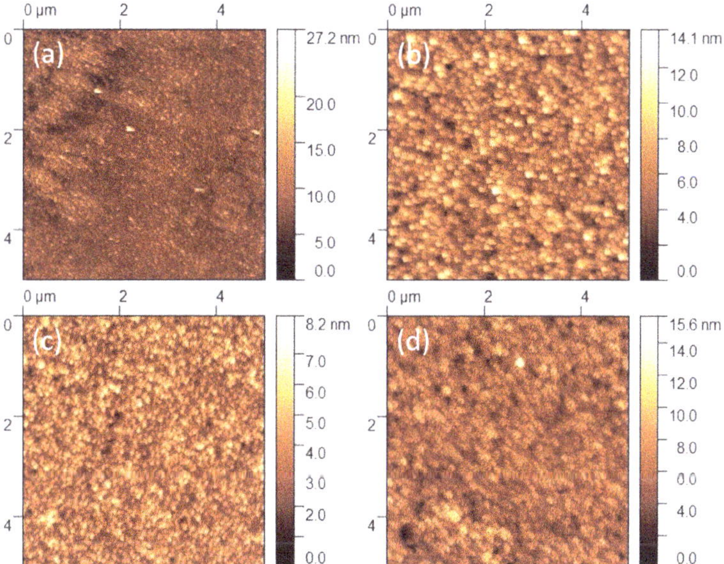

Figure 3. AFM 2D surface topography images of the RF-magnetron sputtered ZnSe thin films, prepared at an RF power of: (**a**) 60 W, (**b**) 80 W, (**c**) 100 W, and (**d**) 120 W.

Table 2. Morphological parameters evaluated by AFM in tapping mode for the ZnSe thin films RF-sputtered at: 60 W, 80 W, 100 W, and 120 W. Corresponding thickness values of the films, as measured by SEM, are indicated as well.

Sample	RF Power (W)	Thickness (nm)	R_A (nm)	RMS (nm)	Ssk	Sku
ZnSe1	60	37	1.9	2.4	0.8	1.4
ZnSe2	80	170	1.5	1.8	0.3	0.2
ZnSe3	100	271	0.8	1.0	0.3	0.1
ZnSe4	120	351	1.3	1.6	0.3	0.5

The smoothness of the prepared ZnSe films has also been interpreted in terms of the determined statistical parameters, i.e., Skewness (Ssk) and Excess Kurtosis (Sku) coefficients (see Table 2), which generally indicate clean and almost flat surfaces of all the sputtered ZnSe films. In particular, an overall positive Skewness with Ssk values close to zero (i.e., 0.3), clearly indicates the presence of a very limited number of height values above the average, although the film grown at 60 W could exhibit a flat surface with tiny protruding features as the slightly larger SSk value of 0.8 suggests. Complementarily, the calculated Sku coefficients statistically suggest a distribution of the heights with a Platykurtic profile of the examined surfaces (Ssk < 3), hence with a small variation of the heights in respect to the horizontal reference plane. As one can easily notice, the sample prepared at 60 W exhibits a higher Sku value (i.e., 1.4), confirming the hypothesis of presenting few extreme protruding heights on the surface. In contrast, the ZnSe film sputtered at 100 W features the lowest Sku value (i.e., 0.1), pointing out again that in this conditions the ZnSe film surface is homogenous and flat without any pit or hillock defects [42,43].

Cross-sectional observations of the ZnSe-coated glass substrates were performed by SEM operating in secondary electron imaging mode. The SEM specimens were carefully prepared by dicing the ZnSe–sputtered glass substrates and partially covering their section with conducting carbon tape to minimize the charging effects, hence to facilitate the SEM analysis. Figure 4 shows cross-sectional SEM micrographs of the ZnSe thin films deposited on glass by magnetron sputtering in an RF plasma engaged in Ar atmosphere kept at 0.86 Pa for 30 min, at different RF powers of 60 W (a), 80 W (b), 100 W (c) and 120 W (d). Several measurements have been acquired for each sample to allow an estimation of the films thickness, as indicated in the images. Average thickness values, as estimated from the SEM analysis, are collected in Table 2. It can be observed that the obtained ZnSe films are compact and sputtered on the glass substrates with good conformality. Experimental data resulted from the SEM analysis are graphically represented in Figure 5, where the black circles represent the SEM-measured thickness values, while on the right red Y-axis, the corresponding calculated sputtering rate is shown. As can be noticed, the ZnSe thickness increases with RF power, while the deposition rate slows-down towards 120 W. The observed behavior in Figure 5 can be obviously attributed to an increase of the kinetic energy and velocity of the particles being sputtered away from the target which results in a higher deposition rate [44]. However, the slowing-down of the increase in the deposition rate observed towards 120 W can be associated with an excessively high energy of the ions injection into the target, leading to an energy and quantity loss of the Ar ions, thus causing further a reduction in the deposition rate, typical behavior also mentioned elsewhere [45].

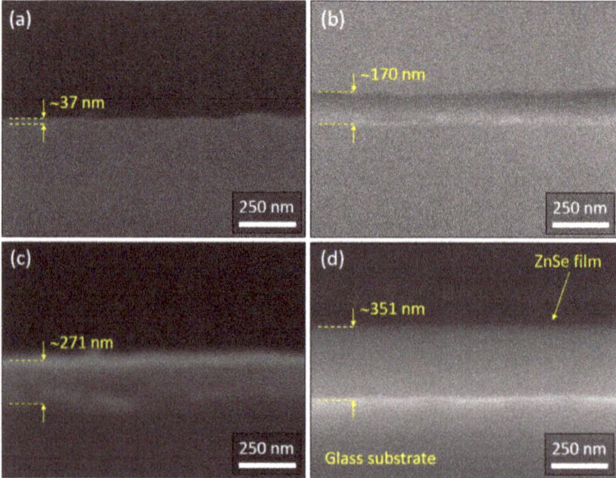

Figure 4. Cross-sectional SEM micrographs of the ZnSe thin films sputtered on glass substrates for 30 min, while keeping a constant gas pressure of 0.86 Pa and applying an RF power of: 60 W (**a**), 80 W (**b**), 100 W (**c**) and 120 W (**d**). Corresponding SEM measurements of the ZnSe films thickness are indicated.

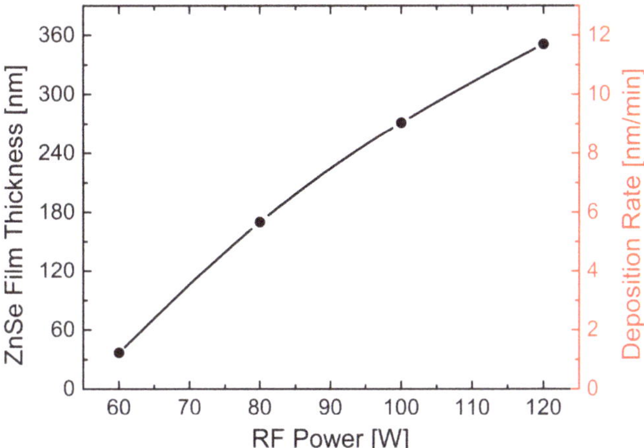

Figure 5. Variation of ZnSe films thickness with RF power, in given sputtering conditions as described in the text. The curve is based on the SEM analysis of the samples. Corresponding calculated sputtering rate is indicated in nm/min on the right red Y-axis.

3.3. Optical Characterizations

Optical properties of ZnSe thin films deposited on glass substrates were investigated by optical transmission and absorption spectroscopy in the spectral range between 200 nm and 1200 nm at room temperature. The optical transmission results are shown in Figure 6. As a reference, the optical transmission of the glass substrate is also presented (the orange curve). Except for the sample ZnSe1, all investigated films show high transmittances in the visible range, with values larger than 80%. The reason behind a reduced optical transmission of the ZnSe film sputtered at 60 W, could be related to its inferior morphological and structural properties, as suggested by both the GIXRD pattern presented in Figure 1a and the AFM parameters collected in Table 2.

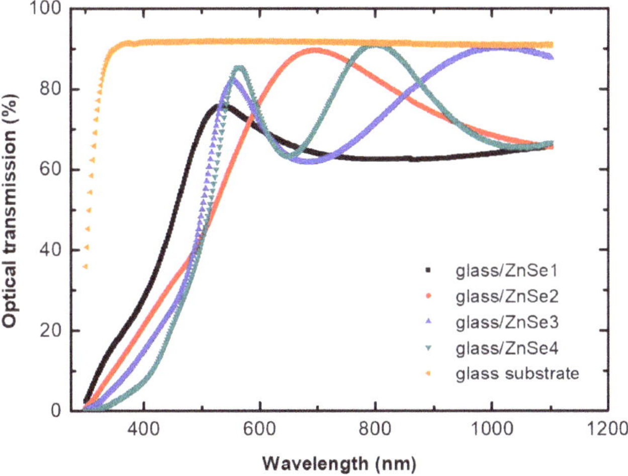

Figure 6. Optical transmission spectra of ZnSe thin films deposited onto optical glass substrates at different RF power. The reference spectrum of the optical glass substrate is depicted in orange.

Figure 7 shows the acquired optical absorbance data of the samples ZnSe1 to ZnSe4. The calculation of the absorption coefficient was performed using the following expression:

$$\alpha = \frac{A}{d} \tag{1}$$

where A is the optical absorbance and d is the thickness of the films. The thickness values that were considered for the computation of the optical band gap of the prepared RF-sputtered ZnSe films were those determined by cross-sectional SEM analysis (see Table 2). Later on, α was used to estimate the optical band gap energy using the Tauc's plot method following the well-known relationship:

$$\alpha = A\frac{\left(\hbar\omega - E_g\right)^{1/2}}{\hbar\omega} \tag{2}$$

Equation (2) gives the dependence of the absorption coefficient on incident photons energies in the case of direct band gap semiconductors (like ZnSe) near the fundamental absorption edge, in which α is the absorption coefficient, A is a constant, $\hbar\omega$ is the energy of incident photons and E_g is the optical bandgap corresponding to Γ point in the first Brillouin zone. The insets of Figure 7a–d present the corresponding Tauc's plots, where the values for the optical band gap energies can be determined by the interceptions over the energy axis of the extrapolated linear portion of the plots. The values of the band gap energies of ZnSe films ($E_{g(OS)}$) as obtained by fitting of the experimental data using Equation (2) are collected in Table 3. The obtained results are similar to those from literature for ZnSe thin films deposited by vacuum thermal evaporation [46].

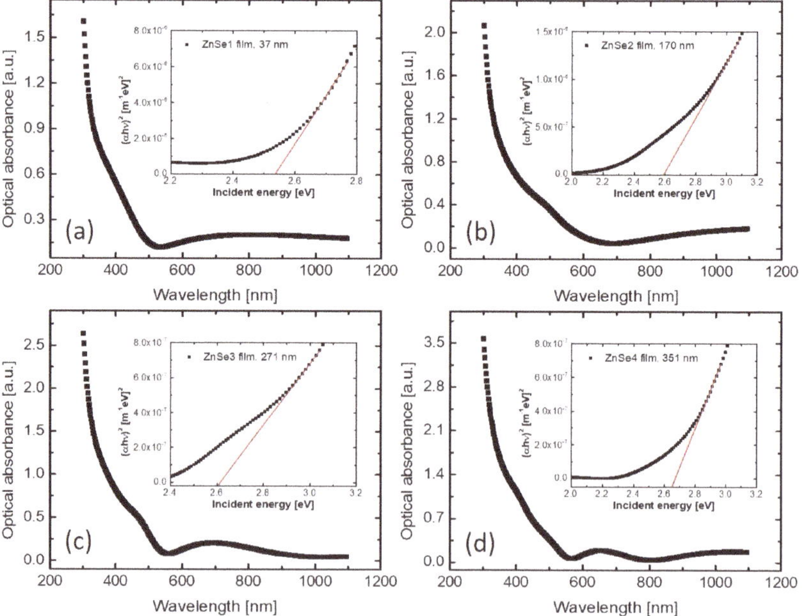

Figure 7. Spectral dependencies of optical absorbance for samples ZnSe1 (**a**), ZnSe2 (**b**), ZnSe3 (**c**), and ZnSe4 (**d**). The inset of graphs shows the $(\alpha\hbar\omega)^2$ vs. $\hbar\omega$ dependencies (Tauc's plots), used to determine the optical band gap energies ($E_{g(OS)}$) of the films.

Table 3. Calculated band gap ($E_{g_{(OS)}}$) values (from OS analysis) of the ZnSe thin films sputtered at an RF power of: 60 W, 80 W, 100 W and 120 W. Corresponding thickness values of the films, as measured by SEM, are indicated as well.

Sample	RF Power (W)	Thickness (nm)	$E_{g_{(OS)}}$ (eV)
ZnSe1	60	37	2.54
ZnSe2	80	170	2.59
ZnSe3	100	271	2.60
ZnSe4	120	351	2.65

Optical constants (refractive index and extinction coefficient) of ZnSe thin films deposited on optical glass substrates were determined using SE. The ellipsometric (Ψ, Δ) spectra were recorded in reflected light at an incident angle of 75° with respect to film surfaces. An optical three layers model (upper rough layer/ZnSe layer/glass substrate) was used while the dispersion model chosen for fitting the ellipsometric spectra was Adachi–New Forouhi (ANF) model [47]. After choosing the optical model and the dispersion function, the fitting parameters were varied by least square regression until a minimum difference between experimental and computed (Ψ, Δ) spectra was obtained. As a risk function, the mean-squared error (MSE) was used and the Levenberg–Marquardt regression algorithm was used in order to minimize MSE function [48]. Spectra of refractive indices (n) and extinction coefficients (k) of the films are plotted in Figure 8, whilst the results obtained by SE are displayed in Table 4. One can observe that the refractive index (Figure 8a) tends to increase with increasing the thickness of the films. The decrease of the refractive index with the increase of photon wavelength in the visible region (VIS) of the electromagnetic spectrum indicates the normal dispersion behavior of ZnSe thin films in this region. In near-infrared (NIR) region the refractive index tends to be relatively constant, the obtained values being in good agreement with previous works [49,50]. Due to the fact that ZnSe films are almost transparent in NIR region, extinction coefficients were displayed in Figure 8b only for VIS region, their values increasing with photon energy. Increasing the thickness of ZnSe films by increasing the RF sputtering power leads to a decrease in extinction coefficients. The values of the extinction coefficients obtained from the SE. data analysis, were then used to compute the absorption coefficients, using the commonly known equation:

$$\alpha = \frac{4\pi k}{\lambda} \tag{3}$$

where α is the absorption coefficient, k is the extinction coefficient and λ represents the wavelength.

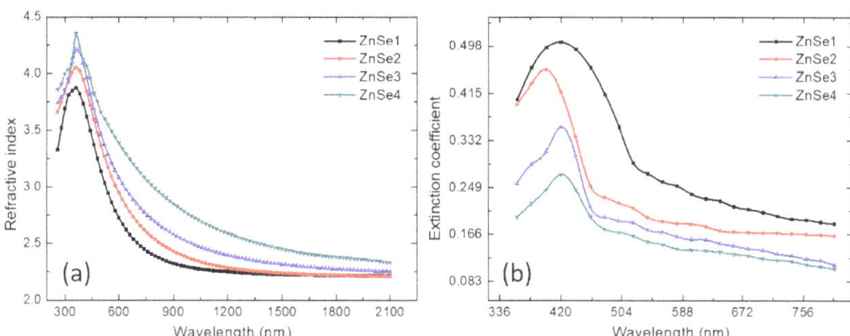

Figure 8. Spectral dependencies of refractive indices (**a**) and extinction coefficients (**b**) of the ZnSe thin films with different thicknesses, prepared by RF magnetron sputtering while varying the RF power.

Further, knowing the absorption coefficients, Tauc's method was similarly used to compute the energy band gap values ($E_{g_{(SE)}}$) using Equation (2). Notably, the values of band gap energies, showing a shift from the standard bulk band gap value of 2.7 eV (Table 4), match very well the values calculated from OS analysis of the ZnSe thin films (Table 3). The observed tendency of the band gap increase with the raise of the RF power can be explained by the domain of nanocrystallite sizes, ranging from 51.8 nm to 107.4 nm when the RF power is increased from 80 to 120 W, respectively (see Table 1), together with a reduction of the number of inter-grain boundaries and an increase of the thin film thickness. Taking into account these changes induced by increasing the RF power, the density of defects, leaving-out energy levels in the semiconductor's band gap and thus being involved in the value of the optical band gap, reduces within whole grains and especially at their central parts. Hence, the optical band gap tends to increase towards the corresponding value of the bulk single crystal semiconductor, here 2.7 eV [51]. As can be seen in Table 4, relatively small values for MSE function were obtained for all investigated ZnSe samples indicating that the fitting procedure of ellipsometric spectra was performed accurately.

Table 4. Refractive indices (n) and extinction coefficients (k), as obtained from the SE analysis. Corresponding band gap ($E_{g_{(SE)}}$) values as calculated form the SE data and the SEM-measured thicknesses are indicated as well.

Sample	RF Power (W)	Thickness (nm)	n (at $\lambda = 600$ nm)	k (at $\lambda = 600$ nm)	$E_{g_{(SE)}}$ (eV)	MSE
ZnSe1	60	37	2.723	0.237	2.56	6.02
ZnSe2	80	170	2.946	0.184	2.59	5.85
ZnSe3	100	271	3.083	0.157	2.61	6.81
ZnSe4	120	351	3.386	0.139	2.64	4.66

3.4. Electrical Characterizations

Ultimately, Au/ZnSe/Au sandwich structures were fabricated via RF magnetron sputtering of ZnSe at 100 W, by using Au electrodes in order to diminish the noise from contacts. The current–voltage (I-V) dark characteristic at room temperature (300 K) is presented in Figure 9a, showing a nonlinear but symmetrical behaviour, as expected when using the two electrodes made of the same metal. Figure 9b depicts the characteristics in logarithmic scale of two linear fits having different slopes, suggesting a transition from one conduction mechanism to another. At low injection levels (low applied voltage-LV) the slope of the first linear fit is around 1 (slope = 1.12) leading to the assumption that the device enters an ohmic conduction regime described by the following equation [52,53]:

$$J_{ohmic} = q n_0 \mu \frac{U}{d} \tag{4}$$

where q is the electronic charge, n_0 is the concentration of thermally generated free electrons in the conduction band at thermal equilibrium, μ is the electron mobility, U is the applied voltage and d is the thickness of ZnSe film (considered 271 nm for the sample sputtered at 100 W). In the LV regime, the transport through the structure is thus realized by the thermal equilibrium charge carriers of n_0 concentration.

With an evaluated value of 0.6 V for the transition voltage ($U_{transition}$), the aforementioned structure has at high injection levels (HV) a slope of 3.83 (Figure 9b), suggesting, as expected for high resistivity materials [54,55], the existence of a conduction mechanism of space charge limited currents (SCLC). The analytical relationship describing the I-V characteristics in the HV regime over $U_{transition}$, is frequently a power function but not always, because it generally depends on the presence of the defect levels in the band gap of the semiconductor, hence changing as a function of defects concentration and their energy distribution [56]. Taking into account the slope value (3.83), in the case of the prepared ZnSe thin films, there is a SCLC in the presence of an exponential trap distribution described by:

$$\rho(E) = \frac{N_t}{k_B T_C} e^{\frac{-E}{k_B T_C}} \tag{5}$$

where $\rho(E)$ is the trap density per unit energy range at an E level of energy below the conduction band, N_t is the total density of trapping levels in the exponential distribution, k_B is the Boltzmann constant and T_C is the characteristic temperature which is a typical parameter for the exponential trap distribution showing how fast the variation of the density of traps is, with the changing of energy measured from the minimum of conduction band towards the middle of the band gap (in the case of traps for electrons). Based on the above-mentioned distribution, the I-V characteristics in the HV regime is described by a power function as it follows [53]:

$$J_{SCLC_{exp}} = q\mu N_C \left(\frac{\varepsilon}{qN_t}\right)^{\gamma} \frac{U^{\gamma+1}}{d^{2\gamma+1}} \quad , \quad \gamma = \frac{T_C}{T} \tag{6}$$

where N_C is the effective density of states in the conduction band, ε is the dielectric constant of ZnSe, γ is the ratio between T_C (characteristic temperature) and T (absolute temperature, i.e., 300 K in this case). As one may see in Figure 9b, the linear fit, i.e., $\log_{10} J_{SCLC_{exp}} = f(\log_{10} U)$, in the HV region has a slope of 3.83 ($\gamma + 1$), therefore the γ coefficient is 2.83 and the characteristic temperature T_C is 849 K. Coupling thus the analytical relationship of the SCLC for an exponential trap distribution in the band gap of ZnSe, i.e., Equation (6), with the obtained experimental results, the above-mentioned parameters have been obtained through fitting the experimental data with the typical SCLC I-V characteristic of an exponential trap distribution. Such a methodological approach was already engaged in many other reports on similar materials [57,58], including our previous studies [16,32,35,42,52].

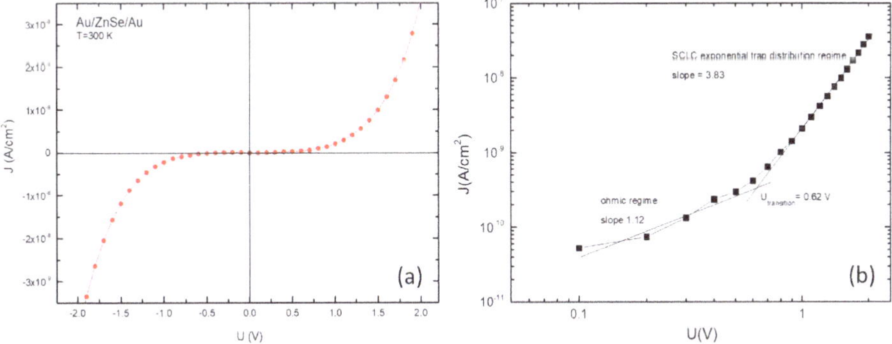

Figure 9. (**a**) Current–voltage (I-V) dark ambipolar characteristics recorded at room temperature for Au/ZnSe/Au sandwich structure, with ZnSe thin film sputtered at 100 W. (**b**) Corresponding I-V characteristics of Au/ZnSe/Au at room temperature, plotted in a double-logarithmic scale.

4. Conclusions

ZnSe thin films with different thicknesses were deposited onto optical glass substrates through RF magnetron sputtering, by varying the deposition power from 60 W to 120 W. The XRD structural characterization emphasizes the polycrystallinity of these thin films with a pronounced (111) texture, showing as well, an increase of the crystallite size with increasing the RF plasma power. Furthermore, the AFM superficial morphology analysis of the films indicates that while raising the RF power, R_A and RMS parameters decrease, the ZnSe layer prepared at 100 W exhibiting an almost flat surface with the smallest values of roughness parameters, i.e., 0.8 nm and 1 nm, respectively. Additional cross-sectional SEM analysis allowed then a direct estimation of the ZnSe films thickness, showing as expected, an increase of the thickness by increasing the incident power, from 37 nm to 351 nm. In a second stage, OS (absorption and transmission) and SE techniques were employed to evaluate the band gap energies and to extract the optical constants of the ZnSe thin films (refractive indices and extinction coefficients). Notably, the band gaps estimated by both optical methods were similar, the values spanning between 2.54 eV and 2.65 eV, and showing a visible tendency of band gap increase with raising the RF power. Finally, the electrical measurements highlighted that the ZnSe thin films have high resistivity and at high-injection levels the conduction mechanism relies on a SCLC with an exponential trap distribution. The results obtained from all our investigations pointed out that in given sputtering conditions, at an RF plasma power of 100 W, the obtained ZnSe thin films exhibit superior properties, in terms of: better crystallinity, appropriate thickness maintaining a high transparency, low R_A and RMS values, an adequate optical band gap of 2.61 eV (slightly larger value compared to the one of the conventional CdS layer), and ultimately good electrical properties. Consequently, these parameters are optimal to prepare excellent ZnSe thin films, perfectly suitable to be used (i) as environmentally-friendly "window" materials for solar cells to reduce the amount of Cd, commonly used for the second generation of solar cells relying on CdTe as main absorber, or alternatively, (ii) as buffer layers in solar cell architectures completely free of Cd.

Author Contributions: Conceptualization, Ş.A.; methodology, Ş.A.; software, A.R. and L.I.; validation, Ş.A.; formal analysis, O.T., V.-A.A., A.-M.P., S.I. and A.-M.R.; investigation, O.T., V.-A.A., A.-M.P., S.I. and A.-M.R.; resources, V.-A.A., S.I. and Ş.A.; writing–original draft preparation, O.T. and V.-A.A.; writing–review and editing, O.T., V.-A.A., S.I. and Ş.A.; supervision, Ş.A.; project administration, V.-A.A., S.I. and Ş.A.; funding acquisition, V.-A.A., S.I. and Ş.A. All authors have read and agreed to the published version of the manuscript.

Funding: This research was funded by the "Executive Unit for Financing Higher Education, Research, Development and Innovation" (UEFISCDI, Romania), through the grants: 115/2020 (PN-III-P1-1.1-TE-2019-0868) and 25/2020 (PN-III-P1-1.1-TE-2019-0846).

Institutional Review Board Statement: Not applicable.

Informed Consent Statement: Not applicable.

Data Availability Statement: Not applicable.

Conflicts of Interest: The authors declare no conflict of interest. The funders had no role in the design of the study; in the collection, analyses, or interpretation of data; in the writing of the manuscript, or in the decision to publish the results.

Abbreviations

The following abbreviations are used in this manuscript:

R&D	Research & Development
MDEO	R&D Center for Materials and Electronic & Optoelectronic Devices
IMCN	Institute of Condensed Matter and Nanosciences
UCLouvain	Université catholique de Louvain
RF	Radio frequency
(GI)XRD	(Grazing incidence) X-ray diffraction
AFM	Atomic force microscope
SEM	Scanning Electron Microscope
OS/SE	Optical spectroscopy/Spectroscopic ellipsometry
PME	Phase modulated spectro-ellipsometer
2D	Two-dimensional
RMS	Root mean square roughness
Ssk/Sku	Skewness/Kurtosis
MSE	Mean-squared error

<table>
<tr><td>VIS</td><td>Visible spectral region</td></tr>
<tr><td>NIR</td><td>Near-infrared spectral region</td></tr>
<tr><td>SCLC</td><td>Space charge limited currents</td></tr>
<tr><td>UEFISCDI</td><td>Executive Unit for Financing Higher Education, Research, Development and Innovation</td></tr>
</table>

References

1. Hu, R.; Xi, W.; Liu, Y.; Tang, K.; Song, J.; Luo, X.; Wu, J.; Qiu, C.W. Thermal camouflaging metamaterials. *Mater. Today* **2021**, *45*, 120–141. [CrossRef]
2. Xi, W.; Liu, Y.; Song, J.; Hu, R.; Luo, X. High-throughput screening of a high-Q mid-infrared Tamm emitter by material informatics. *Opt. Lett.* **2021**, *46*, 888–891. [CrossRef]
3. Hu, R.; Wang, Y.; Zou, Y.; Chen, X.; Liu, S.; Luo, X. Study on phosphor sedimentation effect in white light-emitting diode packages by modeling multi-layer phosphors with the modified Kubelka-Munk theory. *J. Appl. Phys.* **2013**, *113*, 063108. [CrossRef]
4. Xie, B.; Wang, Y.; Liu, H.; Ma, J.; Zhou, S.; Yu, X.; Lan, W.; Wang, K.; Hu, R.; Luo, X. Targeting cooling for quantum dots by 57.3 °C with air-bubbles-assembled three-dimensional hexagonal boron nitride heat dissipation networks. *Chem. Eng. J.* **2022**, *427*, 130958. [CrossRef]
5. Hu, R.; Song, J.; Liu, Y.; Xi, W.; Zhao, Y.; Yu, X.; Cheng, Q.; Tao, G.; Luo, X. Machine learning-optimized Tamm emitter for high-performance thermophotovoltaic system with detailed balance analysis. *Nano Energy* **2020**, *72*, 104687. [CrossRef]
6. Saha, S.; Johnson, M.; Altayaran, F.; Wang, Y.; Wang, D.; Zhang, Q. Electrodeposition Fabrication of Chalcogenide Thin Films for Photovoltaic Applications. *Electrochem* **2020**, *1*, 286–321. [CrossRef]
7. Ou, K.; Wang, S.; Wan, G.; Huang, M.; Zhang, Y.; Bai, L.; Yi, L. A study of structural, morphological and optical properties of nanostructured ZnSe/ZnS multilayer thin films. *J. Alloys Compd.* **2017**, *726*, 707–711. [CrossRef]
8. Godlewski, M.; Guziewicz, E.; Kopalko, K.; Łusakowska, E.; Dynowska, E.; Godlewski, M.M.; Goldys, E.M.; Phillips, M.R. Origin of white color light emission in ALE-grown ZnSe. *J. Lumin.* **2003**, *102–103*, 455–459.
9. Mittal, V.; Sessions, N.P.; Wilkinson, J.S.; Murugan, G.S. Optical quality ZnSe films and low loss waveguides on Si substrates for mid-infrared applications. *Opt. Mater. Express* **2017**, *7*, 712–725. [CrossRef]
10. Vivet, N.; Morales, M.; Levalois, M.; Charvet, S.; Jomard, F. Optimization of the structural, microstructural and optical properties of nanostructured Cr^{2+}:ZnSe films deposited by magnetron co-sputtering for mid-infrared applications. *Thin Solid Film.* **2010**, *519*, 106–110. [CrossRef]
11. Venkatachalam, S.; Agilan, S.; Mangalaraj, D.; Narayandass, S.K. Optoelectronic properties of ZnSe thin films. *Mater. Sci. Semicond. Process.* **2007**, *10*, 128–132. [CrossRef]
12. Lin, T.K.; Chang, S.J.; Su, Y.K.; Chiou, Y.Z.; Wang, C.K.; Chang, S.P.; Chang, C.M.; Tang, J.J.; Huang, B.R. ZnSe MSM photodetectors prepared on GaAs and ZnSe substrates. *Mater. Sci. Eng. B* **2005**, *119*, 202–205. [CrossRef]
13. Feng, G.; Yang, C.; Zhou, S. Nanocrystalline Cr^{2+}-doped ZnSe Nanowires Laser. *Nano Lett.* **2013**, *13*, 272–275. [CrossRef] [PubMed]
14. Holzman, J.F.; Vermeulen, F.E.; Irvine, S.E.; Elezzabi, A.Y. Free-space detection of terahertz radiation using crystalline and polycrystalline ZnSe electro-optic sensors. *Appl. Phys. Lett.* **2002**, *81*, 2294–2296. [CrossRef]
15. Elsaeedy, H.I.; Hassan, A.A.; Yakout, H.A.; Qasem, A. The significant role of ZnSe layer thickness in optimizing the performance of ZnSe/CdTe solar cell for optoelectronic applications. *Opt. Laser Technol.* **2021**, *141*, 107139. [CrossRef]
16. Antohe, S.; Ion, L.; Girtan, M.; Toma, O. Optical and morphological studies of thermally vacuum evaporated ZnSe thin films. *Rom. Rep. Phys.* **2013**, *65*, 805–811.
17. Khurram, A.A.; Jabar, F.; Mumtaz, M.; Khan, N.A.; Mehmood, M.N. Effect of light, medium and heavy ion irradiations on the structural and electrical properties of ZnSe thin films. *Nucl. Instrum. Methods Phys. Res. Sect. B Beam Interact. Mater. Atoms* **2013**, *313*, 40–44. [CrossRef]
18. Bacaksiz, E.; Aksu, S.; Polat, I.; Yılmaz, S.; Altunbaş, M. The influence of substrate temperature on the morphology, optical and electrical properties of thermal-evaporated ZnSe thin films. *J. Alloys Compd.* **2009**, *487*, 280–285. [CrossRef]
19. Lokhande, C.D.; Patil, P.S.; Tributsch, H.; Ennaoui, A. ZnSe thin films by chemical bath deposition method. *Sol. Energy Mater. Sol. Cells* **1998**, *55*, 379–393. [CrossRef]
20. Kale, R.B.; Lokhande, C.D.; Mane, R.S.; Han, S.H. Use of modified chemical route for ZnSe nanocrystalline thin films growth: Study on surface morphology and physical properties. *Appl. Surf. Sci.* **2006**, *252*, 5768–5775. [CrossRef]
21. Wei, A.; Zhao, X.; Liu, J.; Zhao, Y. Investigation on the structure and optical properties of chemically deposited ZnSe nanocrystalline thin films. *Phys. B Condens. Matter* **2013**, *410*, 120–125. [CrossRef]
22. Rumberg, A.; Sommerhalter, C.; Toplak, M.; Jäger-Waldau, A.; Lux-Steiner, M.C. ZnSe thin films grown by chemical vapour deposition for application as buffer layer in CIGSS solar cells. *Thin Solid Films* **2000**, *361–362*, 172–176. [CrossRef]
23. Kumar, V.; Khan, K.L.A.; Singh, G.; Sharma, T.P.; Hussain, M. ZnSe sintered films: Growth and characterization. *Appl. Surf. Sci.* **2007**, *253*, 3543–3546. [CrossRef]
24. Arslan, M.; Maqsood, A.; Mahmood, A.; Iqbal, A. Structural and optical properties of copper enriched ZnSe thin films prepared by closed space sublimation technique. *Mater. Sci. Semicond. Process.* **2013**, *16*, 1797–1803. [CrossRef]

25. Riveros, G.; Gómez, H.; Henríquez, R.; Schrebler, R.; Marotti, R.E.; Dalchiele, E.A. Electrodeposition and characterization of ZnSe semiconductor thin films. *Sol. Energy Mater. Sol. Cells* **2001**, *70*, 255–268. CANCUN 2000. [CrossRef]
26. Kim, T.W.; Jung, M.; Lee, D.U.; Oh, E.; Lee, S.D.; Jung, H.D.; Kim, M.D.; Kim, J.R.; Park, H.S.; Lee, J.Y. Structural and optical properties of undoped and doped ZnSe/GaAs strained heterostructures. *Thin Solid Film.* **1997**, *298*, 187–190. [CrossRef]
27. Williams, J.E.; Camata, R.P.; Fedorov, V.V.; Mirov, S.B. Pulsed laser deposition of chromium-doped zinc selenide thin films for mid-infrared applications. *Appl. Phys. A* **2008**, *91*, 333–335. [CrossRef]
28. Rizzo, A.; Tagliente, M.A.; Caneve, L.; Scaglione, S. The influence of the momentum transfer on the structural and optical properties of ZnSe thin films prepared by r.f. magnetron sputtering. *Thin Solid Film.* **2000**, *368*, 8–14. [CrossRef]
29. Ion, L.; Iftimie, S.; Radu, A.; Antohe, V.A.; Toma, O.; Antohe, S. Physical properties of RF-sputtered ZnSe thin films for photovoltaic applications: Influence of film thickness. *Porc. Rom. Acad. Ser. A* **2021**, *22*, 25–34.
30. Gashin, P.; Focsha, A.; Potlog, T.; Simashkevich, A.V.; Leondar, V.V. n-ZnSe/p-ZnTe/n-CdSe tandem solar cells. *Sol. Energy Mater. Sol. Cells* **1997**, *46*, 323–331. [CrossRef]
31. Eisele, W.; Ennaoui, A.; Schubert-Bischoff, P.; Giersig, M.; Pettenkofer, C.; Krauser, J.; Lux-Steiner, M.; Zweigart, S.; Karg, F. XPS, TEM and NRA investigations of Zn(Se,OH)/Zn(OH)$_2$ films on Cu(In,Ga)(S,Se)$_2$ substrates for highly efficient solar cells. *Sol. Energy Mater. Sol. Cells* **2003**, *75*, 17–26. [CrossRef]
32. Toma, O.; Ion, L.; Iftimie, S.; Antohe, V.A.; Radu, A.; Raduta, A.M.; Manica, D.; Antohe, S. Physical properties of rf-sputtered ZnS and ZnSe thin films used for double-heterojunction ZnS/ZnSe/CdTe photovoltaic structures. *Appl. Surf. Sci.* **2019**, *478*, 831–839. [CrossRef]
33. Saikia, P.; Saikia, D.P.K.; Saikia, D. Fabrication and characterization of ZnSe/ZnTe/CdTe/HgTe multijunction solar cell. *Optoelectron. Adv. Mater. Rapid Commun.* **2011**, *5*, 204–207.
34. Prabhu, M.; Kamalakkannan, K.; Soundararajan, N.; Ramachandran, K. Fabrication and characterization of ZnSe thin films based low-cost dye sensitized solar cells. *J. Mater. Sci. Mater. Electron.* **2015**, *26*, 3963–3969. [CrossRef]
35. Toma, O.; Ion, L.; Iftimie, S.; A.Radu.; Antohe, S. Structural, morphological and optical properties of rf-sputtered CdS thin films. *Mater. Des.* **2016**, *100*, 198–203. [CrossRef]
36. Chan, K.Y.; Teo, B.S. Investigation into the influence of direct current DC power in the magnetron sputtering process on the copper crystallite size. *Microelectron. J.* **2007**, *38*, 60–62. [CrossRef]
37. Ghorannevis, Z.; Akbarnejad, E.; Ghoranneviss, M. Effects of various deposition times and RF powers on CdTe thin film growth using magnetron sputtering. *J. Theor. Appl. Phys.* **2016**, *10*, 225–231. [CrossRef]
38. Langford, J.I.; Delhez, R.; de Keijser, T.H.; Mittemeijer, E.J. Profile Analysis for Microcrystalline Properties by the Fourier and Other Methods. *Aust. J. Phys.* **1988**, *41*, 173–188. [CrossRef]
39. Venkatachalam, S.; Mangalaraj, D.; Narayandass, S.K. Influence of substrate temperature on the structural, optical and electrical properties of zinc selenide ZnSe thin films. *J. Phys. D Appl. Phys.* **2006**, *39*, 4777–4782. [CrossRef]
40. Duta, M.; Anastasescu, M.; Calderon-Moreno, J.M.; Predoana, L.; Preda, S.; Nicolescu, M.; Stroescu, H.; Bratan, V.; Dascalu, I.; Aperathitis, E.; et al. Sol-gel versus sputtering indium tin oxide films as transparent conducting oxide materials. *J. Mater. Sci. Mater. Electron.* **1988**, *27*, 4913–4922. [CrossRef]
41. Zhou, C.; Li, T.; Wei, X.; Yan, B. Effect of the Sputtering Power on the Structure, Morphology and Magnetic Properties of Fe Films. *Metals* **2020**, *10*, 896. [CrossRef]
42. Toma, O.; Ion, L.; Girtan, M.; Antohe, S. Optical, morphological and electrical studies of thermally vacuum evaporated CdTe thin films for photovoltaic applications. *Sol. Energy* **2014**, *108*, 51–60. [CrossRef]
43. Eaton, P.; West, P. *Atomic Force Microscopy*; Oxford University Press: Oxford, UK, 2010; 256p.
44. Andújar, J.L.; Bertran, E.; Canillas, A.; Roch, C.; Morenza, J.L. Influence of pressure and radio frequency power on deposition rate and structural properties of hydrogenated amorphous silicon thin films prepared by plasma deposition. *J. Vac. Sci. Technol. A* **1991**, *9*, 2216–2221. [CrossRef]
45. Wu, S.; Chen, H.; Du, X.; Liu, Z. Effect of deposition power and pressure on rate deposition and resistivity of titanium thin films grown by DC magnetron sputtering. *Spectrosc. Lett.* **2016**, *49*, 514–519. [CrossRef]
46. Yadav, B.K.; Singh, P.; Pandey, D.K. Synthesis and Non-Destructive Characterization of Zinc Selenide Thin Films. *Z. Nat. A* **2019**, *74*, 993–999. [CrossRef]
47. Yoshikawa, H.; Adachi, S. Optical Constants of ZnO. *Jpn. J. Appl. Phys.* **1997**, *36*, 6237–6243. [CrossRef]
48. Erman, M.; Theeten, J.B.; Frijlink, P.; Gaillard, S.; Hia, F.J.; Alibert, C. Electronic states and thicknesses of GaAs/GaAlAs quantum wells as measured by electroreflectance and spectroscopic ellipsometry. *J. Appl. Phys.* **1984**, *56*, 3241–3249. [CrossRef]
49. Ashraf, M.; Akhtar, S.M.J.; Khan, A.F.; Ali, Z.; Qayyum, A. Effect of annealing on structural and optoelectronic properties of nanostructured ZnSe thin films. *J. Alloys Compd.* **2011**, *509*, 2414–2419. [CrossRef]
50. Subbaiah, Y.P.V.; Prathap, P.; Devika, M.; Reddy, K.T.R. Close-spaced evaporated ZnSe films: Preparation and characterization. *Phys. B Condens. Matter* **2005**, *365*, 240–246. [CrossRef]
51. Biswas, S.K.; Chaudhuri, S.; Choudhury, A. Effects of heat treatment on the optical and structural properties of InSe thin films. *Phys. Status Solidi (a)* **1988**, *105*, 467–475. [CrossRef]
52. Antohe, S. Electrical and Photovoltaic Properties of Tetrapyrydilporphyrin Sandwich Cells. *Phys. Status Solidi (a)* **1993**, *136*, 401–410. [CrossRef]

53. Ruxandra, V.; Antohe, S. The effect of the electron irradiation on the electrical properties of thin polycrystalline CdS layers. *J. Appl. Phys.* **1998**, *84*, 727–733. [CrossRef]
54. Iftimie, S.; Majkic, A.; Besleaga, C.; Antohe, V.A.; Radu, A.; Radu, M.; Arghir, I.; Florica, C.; Ion, L.; Bratina, G.; Antohe, S. Study of electrical and optical properties of ITO/PEDOT/P3HT:PCBM(1:1)/LiF/Al photovoltaic structures. *J. Optoelectron. Adv. Mater.* **2020**, *12*, 2171–2175.
55. Antohe, S.; Iftimie, S.; Hrostea, L.; Antohe, V.A.; Girtan, M. A critical review of photovoltaic cells based on organic monomeric and polymeric thin film heterojunctions. *Thin Solid Film.* **2017**, *642*, 219–231. [CrossRef]
56. Lampert, M.A.; Mark, P. *Current Injection in Solids*; Electrical Science, Academic Press: New York, NY, USA, 1970; 351p, ISBN 978-0-12-435350-3.
57. Rao, G.; Bangera, K.; Shivakumar, G.K. Studies on vacuum deposited p-ZnTe/n-ZnSe heterojunction diodes. *Solid-State Electron.* **2010**, *54*, 787–790. [CrossRef]
58. Seyam, M.A.M.; El-Shair, H.T.; Salem, G.F. Electrical properties and transport mechanisms of p-ZnTe/n-Si heterojunctions. *Eur. Phys. J. Appl. Phys.* **2008**, *41*, 221–227. [CrossRef]

 *nanomaterials*

MDPI

Article

Compact SnO$_2$/Mesoporous TiO$_2$ Bilayer Electron Transport Layer for Perovskite Solar Cells Fabricated at Low Process Temperature

Junyeong Lee [1], Jongbok Kim [2], Chang-Su Kim [3] and Sungjin Jo [1,*]

[1] School of Energy Engineering, Kyungpook National University, Daegu 41566, Korea; junyeong112@knu.ac.kr
[2] Department of Materials Science and Engineering, Kumoh National Institute of Technology, Gumi 39177, Korea; jbkim@kumoh.ac.kr
[3] Department of Advanced Functional Thin Films, Surface Technology Division, Korea Institute of Materials Science, Changwon 51508, Korea; cskim1025@kims.re.kr
[*] Correspondence: sungjin@knu.ac.kr

Abstract: Charge transport layers have been found to be crucial for high-performance perovskite solar cells (PSCs). SnO$_2$ has been extensively investigated as an alternative material for the traditional TiO$_2$ electron transport layer (ETL). The challenges facing the successful application of SnO$_2$ ETLs are degradation during the high-temperature process and voltage loss due to the lower conduction band. To achieve highly efficient PSCs using a SnO$_2$ ETL, low-temperature-processed mesoporous TiO$_2$ (LT m-TiO$_2$) was combined with compact SnO$_2$ to construct a bilayer ETL. The use of LT m-TiO$_2$ can prevent the degradation of SnO$_2$ as well as enlarge the interfacial contacts between the light-absorbing layer and the ETL. SnO$_2$/TiO$_2$ bilayer-based PSCs showed much higher power conversion efficiency than single SnO$_2$ ETL-based PSCs.

Keywords: compact SnO$_2$; mesoporous TiO$_2$; oxygen plasma; perovskite solar cell low process temperature

Citation: Lee, J.; Kim, J.; Kim, C.-S.; Jo, S. Compact SnO$_2$/Mesoporous TiO$_2$ Bilayer Electron Transport Layer for Perovskite Solar Cells Fabricated at Low Process Temperature. *Nanomaterials* **2022**, *12*, 718. https://doi.org/10.3390/nano12040718

Academic Editor: Vlad Andrei Antohe

Received: 19 January 2022
Accepted: 16 February 2022
Published: 21 February 2022

Publisher's Note: MDPI stays neutral with regard to jurisdictional claims in published maps and institutional affiliations.

1. Introduction

Perovskite solar cells (PSCs) have received attention because their power conversion efficiency (PCE) has rapidly increased by over 25% [1,2]. Many researchers have tried to enhance the performance of PSCs and translate them from the laboratory to commercial products [3–6].

For the state-of-the-art device configuration, PSCs usually consist of a transparent electrode, an electron transport layer (ETL), a light-absorbing layer, a hole transport layer (HTL), and a metal electrode [7]. In the pursuit of high-performance PSCs, the ETL has become the subject of high interest and one of the most challenging scientific issues [8]. As a conventional ETL material, TiO$_2$ has been widely adopted. However, many attempts have been made to substitute TiO$_2$ with alternative materials that have better optoelectronic properties [9]. SnO$_2$ is the most investigated ETL after TiO$_2$ due to its high electron mobility, high conductivity, wide optical bandgap, and excellent chemical stability [10]. Although SnO$_2$ ETL-based PSCs have made rapid progress recently, their performance is still lower than that of PSCs using mesoporous TiO$_2$ (m-TiO$_2$) as an ETL [11]. In addition, the low conduction band of SnO$_2$ reduces the built-in potential of the Schottky barrier between the perovskite and SnO$_2$, resulting in the voltage loss of the PSCs [12].

To achieve highly efficient PSCs using a SnO$_2$ ETL, SnO$_2$ ETL combination and surface modification techniques that can improve electron injection and suppress electron recombination have been developed [13]. Various inorganic metal oxides, such as ZnO [14], MgO [15], and TiO$_2$ [16,17], as well as organics, including carbon-based materials [18], self-assembled monolayers (SAM) [19], and polymers [20], have been adopted in SnO$_2$

ETL-based PSCs to combine with or modify SnO_2. Among them, the conventional m-TiO_2 layer is the preferable candidate to be combined with the compact SnO_2 (c-SnO_2) layer because the mesoporous scaffold can facilitate sufficient pore filling of the light-absorbing layer and improve electron extraction and transport over a single c-SnO_2 layer [17,21]. Moreover, TiO_2 is a better choice in view of its established cascaded energy-level alignment between the electrode and light-absorbing layer, which results in a significantly improved performance. However, m-TiO_2 generally requires a high-temperature sintering process of up to 450 °C to remove organic additives that cause deterioration in the photovoltaic performance [22]. This high-temperature process restricts the application of m-TiO_2 on the c-SnO_2 layer because the high-temperature process induces not only a large amount of charge traps and a recombination center in the SnO_2 layer but also poor interfacial contact, leading to interface recombination and shunting paths. Therefore, one important challenge is determining how to construct m-TiO_2 on the c-SnO_2 layer to take advantage of SnO_2. We recently achieved low-temperature processed PSCs by employing m-TiO_2 as ETL [23]. To remove the organic additives in the low-temperature-processed TiO_2 (LT-TiO_2), we adopted the oxygen plasma process. The simple and effective method of oxygen plasma treatment enhances charge extraction and transport, thereby improving photovoltaic performance. Therefore, our newly developed oxygen plasma treatment for LT m-TiO_2 is a promising strategy for combining the m-TiO_2 layer with the c-SnO_2 layer to produce an efficient bilayer ETL.

In this work, we demonstrated that the LT m-TiO_2 can be adopted to construct a compact/mesoporous structured bilayer ETL to prevent the degradation of SnO_2 by the high-temperature process. When the conventional m-TiO_2 layer was deposited on the SnO_2 layer and then the bilayer ETL underwent the high-temperature sintering process (BLH), the photovoltaic performance of this bilayer ETL-based PSC (BLH-PSC) deteriorated more than that of a single c-SnO_2 ETL-based PSC (SL-PSC). On the contrary, when the oxygen plasma treatment was applied to the LT m-TiO_2 deposited on the SnO_2 (BLP), the PSC with this bilayer ETL (BLP-PSC) exhibited an excellent PCE of 15.36%, which is higher than that of the SL-PSC (13.68%). Moreover, detailed characterizations demonstrated that the SnO_2/TiO_2 bilayer ETL is beneficial for carrier extraction and transport.

2. Materials and Methods

2.1. Fabrication of the SnO$_2$ Layer

A fluorine-doped tin oxide (FTO) electrode was patterned using zinc powder and diluted HCl solution. Then, the patterned FTO substrate was cleaned with deionized water (DI), acetone, and ethanol in an ultrasonic bath. After ultraviolet–ozone (UVO) treatment for 15 min, 0.05 M $SnCl_2 \cdot 2H_2O$ solution diluted in ethanol was spin-coated on the patterned FTO substrate and sintered at 200 °C for 1 h.

2.2. Fabrication of the Bilayer ETL

The m-TiO_2 solution was prepared by dissolving TiO_2 nanoparticle paste (Dyesol, Queanbeyan, Australia) in ethanol at a ratio of 1:10 (wt %). After 15 min of UVO treatment, the m-TiO_2 solution was spin-coated on the SnO_2 layer and sintered at 150 °C for 4 h for BLL-PSC and BLP-PSC. To remove TiO_2 nanoparticle aggregates, the bilayer ETL substrate was dipped in ethanol and stirred for 15 s, and then the substrate was annealed at 150 °C for 30 min. In contrast, the m-TiO_2 layer was sintered at 450 °C for 1 h and was not rinsed with ethanol for BLH-PSC. The substrate was dipped in 20 mM $TiCl_4$ solution at 90 °C for 15 min and sintered at 150 °C for 30 min.

2.3. Fabrication of the PSC

The $MAPbI_3$ solution was prepared by mixing methylammonium iodide (MAI), PbI_2, dimethyl sulfoxide, and *N,N*-dimethylformamide. In the case of BLP-PSCs, the m-TiO_2 layer was treated by oxygen plasma at a radio frequency (RF) power of 20 W for 10 min. Then, the $MAPbI_3$ layer was spin-coated and annealed at 65 °C for 1 min and at 100 °C for

10 min. The mixed Spiro-OMeTAD solution, which contained Spiro-OMeTAD (Jilin OLED, Changchun, Jilin Sheong, China), lithium salt, 4-tert-butylpyridine, and chlorobenzene, was spin-coated on the $MAPbI_3$ layer. A silver electrode was deposited via a thermal evaporator. All chemicals were purchased from Sigma-Aldrich (St. Louis, MO, USA).

2.4. Fabrication of the Flexible PSC

The indium-doped tin oxide (ITO)/polyethylene naphthalate (PEN) substrate (Peccell Technologies, Yokohama, Japan) was used to fabricate flexible PSCs. All fabrication processes were identical to that for BLP-PSC on FTO substrate, only the SnO_2 layer was annealed at 150 °C for 5 h.

2.5. Measurements

The surface morphologies were characterized using a field-emission scanning electron microscope (SEM) (S-4800, HITACHI, Tokyo, Japan). The photovoltaic characteristics were measured under 100 mW/cm^2 illumination using a solar simulator (Sol2A, Oriel, Irvine, CA, USA) with scan rate of 0.02 V at 25 °C. The internal electrochemical behavior was characterized using electrochemical impedance spectroscopy (EIS) (Compactstat.h, Ivium Technologies, Eindhoven, Netherlands) at a frequency range of 1 Hz to 1 MHz. The bending test was performed at a rate of 1 cycle per 0.5 s and a bending radius of 13 mm using a radius bending tester (JIRBT-620, JUNIL TECH, Daegu, Korea).

3. Results and Discussion

To investigate the possibility of using the conventional high-temperature-processed m-TiO_2 layer as the layer combined with the c-SnO_2 layer, we performed a comparative study of PSCs using both planar- and mesoporous-type PSCs. SL-PSCs in the configuration of FTO/c-SnO_2/perovskite/spiro-OMeTAD/Ag and BLH-PSCs in the configuration of FTO/c-SnO_2/m-TiO_2/spiro-OMeTAD/Ag were fabricated. In the case of the BLH-PSCs, the m-TiO_2 layer was sintered at 450 °C after spin-coating on the SnO_2 layer. The current density–voltage (J-V) curves under an irradiation of 100 mW cm^{-2} (AM 1.5) are shown in Figure 1. The SL-PSCs achieved a PCE of 13.68% with an open-circuit voltage (V_{OC}) of 0.99 V, a short-circuit current density (J_{SC}) of 19.77 mA/cm^2, and a fill factor (FF) of 70.22%. In contrast, the BLH-PSCs only achieved a PCE of 11.93% with a V_{OC} of 0.99 V, a J_{SC} of 19.52 mA/cm^2, and an FF of 61.86%. It is thus clear that SL-PSCs perform much better than BLH-PSCs. The large difference in FF could be primary attributed to the degradation of the SnO_2 layer by the high-temperature process [22].

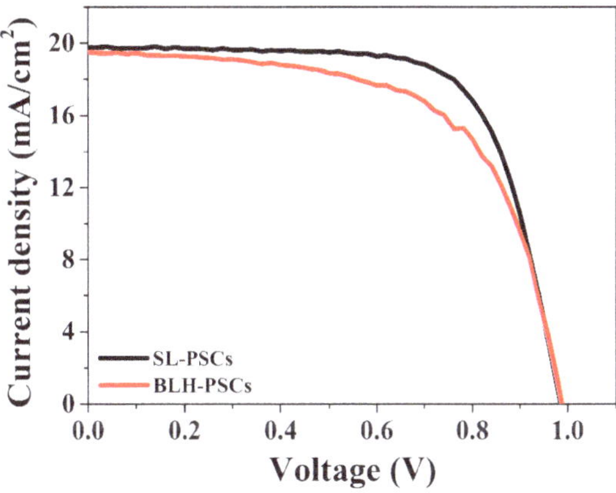

Figure 1. Reverse scan current density–voltage (J-V) curves of SL-PSC and BLH-PSC.

To uncover the underlying reasons for the decreased photovoltaic performance of BLH-PSCs, we fabricated SL-PSCs using a SnO_2 ETL annealed at temperatures from 200–500 °C. Figure 2 shows the dependence of PCE on the annealing temperature of the SnO_2 layer. As the annealing temperature increased, the photovoltaic performance of SL-PSCs decreased. The PECs of SL-PSCs annealed at 200, 300, and 400 °C were 13.68%, 12.19%, and 8.90%, respectively. The SL-PSCs annealed at 400 °C performed poorly, with very low FF and J_{SC}. Moreover, the SL-PSCs annealed at 500 °C did not show any photovoltaic characteristics. The detailed photovoltaic parameters obtained from the J-V curves are summarized in Table S1.

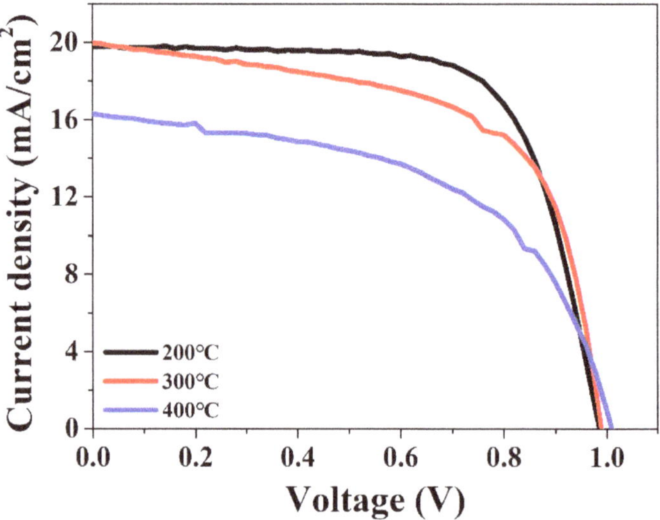

Figure 2. Reverse scan current density–voltage (J-V) curves of perovskite solar cells based on SnO_2 ETL annealed at 200, 300, and 400 °C.

To verify the decreasing trend in PCE, we investigated the morphology change of the SnO_2 layer according to the annealing temperature. Figure 3 shows the top-view SEM images of SnO_2 layers deposited on FTO substrates and annealed at different temperatures. As shown in Figure 3a, the FTO substrate was uniformly covered with the SnO_2 layer, and no pinholes were observed when the SnO_2 layer was annealed at the relatively low temperature of 200 °C. However, as the annealing temperature increased above 300 °C, the SnO_2 nanoparticles agglomerated more and the FTO areas uncovered by SnO_2 increased. The high-temperature-annealed SnO_2 layer could not completely cover the FTO substrate, and thus these pinholes resulted in leakage in the current pathway. Moreover, poor interface contact with the FTO substrate increased the series resistance. Therefore, as shown in Figure 1, BLH-PSCs exhibited lower FF and PCE than SL-PSCs because the high-temperature process of the m-TiO_2 layer caused the degradation of the underlying SnO_2 layer [24]. The above results confirm that the low-temperature processing of the m-TiO_2 layer without causing damage to the SnO_2 layer is important for producing high-performance PSCs using a c-SnO_2/m-TiO_2 bilayer ETL.

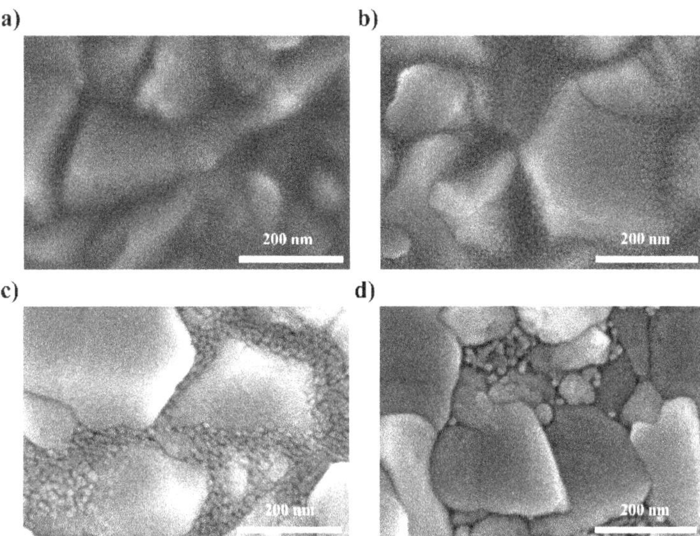

Figure 3. Top SEM image of SnO$_2$ layer after annealing at (**a**) 200, (**b**) 300, (**c**) 400, and (**d**) 500 °C on FTO glass.

To investigate the effectiveness of our strategy for LT m-TiO$_2$, we compared the PCEs of three different PSCs (Figure S2). First, we fabricated SL-PSC without the m-TiO$_2$ layer to determine the role of the m-TiO$_2$ layer. Then, we fabricated two different PSCs based on the FTO/c-SnO$_2$/m-TiO$_2$/spiro-OMeTAD/Ag architecture. The main difference between these two PSCs with a bilayer ETL was the post-treatment of the m-TiO$_2$ layer. In the case of LT m-TiO$_2$-based PSCs (BLL-PSCs), the m-TiO$_2$ layers were only annealed at 150 °C, whereas oxygen plasma treatment was directly performed on the LT TiO$_2$ layers for BLP-PSCs. As shown in Figure 4, the BLL-PSCs exhibited a PCE of only 5.43% with a V_{OC} of 0.85 V, a J_{SC} of 14.81 mA/cm^2, and an FF of 43.01% (Table 1). Although the low-temperature processing of m-TiO$_2$ might not have caused the aggregation of the underlying c-SnO$_2$ layer, the remaining organic additives in m-TiO$_2$ inhibited the full coverage of the perovskite layer on TiO$_2$, which hindered electron transport at the interface between the perovskite and TiO$_2$. According to our previous work, oxygen plasma treatment can successfully remove organic additives from and improve the wettability of the LT TiO$_2$ layer [13]. With oxygen plasma treatment, the performance of BLL-PSCs considerably improved, and the PCE, V_{OC}, J_{SC}, and FF were 15.35%, 1.03 V, 20.65 mA/cm^2, and 72.30%, respectively. Moreover, the PCE of BLP-PSC (15.53%) is higher than that of SL-PSC (13.68%). These results demonstrate that oxygen plasma treatment enables the fabrication of c-SnO$_2$/m-TiO$_2$ bilayer ETL-based PSCs with excellent photovoltaic performance using an LT m-TiO$_2$ layer.

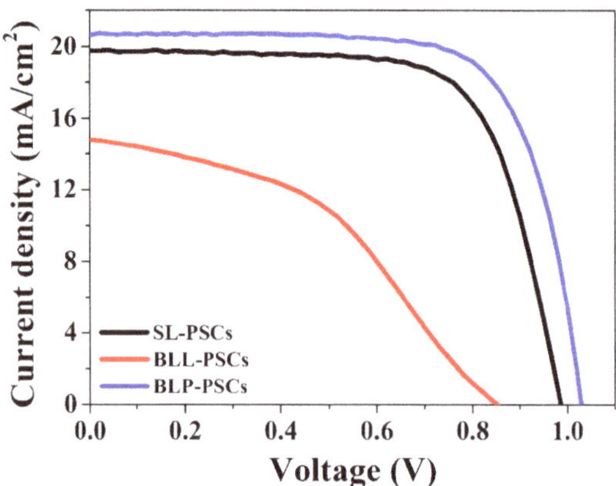

Figure 4. Reverse scan current density–voltage (J-V) curves of SL-PSCs, BLL-PSCs, and BLP-PSCs.

Table 1. Summary of the photovoltaic parameters of PSCs based on different ETLs.

Sample	J_{SC} (mA/cm^2)	V_{OC} (V)	FF (%)	PCE (%)
SL-PSCs	19.77	0.99	70.22	13.68
BLL-PSCs	14.81	0.85	43.01	5.43
BLP-PSCs	20.65	1.03	72.30	15.36

To gain further insight into the effects of the m-TiO$_2$ layer on charge transfer properties at the ETL/perovskite interface, EIS was conducted. The Nyquist plots of different ETLs were obtained in the dark with an applied bias voltage of 0.9 V and are shown in Figure 5. The series resistance (R_s) and charge transport resistance (R_{ct}) were obtained by fitting EIS data according to the relevant equivalent circuit, as shown in the inset of Figure 5. The EIS parameters from the semicircle Nyquist plot are summarized in Table S2. In general, R_s is related to the sheet resistance of electrodes [25], including the contributions from FTO and metal electrodes. In contrast, R_{ct} generally refers to the charge transfer resistance at all the interfaces [26], such as between the carrier selective layer and the perovskite layer, and between the electrode and the carrier selective layer. The R_{ct} value of SL-PSC was 350 Ω, which is slightly higher than that of BLP-PSC (230 Ω). The small R_{ct} value of BLP-PSC further supports that the combination of the m-TiO$_2$ layer with c-SnO$_2$ promotes good interface contact between the ETL and perovskite, leading to an enhanced charge transfer process and the highest PCE [27]. While BLL-PSC exhibited the highest value of R_{ct}, it had the lowest PCE. Figure S1 shows the top-view SEM images of the perovskite layer on c-SnO$_2$ and c-SnO$_2$/m-TiO$_2$ and their corresponding grain size distribution histograms. Both perovskite layers exhibited a similar average grain size with uniform morphology consisting of densely packed grains. Because the grain size, which refers to the density of grain boundaries, is related to the transport of photogenerated carriers and the extension of the charge carrier diffusion length [28], a comparable grain size might not result in different photovoltaic characteristics. However, the mesoporous structure of TiO$_2$ allows the perovskite to infiltrate into TiO$_2$, which increases the interfacial contact between perovskite and TiO$_2$ (Figure S3). Therefore, the improved interfacial contact due to the direct transfer pathway substantially contributed to the increase in PCE of SL-PSCs by combination with the m-TiO$_2$ ETL (Figure S4 and Table S3).

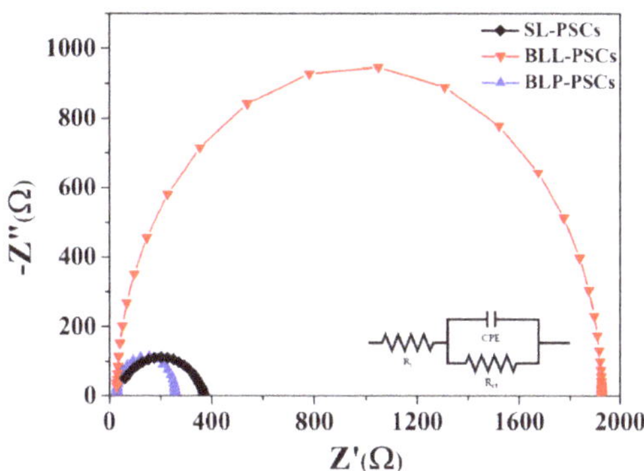

Figure 5. Nyquist plots of SL-PSCs, BLL-PSCs, and BLP-PSCs.

The high performance of c-SnO$_2$/m-TiO$_2$ bilayer ETL-based PSCs was achieved with oxygen plasma treatment at a low temperature, which suggests that high-performance and flexible BLP-PSCs can also be attained via the same procedures. Figure 6 shows the J-V curves of the flexible BLP-PSCs constructed on the PEN/ITO substrate as a function of bending cycles. The flexible cells exhibited a promising PCE of 9.56%, with a V$_{OC}$ of 1.81 V, a J$_{SC}$ of 17.20 mA/cm^2, and an FF of 55.22%. The inferior performance of the flexible cell compared to the rigid cell on the FTO/glass substrate arose from its inferior surface morphology and low transmittance compared to the rigid surface [29]. To investigate the mechanical stability of the flexible cell, a bending durability test was performed. All the photovoltaic characteristics were maintained during 500 cycles of bending without much deterioration.

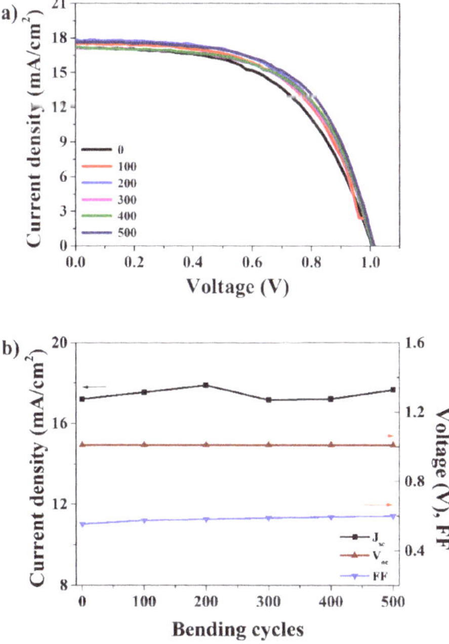

Figure 6. (**a**) Reverse scan current density–voltage (J-V) curves of flexible cells and (**b**) variation in the J$_{SC}$, V$_{OC}$, and FF values of flexible cells as a function of bending cycles.

4. Conclusions

In summary, an LT m-TiO$_2$ layer using oxygen plasma treatment was combined with c-SnO$_2$. The BLP-PSC had a PCE of 15.36%, which is much higher than that of the PSC with a single c-SnO$_2$ ETL. This high efficiency was obtained because the oxygen plasma treatment facilitated the removal of organic additives from LT m-TiO$_2$ and the infiltration of perovskite into m-TiO$_2$, thus enhancing charge transport and extraction. This proves that our strategy to construct a bilayer ETL using LT m-TiO$_2$ is beneficial because it allows the superior characteristics of the underlying c-SnO$_2$ to be maintained, and the mesoporous structure provided increased interfacial contact between the perovskite and the ETL. Consequently, the c-SnO$_2$/m-TiO$_2$ bilayer ETL is thought to be one of the most promising ETL layers for high-efficiency PSCs.

Supplementary Materials: The following supporting information can be downloaded at: https://www.mdpi.com/article/10.3390/nano12040718/s1, Figure S1: Top-view SEM images and grain size distribution histograms (inset) of MAPbI$_3$ layer on SL, BLL, BLP, and BLH; Figure S2: XRD pattern of MAPbI$_3$ layer on BLP, BLL, and SL; Figure S3: Cross-section SEM image of glass/FTO/ETL/MAPbI$_3$; Figure S4: Reverse scan (solid line) and forward scan (dotted line) current density–voltage curves of PSCs based on SL-PSC, BLL-PSC, and BLP-PSC; Table S1: Summary of the photovoltaic parameters of PSCs based on SnO$_2$ ETL annealed at 200, 300, and 400 °C; Table S2: Summary of EIS data; Table S3: Summary of the photovoltaic parameters of SL-PSC, BLL-PSC, and BLP-PSC according to scan direction.

Author Contributions: S.J. conceived and designed the research; J.L. participated in material preparation and device fabrication; J.K. and C.-S.K. participated in data interpretation; S.J. and J.L. wrote the paper; S.J. supervised the project. All authors have read and agreed to the published version of the manuscript.

Funding: This work was supported by the National Research Foundation (NRF) grant funded by the Korea government (MSIT) (2020R1F1A1074743 and 2021R1A4A1031761).

Institutional Review Board Statement: Not applicable.

Data Availability Statement: The data presented in this article is available on request from the corresponding author.

Conflicts of Interest: The authors declare no conflict of interest.

References

1. Yoo, J.J.; Seo, G.; Chua, M.R.; Park, T.G.; Lu, Y.; Rotermund, F.; Kim, Y.-K.; Moon, C.S.; Jeon, N.J.; Correa-Baena, J.-P.; et al. Efficient perovskite solar cells via improved carrier management. *Nature* **2021**, *590*, 587–593. [CrossRef]
2. Green, M.; Dunlop, E.; Hohl-Ebinger, J.; Yoshita, M.; Kopidakis, N.; Hao, X. Solar cell efficiency tables (version 57). *Prog. Photovoltaics: Res. Appl.* **2021**, *29*, 3–15. [CrossRef]
3. Hu, Y.; Niu, T.; Liu, Y.; Zhou, Y.; Xia, Y.; Ran, C.; Wu, Z.; Song, L.; Müller-Buschbaum, P.; Chen, Y.; et al. Flexible Perovskite Solar Cells with High Power-Per-Weight: Progress, Application, and Perspectives. *ACS Energy Lett.* **2021**, *6*, 2917–2943. [CrossRef]
4. Ma, S.; Yuan, G.-Z.; Zhang, Y.; Yang, N.; Li, Y.; Chen, Q. Development of encapsulation strategies towards the commercialization of perovskite solar cells. *Energy Environ. Sci.* **2021**, *15*, 13–55. [CrossRef]
5. Dai, T.; Cao, Q.; Yang, L.; Aldamasy, M.; Li, M.; Liang, Q.; Lu, H.; Dong, Y.; Yang, Y. Strategies for High-Performance Large-Area Perovskite Solar Cells toward Commercialization. *Crystals* **2021**, *11*, 295. [CrossRef]
6. Cai, L.; Liang, L.; Wu, J.; Ding, B.; Gao, L.; Fan, B. Large area perovskite solar cell module. *J. Semicond.* **2017**, *38*, 014006. [CrossRef]
7. Park, N.-G. Perovskite solar cells: An emerging photovoltaic technology. *Mater. Today* **2015**, *18*, 65–72. [CrossRef]
8. Noh, M.F.M.; Teh, C.H.; Daik, R.; Lim, E.L.; Yap, C.C.; Ibrahim, M.A.; Ludin, N.A.; Yusoff, A.R.B.M.; Jang, J.; Teridi, M.A.M. The architecture of the electron transport layer for a perovskite solar cell. *J. Mater. Chem. C* **2018**, *6*, 682–712. [CrossRef]
9. Anaraki, E.H.; Kermanpur, A.; Steier, L.; Domanski, K.; Matsui, T.; Tress, W.; Saliba, M.; Abate, A.; Grätzel, M.; Hagfeldt, A.; et al. Highly efficient and stable planar perovskite solar cells by solution-processed tin oxide. *Energy Environ. Sci.* **2016**, *9*, 3128–3134. [CrossRef]
10. Tiwana, P.; Docampo, P.; Johnston, M.B.; Snaith, H.J.; Herz, L.M. Electron Mobility and Injection Dynamics in Mesoporous ZnO, SnO2, and TiO2 Films Used in Dye-Sensitized Solar Cells. *ACS Nano* **2011**, *5*, 5158–5166. [CrossRef]
11. Min, H.; Lee, D.Y.; Kim, J.; Kim, G.; Lee, K.S.; Kim, J.; Il Seok, S.; Paik, M.J.; Kim, Y.K.; Kim, K.S.; et al. Perovskite solar cells with atomically coherent interlayers on SnO2 electrodes. *Nature* **2021**, *598*, 444–450. [CrossRef]

12. Xiong, L.; Guo, Y.; Wen, J.; Liu, H.; Yang, G.; Qin, P.; Fang, G. Review on the Application of SnO2 in Perovskite Solar Cells. *Adv. Funct. Mater.* **2018**, *28*, 1802757. [CrossRef]
13. Nam, J.; Kim, J.-H.; Kim, C.S.; Kwon, J.-D.; Jo, S. Surface Engineering of Low-Temperature Processed Mesoporous TiO2 via Oxygen Plasma for Flexible Perovskite Solar Cells. *ACS Appl. Mater. Interfaces* **2020**, *12*, 12648–12655. [CrossRef]
14. Song, J.; Zheng, E.; Wang, X.-F.; Tian, W.; Miyasaka, T. Low-temperature-processed ZnO–SnO2 nanocomposite for efficient planar perovskite solar cells. *Sol. Energy Mater. Sol. Cells* **2016**, *144*, 623–630. [CrossRef]
15. Ma, J.; Yang, G.; Qin, M.; Zheng, X.; Lei, H.; Chen, C.; Chen, Z.; Guo, Y.; Han, H.; Zhao, X.; et al. MgO Nanoparticle Modified Anode for Highly Efficient SnO2 -Based Planar Perovskite Solar Cells. *Adv. Sci.* **2017**, *4*, 1700031. [CrossRef]
16. Han, G.S.; Chung, H.S.; Kim, D.H.; Kim, B.J.; Lee, J.-W.; Park, N.-G.; Cho, I.S.; Lee, J.-K.; Lee, S.; Jung, H.S. Epitaxial 1D electron transport layers for high-performance perovskite solar cells. *Nanoscale* **2015**, *7*, 15284–15290. [CrossRef]
17. Kogo, A.; Ikegami, M.; Miyasaka, T. A SnOx–brookite TiO2 bilayer electron collector for hysteresis-less high efficiency plastic perovskite solar cells fabricated at low process temperature. *Chem. Commun.* **2016**, *52*, 8119–8122. [CrossRef]
18. Tang, H.; Cao, Q.; He, Z.; Wang, S.; Han, J.; Li, T.; Gao, B.; Yang, J.; Deng, D.; Li, X. SnO2–Carbon Nanotubes Hybrid Electron Transport Layer for Efficient and Hysteresis-Free Planar Perovskite Solar Cells. *Sol. RRL* **2020**, *4*, 1900415. [CrossRef]
19. Yang, G.; Wang, C.; Lei, H.; Zheng, X.; Qin, P.; Xiong, L.; Zhao, X.; Yan, Y.; Fang, G. Interface engineering in planar perovskite solar cells: Energy level alignment, perovskite morphology control and high performance achievement. *J. Mater. Chem. A* **2017**, *5*, 1658–1666. [CrossRef]
20. Xiong, Z.; Lan, L.; Wang, Y.; Lu, C.; Qin, S.; Chen, S.; Zhou, L.; Zhu, C.; Li, S.; Meng, L.; et al. Multifunctional Polymer Framework Modified SnO2 Enabling a Photostable α-FAPbI3 Perovskite Solar Cell with Efficiency Exceeding 23%. *ACS Energy Lett.* **2021**, *6*, 3824–3830. [CrossRef]
21. Kogo, A.; Sanehira, Y.; Numata, Y.; Ikegami, M.; Miyasaka, T. Amorphous Metal Oxide Blocking Layers for Highly Efficient Low-Temperature Brookite TiO2-Based Perovskite Solar Cells. *ACS Appl. Mater. Interfaces* **2018**, *10*, 2224–2229. [CrossRef]
22. Jung, K.-H.; Seo, J.-Y.; Lee, S.; Shin, H.; Park, N.-G. Solution-processed SnO2thin film for a hysteresis-free planar perovskite solar cell with a power conversion efficiency of 19.2%. *J. Mater. Chem. A* **2017**, *5*, 24790–24803. [CrossRef]
23. Nam, J.; Nam, I.; Song, E.-J.; Kwon, J.-D.; Kim, J.; Kim, C.S.; Jo, S. Facile Interfacial Engineering of Mesoporous TiO2 for Low-Temperature Processed Perovskite Solar Cells. *Nanomaterials* **2019**, *9*, 1220. [CrossRef]
24. Parida, B.; Singh, A.; Oh, M.; Jeon, M.; Kang, J.-W.; Kim, H. Effect of compact TiO2 layer on structural, optical, and performance characteristics of mesoporous perovskite solar cells. *Mater. Today Commun.* **2019**, *18*, 176–183. [CrossRef]
25. Bu, T.; Wu, L.; Liu, X.; Yang, X.; Zhou, P.; Yu, X.; Qin, T.; Shi, J.; Wang, S.; Li, S.; et al. Synergic Interface Optimization with Green Solvent Engineering in Mixed Perovskite Solar Cells. *Adv. Energy Mater.* **2017**, *7*, 1700576. [CrossRef]
26. Chen, C.; Jiang, Y.; Guo, J.; Wu, X.; Zhang, W.; Wu, S.; Gao, X.; Hu, X.; Wang, Q.; Zhou, G.; et al. Solvent-Assisted Low-Temperature Crystallization of SnO2 Electron-Transfer Layer for High-Efficiency Planar Perovskite Solar Cells. *Adv. Funct. Mater.* **2019**, *29*, 1900557. [CrossRef]
27. Li, M.; Wang, Z.-K.; Yang, Y.-G.; Hu, Y.; Feng, S.-L.; Wang, J.-M.; Gao, X.-Y.; Liao, L.-S. Copper Salts Doped Spiro-OMeTAD for High-Performance Perovskite Solar Cells. *Adv. Energy Mater.* **2016**, *6*, 1601156. [CrossRef]
28. Hidayat, R.; Nurunnizar, A.A.; Fariz, A.; Herman; Rosa, E.S.; Shobih; Oizumi, T.; Fujii, A.; Ozaki, M. Revealing the charge carrier kinetics in perovskite solar cells affected by mesoscopic structures and defect states from simple transient photovoltage measurements. *Sci. Rep.* **2020**, *10*, 19197. [CrossRef] [PubMed]
29. Huang, K.; Peng, Y.; Gao, Y.; Shi, J.; Li, H.; Mo, X.; Huang, H.; Gao, Y.; Ding, L.; Yang, J. High-Performance Flexible Perovskite Solar Cells via Precise Control of Electron Transport Layer. *Adv. Energy Mater.* **2019**, *9*, 1901419. [CrossRef]

nanomaterials

Article

Tantalum-Doped TiO$_2$ Prepared by Atomic Layer Deposition and Its Application in Perovskite Solar Cells

Chia-Hsun Hsu [1], Ka-Te Chen [1], Ling-Yan Lin [2,3], Wan-Yu Wu [4], Lu-Sheng Liang [5], Peng Gao [5], Yu Qiu [2,3], Xiao-Ying Zhang [1], Pao-Hsun Huang [6], Shui-Yang Lien [1,4,7,* and Wen-Zhang Zhu [1,7]

[1] School of Opto-Electronic and Communication Engineering, Xiamen University of Technology, Xiamen 361024, China; chhsu@xmut.edu.cn (C.-H.H.); 1922031007@stu.xmut.edu.cn (K.-T.C.); xyzhang@xmut.edu.cn (X.-Y.Z.); wzzhu@xmut.edu.cn (W.-Z.Z.)

[2] Key Laboratory of Green Perovskites Application of Fujian Province Universities, Fujian Jiangxia University, Fuzhou 350108, China; lingyanlin@fjjxu.edu.cn (L.-Y.L.); yuqiu@fjjxu.edu.cn (Y.Q.)

[3] College of Electronics and Information Science, Fujian Jiangxia University, Fuzhou 350108, China

[4] Department of Materials Science and Engineering, Da-Yeh University, Changhua 51591, Taiwan; wywu@mail.dyu.edu.tw

[5] CAS Key Laboratory of Design and Assembly of Functional Nanostructures, Fujian Provincial Key Laboratory of Nanomaterials, Fujian Institute of Research on the Structure of Matter, Chinese Academy of Sciences, Fuzhou 350002, China; lushengliang@fjirsm.ac.cn (L.-S.L.); peng.gao@fjirsm.ac.cn (P.G.)

[6] School of Information Engineering, Jimei University, Xiamen 361021, China; ph.huang@jmu.edu.cn

[7] Fujian Key Laboratory of Optoelectronic Technology and Devices, Xiamen University of Technology, Xiamen 361024, China

* Correspondence: sylien@xmut.edu.cn

Citation: Hsu, C.-H.; Chen, K.-T.; Lin, L.-Y.; Wu, W.-Y.; Liang, L.-S.; Gao, P.; Qiu, Y.; Zhang, X.-Y.; Huang, P.-H.; Lien, S.-Y.; et al. Tantalum-Doped TiO$_2$ Prepared by Atomic Layer Deposition and Its Application in Perovskite Solar Cells. *Nanomaterials* **2021**, *11*, 1504. https://doi.org/10.3390/nano11061504

Academic Editor: Vlad Andrei Antohe

Received: 30 April 2021
Accepted: 4 June 2021
Published: 7 June 2021

Publisher's Note: MDPI stays neutral with regard to jurisdictional claims in published maps and institutional affiliations.

Abstract: Tantalum (Ta)-doped titanium oxide (TiO$_2$) thin films are grown by plasma enhanced atomic layer deposition (PEALD), and used as both an electron transport layer and hole blocking compact layer of perovskite solar cells. The metal precursors of tantalum ethoxide and titanium isopropoxide are simultaneously injected into the deposition chamber. The Ta content is controlled by the temperature of the metal precursors. The experimental results show that the Ta incorporation introduces oxygen vacancies defects, accompanied by the reduced crystallinity and optical band gap. The PEALD Ta-doped films show a resistivity three orders of magnitude lower than undoped TiO$_2$, even at a low Ta content (0.8–0.95 at.%). The ultraviolet photoelectron spectroscopy spectra reveal that Ta incorporation leads to a down shift of valance band and conduction positions, and this is helpful for the applications involving band alignment engineering. Finally, the perovskite solar cell with Ta-doped TiO$_2$ electron transport layer demonstrates significantly improved fill factor and conversion efficiency as compared to that with the undoped TiO$_2$ layer.

Keywords: tantalum; titanium oxide; atomic layer deposition; bubbler temperature; perovskite solar cell; electron transport layer

1. Introduction

Titanium dioxide (TiO$_2$) thin films are widely studied owing to their optoelectrical properties that fulfill the requirements of numerous applications such as dye-sensitized photovoltaic devices [1], photocatalysis [2], and perovskite solar cells (PSCs) [3]. In order to enhance properties, the TiO$_2$ films are commonly doped with transition metal, rare metal, or noble metal ions [4–9]. Some studies indicate that the electron-hole recombination could be reduced due to the generation of the charge trapping centers by foreign ions [10]. A moderate amount of metal dopants promotes the excitons separation and thus improves the mobility of photogenerated carriers [11–13]. Due to the wide forbidden energy gap and the absorption only at UV-light region, the TiO$_2$ films are intentionally doped in order to reduce the energy gap and extend the light absorption to visible light in some applications [14]. The reduced band gap results from the generation of oxygen vacancies produced simultaneously through doping, introducing shallow energy levels below the conduction band edge [15,16]. Zhao et al. reported that doped TiO_2 with suitable metal ions can effectively improve the electron transport of TiO_2 [17]. Among various dopants, Ta^{5+} has an ionic radius (0.64 Å) very close to Ti^{4+} (0.61 Å), and thus it is able to be incorporated into the TiO_2 lattice without severe strain or secondary phase production [18]. Doping with Ta is also more effective than with niobium, and has lower formation energy consumption and superior thermodynamic stability compared to doping with nitrogen [13]. Ta-doped TiO_2 thin films can be prepared by various methods such as sputtering [19], sol-gel spin-coating [13], and chemical vapor deposition [20]. Recently, due to the demands of slim and thin electronics or the requirements of depositing films on high aspect ratio substrates, atomic layer deposition (ALD) technique receives great attention because of its pinhole-free deposition process, accurate thickness control at sub-nanometer level, and high conformality [21]. The Ta incorporation into ALD TiO_2 films is mostly carried out by using the Ta_2O_5/TiO_2 multilayer structure, and controlled by the ALD cycle ratio of the two metal oxides [18,22].

As for planar perovskite solar cells, the electron transport layer (ETL) plays a crucial role in electron transport and hole blocking. Due to the fact that ETLs are mostly prepared using the sol-gel process, the layer may contain pinholes or have low density, rendering holes possible to pass the ETL to increase the recombination rate. A compact layer is usually inserted at the sol-gel ETL/transparent electrode interface to improve the hole blocking ability. Thanks to the pinhole-free structure and highly conformal coverage, a thin ALD layer is able to be used as both ETL and compact layer for the perovskite solar cells [23]. However, the conduction band edge of TiO_2 (about -3.9 eV) is shallower than that of cesium formamidinium-based $Cs_xFA_{1-x}Pb(I_{1-y}Br_y)_3$ (about -4.2 eV), which is one of the most promising classes of perovskite light absorber for solar cells due to its wide band gap and higher durability compared to $MAPbI_3$ [24]. The conduction band mismatch at the TiO_2 ETL/perovskite interface causes a barrier that is theoretically unfavorable for electron transport. Development of doped TiO_2 ETLs is one possible way to have a better band alignment and an improved PSC performance.

In the present study, Ta-doped TiO_2 films are grown by using plasma enhanced ALD (PEALD), and used as an ETL for PSCs. The metal precursors tantalum ethoxide ($Ta(OEt)_5$) as Ta dopant source and titanium isopropoxide (TTIP) as Ti source are simultaneously fed into the deposition chamber. The precursor output flows are controlled by the bubbler temperature, which is varied from 70–90 °C to obtain different Ta doping ratio. The effects of the bubbler temperature on the structural, electrical, and optical properties of the Ta-doped TiO_2 films are investigated. The adjustments of the band gap, conduction band, and valence band positions are presented. Finally, the Ta-doped TiO_2 ETL is applied to PSC fabrication, and the improved conversion efficiency and hysteresis is demonstrated and discussed.

2. Materials and Methods

2.1. Thin Film Preparation

Glass substrates with a size of 20 mm × 20 mm and a thickness of 1 mm were cleaned in an ultrasonic bath with deionized water, acetone, and ethanol for 15 min, and then dried in an oven at 70 °C for 30 min. The substrates were placed on the substrate holder in the deposition chamber of PEALD (R200 advance, Picosun Oy, Espoo, Finland). The used metal precursors were $Ta(OEt)_5$ and TTIP (Aimou Yuan Scientific, Nanjing, China) as the Ta and Ti sources, respectively. Nitrogen with a purity of 99.999% was introduced to the precursor bubblers to transport the precursor vapors to the deposition chamber. The oxidant was provided by inductively coupled oxygen-argon mixed plasma generated by an RF power of 1500 W. The substrate temperature was 250 °C. The temperatures of the $Ta(OEt)_5$ and TTIP bubblers were controlled from 70 to 90 °C by means of a heating jacket. The thickness of the Ta-doped TiO_2 films was 60 nm. The undoped films were prepared under identical conditions, except for the use of the TTIP metal precursor alone. The detailed deposition parameters for the PEALD Ta-doped TiO_2 films are summarized in Table 1.

Table 1. Deposition parameters of the PEALD Ta-doped TiO_2 films.

Parameter	Value
Substrate temperature ($^{\circ}$C)	250
Bubbler temperature ($^{\circ}$C)	70–90
Nitrogen carrier flow rate (sccm)	120
Metal precursor pulse time (s)	1.6
Metal precursor purge time (s)	6
O_2 flow rate (sccm)	150
O_2 pulse time (s)	11
O_2 purge time (s)	5
O_2 plasma power (W)	1500
Ar flow rate (sccm)	80
Post annealing temperature ($^{\circ}$C)	500

2.2. Perovskite Solar Cell Fabrication

The 15 nm-thick ALD TiO_2 films were deposited on the cleaned fluorine-doped tin oxide (FTO) substrates. Lead iodide, lead bromide, methylammonium bromide, and formamidinium iodide were dissolved with a molar ratio of 1:1.15:0.2:0.2 in *N,N*-dimethylformamide (DMF) and dimethyl sulfoxide (DMSO) with 4:1 volume ratio. The CsI dissolved in DMSO as 1.5 mol stock solution was further poured to give $Cs_{0.1}(FA_{0.83}MA_{0.17})_{0.9}Pb(I_{0.83}Br_{0.17})_3$. The perovskite precursor solution was then spin-coated on ALD doped or undoped TiO_2 ETL in two steps, the first spreading step with 1000 rpm for 10 s and the second step with 6000 rpm for 25 s. Chlorobenzene of 110 μL was sprayed on the spinning substrate at five seconds before the end of the second-step spin-coating process. Afterwards, an annealing process at 100 $^{\circ}$C was performed for 60 min. Spiro-OMeTAD (Lumtec, New Taipei, Taiwan) of 50 uL was spin-coated on the perovskite with 4000 rpm for 30 s following cool down of the substrate back to room temperature. Finally, the cell fabrication was finished by evaporating gold films on the Spiro-OMeTAD. The devices had an active area of 0.1 cm^2.

2.3. Characterization

The thickness of the films was determined using an ellipsometer (M-2000, J. A. Woollan Co., Lincoln, NE, USA). The ultraviolet-visible spectrometer (MFS630, Hong-Ming, New Taipei City, Taiwan) was employed to obtain the transmittance spectra. The X-ray diffraction apparatus (XRD, Rigaku TTRAXIII, Ibaraki, Japan) was used for characterizing the crystalline structure of the films. The X-ray photoelectron spectroscopy (XPS, ESCALAB 250Xi, Thermo Fisher, Waltham, MA, USA) was used in order to investigate the chemical states and elemental composition of the films. The Fermi-level and valance band position of the Ta-doped TiO_2 films were obtained using UV photoelectron spectroscopy (UPS, ESCALAB Xi$^+$, Thermo Fisher Scientific, Gloucester, UK) with He I source (photon energy of 21.2 eV). The resistivity of the films was examined by using a four-point probe (T2001A3, Ossila, Sheffield, UK). The current density–voltage (J–V) curves and solar cell external parameters such as open-circuit voltage (V_{oc}), short-circuit current density (J_{sc}), fill factor (FF), and conversion efficiency (η) were measured at AM1.5G (100 mW/cm^2) using a solar simulator (Newport Oriel, Irvine, CA, USA).

3. Results and Discussion

In this work, $Ta(OEt)_5$ precursor is used as the Ta dopant source, and co-injected with TTIP precursor into the deposition zone. The amount of the output precursor molecules is determined by the bubbler temperature controlling the vapor pressure of the metal precursors. Figure 1a shows the temperature-dependent vapor pressures of TTIP and Ta(OEt)$_5$ as given by:

$$P_{\text{Ta(OEt)5}} = 10^{9.66 - \frac{4288}{T}},$$ (1)

$$P_{\text{TTIP}} = 10^{10.12 - \frac{3425}{T}},$$ (2)

where P is the vapor pressure (unit in mmHg), and T is the bubbler temperature. Both of the metal vapor pressures increase with the bubbler temperature. It is believed that the film growth is related to the output vapors of metal precursors, which can be controlled by either varying the carrier gas flows or the bubbler temperature. The latter is adopted in this study. Figure 1b shows the deposition rates of the PEALD Ta-doped TiO$_2$ films prepared at different bubbler temperatures. The value of the undoped film is also indicated for comparison. It is seen that the TiO$_2$ has a deposition rate of 0.252 Å/cycle, which is similar to that reported by other groups [25,26]. The deposition rate is not much affected at the bubbler temperature of 70 °C, and then increases from 0.252 to 0.284 Å/cycle when the bubbler temperature increases from 70 to 90 °C. It is reported that the deposition rate of the ALD tantalum oxide is 0.5–1 Å/cycle [27,28], which is nearly two to four times higher than that of TiO$_2$. This may be one explanation for the increased deposition rate for the Ta-doped TiO$_2$. Another reason is that Ta incorporation leads to a less dense film structure due to the donor doping induced oxygen vacancies defects to maintain charge neutral as commonly seen in doped metal oxides [16]. The relatively loose film structure of the Ta-doped TiO$_2$ contributes to a higher film thickness.

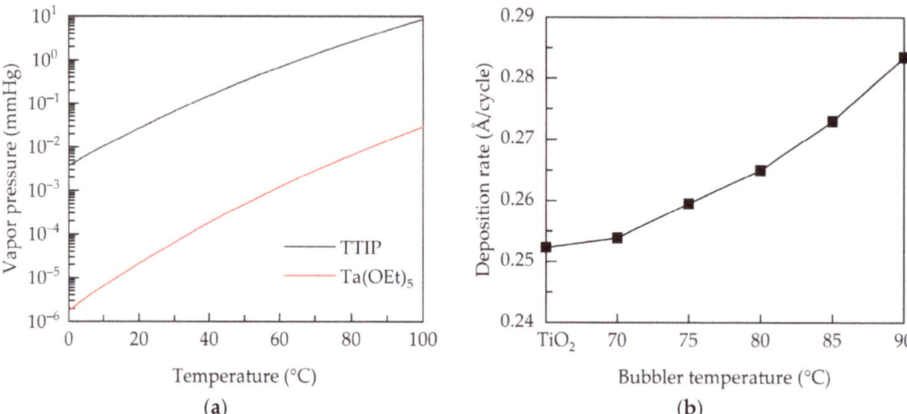

Figure 1. (**a**) Temperature-dependent vapor pressures of the TTIP and Ta(OEt)$_5$ metal precursors. (**b**) Deposition rate of the PEALD films prepared at various bubbler temperatures.

Figure 2a shows the transmittance over a wavelength range from 300 to 800 nm for the PEALD films prepared at various bubbler temperatures. The lower transmittance at the short wavelength region is related to the band-to-band absorption of the TiO$_2$ films. The transmittance decreases with increasing the bubbler temperature, indicating that the increase of Ta concentration enhances the absorption. The reduced transmittance could also be a consequence of the light scattering caused by the oxygen vacancies [29]. The inset shows the transmittance in the short-wavelength region (300–400 nm) to evaluate the absorption edge. It can be seen that the edge shifts towards longer wavelength direction when the bubbler temperature increases, inferring the reduced band gap. Figure 2b shows the band gap deduced from Tauc's equation [30]:

$$(\alpha h v)^n = A (h v - E_g)$$ (3)

where α is the absorption coefficient, hv is the photon energy, A is the material-dependent constant, and E_g is the band gap. The n value is taken as 1/2 for indirect band gap anatase TiO_2 [31]. The band gap values of the PEALD films are evaluated by plotting $(\alpha hv)^{1/2}$ versus hv and extrapolating the linear region of the resultant curves to obtain an interception with the hv-axis. The band gap of the TiO_2 is 3.17 eV, and it decreases from 3.09 to 3 eV by increasing the bubbler temperature from 70 to 90 °C. This is possibly related to the band gap narrowing effect due to the formation of the oxygen vacancies defects that introduce shallow donor levels below the conduction band. Similar effect can also be observed elsewhere [32].

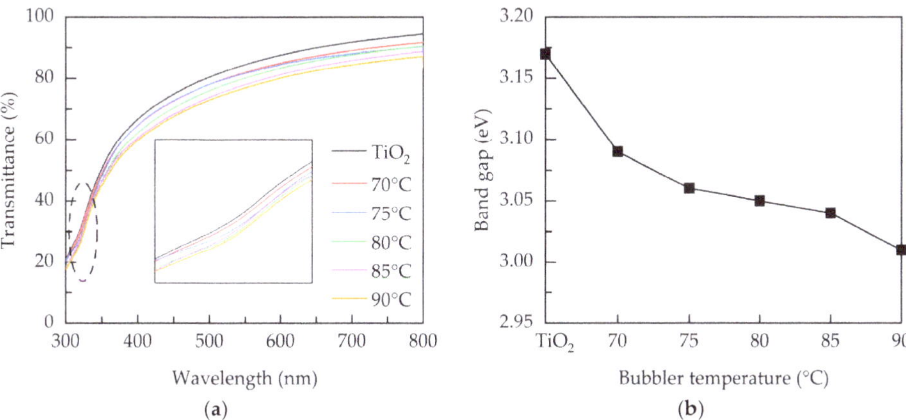

Figure 2. (**a**) Transmittance spectra and (**b**) band gap of the PEALD films prepared with various bubbler temperatures.

Figure 3a shows the XRD patterns of the PEALD undoped and Ta-doped TiO_2 films. The peaks labeled are well matched to anatase TiO_2 (JCPSD#83-2243). With Ta doping, the (103), (004), and (112) peaks vanish, but no additional peak such as Ta or Ta_2O_5 is found, indicating that Ta atoms are well-incorporated into the TiO_2 crystalline structure by replacing Ti^{4+} at random lattice sites. The full width at half maximum (FWHM) of the most intense (101) orientation is extracted to calculate crystallite size according to the Scherrer equation:

$$L = \frac{K\lambda}{\beta cos\theta} \tag{4}$$

with L the crystallite size, K the shape factor taken as 0.9, λ the X-ray wavelength (0.154 nm), β is the width of the observed diffraction line at its half intensity maximum, and θ is the Bragg angle. Figure 3b illustrates the FWHM and crystallite size as function of bubbler temperature. It is seen that the FWHM increases when Ta is substituted into TiO_2, suggesting that the crystallinity is diminished. The crystallite size reduces from 28 nm for undoped TiO_2 to 23 nm for the film deposited at the bubbler temperature of 70 °C. Further increasing the bubbler temperature up to 90 °C results in a slight reduction in crystallite size. Although Ta^{5+} has a similar ionic radius to Ti^{4+}, the electron density around Ta and the doping induced oxygen vacancies cause the rearrangement of the nearby atoms [16,33], which limits crystal growth.

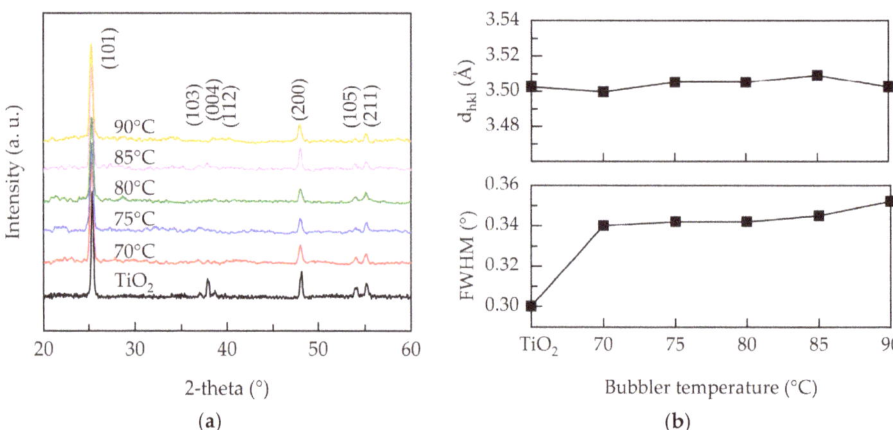

Figure 3. (**a**) XRD pattern, (**b**) interplanar distance and FWHM of the PEALD Ta-doped TiO$_2$ films prepared at various bubbler temperatures.

Figure 4 shows the composition and chemical states measured by XPS for the PEALD Ta-doped TiO$_2$ films prepared at different bubbler temperatures. Figure 4a shows the high-resolution Ta 4f for the PEALD films, evidencing the Ta incorporation into the TiO$_2$ films. The Ta peaks can be deconvoluted into two Gaussian components at 21.7 and 23.7 eV, corresponding to TaO$_x$ (x < 2.5) [34].The atomic ratios of the Ti, O, and Ta elements are depicted in Figure 4b. The TiO$_2$ film has an oxygen-deficient structure. The Ta content is varied from 0.82 to 0.95 at.% with increase in bubbler temperature from 70 to 90 °C. The XPS spectra over the biding energy of 450 to 470 eV corresponding to Ti 2p are shown in Figure 4c. The binding energies of around 453–460 eV and 460–466 eV belong to Ti 2p$_{3/2}$ and Ti 2p$_{1/2}$, respectively [35]. Each spin state can further be deconvoluted into Ti^{4+}, Ti^{3+} and Ti^{2+} peaks [36]. The Ti 2p$_{3/2}$ peaks are analyzed, and the peak ratios of Ti^{2+}, Ti^{3+} and Ti^{4+} to total are shown in Figure 4d. The TiO$_2$ has the highest T^{i4+} peak ratio of about 60%. At the bubbler temperature of 70 °C, the Ti^{4+} ratio decreases and Ti^{3+} ratio increases, indicating that Ta dopants create oxygen vacancies and reduce Ti^{4+} to Ti^{3+}. At the bubbler temperature of 90 °C, the Ti^{4+} ratio further reduces and the Ti^{2+} ratio increases. This result suggests that the higher Ta incorporation causes Ti^{4+} to reduce to not only Ti^{3+} but also Ti^{2+}. Figure 4e shows the high-resolution O 1s spectra for the PEALD films. The peaks are further split into two peaks, one at the lower binding energies (530–531 eV) assigned to lattice oxygen and the other at the higher binding energies of 531–532 eV ascribed to oxygen ions in oxygen-deficient regions in the lattice [37]. The latter is usually deemed to reflect the presence of oxygen vacancies. The ratio of each component to total is calculated as shown in Figure 4f. The lattice oxygen shows an opposite trend to defective oxygen. The increased defective oxygen ratio confirms that the Ta doping leads to the generation of oxygen vacancies defects.

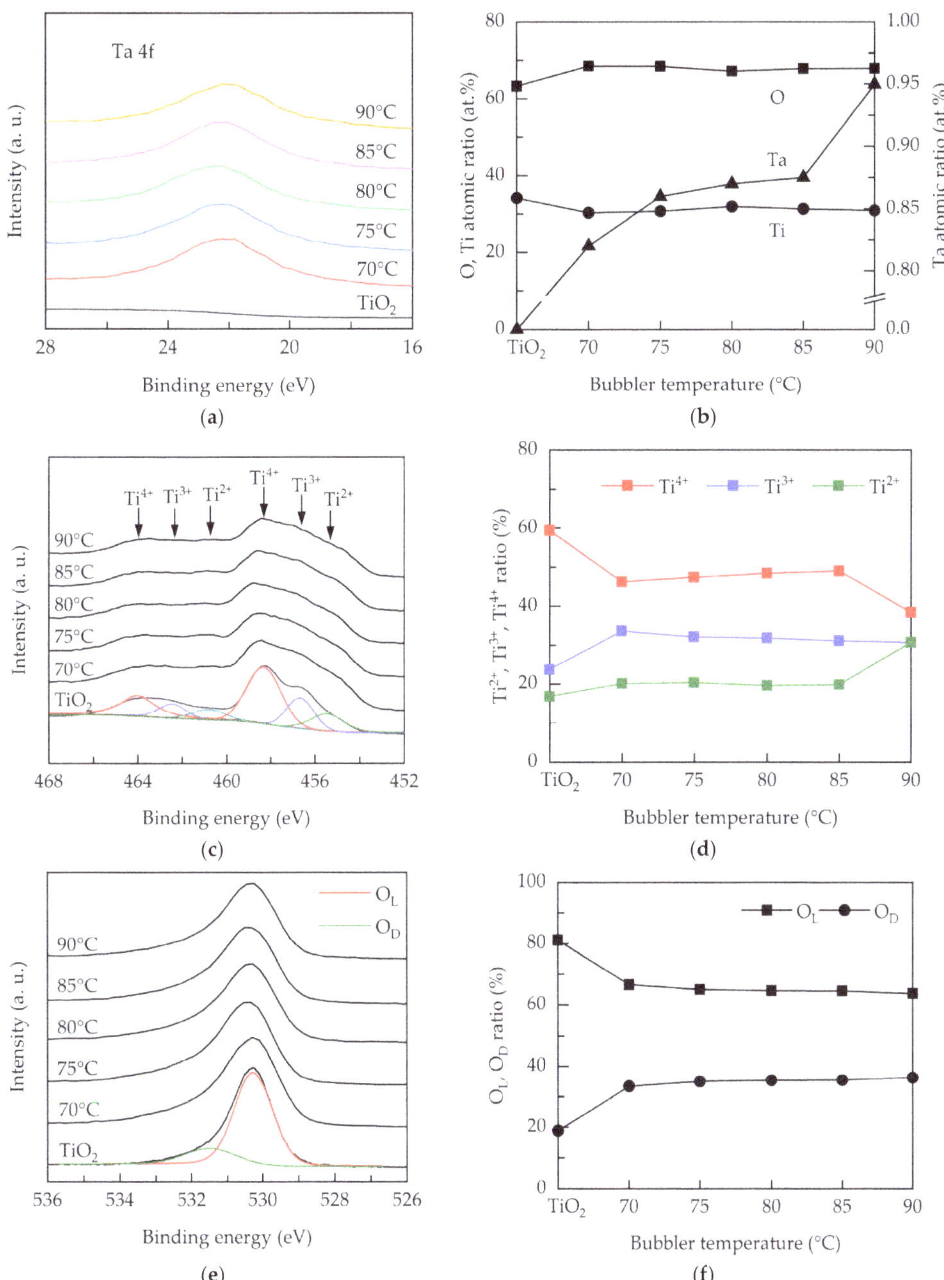

Figure 4. (**a**) Ta 4f spectra, (**b**) O, Ti and Ta atomic ratios, (**c**) O 1s spectra, (**d**) O_L and O_D peak area ratios, (**e**) Ti 2p$_{3/2}$ spectra, and (**f**) Ti^{2+}, Ti^{3+} and Ti^{4+} peak area ratios for the PEALD Ta-doped TiO$_2$ films.

Figure 5 shows the UPS spectra for the PEALD Ta-doped TiO$_2$ films prepared at different bubbler temperatures. As shown in Figure 5a, the extrapolation of the linear region of the curves to the binding energy axis gives the maximum valance band energy

(E_{VBM}) with respective to the Fermi energy. The curves in the high binding energy region are shown in Figure 5b, and similarly, the extrapolation of the linear region of the curves to the x-axis corresponds to the cut-off energy (E_{cutoff}). Accordingly, the positions of Fermi level (E_F) and valance band (E_v) can be calculated as given by [38]:

$$E_F = E_{cutoff} - 21.2, \tag{5}$$

$$E_v = E_F - E_{VBM} \tag{6}$$

where the value of 21.2 eV corresponds to the energy of the used UV light source. The position of the conduction band (E_c) is further obtained using: $E_c = E_v + E_g$ with the band gap values shown previously. The calculated E_c and E_v are summarized in Table 2. It is found the difference between the doped films is relatively small; however, the Ta-doped TiO$_2$ films have both deeper conduction band and valence band edges, as compared to the undoped TiO$_2$ film. The deep valance band positions favor the applications requiring hole blocking ability such as the electron transport layer in solar cells [39,40]. The lowered conduction band edges of the Ta-doped films indicate that Ta incorporation can adjust the conduction band position, which is benefit for interfacial band alignment engineering and a very important feature for being used as a selective charge transport layer of optoelectronics.

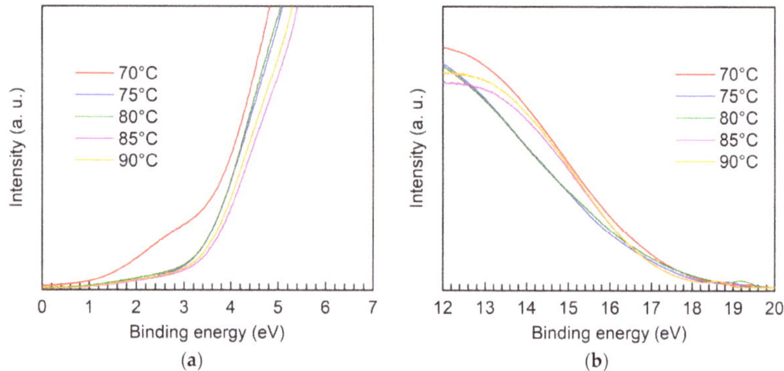

Figure 5. UPS spectra in the (**a**) low binding energy region and (**b**) high binding energy region for the PEALD Ta-doped TiO$_2$ films prepared at different bubbler temperatures.

Table 2. Band gap, valance band position, and conduction band position for the PEALD Ta-doped TiO$_2$ films with different bubbler temperatures.

Bubbler Temperature (°C)	Band Gap (eV)	E_v (eV)	E_c (eV)
TiO$_2$	3.17	−7.07	−3.9
70	3.09	−7.19	−4.1
75	3.06	−7.19	−4.13
80	3.05	−7.22	−4.17
85	3.04	−7.24	−4.2
90	3	−7.24	−4.23

Figure 6 shows the resistivity measured by a four-point probe for the PEALD Ta-doped TiO$_2$ films prepared at different bubbler temperatures. The TiO$_2$ exhibits a considerably high resistivity of 3.4×10^2 Ω·cm. The Ta incorporation leads to significant reduction in resistivity. All the Ta-doped TiO$_2$ films have resistivities at the order of 10^{-1} Ω·cm. The minimum resistivity is 3×10^{-1} Ω·cm at the bubbler temperature of 90 °C. The decrease in resistivity is due to the Ta substitution for Ti atoms in the lattice sites, contributing free electrons. In addition, the Ta incorporation also results in creation of oxygen vacancies (as suggested by XPS), which are widely known as one of the sources of free electrons. Similar resistivity reduction due to the Ta doping is reported by other studies. For example, the resistivity of the Ta-doped TiO_2 films prepared using metal organic chemical vapor deposition decreases four to six orders for at a high Ta doping level of 4–5% [17,41]. In the present work, the Ta content for the bubbler temperatures of 70–90 °C is between 0.82 and 0.95 at.%, leading to resistivities three orders lower than that of the undoped TiO_2 film.

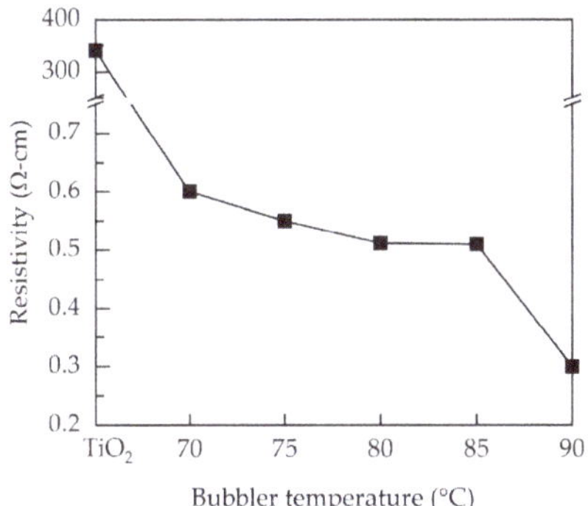

Figure 6. Four-point probe resistivity of the PEALD Ta-doped TiO_2 films prepared at different bubbler temperature.

Figure 7 depicts the isolated energy band diagram of conventional TiO_2 ETL-based PSCs. The E_c and E_v values of each layer are experimentally determined. Similar values can be found in other studies [42–45]. The photogenerated electrons and holes move towards different directions. A downward shift of CBM is generally favorable in the viewpoint of electron transport. However, as shown in the inset of Figure 7, the undoped TiO_2 layer has a shallower CBM than the perovskite absorber, which causes the band mismatch and obstacles for electron transport. The charge accumulation at the perovskite/TiO_2 interface could resist electrons to flow into FTO. In the case of the adoption of Ta-doped TiO_2, with the experimental E_c values around $-(4.1–4.23)$ eV shown in the UPS result, the band mismatch is supposed to be mitigated or absent due to the deeper conduction band edge. Considering the conduction band of the perovskite layer (-4.21 eV), the Ta-doped TiO_2 prepared at the bubbler temperature of 85 °C is used for device fabrication due to the least conduction band offset.

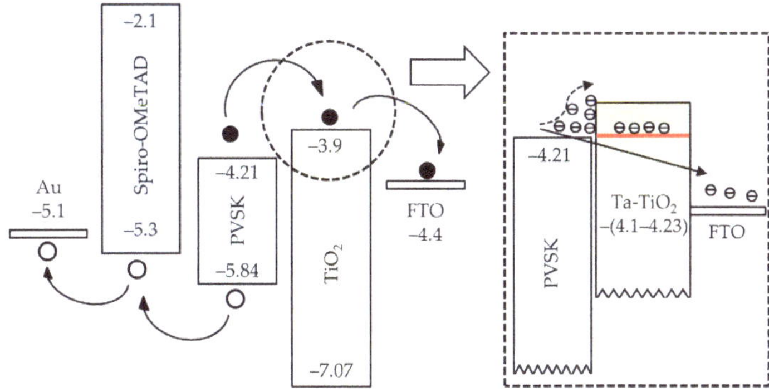

Figure 7. Isolated energy band diagram of each material involved in a TiO_2 ETL-based PSC. The inset indicates the interfaces of perovskite/TiO_2 and perovskite/Ta-doped TiO_2 ETL.

Figure 8a shows the band-bending of energy levels in the PSCs with undoped TiO_2 and Ta-doped TiO_2 ETL deposited at the bubbler temperature of 85 °C. The major difference is at the ETL/perovskite interface. A conduction band barrier is clearly presented at the undoped TiO_2/perovskite absorber interface; however, the barrier disappears in the case of using Ta-doped TiO_2 ETL due to the very little conduction band mismatch. This is expected to be reflected on the improvement in FF of the devices. In addition, as shown in the inset, the slope of the conduction band of the Ta-doped TiO_2 device is steeper than that of undoped TiO_2 device, and this implies a greater built-in electric field, which could enhance the separation of photo-generated electron-hole pairs. The J–V characteristics of the fabricated PSCs in the forward and reverse scans are shown in Figure 8b, and the corresponded external photovoltaic parameters are listed in Table 3. Apart from the slightly improved V_{oc} and J_{sc}, the cell with the Ta-doped TiO_2 ETL has a considerably enhanced FF. The SEM images of the PEALD TiO_2 and Ta-doped TiO_2 films are shown in Figure S1. The two samples are compact without observable pinholes, and do not show much difference in surface morphology likely due to the low-Ta doping level. It is believed that the FF improvement of the perovskite solar cells with the Ta-doped TiO_2 ETL is due to the reduction of the conduction band mismatch, thereby decreasing the transport barrier of electrons. This demonstrates the significance of the conduction band alignment of the ETL. In addition to the absence of the transport barrier at perovskite/ETL interface, the improved FF of the device with the Ta-doped TiO_2 ETL is partially owing to the smaller series resistance as assessed from the inverse slope of the J–V curves at the V_{oc}-point. The hysteresis index of the PSCs is investigated, which is given as [46]:

$$Hysteresis = \frac{\int_{OC}^{SC} J_R(V)\varphi(V)dV - \int_{SC}^{OC} J_F(V)\varphi(V)dV}{\int_{OC}^{SC} J_R(V)\varphi(V)dV + \int_{SC}^{OC} J_F(V)\varphi(V)dV},\tag{7}$$

where the subscript R represents reverse scan, F denotes the forward scan, and H is the unit step function. The result reveals the hysteresis of 1.06% for the device with undoped TiO_2 ETL and 0.21% for the cell with Ta-doped TiO_2 ETL. This is in agreement with the findings reported by Kim et al. [47], where PSCs with doped metal oxide ETLs exhibit less hysteresis compared to undoped metal oxides. It is worth mentioning that the Ta-doped TiO_2 film prepared at the bubbler temperature of 90 °C has been also applied to PSC fabrication as it seems to have the lowest resistivity as well as an appropriate conduction band position. However, the resultant device conversion efficiency is lower than in the case of 85 °C-bubbler temperature. The J–V curve and the corresponded photovoltaic performance of the device at the bubbler temperature of 90 °C are shown in Figure S2. The device has conversion efficiency of 17.3% (V_{oc} of 1.1 V, J_{sc} of 22.5 mA/cm^2, FF of 0.7), lower than that (18.09%) of the device with the Ta-doped TiO_2 ETL deposited at 85°C. It is speculated that the performance reduction might be related to the increased defect formation when the impurity doping level increases. This compromises the cell performance, and thus the optimal Ta bubbler temperature occurs at 85 °C.

Table 3. Performance of PSCs measured in the reverse and forward directions.

Parameter	J_{sc} (mA/cm^2)	V_{oc} (V)	FF	η (%)
Undoped TiO_2 (Forward)	22.9	1.09	0.55	13.02
Undoped TiO_2 (Reverse)	22.3	1.10	0.54	13.42
Ta-doped TiO_2 (Forward)	22.8	1.11	0.70	17.91
Ta-doped TiO_2 (Reverse)	23.0	1.11	0.71	18.09

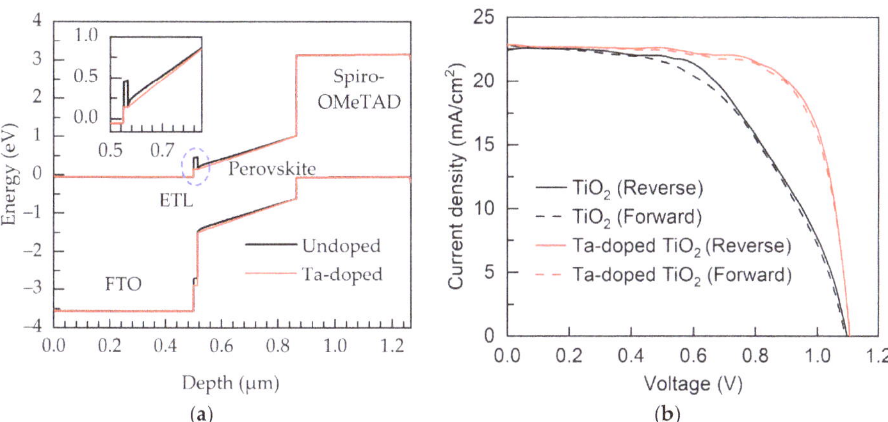

Figure 8. (**a**) Band bending of energy levels in PSCs. (**b**) J–V curves of the PSCs measured in reverse and forward directions. The inset shows the enlarged view at the ETL/perovskite interface.

The performance of 12 devices under reverse scan is shown in Figure 9 for TiO$_2$ and Ta-doped TiO$_2$ perovskite solar cells. The results demonstrate small standard deviations of the device photovoltaic parameters. For instance, the standard deviation of the device efficiency is less than 0.4% for TiO$_2$ devices and 0.46% for Ta-doped TiO$_2$ devices, indicating high reproducibility. Moreover, the average conversion efficiency has the same trend shown in Figure 9, inferring that the comparison of efficiency among different groups is reliable.

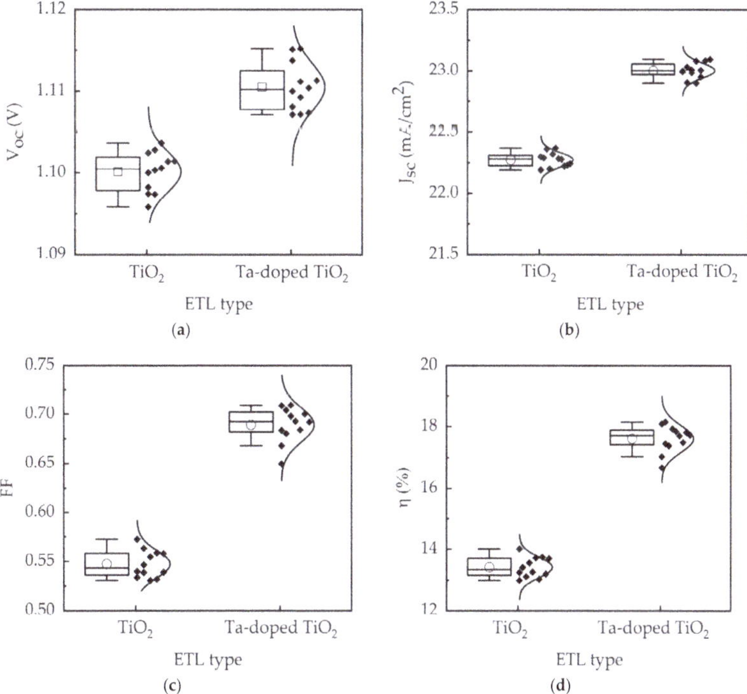

Figure 9. Performance of 12 devices for TiO$_2$ and Ta-doped TiO$_2$ perovskite solar cells: (**a**) V_{OC}, (**b**) J_{SC}, (**c**) FF, and (**d**) η.

4. Conclusions

The Ta-doped TiO_2 thin films are prepared using PEALD with co-injected $Ta(OEt)_5$ and TTIP metal precursors as Ta and Ti sources, respectively. The bubbler temperature varies to obtain different Ta content in the films. The XRD result shows that the Ta incorporation does not lead to additional crystalline phases but reduces crystallinity. As confirmed by XPS, the Ta dopants cause Ti^{4+} ions reducing to Ti^{3+} and Ti^{2+}, due to the creation of the oxygen vacancies defects. The resistivity significantly reduces by three orders at the Ta content of 0.82–0.95 at.%, as compared to that of the undoped TiO_2. The band gap, conduction band, and valence band position can be tailored by the Ta incorporation, and this is favorable for the applications involving interfacial band structure engineering. Finally, the PSC with Ta-doped TiO_2 ETL demonstrates a significantly improved FF from 0.54 to 0.71 and conversion efficiency from 13.42% to 18.09% as compared to the PSC with conventionally undoped TiO_2 ETL.

Supplementary Materials: The following are available online at https://www.mdpi.com/article/10.3390/nano11061504/s1, Figure S1: SEM images of PEALD (a) TiO_2 and (b) Ta-doped TiO_2 deposited at the bubbler temperature of 85 °C, Figure S2: J–V curve of the perovskite solar cell with the Ta-doped TiO_2 ETL deposited at the bubbler temperature of 90 °C.

Author Contributions: Conceptualization, C.-H.H. and S.-Y.L.; methodology, K.-T.C.; formal analysis, C.-H.H., K.-T.C., L.-Y.L., W.-Y.W., L.-S.L., P.G., Y.Q. and W.-Z.Z.; investigation, K.-T.C., X.-Y.Z. and P.-H.H.; writing—original draft preparation, C.-H.H.; writing—review and editing, C.-H.H. and S.-Y.L.; supervision, S.-Y.L.; funding acquisition, C.-H.H., S.-Y.L. and P.G. All authors have read and agreed to the published version of the manuscript.

Funding: This work is supported by a scientific and technological project in Xiamen (no. 3502ZCQ20191002), scientific research projects of Xiamen University of Technology (no. 405011904, 40199029, YKJ19001R, and XPDKQ19006), the Natural Science Foundation of Fujian Province (no. 2020H0025, 2019J01877, 2020J01930). P.G. knowledge 'the National Natural Science Foundation of China (Grant No. 21975260), and the Recruitment Program of Global Experts (1000 Talents Plan) of China'. This study is also sponsored by the Fujian Association for Science and Technology.

Data Availability Statement: Data sharing is not applicable to this article as no new data were created or analyzed in this study.

Conflicts of Interest: The authors declare no conflict of interest.

References

1. Liu, Q.; Wang, J. Dye-sensitized solar cells based on surficial TiO_2 modification. *Sol. Energy* **2019**, *184*, 454–465. [CrossRef]
2. Guo, Q.; Zhou, C.; Ma, Z.; Yang, X. Fundamentals of TiO_2 photocatalysis: Concepts, mechanisms, and challenges. *Adv. Mater.* **2019**, *31*, 1901997. [CrossRef] [PubMed]
3. Wang, S.; Liu, B.; Zhu, Y.; Ma, Z.; Liu, B.; Miao, X.; Ma, R.; Wang, C. Enhanced performance of TiO_2-based perovskite solar cells with ru-doped TiO2 electron transport layer. *Sol. Energy* **2018**, *169*, 335–342. [CrossRef]
4. Wu, M.-C.; Chan, S.-H.; Lee, K.-M.; Chen, S.-H.; Jao, M.-H.; Chen, Y.-F.; Su, W.-F. Enhancing the efficiency of perovskite solar cells using mesoscopic zinc-doped TiO_2 as the electron extraction layer through band alignment. *J. Mater. Chem. A* **2018**, *6*, 16920–16931. [CrossRef]
5. Khlyustova, A.; Sirotkin, N.; Kusova, T.; Kraev, A.; Titov, V.; Agafonov, A. Doped TiO_2: The effect of doping elements on photocatalytic activity. *Mater. Adv.* **2020**, *1*, 1193–1201. [CrossRef]
6. El Ruby Mohamed, A.; Rohani, S. Modified TiO_2 nanotube arrays (TNTAs): Progressive Strategies towards visible light responsive photoanode, a review. *Energy Environ. Sci.* **2011**, *4*, 1065. [CrossRef]
7. Singh, S.; Srivastava, P.; Bahadur, L. Neetu Hydrothermal synthesized Nd-doped TiO_2 with anatase and brookite phases as highly improved photoanode for dye-sensitized solar cell. *Sol. Energy* **2020**, *208*, 173–181. [CrossRef]
8. Sun, J.; Yu, S.; Zheng, Q.; Cheng, S.; Wang, X.; Zhou, H.; Lai, Y.; Yu, J. Improved performance of inverted organic solar cells by using la-doped TiO_2 film as electron transport layer. *J. Mater Sci. Mater Electron.* **2017**, *28*, 2272–2278. [CrossRef]
9. Qu, X.; Hou, Y.; Liu, M.; Shi, L.; Zhang, M.; Song, H.; Du, F. Yttrium doped TiO_2 porous film photoanode for dye-sensitized solar cells with enhanced photovoltaic performance. *Results Phys.* **2016**, *6*, 1051–1058. [CrossRef]
10. Akpan, U.G.; Hameed, B.H. The advancements in sol–gel method of doped-TiO_2 photocatalysts. *Appl. Catal. A Gen.* **2010**, *375*, 1–11. [CrossRef]

11. Cui, Q.; Zhao, X.; Lin, H.; Yang, L.; Chen, H.; Zhang, Y.; Li, X. Improved efficient perovskite solar cells based on ta-doped TiO_2 nanorod arrays. *Nanoscale* **2017**, *9*, 18897–18907. [CrossRef] [PubMed]
12. Carey, J.J.; McKenna, K.P. Screening doping strategies to mitigate electron trapping at anatase TiO_2 surfaces. *J. Phys. Chem. C* **2019**, *123*, 22358–22367. [CrossRef]
13. Ranjan, R.; Prakash, A.; Singh, A.; Singh, A.; Garg, A.; Gupta, R.K. Effect of tantalum doping in a TiO_2 compact layer on the performance of planar Spiro-OMeTAD free perovskite solar cells. *J. Mater. Chem. A* **2018**, *6*, 1037–1047. [CrossRef]
14. Xu, C.; Lin, D.; Niu, J.-N.; Qiang, Y.-H.; Li, D.-W.; Tao, C.-X. Preparation of Ta-Doped TiO_2 Using Ta_2O_5 as the doping source. *Chin. Phys. Lett.* **2015**, *32*, 088102. [CrossRef]
15. Alim, M.A. Photocatalytic Properties of Ta-Doped TiO_2. *Ionics* **2017**, *23*, 1–15. [CrossRef]
16. Pan, X.; Yang, M.-Q.; Fu, X.; Zhang, N.; Xu, Y.-J. Defective TiO_2 with oxygen vacancies: Synthesis, properties and photocatalytic applications. *Nanoscale* **2013**, *5*, 3601. [CrossRef]
17. Zhao, W.; He, L.; Feng, X.; Luan, C.; Ma, J. Structural, electrical and optical properties of epitaxial ta-doped titania films by MOCVD. *Cryst. Eng. Comm.* **2018**, *20*, 5395–5401. [CrossRef]
18. Choi, J.-H.; Kwon, S.-H.; Jeong, Y.-K.; Kim, I.; Kim, K.-H. Atomic layer deposition of Ta-doped TiO_2 electrodes for dye-sensitized solar cells. *J. Electrochem. Soc.* **2011**, *158*, B749. [CrossRef]
19. Hitosugi, T.; Furubayashi, Y.; Ueda, A.; Itabashi, K.; Inaba, K.; Hirose, Y.; Kinoda, G.; Yamamoto, Y.; Shimada, T.; Hasegawa, T. Ta-doped anatase TiO_2 epitaxial film as transparent conducting oxide. *Jpn. J. Appl. Phys.* **2005**, *44*, L1063–L1065. [CrossRef]
20. Bawaked, S.M.; Sathasivam, S.; Bhachu, D.S.; Chadwick, N.; Obaid, A.Y.; Al-Thabaiti, S.; Basahel, S.N.; Carmalt, C.J.; Parkin, I.P. Aerosol assisted chemical vapor deposition of conductive and photocatalytically active tantalum doped titanium dioxide films. *J. Mater. Chem. A* **2014**, *2*, 12849. [CrossRef]
21. Dasgupta, N.P.; Lee, H.-B.-R.; Bent, S.F.; Weiss, P.S. Recent advances in atomic layer deposition. *Chem. Mater.* **2016**, *28*, 1943–1947. [CrossRef]
22. Hajibabaei, H.; Zandi, O.; Hamann, T.W. Tantalum nitride films integrated with transparent conductive oxide substrates via atomic layer deposition for photoelectrochemical water splitting. *Chem. Sci.* **2016**, *7*, 6760–6767. [CrossRef] [PubMed]
23. Jeong, S.; Seo, S.; Park, H.; Shin, H. Atomic layer deposition of a SnO_2 electron-transporting layer for planar perovskite solar cells with a power conversion efficiency of 18.3%. *Chem. Commun.* **2019**, *55*, 2433–2436. [CrossRef] [PubMed]
24. Niu, G.; Li, W.; Li, J.; Liang, X.; Wang, L. Enhancement of thermal stability for perovskite solar cells through cesium doping. *RSC Adv.* **2017**, *7*, 17473–17479. [CrossRef]
25. Pfeiffer, K.; Schulz, U.; Tünnermann, A.; Szeghalmi, A. Antireflection coatings for strongly curved glass lenses by atomic layer deposition. *Coatings* **2017**, *7*, 118. [CrossRef]
26. Zhao, C.; Hedhili, M.N.; Li, J.; Wang, Q.; Yang, Y.; Chen, L.; Li, L. Growth and characterization of titanium oxide by plasma enhanced atomic layer deposition. *Thin Solid Films* **2013**, *542*, 38–44. [CrossRef]
27. Heil, S.B.S.; Roozeboom, F.; van de Sanden, M.C.M.; Kessels, W.M.M. Plasma-assisted atomic layer deposition of Ta_2O_5 from alkylamide precursor and remote O_2 plasma. *J. Vac. Sci. Technol. A Vac. Surf. Film.* **2008**, *26*, 472–480. [CrossRef]
28. Maeng, W.J.; Park, S.-J.; Kim, H. Atomic layer deposition of Ta-Based thin films: Reactions of alkylamide precursor with various reactants. *J. Vac. Sci. Technol. B* **2006**, *24*, 2276. [CrossRef]
29. Chen, H.; Gu, H.; Xing, J.; Wang, Z.; Zhou, G.; Wang, S. Correlated evolution of dual-phase microstructures, mutual solubilities and oxygen vacancies in transparent La_2-Lu Zr_2O_7 ceramics. *J. Mater.* **2021**, *7*, 185–194. [CrossRef]
30. Tauc, J. Optical properties and electronic structure of amorphous Ge and Si. *Mater. Res. Bull.* **1968**, *3*, 37–46. [CrossRef]
31. Deák, P.; Aradi, B.; Frauenheim, T. Band lineup and charge carrier separation in mixed rutile-anatase systems. *J. Phys. Chem. C* **2011**, *115*, 3443–3446. [CrossRef]
32. Wang, J.; Wang, Z.; Huang, B.; Ma, Y.; Liu, Y.; Qin, X.; Zhang, X.; Dai, Y. Oxygen vacancy induced band-gap narrowing and enhanced visible light photocatalytic activity of ZnO. *ACS Appl. Mater. Interfaces* **2012**, *4*, 4024–4030. [CrossRef] [PubMed]
33. Liu, Y.; Zhou, W.; Wu, P. Electronic Structure and optical properties of Ta-doped and (Ta, N)-codoped $SrTiO_3$ from hybrid functional calculations. *J. Appl. Phys.* **2017**, *121*, 075102. [CrossRef]
34. Rahaman, S.; Maikap, S.; Tien, T.-C.; Lee, H.-Y.; Chen, W.-S.; Chen, F.T.; Kao, M.-J.; Tsai, M.-J. Excellent resistive memory characteristics and switching mechanism using a Ti nanolayer at the Cu/TaO_x interface. *Nanoscale Res. Lett.* **2012**, *7*, 345. [CrossRef] [PubMed]
35. Vikraman, D.; Park, H.J.; Kim, S.-I.; Thaiyan, M. Magnetic, structural and optical behavior of cupric oxide layers for solar cells. *J. Alloy. Compd.* **2016**, *686*, 616–627. [CrossRef]
36. Kruse, N.; Chenakin, S. XPS Characterization of Au/TiO_2 catalysts: Binding energy assessment and irradiation effects. *Appl. Catal. A Gen.* **2011**, *391*, 367–376. [CrossRef]
37. Hannula, M.; Ali-Löytty, H.; Lahtonen, K.; Sarlin, E.; Saari, J.; Valden, M. Improved stability of atomic layer deposited amorphous TiO_2 photoelectrode coatings by thermally induced oxygen defects. *Chem. Mater.* **2018**, *30*, 1199–1208. [CrossRef] [PubMed]
38. Wang, J.; Qin, M.; Tao, H.; Ke, W.; Chen, Z.; Wan, J.; Qin, P.; Xiong, L.; Lei, H.; Yu, H.; et al. Performance enhancement of perovskite solar cells with Mg-doped TiO_2 compact film as the hole-blocking layer. *Appl. Phys. Lett.* **2015**, *106*, 121104. [CrossRef]
39. Zhu, L.; Lu, Q.; Lv, L.; Wang, Y.; Hu, Y.; Deng, Z.; Lou, Z.; Hou, Y.; Teng, F. Ligand-free rutile and anatase TiO_2 nanocrystals as electron extraction layers for high performance inverted polymer solar cells. *RSC Adv.* **2017**, *7*, 20084–20092. [CrossRef]

40. Man, G.; Schwartz, J.; Sturm, J.C.; Kahn, A. Electronically passivated hole-blocking titanium dioxide/silicon heterojunction for hybrid silicon photovoltaics. *Adv. Mater. Interfaces* **2016**, *3*, 1600026. [CrossRef]
41. Mazzolini, P.; Gondoni, P.; Russo, V.; Chrastina, D.; Casari, C.S.; Li Bassi, A. Tuning of electrical and optical properties of highly conducting and transparent Ta-doped TiO_2 polycrystalline films. *J. Phys. Chem. C* **2015**, *119*, 6988–6997. [CrossRef]
42. Lin, L.; Jiang, L.; Li, P.; Li, X.; Qiu, Y. Numerical modeling of inverted perovskite solar cell based on CZTSSe hole transport layer for efficiency improvement. *J. Photon. Energy* **2019**, *9*, 1. [CrossRef]
43. Anwar, F.; Mahbub, R.; Satter, S.S.; Ullah, S.M. Effect of different HTM layers and electrical parameters on ZnO nanorod-based lead-free perovskite solar cell for high-efficiency performance. *Int. J. Photoenergy* **2017**, *2017*, 9846310. [CrossRef]
44. Lin, L.; Jiang, L.; Li, P.; Xiong, H.; Kang, Z.; Fan, B.; Qiu, Y. Simulated development and optimized performance of $CsPbI_3$ based all-inorganic perovskite solar cells. *Sol. Energy* **2020**, *198*, 454–460. [CrossRef]
45. Liang, L.; Luo, H.; Hu, J.; Li, H.; Gao, P. Efficient perovskite solar cells by reducing interface-mediated recombination: A bulky amine approach. *Adv. Energy Mater.* **2020**, *10*, 2000197. [CrossRef]
46. Lee, J.-W.; Kim, S.-G.; Bae, S.-H.; Lee, D.-K.; Lin, O.; Yang, Y.; Park, N.-G. The interplay between trap density and hysteresis in planar heterojunction perovskite solar cells. *Nano Lett.* **2017**, *17*, 4270–4276. [CrossRef]
47. Kim, J.K.; Chai, S.U.; Ji, Y.; Levy-Wendt, B.; Kim, S.H.; Yi, Y.; Heinz, T.F.; Nørskov, J.K.; Park, J.H.; Zheng, X. Resolving hysteresis in perovskite solar cells with rapid flame-processed cobalt-doped TiO_2. *Adv. Energy Mater.* **2018**, *8*, 1801717. [CrossRef]

nanomaterials

Review

Recent Advances in Hole-Transporting Layers for Organic Solar Cells

Cinthya Anrango-Camacho [1], Karla Pavón-Ipiales [1], Bernardo A. Frontana-Uribe [2,3] and Alex Palma-Cando [1,*]

[1] Grupo de Investigación Aplicada en Materiales y Procesos (GIAMP), School of Chemical Sciences and Engineering, Yachay Tech University, Hda. San José s/n y Proyecto Yachay, Urcuqui 100119, Ecuador; cinthya.anrango@yachaytech.edu.ec (C.A.-C.); karla.pavon@yachaytech.edu.ec (K.P.-I.)

[2] Centro Conjunto de Investigación en Química Sustentable UAEMex-UNAM, Carretera Toluca Atlacomulco, Km 14.5, Toluca 50200, Mexico; bafrontu@unam.mx

[3] Instituto de Química, Universidad Nacional Autónoma de México, Circuito Exterior, Ciudad Universitaria, Ciudad de México 04510, Mexico

* Correspondence: apalma@yachaytech.edu.ec

Abstract: Global energy demand is increasing; thus, emerging renewable energy sources, such as organic solar cells (OSCs), are fundamental to mitigate the negative effects of fuel consumption. Within OSC's advancements, the development of efficient and stable interface materials is essential to achieve high performance, long-term stability, low costs, and broader applicability. Inorganic and nanocarbon-based materials show a suitable work function, tunable optical/electronic properties, stability to the presence of moisture, and facile solution processing, while organic conducting polymers and small molecules have some advantages such as fast and low-cost production, solution process, low energy payback time, light weight, and less adverse environmental impact, making them attractive as hole transporting layers (HTLs) for OSCs. This review looked at the recent progress in metal oxides, metal sulfides, nanocarbon materials, conducting polymers, and small organic molecules as HTLs in OSCs over the past five years. The endeavors in research and technology have optimized the preparation and deposition methods of HTLs. Strategies of doping, composite/hybrid formation, and modifications have also tuned the optical/electrical properties of these materials as HTLs to obtain efficient and stable OSCs. We highlighted the impact of structure, composition, and processing conditions of inorganic and organic materials as HTLs in conventional and inverted OSCs

Keywords: hole transporting layer; organic solar cells; photoconversion efficiency; stability; metal oxides; metal sulfides; nanocarbon materials; conducting polymers; conjugated polyelectrolyte; small organic molecules

Citation: Anrango-Camacho, C.; Pavón-Ipiales, K.; Frontana-Uribe, B.A.; Palma-Cando, A. Recent Advances in Hole-Transporting Layers for Organic Solar Cells. *Nanomaterials* **2022**, *12*, 443. https://doi.org/10.3390/nano12030443

Academic Editor: Vlad Andrei Antohe

Received: 13 December 2021
Accepted: 24 January 2022
Published: 28 January 2022

Publisher's Note: MDPI stays neutral with regard to jurisdictional claims in published maps and institutional affiliations.

1. Introduction

Solar energy has enough power capacity to satisfy the whole world's demand [1,2]. According to Luqman et al. [3], the amount of solar energy irradiated at the Earth's atmosphere ranges from 200 to 250 Wm^{-2} per day, of which ca. 70% is available for conversion into power generation [4,5]. Research on solar energy technology, which aims to convert sunlight directly into electrical energy, is vital to switch into low-carbon energy systems [6,7]. The intense developments concerning solar energy have boosted the investigations to optimize the efficiency and stability of emerging photovoltaic technology, such as dye-sensitized solar cells, organic solar cells (OSCs), perovskite solar cells, quantum dot solar cells, and so on, of which OSCs are one of the most promising technologies [8–14]. Since Kearns and Calvin's pioneering work on OSCs in 1958, one significant breakthrough in solar energy technology has been the efficient electron transfer between a conjugated polymer and fullerene derivative [15,16]. It encouraged the interest in the light-harvesting of OSCs from the structure device into the materials used for their construction [16–19]. OSCs are based on organic semiconductors as active layers with unique advantages to achieve low-cost renewable energy harvesting, owing to their material and manufacturing advances [20,21]. OSCs have some advantages: low-cost fabrication, solution processes, light weight, flexibility, a great opportunity for large-scale roll-to-roll production, and a low environmental disposable impact. Furthermore, OSCs showed shortened energy payback time, that is, the time needed to recover the device fabrication energy [22]. OSCs' record efficiency is over 18.2% in single cells [23,24] and over 18.6% in tandem cells [25]. In the last ten years, extensive research and development have been conducted in OSCs to improve lifetimes (7–10 years) and the power conversion efficiency (PCE) over 10% in roll-to-roll industrial manufacturing [23,26–28]. Some companies have developed OSCs commercially, such as Heliatek, infinityPV, and OPVIUS GmbH, which manufacture flexible OSC modules [29] representing 5% of the market [30]. OSCs have been used in small-scale applications as building integrated photovoltaics, e.g., incorporated on roofs and walls of storage buildings and solar parks [31].

The core of the OSCs is a blend of electron-donor materials (e.g., conjugated polymers) and fullerene-based or non-fullerene-based electron-acceptor materials [32]. This central layer is called the photoactive layer and absorbs solar radiation. A typical OSC has a bulk heterojunction (BHJ) structure that is a mixed-blend of donor and acceptor materials, which constitute the photoactive layer [33]. When the solar cell is irradiated, the photoactive layer absorbs photons to generate excitons (bound electron-hole pairs), which dissociate into free charge carriers in the donor-acceptor interface, producing separated holes and electrons. These free charges are then extracted and transported to the corresponding electrodes [34,35]. Interfacial layers are generally utilized to tailor the work function (WF) of electrodes for the maximization of charge carrier (e.g., electrons and holes) collection. They modify the interface to alter the photoactive layer morphology and minimize charge carrier recombination (improving the charge selectivity) at the interface between the active layer and transport layer [36]. Moreover, the interfacial layers help to form an ohmic contact between electrodes and active layers as well as tune the energy level alignment to facilitate the charge extraction [37,38]. Hole-transporting layers (HTLs), also called anode interfacial layers (AILs), facilitate hole extraction and transportation while blocking electron flux. Hole-transport materials are deposited between the photoactive layer and the anode, improving the device performance. HTLs, used in conventional polymer solar cells (PSCs), were first reported in the late 1990s after a similarly reported experimentation in organic light-emitting diodes (OLEDs) [39,40]. Some important characteristics are required for hole-transport materials such as a high conductivity, high transparency (since the sunlight is absorbed by the photoactive layer through the HTL on anode), solution processability and favorable stability, high WF (since the energy level of materials should be appropriate for charge collection), and predominantly good hole mobility [39].

Over the past five years, the research community has been working on achieving high efficiency and stability and low cost of production on emerging clean energy sources, such as OSCs, with a priority on interfacial layer engineering. Tian et al. analyzed the diverse molecular structures employed as HTL and electron transporting layers (ETLs) to minimize energy losses in non-fullerene OSCs [41]. Palilis et al. discussed the relationship between the optoelectronic and physical properties of inorganic materials and their functionality at the interface [42]. Gusain et al. showed the physical mechanisms involved with the interfacial issues and the routes adopted to address them [43]. Amollo et al. explored the physical and optical properties of polymers and metal oxides together with their hybrids and graphene to guide the choice of suitable interfacial materials [44]. Wu et al. showed the impact of nanotechnology and nanomaterials in manufacturing multifunctional interfacial layers to enhance OSCs' performance [45]. Huang et al. reviewed the feasibility of tuning the optical and electrical properties of solution-processed ternary oxides, as potential carrier transports layers, from the large range of crystal structures and adjustable atomic ratio [46]. Herein, we presented an extensive state-of-the-art review about the advances in HTLs that show great potential for enhancing the efficiency (e.g., PCE) and stability of OSCs. The progress made on improving HTL properties of inorganic (metal oxides and sulfides), nanocarbon materials, conjugated polymers, and small organic molecules as HTLs in OSCs was discussed, focusing on solution-processing conditions, deposition methods, doping, composite/hybrid formation, and chemical modifications. Considering the numerous and highly dispersed literature, we tried to include relevant information reported in scientific journals. This review included a short section on the structure and characterization of OSCs and some remarks on HTLs followed by reports of the last five years in the use of hole-transporting materials as HTLs in OSCs. Summary tables of the photovoltaic device architecture and their performance are presented at the end of Sections 4.3 and 4.5.

2. Structure and Characterization of Organic Solar Cells

A conventional OSC consists of an active layer sandwiched between two electrodes with their respective extracting layers to ensure mobility, collection, and transport of the charge carriers [47]. At the bottom, the anode electrode is a transparent conductive oxide, such as indium tin oxide (ITO), and at the top, the cathode is a low WF metal, such as Ca and Al (see Figure 1) [48]. The OSCs based on two organic semiconductors in the active layer can have two architectures: the bilayer and the BHJ devices. Tang et al. presented the sequential stacking of donor and acceptor semiconductors to form the bilayer planar heterojunction in 1986 [17]. However, it has limitations, such as the small surface area between the donor/acceptor interface and the poor excitons' dissociation. Then, the introduction of BHJ devices in 1990 solved bilayer devices' issues [19]. They involve mixing donor and acceptor materials in the bulk body of an OSC to reduce phase separation. Donor and acceptor domains are twice the size of the exciton diffusion length (~10 nm). To expand the active layer's absorption range, tandem OSCs have been proposed to stack two single-junctions with different absorption ranges [49,50]. According to the charge flow direction, OSCs can be divided into conventional and inverted devices (see Figure 1) [51]. Under light irradiation, photons are absorbed by the donor material in the active layer to form excited states, called excitons, which are bound electron-hole pairs. Excitons diffuse towards the donor/acceptor material interface and separate into free charge carriers. Holes and electrons move apart in the highest occupied molecular orbital (HOMO) and the lowest unoccupied molecular orbital (LUMO) levels, respectively (see Figure 1). Then, the separated charge carriers are transported and collected at the electrodes supplying a photocurrent [52].

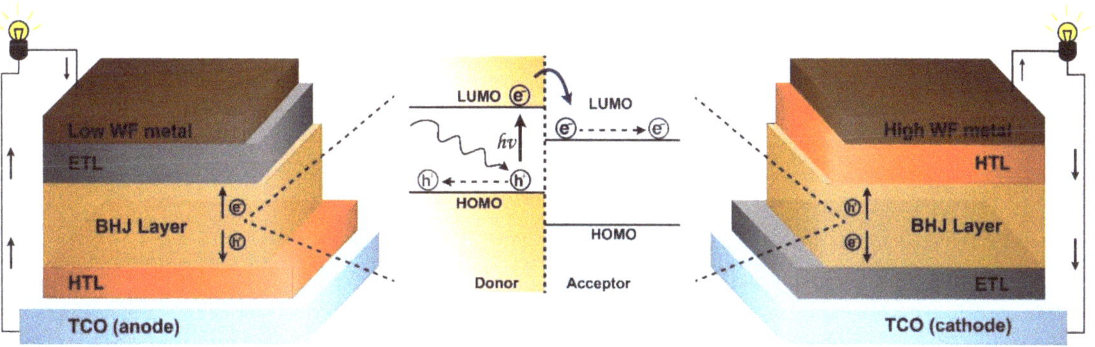

Figure 1. Schematic structure of conventional and inverted OSCs, and a simplified view of the operating principle in the active layer.

The current-voltage (J-V) curve of an OSC is characterized under 1000 W/m^2 light of AM 1.5 solar spectrum [53]. Figure 2a shows a J-V curve of an OSC under darkness (dashed line) and illumination (solid line) conditions. Photocurrent is not flowing through the electrodes under dark conditions, just the current by the forward bias of contacts as a diode. Under irradiation, photocurrent is generated. PCE is determined by the product of three parameters: short-circuit current density (J_{sc}), open-circuit voltage (V_{oc}), and fill factor (FF) over the incident light power density (P_{in}) as follows [54]:

$$PCE = \frac{V_{oc} \times J_{sc} \times FF}{P_{in}} \qquad (1)$$

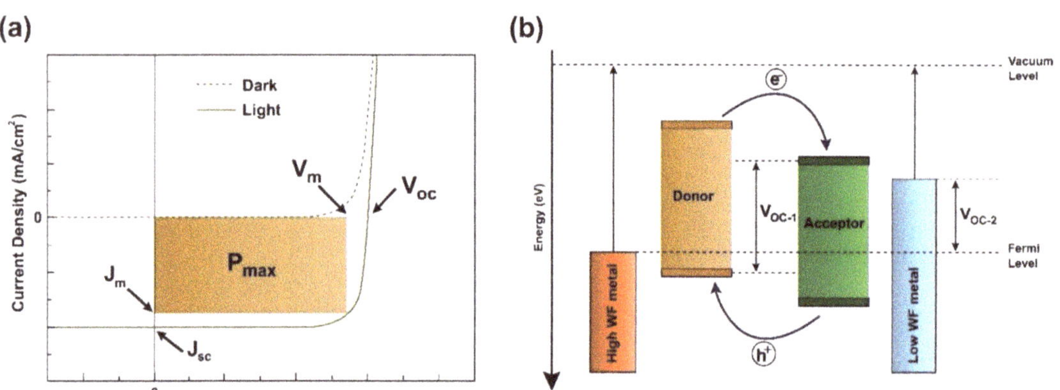

Figure 2. (**a**) Typical J-V curve and (**b**) schematic representation of the V_{OC} in a conventional OSC.

The ratio of collected photogenerated charges and the number of incident photons is related to the external quantum efficiency (EQE) of the OSC. V_{oc} is the main driving force for charge separation once the exciton reaches the donor/acceptor interface [55,56]. V_{oc} is the difference of WFs between the quasi-Fermi levels of holes ($E_{F,h}$) in the HOMO level of the donor and the quasi-Fermi levels of electrons ($E_{F,e}$) in the LUMO level of the acceptor in a BHJ under the formation of ohmic contacts with the cathode and anode (depicted as V_{oc-1} in Figure 2b). If a Schottky contact appears in both BHJ/electrode interfaces, the V_{oc} would decrease and would depend on the difference between the WFs of the two metal contacts (depicted as V_{oc-2} in Figure 2b) [57,58]. FF is the ratio between the maximum power output (P_{max}) and the maximum attainable power output ($J_{sc} V_{oc}$). P_{max} describes the maximum power drawn from the device and is the product of the maximum current (J_m) and voltage (V_m) (see Figure 2a) [59], as follows:

$$FF = \frac{P_{max}}{J_{sc} V_{oc}} = \frac{J_m V_m}{J_{sc} V_{oc}} \qquad (2)$$

The main factors that influence the FF are the series resistance (R_s) and the shunt resistance (R_{sh}). Their interaction determines the current flow. R_s is attributed to the conductivity of electrodes, BHJ and extracting interface layers, as well as the contact resistance between them [60]. A small R_s increases the mobility of the charge carriers and the performance of OSCs. R_{sh} reflects the current losses from the pinholes and traps in the film. Established relationships describing J-V behavior in OSCs and directly accounting for resistance effects on cell performance are the following [61–64]:

$$J = J_d + J_{sh} - J_{ph} = J_0 \left\{ \exp\left[\frac{e(V - JR_s)}{nk_BT} - 1\right] \right\} + \frac{V - J R_s}{R_{sh}} - J_{ph} \qquad (3)$$

where, J_d is the diode current density, J_{sh} is the leakage current density, J_{ph} is the photogenerated current density, J_0 is the reverse saturation current, e is the elementary charge, n is the diode ideality factor, k_B is Boltzmann's constant, and T is temperature. J_{sh} is an undesirable current injected from the electrodes in the opposite direction to J_{sc}. A suitable interface morphology decreases J_{sh} and increases R_{sh} independently of the light intensities [65]. Thus, the contact quality at the active layer/electrode interface is critical to optimize FF, V_{oc}, and J_{sc}. Interface transporting layers enhance all these parameters because they tune the energy level alignment at the active layer and electrodes, the surface morphology, and the contact to boost the efficiency and stability of the OSCs [66].

3. Hole-Transporting Layers

Interfacial layers are critical components of OSCs to enhance the collection efficiency of holes and electrons toward the anode and cathode electrodes. In photovoltaic devices, including OSCs, there are barriers to charge extraction by the non-ideal contact between the active layer and the electrodes [39]. This limited interfacial energy alignment inhibits the spontaneous charge transport, resulting in charge accumulation at the interface, thus decreasing Voc, FF, and PCE [67]. Interfacial layers with suitable WFs contribute to match the energy levels of donor and acceptor materials with the electrodes, favoring the charge transport and stability [38]. The interfacial layers must be charged selectively to avoid charge recombination at the electrodes in addition to the tuning of the energy levels. HTLs and ETLs increase the hole and electron mobility in the opposite direction to collect only one type of charge on each electrode [68]. In the 1990s, HTLs were introduced to the organic electronics field by Tokito et al., who showed that hole-injection increased from inserting vanadium, molybdenum, and ruthenium oxides layers into OLEDs [69]. HTL's central role is the efficient hole extraction and transport from the HTL/active interface to the anode/HTL interface, increasing power generation [70]. To achieve high-performance OSCs, the materials used for HTLs need to show (i) high WF that matches with the HOMO energy level of the donor material and the anode energy level, (ii) transparency to increase the light absorption by the active layer, (iii) high hole mobility to lower the charge accumulation and recombination, (iv) a large band gap to block electron carriers, and (v) chemical resistance to external factors [38,71,72]. The first materials used as HTLs in OSCs were inorganic p-type transition metal oxides (MoO_3, WO_3, NiO, Fe_3O_4) or metal sulfides (MoS_2), which showed high stability and performance [73–77]. Most of them required high vacuum for deposition, which, compared with organic materials, might be costly for industrial and large-scale processing [78]. Poly(3,4-ethylenedioxythiophene)-poly(styrene sulfonate) (PEDOT:PSS) is still the standard conducting polymer used as HTL in OSCs because of the low costs, minimal toxicity, facile solution processing, and high WF. However, it is not stable at standard conditions owing to its hygroscopic and acidic nature [79]. Currently, there is an excellent development of cost-effective low-temperature deposition strategies for industrial scaling to avoid the traditional vacuum method used in the manufacture of HTLs. Casting process deposits the material dissolved in liquid form in a solvent on the underlying substrate, followed by drying. Spray casting solves the lack of control in film morphology and uniformity [80]. Spin coating is the most common deposition method of PEDOT:PSS due to its high reproducibility in film thickness and morphology. It applies the spinning at a certain rotation speed of the substrate to dry the deposited liquid material. However, neither large area applicability nor film patterning are achievable by this technique [81]. Electrochemical deposition or electrodeposition allows depositing polymers and inorganic materials through an electric field [82]. The control on deposition has broader applicability for the formation of composites [83]. The roll-to-roll technique is usually utilized in flexible OSCs because the flexible substrate is unwound to pass through printing or coating machines, followed by being rewound on a roll. It opens the applicability for large-area production because substrates are not handled individually but instead in rolls [84,85]. Compared with the vacuum method, these deposition techniques offer the advantage of a continuous and large-area process at mild conditions, avoiding wasting raw materials.

4. Hole-Transporting Materials as HTLs in OSCs

4.1. Metal Oxides

4.1.1. Molybdenum Oxide

MoO_x is an *n*-type material with a valence band edge around 2.5–3 eV below the Fermi level and a conduction band closer to the Fermi level [70]. MoO_3 has a high WF (6.9 eV) and conductivity of 1.2×10^{-7} Sm^{-1} due to the different states of O and the multivalence of Mo in its three crystal phases (α-MoO_3, β-MoO_3, h-MoO_3) [86–88]. MoO_3 is a promising HTL due to its electronic structure, transparency, conductivity, and stability, enhancing the hole extraction and thus the efficiency of OSCs, compared with PEDOT:PSS [89]. Lee et al. reported that MoO_x HTL-based OSCs are more stable at a high operating temperature near 300–420 K than PEDOT:PSS [90]. Therefore, there is much research in strategies to optimize the solution-processing methods and the film properties of MoO_3 [91–93]. Bortoti et al. obtained the orthorhombic phase of MoO_3 (α-MoO_3) by refluxing MoS_2 in HNO_3 and H_2SO_4 as the oxidant media, followed by heating at 120 °C for 10 min to evaporate the solvent [94]. The energy level of α-MoO_3 well-matched with that of the P3HT. A PCE of 1.55% was obtained in a FTO/ZnO/P3HT:PC_{60}BM/MoO_3/Ag cell structure. Ji et al. used ammonium heptamolybdate (AHM) as the precursor solution to prepare a solution-processed MoO_3 array on P3HT:PC_{61}BM by the ultrasonic spray-coating method at 80 °C [95]. The solution-processed MoO_3 micro arrays improved the charge transport between the active layer and the anode. Thus, the V_{oc} and FF increased to 0.59 V and 59.2%, and a higher PCE of 3.40% was achieved. MoO_3 is adequate to attain a high built-in potential and V_{oc} because it can suppress interfacial reactions at the HTL/BHJ interface. MoO_3 nanoparticles (NPs) can be added at the interface between the active layer and the PEDOT:PSS to take advantage of the localized surface-resonance plasmon (LSRP) effect of NPs and the electronic structure of MoO_3 [96]. MoO_3 NPs increased the path length of the absorbed light and blocked the electrons flow to the anode, resulting in a higher J_{sc} and FF, and thus a PCE of 4.11% was reached over a long period of 30 days [97]. The high transparency of MoO_x allows an enhanced back-reflected light into the active layer to enhance the photocurrent, as shown in the EQE curves (see Figure 3a) [98]. At low temperatures of 80–200 °C, Jagadamma et al. prepared an alcohol-based MoO_x nanocrystalline suspension processed directly over temperature-sensitive active layers (see Figure 3b) [99]. The water-free solvent and the fine MoO_x nanocrystal diameter (<5 nm) resulted in a compact and smooth film with a thickness around ~5–10 nm. All inverted OSCs reached a PCE above 9%, retaining 90% of their efficiency after five months of aging. MoO_3 nanocrystals (NCs) with a size greater than 5 nm can form a composite of MoO_x with Ag nanowires (NWs) to lower the nanowire junction resistance by close packing Ag NWs. The Ag NWs/MoO_x composite also served as a barrier for Ag diffusion into the active layer's bulk. Wang et al. added AgAl NPs into MoO_x HTL to prevent the Ag diffusion by forming AlO_x [100]. The PTB7-Th:PC_{71}BM cell retained 60% of the initial PCE (9.28%) over 120 days. Cong et al. used ammonium molybdate and citric acid in 2-methoxyethanol as the precursor to prepare MoO_x, followed by 10 vol.% of H_2O_2 to form a stable conductive film [101]. The presence of H_2O_2 induced oxygen vacancies to help in the polyvalence and conductivity of the MoO_x film. Jung et al. prepared a solution-processed MoO_x from the dissolution of MoO_x powder in ammonium hydroxide (NH_4OH) and isopropanol solvent [102]. The Mo^{5+}-OH bonds induced by hydroxyl radicals facilitated the charge transport with higher hole mobilities, of 2.3×10^{-6} $cm^2V^{-1}s^{-1}$, than PEDOT:PSS, of 2.1×10^{-6} $cm^2V^{-1}s^{-1}$. The gap states induced in the bandgap by the oxygen defects tuned the Fermi level of MoO_x with the HOMO of PBDB-T as the donor material, showing overall improvement in FF and J_{sc} with a PCE of 10.86%. The excess of oxygen vacancies during the film formation results in recombination sites which compromise the performance and stability of the OSC [37]. Kobori et al. improved the J_{sc} and FF when the as-deposited solution-processed MoO_x film was annealed at 160 °C for 2 min [103]. The enhancement in the efficiency from 1.40% to 6.57% is because of surface passivation of MoO_x HTL by annealing treatment, resulting in a reduction of oxygen vacancies in the MoO_x film (see Figure 3c). It helps the fabrication of OSCs with temperature-sensitive low-bandgap polymers, such as PTB7-Th:PC$_{71}$BM and PCPDTBT:PC$_{71}$BM. Li et al. reported that low-temperature annealing treatment could also enhance the preparation of solution-processed MoO$_x$ films from peroxomolybdic acid organosol precursor solution at 150 °C, while also achieving passivation of the surface [104].

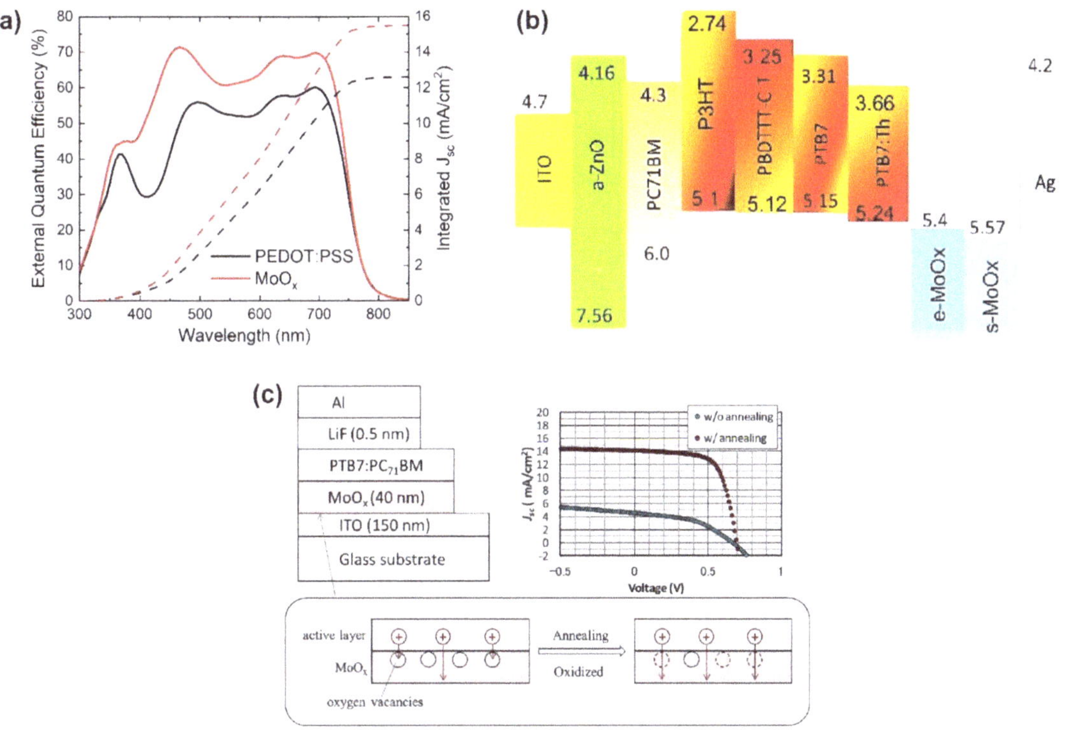

Figure 3. (**a**) External quantum efficiency of inverted OSCs with MoO$_3$ and PEDOT:PSS HTLs. Adapted with permission from [98]. (**b**) Energy level diagram of inverted OSCs with different donor polymers. Adapted with permission from [99]. Copyright 2016, Elsevier. (**c**) Reduction of oxygen vacancies by annealing treatment. Adapted with permission from [103]. Copyright 2016, Elsevier.

Ultraviolet (UV) annealing can retain a higher PCE over a longer period if compared with OSCs' efficiency under no annealing or under thermal annealing (100 °C) [105]. UV annealing removed the adhered organic contaminants on the MoO$_3$ film surface by two short wave UV lights at 185 nm and 285 nm. This radiation decomposed O$_3$ into O$_2$ and active O, which oxidized and removed any organic contaminant by transformation into volatile gases. Cai et al. achieved a PCE of 9.27% in the PBDB-T:ITIC BHJ cell using an ultraviolet-deposited MoO$_3$ film [106]. Tan et al. developed a solution-processed, annealing-free aqueous MoO$_x$ for non-fullerene OSCs [107]. By adding a small amount of water to MoO$_2$(acac)$_2$, the ligand of MoO$_2$(acac)$_2$ was removed from the MoO$_x$ film, avoiding thermal treatments, and enhancing the PCE of PBDB-T-2F:Y6 cell up to 17.0%. In addition to the impurities in the precursor solution, external factors, such as air, create oxygen defects in the MoO$_x$ film lattice, which change the electric properties (e.g., WF, energy levels) and the performance of the OSC [108,109]. Soultati et al. reported the microwave (MW) air annealing approach for recovering the WF in stoichiometric MoO$_x$ and the efficiency of the FTO/MW-MoO$_x$/P3HT:PC$_{71}$BM/Al cell up to 5.0% [110]. The WF recovery resulted in the formation of a large interfacial dipole at the FTO/MW-MoO$_x$/P3HT:PC$_{71}$BM interfaces, favoring hole extraction via gap states.

In addition to post-treatments, the film properties of the MoO_3 HTL in OSCs also improve through strategies involving doping, composite/hybrid formation, multilayers, and deposition techniques. Chang et al. reported vanadium-doped MoO_x films at different ammonium metavanadate concentrations. The smallest band offset (1.13 eV) between the valence band edge of $V_{0.05}MoO_x$ and P3HT HOMO level favored the hole transport due to having the lowest resistance among all V-MoO_x films [111]. Marchal et al. reported a decrease of 3 nm in the surface roughness of MoO_x HTL by adding 0.5 mol% of Zr and Sn via a combustion chemical deposition method at low temperatures [112]. The Zr and Sn atoms also covered the surface defects of MoO_x, forming a uniform and well-covered HTL film on the ITO electrode (see Figure 4a). Bai et al. employed a small amount of p-type NiO_x into n-type MoO_3 in one step [113]. Since MoO_3:NiO_x was highly transparent and had a conduction band of 3.25 eV and a WF of 5.10 eV (see Figure 4b), the MoO_3:NiO_x film was able to block electrons while enhancing the contact to charge transport toward the anode, achieving a PCE of 10.81% in PBDB-T:IT-M BHJ OSCs. Li et al. showed the feasibility of the work function tuning of MoO_x to use as both HTL and ETL through the Cs intercalation approach [114]. MoO_x and the intercalated mole ratio MoO_x:Cs (1:0.5) tested in P3HT-based conventional and inverted OSCs as HTL and ETL achieved PCEs of 3.50% and 3.20%. Besides, high PCEs of 7.35% and 6% were obtained in the PBDTDTTT-S-T-based conventional and inverted OSCs. The Cs-intercalation within the MoO_x acts as an n-type semiconductor [115] to tune the work function from 5.30 to 4.16, which favors the energy alignment at the interface and the reduction in the charge carrier losses. Yoon et al. synthesized a dual-HTL by mixing solution-processed copper iodide (CuI) and thermally evaporated MoO_3 [116]. The interaction between MoO_3 and the CuI increased the forbidden gap states in the MoO_3 layer for the hole transport by forming small oxygen vacancies and Mo^{5+} defect states. Zhiqui et al. reported a composite of copper bromide ($CuBr_2$) and molybdenum trioxide (MoO_3) as the HTL for OSCs [117]. CuBr optimized interfacial contact to increase charge carriers, and MoO_3 blocked electron transport, resulting in improved FF (65.20%), J_{sc} (19.65 mAcm^2), and an increase in the PCE from 7.30 to 9.56%. Li et al. prepared CTAB-modified MoO_3 nanocomposites by adding a small amount of cetyltrimethylammonium bromide (CTAB) solution into ammonium molybdate and annealing it at 200 °C in a glovebox [118]. CTAB passivated the surface traps of MoO_3 films to avoid the recombination sites, resulting in a film with PCEs of $5.80 \pm 0.13\%$ in P3HT:ICBA and $8.34 \pm 0.13\%$ in PTB7:PC_{71}BM OSCs. The formation of polynuclear metal-oxo clusters (PMC) of tungsten/molybdenum as HTLs showed PCEs of 14.3% and higher stability than PEDOT:PSS [119]. The variation in the W/Mo ratio allowed the increase of the hole transport from the polymer donor (PBDBT-2F) toward the anode due to the formation of an inorganic-organic charge transfer complex with a barrier-free interface. This unique characteristic of PCM clusters in OSCs might promote new insights for its utility in high-performance optoelectronic devices. Kwon et al. also boosted the efficiency by developing an alloy of molybdenum-tungsten disulfides films as HTL to replace PEDOT:PSS efficiently [120]. As was mentioned before, Ag NPs can be incorporated into MoO_3 to enhance the electrical and optical properties of the HTL. Indeed, it can form a MoO_3/AgNPs/MoO_3 structure as HTL to improve the J_{sc} and reduce the recombination by the backscattering and surface plasmon effects of AgNPs [121]. Zhang et al. prepared a solution-processed MoO_3/AgNPs/MoO_3 (MAM) HTL in PTB7:PC70BM cells [122]. The MAM multilayer enabled an enhanced charge collection by suppressing charge recombination. The efficiency of the OSC was superior (7.68%) to that of the s-MoO_3 (6.72%). The manufacture of OSCs has also been limited by the material's finite availability, such as the transparent anode electrode, ITO [123,124]. An ITO-free flexible OSC obtained by Chen et al. used multiple layers of molybdenum oxide MoO_3/LiF/MoO_3/Ag/MoO_3 as transparent electrodes, facilitating the transmittance and charge transport [125]. Lee et al. reported a reduced atomic percentage of In and Sn at the surface of ITO electrodes by graded sputtering of MoO_3 HTLs (see Figure 4c) [126]. The MoO_3-graded ITO (MGI) electrode formed three regions: (i) the bottom ITO region, providing high transparency

(83.8%), (ii) the Mo-In-Sn-O graded interlayer, and (iii) the MoO_3 region, which served as the HTL. For thin HTLs, the deposition method might cause defects or form compacted layers depending on the working conditions. Uniform s-MoO_x HTLs prepared by direct current (DC) magnetron sputtering showed enhanced charge transport with a FF of 50%, as the s-MoO_x film's surface was smoother and controlled by DC in comparison with the conventional evaporated approach [127]. Chaturvedi et al. applied a DC voltage of 1 kV during the spray deposition of MoO_3 HTL, obtaining a PCE of 2.71% [128]. The applied electric field controlled both the optical and electrical properties of the thin MoO_3 film. Dong et al. used a laser-assisted method to obtain a hydrogenated molybdenum oxide H_yMoO_{3-x} film for flexible OSCs (see Figure 4d) [129]. By controlling the energy of the KrF laser (λ = 248 nm) during the irradiation of photons on the AHM precursor solution, the WF (5.6 eV) and the hole transport of H_yMoO_{3-x} film increased, allowing higher PTB7:PC_{70}BM cell performance. The laser processing time lasts only 30 ns, so it is suitable in time and economically compared with the thermal evaporation method.

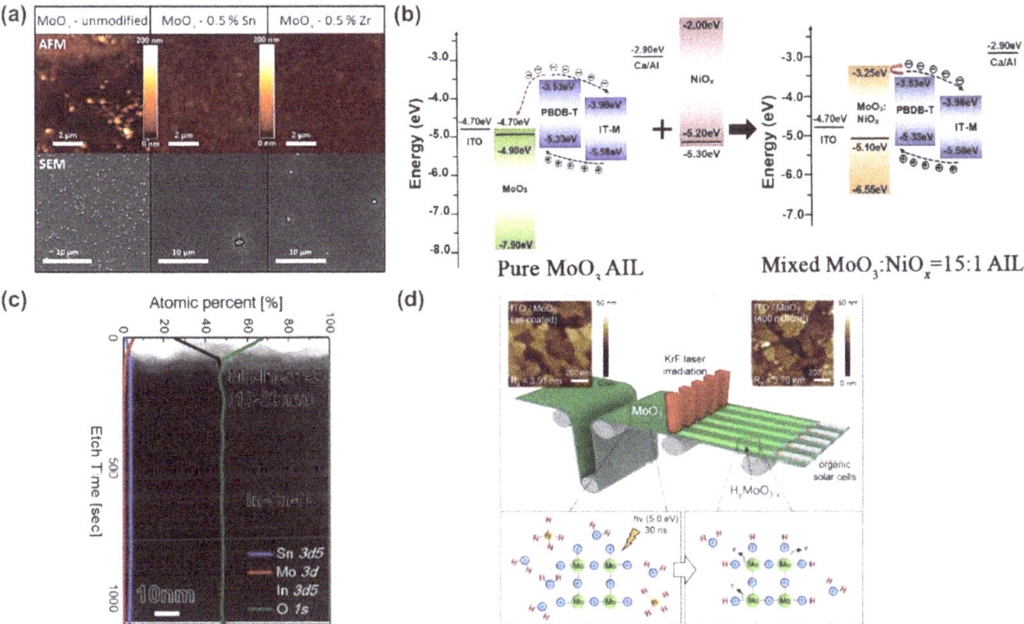

Figure 4. (**a**) AFM and SEM images of unmodified and modified MoOx HTL. Adapted with permission from [112]. (**b**) Energy levels of pristine and doped MoO_3 with a NiOx layer. Adapted with permission from [113]. Copyright 2019, Elsevier. (**c**) XPS depth profile of MGI electrode. Adapted with permission from [126]. Copyright 2016, Elsevier. (**d**) Scheme of laser-assisted synthesis of H_yMO_{3-x}. Adapted with permission from [129]. Copyright 2016, Royal Society of Chemistry.

4.1.2. Tungsten Oxide

Tungsten oxide is an *n*-type material with a WF ranging from 4.7 to 6.4 eV depending on the film preparation [130–133]. Tungsten oxide is a hole extracting layer that can work efficiently in conventional and inverted OSCs using vacuum and solution-processing methods [74,134]. WO_x is an amorphous structure that (i) forms smooth surface morphologies, (ii) increases the charge mobility in the active layer, and (iii) enhances the charge collection because V_{oc} depends linearly on the anodic WF when there is not ohmic contact at the anode/donor interface [135]. Thus, the enhancement in solution-processing WO_x-based OSCs is particularly focused on increased light absorption. Lee et al. designed an Au@SiO_2-WO_3 nanocomposite (NC) which works as a photon antenna for high light absorption [136]. The localized surface-plasmon resonance (LSPR) effect of AgNPs enhances the intensity of photon absorption in the P3HT:PC$_{61}$BM BHJ cell, resulting in increased J$_{sc}$. The favorable plasmonic effect is comparable to some reported literature for plasmonic nanomaterials-based optoelectronic devices [137–139]. Moreover, high hole mobility of WO$_x$ NPs boosted the device PCE by up to 1.6%. The surface morphology of the Au@SiO$_2$-WO$_3$ NC film was kept uniform due to the SiO$_2$ shell avoiding the aggregation effect of the Au NPs. Instead of SiO$_2$, the aggregation effect can be avoided by controlling the concentration of Au NPs. Using 10 wt% of Au NPs, the Au-WO$_3$ NC HTL decreased the surface morphology's roughness, achieving a PCE of 60.37% [140]. Shen et al. enhanced the light absorption and the PCE of OSCs based on the LSRP effect of structure-differentiated silver nano-dopants in solution-processed WO$_x$ HTL [141]. Three silver nano-dopants, (i) naked Ag NPs (nAgp), (ii) SiO$_2$-covered Ag NPs (SiAgp), and (iii) naked Ag nanoplates (nAgPl), were synthesized. The triangular nAgPl reached the highest PCE of 4.6% while spherical nAgp reached the lowest. The spherical nAgp surface decreased the PCE because its surface can directly contact the donor/acceptor material of the active layer, resulting in excitons quenching and thus weakening LSRP effects (see Figure 5). The shape of NPs affects the overall performance of OSCs by tuning plasmon-electrical [142], plasmon-optical [143,144], and charge-storage effects [145]. Ren et al. reported the high efficiency of OSCs by incorporating gold nanostars (Au NSs) between HTL and the active layer [146]. The plasmonic asymmetric modes of Au NSs enhanced the optical absorption of the active layer and the balance of photogenerated charges by shortening transport path length in the HTL. The localized plasmonic effect of NPs manipulates transport paths of photogenerated carriers in bulk heterojunction OSCs and thus reduces the charge recombination sites and the space-charge-limit effect [147,148]. Li et al. reported comparable results by applying Ag nanoprisms to achieve higher PCE through the improvement in the broadband absorption [149]. Remya et al. performed a study between dehydrated and di-hydrated WO$_3$ films as HTL in the inverted P3HT:PC$_{61}$BM and PTB7:PC$_{71}$BM cells [150]. The hydrated phase of WO$_3$ enabled a suitable energy level alignment with the active layer by tuning the water coordination, resulting in a higher PCE of 5.1% and 7.8%, respectively.

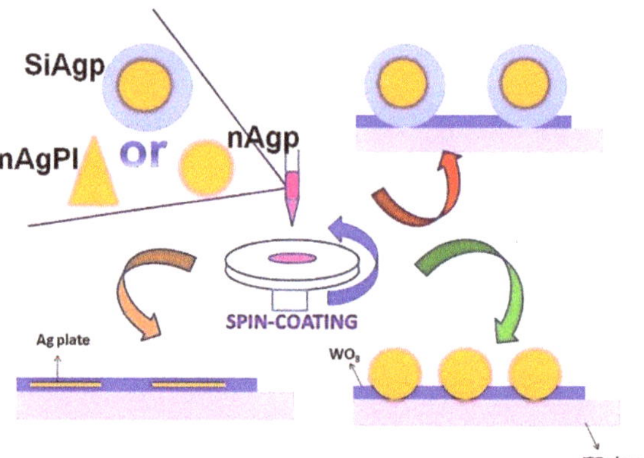

Figure 5. Schematic representation of different silver nanoparticles structures/WO$_3$ layers by spin coating. Adapted with permission from [141]. Copyright 2016, Elsevier.

4.1.3. Vanadium Oxide

V$_2$O$_5$ is a hole-transporting/electron-blocking layer that acts as a protecting layer [151] avoiding surface reactions by the moisture from the working conditions, resulting in improved efficiency and stability [152]. The electronic structure of V$_2$O$_5$ corresponds to an *n*-type material with deep electronic states and WF ranging from 4.7 eV to 7.0 eV, depending on the processing method [73,153]. Li et al. reported the Cs-intercalation method to tune the work function of V$_2$O$_x$ and used Cs-intercalated V$_2$O$_x$ and V$_2$O$_x$ as both ETLs and HTLs in organic optoelectronic devices [114]. The work function tuning and the reduction in the interfacial barrier of Cs-intercalated V_2O_x allowed for obtaining PCEs in P3HT and PBDTDTTT-S-T-based conventional and inverted OSCs up to 3.59% and 7.44%. Xu et al. reported a low-temperature solution-processed V_2O_5 by dissolving V_2O_5 powder into water at room temperature [154]. V_2O_5-based HTLs showed a PCE of 8.05% in $ITO/V_2O_5/PTB7:PC_{70}BM/LiF/Al$ OSCs compared with PEDOT:PSS-based HTLs with a PCE of 7.46%. V_2O_5 served as an optical spacer that increased light absorption, leading to a higher photocurrent. V_2O_5 powder can also be treated directly from the melting-quenching sol-gel method to obtain an easy tunable $V_2O_5 \cdot nH_2O$ HTL [155]. The energy positioning of the $V_2O_5 \cdot nH_2O$ HTL (with $n = 1$) was closer to PEDOT:PSS [67], allowing an ohmic contact with the novel conjugated polymer donor (PBDSe-DT2PyT) and the acceptor of $P_{71}CBM$; thus, a large V_{oc} and a PCE of 5.87% were obtained. The layered and hydrated phase of V_2O_5 is an affordable and tunable charge transport material. Although $V_2O_5 \cdot H_2O$-based HTLs exhibit better performance than PEDOT:PSS-based HTLs, the melting–quenching sol-gel method might be an expensive method owing to the high melting temperature of V_2O_5 (~800°C). Cong et al. applied a green method to prepare vanadium oxide hydrate layers ($VO_x \cdot nH_2O$) to enhance the PCE in organic $PTB7$-$Th:PC_{71}BM$- and $P3HT:PC_{61}BM$-based polymer solar cells up to 8.11% and 3.24% [156]. The combined H_2O_2 and ultraviolet ozone (UVO) in-situ treatments allowed for a smooth surface and improved wettability with the presence of dangling bonds on the HTL surface to enhance interfacial contact. The presence of V^{4+} in the composition analysis of $VO_x \cdot nH_2O$ accounted for a small amount of oxygen vacancies, causing n-type doping, which is essential to hole transport by extracting electrons through its conduction band [157]. Vishnumurthy et al. reported that V_2O_5 HTL optimized the efficiency of thienothiophene-diketopyrrolopyrole-based OSCs by up to 1.02% [158]. Remya et al. prepared an efficient hole-transport/electron-blocking hydrated vanadium oxide (HVO) from V_2O_5 powder with hydrogen peroxide [159]. In the $P3HT:PC_{61}BM$ and $PBDTT$-$FTTE:PC_{71}BM$ BHJ cells, HVO HTL performance was superior to PEDOT:PSS, obtaining 56% enhancement (7.12–11.14%) in the PCE for the $PBDTT$-$FTTE:PC_{71}BM$-based inverted OSC with a lower degradation of 1.4% over 20 weeks. In addition to the V_2O_5 powder, V_2O_5 HTLs can be prepared by other precursors. Xu et al. reported an ammonium metavanadate ammonal water solution for processing VO_x HTLs in $PTB7:PC_{71}BM$ BHJ cells with a PCE of 7.7% [160]. This HTL showed a WF of 5.3 eV and high conductivity by air-annealing treatment at 210 °C for 5 min. The thermal treatment smoothed the surface film to reduce the leakage current, obtaining a higher J_{sc}. Although the stability was better than PEDOT:PSS with a remaining 83% efficiency after four days, it was still low compared with other inorganic HTLs. Shafeeq et al. reported the formation of uniform and crystalline V_2O_5 nanorods by thermal decomposition of ammonium metavanadate NH_4VO_3 to enhance surface morphology and efficiency of OSCs [161]. Alsulami et al. obtained a stable V_2O_x HTL by using vanadium (V) oxytriisopropoxide as the precursor, which converted into V_2O_x by hydrolysis in air [162]. The PCE of the V_2O_x HTL was insensitive to thermal annealing at 100 °C and 200 °C because its optical and electronic properties were comparable to the vacuum-deposited V_2O_5. Besides, the highly tunable V_2O_5 thin films prepared by the solution-processing method boost inverted OSCs because of their higher stability under air conditions [163]. To optimize the interface properties and OSC performance, VO_x NP can efficiently be mixed with PEDOT:PSS solution, resulting in a stable $VO_x:PEDOT:PSS$ HTL by the uniform molecular distribution of VO_x with PEDOT:PSS as reported by Teng et al. [164]. They achieved a PCE, of 10.2%, compared with PEDOT:PSS, of 5.27%, when $VO_x:PEDOT:PSS$ was used as HTL in the TPD-3F:IT-4F cells. Xia et al. reported a nanoparticulate compact V_2O_5 film as HTL using a facile metal-organic decomposition method to replace the traditional HTLs [165]. By adding polyethylene glycol (PEG) as an additive in the precursor, a uniform and compact film of V_2O_5 served as HTL in the PTB7:PC70BM, improving the interface contact, J_{sc}, and the FF. Compared with the spin coating, the spray coating of V_2O_5 HTL has allowed the large-scale production of flexible OSCs in a roll coater [166]. Using a precursor solution of vanadium oxytriisopropoxide (VTIP) diluted in ethanol (1:100), V_2O_5 HTLs exhibited improved electrical properties. The mechanical stress on V_2O_5 HTL was mitigated by introducing a PEDOT:PSS binding-interfacial layer between V_2O_5 HTL and the Ag electrode in the inverted P3HT:PC_{60}BM and PBDTTTz-4:PC_{60}BM BHJ cells. Arbab and Mola also explored electrochemical deposition that resulted in 80% enhancement in PCE (2.43%) compared with PEDOT:PSS-based OSCs [167]. Kavuri et al. reported electrospray deposition (ESD) for V_2O_5 HTL in PTB7:PC_{71}BM-based OSCs with a PCE of 7.61% [168]. Compared to the spin-coating, the ESD allowed more control in the deposition conditions and reduced the manufacturing costs of V_2O_5-based OSCs. Surface morphology, charge mobility, and interfacial contact were adjusted as a function of the solvent evaporation rate. V_2O_5 HTL has also been effective in ITO-free polymer solar cells with an optimized precursor solution (VTIP) of 0.005% [169]. The deposition of V_2O_5 HTL on PEDOT:PSS, as the anode, led to increase R_{sh} and conductivity with the active layer of P3HT:PC_{61}BM by the hydrophobic surface of V_2O_5, resulting in an uniform and compact HTL with a PCE of 3.33%. V_2O_5 is also a potential material that increases the anode's WF of indium zinc oxide (IZO), exhibiting a higher PCE of 2.8% than that flexible OSCs with only IZO [170].

4.1.4. Nickel Oxide

Non-stoichiometric NiO_x is a wide bandgap p-type semiconductor [171]. NiO_x is an efficient electron-blocking layer to the anode due to its conduction band minimum, 1.8 eV, which is above the LUMO of the organic donor P3HT (3.0 eV) [75]. Due to the conduction band of NiO_x being closer to the vacuum level, it is able to suppress electron recombination at the anode [172]. The ohmic contact between NiO and P3HT allows holes to freely transport from the active layer to the anode through the Ni^{2+} vacancy-based hole-conducting anode band [173]. Parthiban et al. demonstrated an enhancement in OSC performance with a NiO HTL deposited via spin coating [174]. Using the precursor solution of nickel acetate and a simple post-annealing process (>300 °C) to reduce roughness, NiO HTL achieved a FF of 63.0% and a corresponding PCE of 4.45% in RP(BDT-PDBT):PC_{70}BM solar cells. Although NiO-based HTLs exhibit better performance and stability than PEDOT:PSS, the high annealing temperature required to convert the nickel precursors into the NiO thin films make it expensive and not compatible with flexible substrates. Chavhan et al. reported a room-temperature approach to manufacture NiO_x films from a nickel formate precursor solution via UV-ozone treatment [175]. In terms of efficiency, the UV-ozone treatment results were ideal for increasing the WF by creating hydroxides at the surface, avoiding high processing temperatures. A high PCE of 6.1% in NiO_x HTL treated with UV-ozone was related to increased presence of NiO(OH) at the surface. Besides the precursor method, Jiang et al. used chemical precipitation to obtain non-stoichiometric NiO_x NPs at room temperature without any post-treatment [176]. The atomic ratio between Ni and O (1:1.14) reduced the R_s of the opto-electronic device as the p-type conductivity was enhanced by the presence of two oxidation states (Ni^{2+} and Ni^{3+}) that favor Ni^{2+} vacancies. Thus, the FF and the J_{sc} increased up to 67.20% and 9.67 $mAcm^{-2}$ to yield a PCE of 3.81% in P3HT-based conventional OSCs. The high performance of NiO_x NPs HTL-based OSCs was also demonstrated for low-bandgap polymers. Alternatively, p-type ternary metal oxides are promising candidates for enhancing electron-blocking ability due to their tunable electronic and optical properties through the hypocrystalline hydroxide-based method [177]. To date, the high surface roughness of fluorine-doped tin oxide (FTO) has limited its application in OSCs; however, the surface roughness can be decreased from 10.36 nm to 6.74 nm by fully covering it with an optimized NiO layer (see Figure 6) [178]. A polyethylene glycol (PEG) assisted sol-gel process altered the c-NiO/FTO surface because it has a stabilizing effect on NiO NPs, so it allowed the crystallization of a close-packed structure of NiO film. The further deposition of PEDOT:PSS led to the formation of a free-pinhole layer with an RMS roughness of 2.44 nm and selective hole transport, increasing the PCE from 5.68% to 7.93%. Although organic devices based on spin-coated NiO HTLs have emerged successfully in the photo-electronic field, it is vital to focus research efforts for printing technologies for large-area roll-to-roll production. Printing technology usually results in thick NiO films, increasing the interfacing between the active and HTL layers and shortening hole carriers' migration due to its short lifetime [179]. Singh et al. obtained a thin film of NiO_x by controlling substrate-processing conditions and inkjet printing [180]. Optimal conditions of UVO pretreatment, drop spacing, and substrate temperature at 25 °C resulted in a PCE of 2.60% in the P3HT:PC_{60}BM cell with superior environmental stability. Huang et al. used copper (5.0 at.%) as a dopant to increase the electrical conductivity of NiO_x film, resulting in a reduction of R_s from 11.25 to 9.98 Ωcm^2 [181]. The Cu-doped NiO_x (Cu:NiO_x) also improved the interface contact with the active layer and facilitated the charge transport, resulting in a higher PCE of 7.1% in PCDTBT:PC_{71}BM-based cells. The enhancement in the optoelectronic properties, surface morphology, and stability of NiO_x HTL by doping is comparable with reported literature, as observed in the co-doping of NiO_x NPs with Li and Cu [182]. The co-doping favored the conductivity by increasing the Ni^{3+}/Ni^{2+} ratio and kept the high transparency in the well-dispersed solution based on NiO_x NPs.

Figure 6. AFM and SEM images of bare nickel oxide on FTO (without PEG) and compact nickel oxide (c-NiO) on FTO (with PEG). Adapted with permission from [178].

4.1.5. Other Oxides

CuO_x are *p*-type semiconductors with narrow band gaps of 1.3–2.0 eV for CuO and 2.1–2.3 eV for Cu_2O [183–186]. HTLs of CuO_x spin-coated on ITO decreased the interfacial barrier using a green solvent of copper acetylacetonate ($Cu(C_5H_7O_2)_2$), improving cell efficiency of PTB7:PC71BM cell up to 8.68% [187]. After H_2O_2 and UVO treatment, CuO_x HTLs increased the WF to 5.45 eV, forming an excellent ohmic contact, while the V_{oc} increased to 0.74 V. Furthermore, the oxidation of CuO_x by UVO treatment enhanced the interfacial contact and the light absorption in the visible range, obtaining a high transmittance of 88%, low R_s of 2 Ωcm^2, and higher hole transport to the anode. The OSCs' initial performance (8.68%) dropped down to 47% over 50 h of storage in the air. The *p*-type $CuCrO_2$ is a semiconductor that belongs to the delafossite compounds [188]. $CuCrO_2$ HTLs are of great interest in optoelectronic applications due to their high transparency, large hole diffusion coefficient, high WF, and ionization energy, which are essential in the manufacture of OSCs [189–191]. Other strategies to boost the potential of $CuCrO_2$ HTL involve In doping, in which optical transmittance and hole conductivity are increased [192]. New alternative techniques to produce efficient and cost-effective $CuCrO_2$ HTLs for roll-to-roll manufacturing are developing, such as microwave assisted-heating to produce $CuCrO_2$ nanocrystals with an efficient PCE of 4.9% [193], or the combustion synthesis to produce $CuCrO_2$ thin films by low-temperature processing at 180 °C with a PCE of 4.6% (see Figure 7) [194]. Both methods are highly efficient and represent advances for lowering fabrication costs. UV-ozone post-treatment or annealing increases the metallic copper oxidation to Cu^{+2} to promote the electronic conduction by the hopping mechanism between Cu^{1+} and Cu^{+2} species. The higher oxidation state of Cu^{2+} enhanced the electronic properties, exhibiting deeper ionization energy (IE) and Fermi energy (E_F). The Cu doping favored the surface-roughness reduction, resulting in an improved interfacial contact, and thus favored J_{sc}, FF, and PCE.

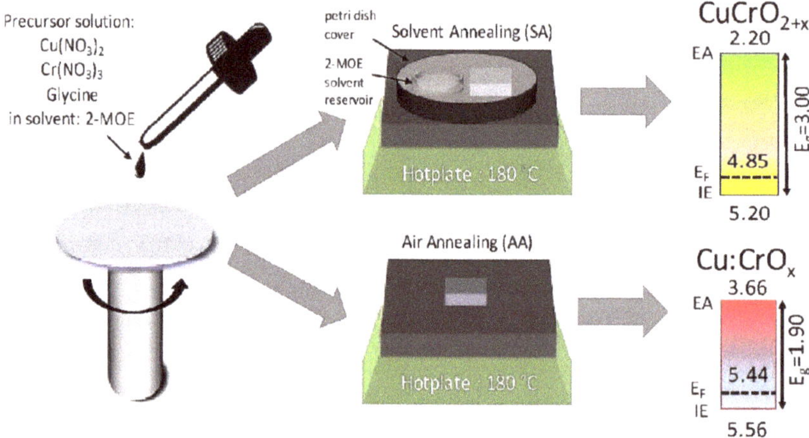

Figure 7. Schematic representation of the combustion synthesis. Adapted with permission from [194]. Copyright 2018, American Chemical Society.

Wahl et al. reported the first HTL based on ITO NPs in inverted OSCs [195]. The addition of ethylenediamine into ITO NPs stabilized it to deposit uniform HTLs on the underlying absorber layer. The deposition of the ITO NPs HTLs by doctor blading allowed controlling the thickness between 15 and 20 nm. Post-treatments of thermal annealing and plasma were beneficial for the film's electronic properties, achieving a PCE of 3%. However, plasma application needs to be mild to avoid OSCs' detrimental performance. The doping method using high-WF metals might be a good alternative over plasma treatments to develop high-quality films in OSCs. The solubility of metal oxides in common solvents such as DMF or water is another main factor for its application in the roll-to-roll manufacturing of OSCs. Bhargav et al. reported the suitability of DMF-soluble Co_3O_4 as HTLs in $PCDTBT:PC_{71}BM$ BHJ [196]. Co_3O_4 HTLs showed transparency around 81% and a smooth surface, allowing for a remarkably high FF of 49.1% and higher PCE (3.21%) compared with PEDOT: PSS-based OPVs.

4.2. Metal Sulfides

4.2.1. Molybdenum Disulfide

MoS_2 with a layered structure is a metal dichalcogenide (TMD) semiconductor that can display two phases under normal conditions, the traditional trigonal prismatic $H-MoS_2$ phase and the distorted octahedral $ZT-MoS_2$ phase with hole mobilities of 3.8×10^2 $cm^2V^{-1}s^{-1}$ and 5.7×10^4 $cm^2V^{-1}s^{-1}$, respectively [197]. Instead of using a vacuum or temperature-dependent process to prepare the traditional MoS_2 HTL, Barrera et al. prepared suspensions of MoS_2 via liquid exfoliation at room temperature [198]. The high WF of MoS_2 resulted in enhanced charge mobility; however, the low transmittance of the film affected the J_{sc}. An effective way to address films' low transmittance is by using composites or hybrid layers with tunable transparency. Martinez-Rojas et al. reported a hybrid layer of MoS_x:MoO_3 on FTO substrates with high transmittance by a pulsed electrochemical method [199]. After 150 cycles of depositing MoS_x on the MoO_3, the percentage of transmitted light decreased significantly due to the agglomeration of MoS_x (see Figure 8a). A hybrid layer with 100 cycles of deposition resulted in 10% higher PCE than the one obtained using MoO_3 or MoS_x HTL. MoS_x was an efficient electron-blocking layer, while MoO_3 increased conductivity, resulting in enhanced hole-transporting properties. The effectiveness of UVO treatment to form homogeneous films and increase the WF was tested in a layer of MoS_2 quantum dots (QDs), showing a PCE of 2.62% and 8.7% for P3HT and PTB7-Th donor systems [200]. The solar cell efficiency increased after 30 min of UVO exposure, but longer UVO treatment periods degraded the HTL, resulting in decreased PCEs (see Figure 8b). The UVO-MoS_2 QDs showed compact and uniform layers with a lower surface roughness of 1.19 nm than UVO-MoS_2 nanosheets of 2.03 nm. The OSC achieved long-term durability due to the improved interfacial contact, showing 64% of its initial PCE after 47 days (see Figure 8c). Annealing treatments can also decrease the surface roughness and favor the optoelectronic properties of the film. At 300 °C, MoS_x flatted the surface morphology, enhancing the PCE by up to 7.5%, 52% of which was retained after two months [201]. However, an annealing treatment is not as efficient as a UVO treatment for temperature-sensitive devices.

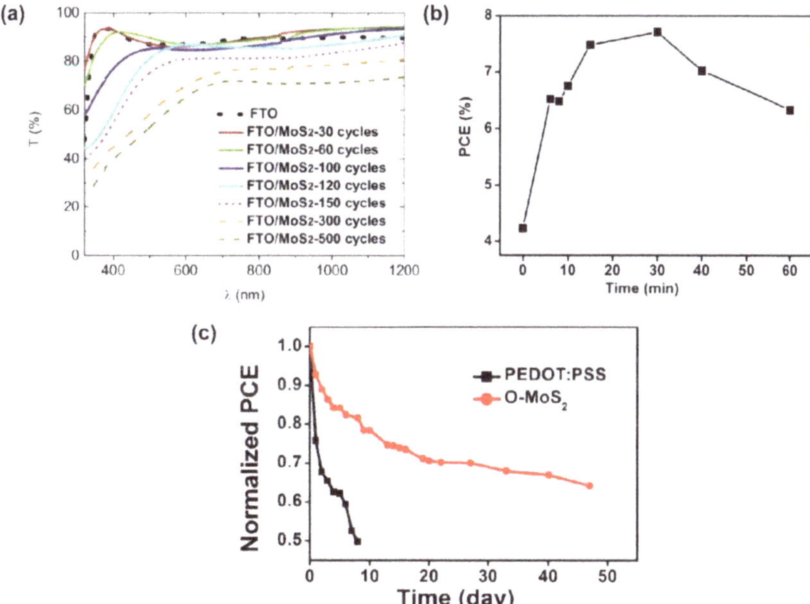

Figure 8. (**a**) Transmission spectra of FTO/MoS_2 at various numbers of scan cycles. Adapted with permission from [199]. Copyright 2017, Elsevier. (**b**) Dependence of PCE on the duration of the UVO treatment and (**c**) stability of devices using UVO-MoS_2 QDs compared with PEDOT:PSS. Adapted with permission from [200]. Copyright 2016, American Chemical Society.

4.2.2. Tungsten Disulfide

Adilbekova et al. used a liquid-phase exfoliation technique to manufacture WS_2 HTL using aqueous ammonia that does not require high-temperature post-treatments [202]. Stabilizers or post-processing treatments were excluded from obtaining WS_2 nanosheets since stoichiometric quality and structural properties were unchanged after performing the top-down method. Due to the *p*-type character of the 2D nanosheets, the HTLs were selective to hole transport toward the anode, achieving a PCE of 15.6% in the PBDBT-2F:PC_{71}BM BHJ cells. Following the same line, Lin et al. fabricated uniform WS_2 layers on

ITO [203]. WS$_2$ flakes were wider and thinner than MoS$_2$, covering the whole surface of ITO. The surface coverage was dependent on the shape and size of the selected material obtained by the exfoliation procedure and its interaction with the substrate. WS$_2$-based HTL in ternary BHJ OSCs (PBDB-T-2F:Y6:PC$_{71}$BM) increased PCE by up to 17%. Ram et al. demonstrated that the use of WS$_2$ as HTL increased the PCE of PBDB-T-2F:Y6:SF(BR)$_4$ ternary cells by 20.87% [204]. The low hygroscopic nature and low acidity of WS$_2$ reduced the contact resistance between the active layer and the ITO.

4.2.3. Nickel Sulfide

Taking advantage of the dependence of the phase diagrams of NiS with the sulfur content, Hilal and Han synthesized the hexagonal phase of NiS as HTL in OSCs processed by the simple solvothermal method at room temperature [205]. The surface morphology of NiS was smoothened by increasing the sulfur content to 2 g, forming a globular flower-like NiS morphology with increased surface area (see Figure 9). In addition to the enhancement in the hole transport, NiS stabilized the OSC; hence, P3HT:PCBM-based cells retained 26% of their initial efficiency value after 15 days.

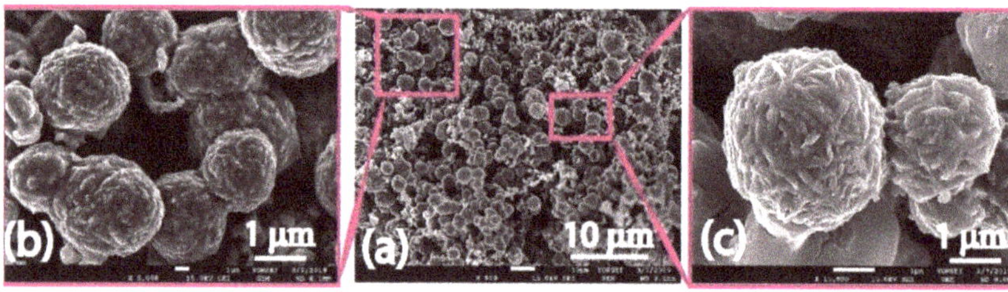

Figure 9. FE-SEM images of globular flower-shaped NiS at magnifications of (**a**) 10 μm and (**b,c**) 1 μm. Adapted with permission from [205]. Copyright 2019, Elsevier.

4.2.4. Other Sulfides

An efficient OSC was achieved by Bhargav et al. using an inorganic HTL made of CuS by a low-cost and efficient manufacturing process [206]. CuS thin films were deposited onto ITO by a solution process instead of vacuum deposition, resulting in a high transparency of 84%. Due to the decreased ohmic resistance, the device structure ITO/CuS/PTB7:P$_{71}$BM/Al reached a high PCE of 4.32% due to the improved FF of 50.1%. A new room-temperature method known as Successive Ionic Layer Adsorption and Reaction (SILAR) was reported by Jose et al. to produce efficient *p*-type Zn-doped CuS HTLs [207]. Due to the high conductivity and low light absorption in the visible region, a PCE of 1.87% was obtained with enhanced charge mobility of 1.5 cm^2 V^{-1}s^{-1}. The use of 2D materials like antimonene quantum dots (AMQS) in HTLs has emerged in OSCs production due to their facile synthesis and unique properties [208]. Wang et al. reached an enhanced PCE of 8.8% by the surface passivation of copper(I) thiocyanate (CuSCN) HTL with AMQSs [209]. The AMQSs smoothened the film surface of CuSCN, tuned the WF, and raised the exciton generation rate from 8.79 × 10^{27} m^{-3}S^{-1} to 9.95 × 10^{27} m^{-3}S^{-1}. Compared with PEDOT:PSS HTLs, CuSCN/AMQSs HTLs were more stable at room temperature, retaining 68% of the initial PCE over 1 month not only in fullerene systems such as PTB7- Th:PC$_{71}$BM, but also in non-fullerene systems. Other strategies involving triple-interface passivation [210], multifunctional interface layer using lead sulfide quantum dots (QDs) [211], and self-polymerization of the monomer have been also reported to passivate surface roughness and interface defects [212]. The surface passivation is key in the construction of OSCs to reduce non-radiative recombination losses which in turns affect the charge separation rate once excitons achieve the donor/acceptor interface, resulting in a low V$_{oc}$ and FF. The *p*-doping of CuSCN with C$_{60}$F$_{48}$, an electron acceptor, is an effective method to obtain highly conductive HTLs for its application in OSC devices [213]. By adding 0.5 mol% of C$_{60}$F$_{48}$ that also acts as a nucleating agent, the CuSCN:C$_{60}$F$_{48}$ film was more dense than the pristine

CuSCN surface. Moreover, reduced surface roughness, leakage current (see Figure 10a), and improved hole mobility of 0.18 cm^{2}V^{-1}s^{-1} were attributed to the percolation conduction mechanism, resulting in a PCE of 6.6% in the PCDTBT:PC$_{70}$BM-based OSCs. Wang et al. achieved a PCE of 15.28% in OSCs based on the non-fullerene PM6:Y6 blend by doping CuSCN film with 1% of TFB (see Figure 10b) [214]. Worakajit et al. increased the hole mobility in CuSCN from 0.01 to 0.05 cm^{2}V^{-1}s^{-1} by passivating surface morphology and the crystallinity with diethyl sulfide (DES) molecules and acetone as antisolvent treatment [215]. Suresh Kumar et al. succeeded in fabricating Cu$_2$CdSnS$_4$ (CCTS) HTLs over ITO substrates deposited by spin coating at room temperature [216]. A PCE of 3.63% in the P3HT:PC$_{71}$BM blend was achieved by controlling the distribution particle size. The bandgap decreases with an increase in the size of CCTS and layer thickness. Minimum surface roughness of 11.07 nm was found after deposition of three layers of CCTS thin films, improving the thin film's compactness, hole-transport efficiency, and stability in environmental conditions.

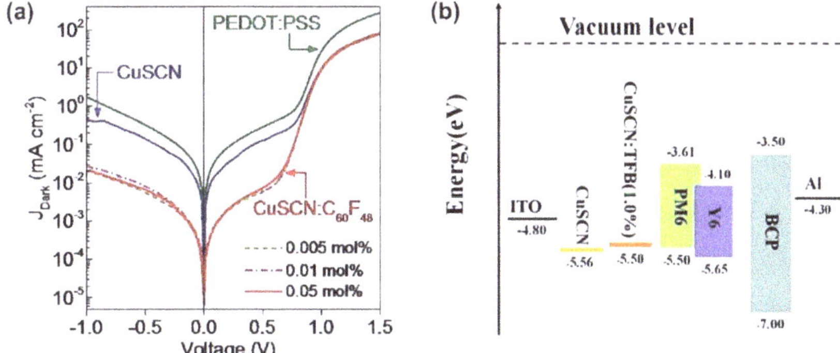

Figure 10. (a) J-V curves of CuSCN:C$_{60}$F$_{48}$ under dark conditions. Adapted with permission from [213]. Copyright 2018, John Wiley and Sons. (b) Energy level alignment of ITO/CuSCN, CuSCN:TFB(1.0%)/PM6:Y6/BCP/Al OSCs. Adapted with permission from [214]. Copyright 2020, American Chemical Society.

4.3. Nanocarbons

4.3.1. Graphene Oxide

Nanocarbon materials like graphene have been applied as HTLs in OSCs due to their unique electrical, optical, and structural properties [217]. Due to the low water dispersibility caused by the nonpolar sp^2 hybridized carbon structure, the oxidized form of graphene, graphene oxide (GO), has also been used in OSCs by the high solubility in eco-friendly water solvents [218]. The hydroxyl groups and epoxy groups located in the basal plane of the graphene sheet and carboxylic acids at the edge limit the conductivity of GO [219]. In fact, an excess of 25% of oxygen atoms on the GO sheet's surface reduced its conductivity until it became an insulator material [220]. Thus, it is crucial to control the concentration and thickness of GO for suitable performance as HTLs. Rafique et al. tested the thickness and concentrations of spin-coated GO, selecting 1 mg/mL to form thin conductive films in BHJ OSCs with a PCE of 2.73% [221]. The reduction process is another feasible way to increase the conductivity of GO layers. The reduction removes the excess of oxygen atoms from the GO surface and recovers the conjugated honeycomb structure [222]. Huang et al. succeeded in synthesizing eco-friendly reduced graphene oxide (rGO) by using a modified Hummer's method to produce GO and thermal treatment to reduce it [223]. A mild temperature of 280 °C was used to obtain rGO and enhance OSCs' conductivity based on P3HT:PC$_{71}$BM and PTB7:PC$_{71}$BM with a PCE of 3.39% and 7.62%, respectively. The dispersibility must be controlled to ensure good coverage of the underlying substrate. Lee et al. mixed highly dispersible semiconducting fullerenol surfactant with GO, obtaining water-dispersible and conductive films [224]. The conductivity increased from 5×10^{-4} Scm^{-1} for the pristine GO layer to 1×10^{-2} Scm^{-1} for the fullerenol-GO layer, resulting in a PCE of 3.15%. Chemical and physical methods involving the reduction of GO seek to tune the WF, improve electrical properties, reduce absorption, and increase hole mobility and charge collection capability. Kwon et al. obtained rGO by electron-beam irradiation with shorter processing times than reported gamma (γ)-rays [225]. Following the same line, Fakharan et al. applied a YAG-pulsed laser to produce rGO in formic acid for OPVs with a PCE of 4.02% [226]. They also highlighted the solvent's role in manufacturing devices to achieve an rGO with superior physical and electrical features. Unlike the traditional chemical-reduction methods, pulsed laser or electron-beam allowed the reduction of graphene over the in-situ formation of reducing species selectively. Dericiler et al. reported graphene nanosheets prepared from the electrochemical exfoliation of graphene powder followed by dispersion in DMF solvent [227]. They used the graphene nanosheets suspension as an additive to PEDOT:PSS HTLs to enhance the stability and charge mobility in the P3HT:PC$_{60}$BM, achieving 66% enhancement in the PCE compared with the reference cell based on pure PEDOT:PSS HTLs.

The application of UVO irradiation has shown excellent efficiency in reducing GO in large-scale OSCs manufacturing. Xia et al. [228] and Rafique et al. [229] exposed GO to UVO treatment, resulting in optimized performance in P3HT:PC$_{71}$BM and PCDTBT:PC$_{71}$BM blend systems. UVO oxidizes the surface of GO and removes CO_2 molecules, leaving a uniform, smoothed, and conductive film. Ultraviolet irradiation was controlled to remove only C-O bonds from the GO surface. UVO-treated GO films allowed for exceeding the value of FF and J$_{sc}$ obtained from PEDOT: PSS. Taking advantage of graphene's chemical structure, the functionalization is very promising for obtaining desirable properties in HTLs, such as high hole mobility, charge collection, transparency, and stability, among others. Zhao et al. fabricated highly stable P3HT:PC$_{71}$BM-based OSCs with a PCE of 3.56% by forming covalent bonds between graphene and sulfanilic acid through C-N linkers [230]. The covalent functionalization increased the WF that enhance the interface's charge transport and the overall photovoltaic characteristics (see Figure 11a). Ali et al. confirmed the potential for tuning the bandgap and electrical properties when reduced and sulfonated GO films were applied as HTLs for a wide range of donor-acceptor systems [231]. Other approaches like non-covalent phosphorylation and fluorination have been remarkably effective in enhancing the charge collection and transport via inducing low ohmic contact [232,233]. The presence of the phosphate ester or fluor in the surface of GO increased the WF of ITO/GO and tuned the HOMO level of the donor by the p-doping effect. Fluorinated GO (F$_5$-GO) was reported to work as an interlayer between ITO and PEDOT:PSS [234]. This material improved hole transport, resulting in a low R$_s$ of 2 Ωcm^2 and a PCE of 7.67% for PTB7:PC$_{71}$BM-based OSCs. Park et al. reported an orthogonal printable HTL by spray casting a highly stable dispersion of fluorine-functionalized reduced graphene oxide (FrGO) [235]. By decreasing the sheet size to 0.3 μm, the PCE increased to 9.27 and 9.02% for PTB7-Th:EH-IDTBR and PTB7-Th:PC$_{71}$BM-based OSCs, respectively. This improvement was attributed to the hole-transport efficiency, decreased leakage current, and higher conductivity of the FrGO layers. Zhen et al. reported graphene-MoS$_2$ hybrid thin films via liquid-phase graphene exfoliation, improving the charge transportation as an interlayer to achieve a PCE 9.5% [236]. This interlayer increased the device stability by retaining 93% of the initial PCE after 1000 h at room temperature. Shoyiga et al. reported reduced graphene oxide-anatase titania (RGOT) nanocomposites by hydrothermal synthesis [237]. RGOT HTL is an efficient charge-transport channel whose higher conductivity and exciton dissociation efficiency decreased the rates of electron-hole recombination (see Figure 11b), resulting in a high J$_{sc}$, low R$_s$, and, thus, improved photovoltaic performance.

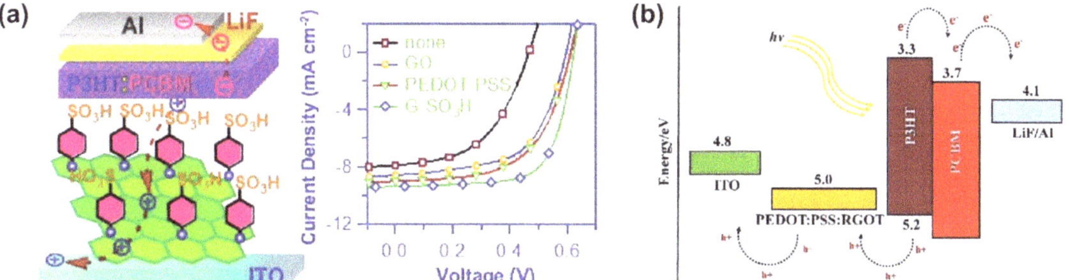

Figure 11. (**a**) J-V curve of the OSCs with G–SO$_3$H films as HTL. Adapted with permission from [230]. Copyright 2018, American Chemical Society. (**b**) Energy level diagram of an OSC with RGOT-modified HTL. Adapted with permission from [237]. Copyright 2020, John Wiley and Sons.

Lee et al. improved the performance of rGO by chemical doping with tetrafluorote-tracyanoquino-dimethane (F$_4$TCNQ) [238]. The *p*-doping of rGO with F$_4$TCNQ increased the WF by 0.2 eV and the conductivity by inducing charge transfer between the F$_4$TCNQ and the graphene layer. F$_4$TCNQ enhanced the interchain interaction and crystalliza-tion of the P3HT film to improve the hole mobility from the active layer to the anode. Lee et al. [239] and Sun et al. [240] reported that GO modified with alkali chlorides such as AuCl$_3$ or CuCl$_2$ dopants in a conventional architecture exhibited an average PCE of 3.77% and 7.68%, respectively. The AuCl$_3$-doped graphene increased the electrical conductivity ($\sim$2.0 $\times$ 10^5 Sm^{-1}) compared with the reported fullerenol-rGO layer (1 $\times$ 10^{-2} Scm^{-1}). GO:CuCl$_2$ layers formed a uniform and continuous film. Although the efficiency achieved by the dopants is even comparable to that of the control devices with PEDOT:PSS, the sta-bility was superior. Graphene-based derivatives (GBD) do not corrode the metal substrate as PEDOT:PSS, to leading the development of OSCs' efficient performance by controlling the properties and deposition conditions of GBD as reported by Capasso et al. [241]. Sarkar et al. embedded Au NPs into GO for increasing the light trapping in the active layer [242]. The exerted plasmonic effect and the plasmon-exciton interaction of NPs increased the light harvested by the active layer, resulting in enhanced J$_{cc}$ and PCE. Besides, the enhanced conductivity of GO helped to reduce the leakage current, thereby improving the photogen-erated current, R$_s$, and FF of the device. A composite of 1 wt% of graphene nanosheet and water-dispersible polyaniline-poly(2-acrylamido-2-methyl-1-propanesulfonic acid) com-plex was used as HTL in OSCs [243]. The graphene nanostacks (GN) from the composite penetrated the BHJ of the OSC and facilitated the charge transport by forming additional pathways (see Figure 12a). The electric field generated from the edges of the GN increased the exciton dissociation. As a result, the composite performance raised the PCE from 2.12% (PANI) to 2.92% (G-PANI) in the P3HT:PC$_{70}$BM cells. Aatif et al. also reported the surface morphology's planarization after applying GO/molybdenum composite, resulting in a PCE of 5.1% with the PCDTBT:PC$_{71}$BM-based OSCs [244]. Quasi-3D GO:NiO$_x$ nanocomposites are potential *p*-type HTLs in ITO/ZnO/PTB7-Th:PC$_{71}$BM/HTL/Ag architectures [245]. Using the solvothermal method, NiO$_x$ NPs interacted with the low oxidized form of GO by hydrogen bonds to form the quasi 3-D arrangement (see Figure 12b). The high performance of these nanocomposite HTLs is due to the enhanced vertical conductivity with low recom-bination rates and enhanced electron-blocking ability by the small conduction band of NiO$_x$ NPs (1.55 eV) (see Figure 12c). The metallic nature of NiO$_x$ NPs improved the stability by retaining half of the initial PCE (12.3%) in environmental conditions. Dang et al. reported a solution-processed hybrid graphene-MoO$_3$ (G-MoO$_3$), via the hydrothermal method, to apply as HTL in OSCs [246]. The G-MoO$_3$ exhibited higher transparency in the visible region compared with the thermal-evaporated MoO$_3$. Moreover, the low injection barrier (0.2 eV) and the higher hole mobility in G-MoO$_3$ (4.16 $\times$ 10^{-5} cm^{2}V^{-1}s^{-1}) than in MoO$_3$ (1.25 $\times$ 10^{-5} cm^{2}V^{-1}s^{-1}) were beneficial to achieve a PCE of 7.07%. The rGO and perylene derivative 3,4,9,10-perylenetetracarboxylic dianhydride (PTCDA) nanohybrid HTL showed an increased cell performance up 4.70% in PBDTTT-CT:PC$_{71}$BM-based cells [247]. The rGO:PTCDA nanohybrid HTL formed permanent dipoles by the PTCDA_rGO bond formation, increasing the hole extraction, electrical conductivity, and tuning the WF.

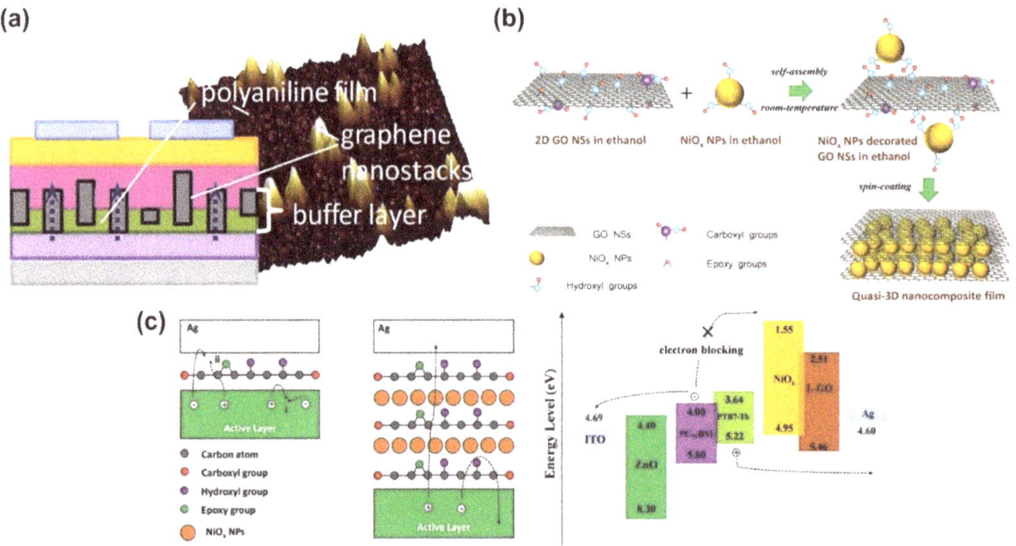

Figure 12. (**a**) Topography AFM-images of G/PANI–PAMPSA nanocomposite layer. Adapted with permission from [243]. Copyright 2018, Elsevier. (**b**) Schematic representation of the preparation at room temperature of self-assembled quasi-3D GO:NiO$_x$ nanocomposite and (**c**) hole extraction properties and dynamics at the interface. Adapted with permission from [245]. Copyright 2018, John Wiley and Sons.

4.3.2. Other Nanocarbons

The production efficiency of graphene quantum dots (GQDs) in the photovoltaic field has been limited by the expensive manufacturing methods, materials availability, and the time-consuming manufacturing [248,249]. However, the development of green and low-cost methods, such as the synthesis of GQDs from carbon fibers by acid treatment and chemical exfoliation or doping with nitrogen, has boosted its potential application in the fabrication of large-area OSCs [250]. Hoang et al. succeeded in the green synthesis of GQDs from graphene using the microwave-assisted hydrothermal method for 10 min [251]. An enhancement of 44% in PCE was achieved by doping the active layer with 2 mg of GQDs. The GQDs filled the interstitial positions between P3HT and PC$_{60}$BM to increase the charge transport of holes and electrons and the photocurrent generation. Zhang et al. reported amino-functionalized multi-walled carbon nanotubes (a-MWNTs), via hydrothermal synthesis, as HTLs in conventional OSCs with the configuration ITO/a-MWNTs/PCDTBT/PC$_{71}$BM/LiF/Al [252]. Compared with the carboxylic acids, the amino functionalization reduced the defects and the resistivity of a-MWNTs (see Figure 13a). The a-MWNTs enhanced the device's charge mobility, collection, and performance by 6.9%. Single-walled carbon nanotubes (SWCNTs) are promising *p*-type transparent conductors owing to their superior hole mobility, conductivity, and facile tuning of the WF by doping method [253]. A highly-conductive composite of unzipped single-walled carbon nanotubes (u-SWNTs) and PEDOT:PSS was synthesized by a facile solution processing method as reported by Zhang et al. (see Figure 13b) [254]. The hybrid PEDOT:PSS doped with u-SWNTs decreased the surface roughness. Oxygen-containing groups of u-SWNTs improved the compatibility between u-SWNTs and PEDOT:PSS to block electrons and increase the hole transport. Using 0.1 mg mL^{-1} of u-SWNTs, the conductivity of the uSWNTs/PEDOT:PSS increased to 2.08 Scm^{-1}. The R$_s$ was insensitive to the layer thickness, resulting in improved charge-carrier transport through the gap of u-SWNTs (see Figure 13c). Thus, PBDB-T-2F:IT-4F devices with u-SWNTs/PEDOT:PSS HTLs exhibited an enhancement in the PCE from 13.72% to 14.60% (see Figure 13d).

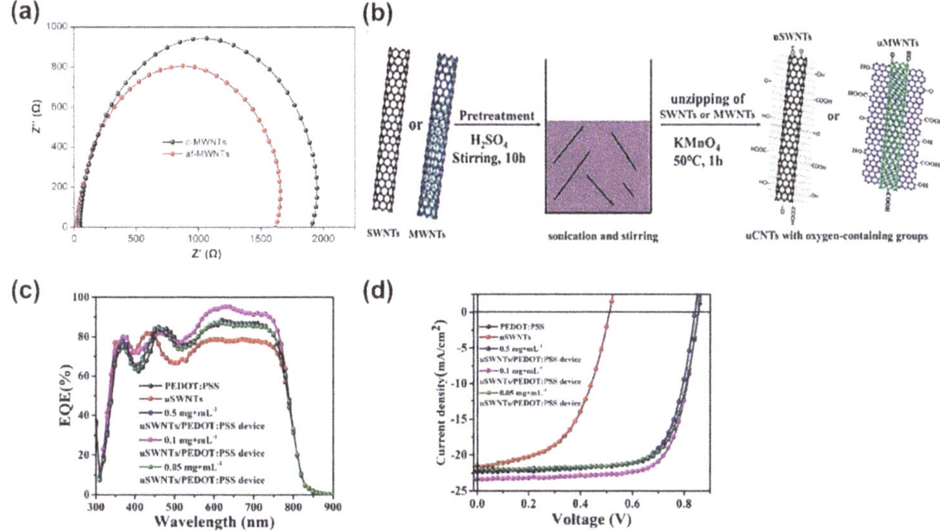

Figure 13. (**a**) Impedance spectra of OSCs with c-MWNTs and af-MWNTs HTLs. Adapted with permission from [252]. Copyright 2019, Elsevier. (**b**) Schematic representation of u-SWNTs and u-MWNTs synthesis, (**c**) EQE of u-SWNTs, SWNTs, and PEDOT:PSS, and (**d**) J-V curves of the OSCs with u-SWNTs HTLs. Adapted with permission from [254]. Copyright 2020, Royal Society of Chemistry.

Table 1 enlists a series of HTLs reviewed up to this point. The anode configuration with its work function, deposition technique for the HTL, active layer composition, and the OSCs performance parameter such as V$_{OC}$, J$_{SC}$, FF, and PCE are provided as well as the reference where the information is available.

Table 1. Performance parameters of OSCs with inorganic films as HTLs.

Anode Configuration and WF (eV)	Deposition Technique	Active Layer	V$_{OC}$ (V)	J$_{SC}$ (mA cm^{-2})	FF (%)	PCE (%)	Ref.
Metallic oxides							
ITO/s-MoO$_3$ (4.92)	spin coating	PBDB-T-2F:Y6	0.84	27.53	73.10	17.00	[107]
ITO/PCM4 (5.4)	blade coating	PBDB-T-2F:Y6	0.83	16.06	68.28	14.30	[119]
HVO/Ag (6.7)	spin coating	PBDTT-FTTE:PC71BM	0.82	22.51	60.19	11.14	[159]
ITO/MoO$_3$ (5.3)	spin coating	PBDB-T:PC71BM	0.88	17.48	71.00	10.86	[102]
MoO$_3$:NiO$_x$ (5.1)	spin coating	PBDB-T:IT-M	0.94	17.26	66.63	10.81	[113]
e-MoO$_x$/Ag (5.4)	spin coating	PTB7-Th:PC71BM	0.79	18.70	69.20	10.42	[99]
VO$_x$:PEDOT:PSS/Ag (5.28)	spin coating	TPD-3F:IT-4F	0.87	16.80	69.10	10.10	[164]
MoO$_3$/AgAl/MoO$_3$/AgAl	thermal evaporation	PTB7-Th:PC71BM	0.78	19.60	61.90	9.79	[100]
CuBr-MoO$_3$/Ag (5.03)	thermal evaporation	PTB7:PC71BM	0.75	19.65	65.20	9.56	[117]
MoO$_x$ NPs/Ag (5.40)	spin coating	PTB7- Th:PC 71 BM	0.79	18.05	65.20	9.50	[99]
ITO/MoO$_3$ (5.29)	spin coating	PBDB-T:ITIC	0.91	15.19	66.59	9.17	[106]
ITO/NiO$_x$ NPs(5.25)	spin coating	PTB7-Th:PC71BM	0.79	18.32	63.10	9.16	[176]
ITO/s-MoO$_3$ (4.92)	spin coating	PTB7-Th:PC71BM	0.79	16.69	67.10	8.90	[107]
ITO/CuO$_x$ (5.06)	spin coating	PTB7:PC71BM	0.74	16.44	71.00	8.68	[187]
MoO$_3$/Ag (5.52)	thermal evaporation	PTB7-Th:PC70BM	0.81	15.90	67.80	8.67	[98]

Table 1. *Cont.*

Anode Configuration and WF (eV)	Deposition Technique	Active Layer	V_{OC} (V)	J_{SC} (mA cm^{-2})	FF (%)	PCE (%)	Ref.
ITO/p-MoO$_3$ (5.26)	spin coating	PTB7:PC71BM	0.73	17.02	68.10	8.46	[104]
ITO/CTAB-MoO$_3$ (5.18)	spin coating	PTB7:PC71BM	0.72	16.88	68.10	8.34	[118]
ITO/VO$_x$·nH$_2$O (5.0)	spin coating	PTB7-th:PC71BM	0.78	15.76	64.62	8.11	[156]
ITO/V$_2$O$_5$	spin coating	PTB7:PC70BM	0.71	17.35	65.00	8.05	[154]
ITO/NiO$_x$ NPs(5.25)	spin coating	PTB7:PC71BM	0.74	16.10	66.42	7.96	[176]
ITO/NP-V$_2$O$_5$ (4.7)	spin coating	PTB7:PC70BM	0.72	15.81	69.01	7.89	[165]
WO$_x$ nanosheets/Ag	spin coating	PTB7: PC71BM	0.81	16.42	58.19	7.76	[150]
ITO/s-VO$_x$ (5.3)	spin coating	PTB7:PC71BM	0.73	15.79	66.82	7.70	[160]
ITO/MoO$_3$/AuNPs/MoO$_3$ (5.6)	spin coating	PTB7:PC70BM	0.73	14.40	73.00	7.68	[122]
ITO/ESD-VO$_x$ (5.6)	spray casting	PTB7:PC71BM	0.74	15.30	67.00	7.61	[168]
ITO/V$_2$O$_x$ (5.42)	spin coating	PBDTDTTT-S-T:PC71BM	0.68	16.29	67.21	7.44	[114]
ITO/MoO$_x$ (5.30)	spin coating	PBDTDTTT-S-T:PC71BM	0.69	16.14	66.02	7.35	[114]
ITO/sMoO$_x$	spin coating	PV10:PC70BM	0.73	13.57	72.55	7.19	[101]
FTO/Cu:NiO$_x$	spin coating	PCDTBT:PC71BM	0.89	12.40	63.85	7.05	[181]
FTO/c-NiO (5.0)	spin coating	PTB7:PC71BM	0.72	14.28	66.98	6.91	[178]
ITO/MoO$_x$	thermal evaporation	PTB7:PC71BM	0.67	14.00	67.00	6.57	[103]
ITO/H$_y$MoO$_{3-x}$ (5.6)	spin coating	PTB7:PC70BM	0.77	13.90	61.20	6.55	[129]
ITO/NiO$_x$ NPs(5.25)	spin coating	PCDTBT:PC71BM	0.90	11.36	62.35	6.42	[176]
ITO/NiO$_x$ (5.6)	spin coating	TQ1:PC70BM	0.87	10.30	71.30	6.39	[175]
ITO/s-V$_2$O$_x$ (5.25)	spin coating	PFDT2BT-8:PC70BM	0.87	10.20	67.10	6.30	[162]
V$_2$O$_x$/Ag (5.42)	spin coating	PBDTDTTT-S-T:PC71BM	0.63	15.81	61.02	6.08	[114]
MoO$_x$/Ag (5.30)	spin coating	PBDTDTTT-S-T:PC71BM	0.61	15.68	62.78	6.00	[114]
ITO/V$_2$O$_5$·H$_2$O (5.04)	spin coating	PBDSe-DT2PyT:PC71BM	0.72	13.96	59.00	5.87	[155]
ITO/CTAB-MoO$_3$ (5.18)	spin coating	P3HT:ICBA	0.82	10.40	67.40	5.80	[118]
ITO/MoO$_x$	thermal evaporation	PTB7-Th:PC71BM	0.74	14.50	44.80	5.52	[103]
FTO/s-MoO$_3$ (5.3)	spin coating	P3HT:ICBA	0.82	11.50	58.00	5.40	[105]
FTO/MoO$_x$ (5.6)	spin coating	P3HT:PC71BM	0.65	12.72	61.00	5.00	[110]
		Metallic sulfides					
ITO/WS$_2$ (5.3)	spin coating	PBDB-T-2F:Y6:SF(BR)$_4$	0.89	29.31	80.00	20.87	[204]
ITO/WS$_2$ (5.5)	spin coating	PBDB-T-2F: Y6: PC71BM	0.84	26.00	78.00	17.00	[203]
ITO/WS$_2$ (5.1)	spin coating	PBDB-T-2F:Y6:PC71BM	0.83	26.00	72.00	15.60	[202]
ITO/CuSCN:TFB (1.0%) (5.50)	spin coating	PM6:Y6	0.85	24.35	73.84	15.28	[214]
ITO/MoS$_2$ (5.04)	spin coating	PBDB-T-2F:Y6:PC71BM	0.81	25.30	71.00	14.90	[202]
ITO/CuSCN:AMQS	spin coating	PBDBT-2F:IT-4F	0.80	18.70	67.80	10.14	[209]
ITO/CuSCN:AMQS	spin coating	PTB7-Th:PC71BM	0.79	17.10	65.20	8.80	[209]
ITO/O-MoS$_2$ QDs (5.2)	spin coating	PTB7-Th: PC71BM	0.79	16.90	65.00	8.66	[200]
ITO/MoS$_x$ (5.10)	spin coating	PTB7-Th: PC71BM	0.77	18.16	53.56	7.50	[201]
ITO/CuSCN:AMQSs	spin coating	PTB7-Th:ITIC	0.82	15.07	59.06	7.15	[209]
ITO/CuSCN:C$_{60}$F$_{48}$ (5.40)	spin coating	PCDTBT:PC70BM	0.92	11.50	61.00	6.60	[213]
		Nanocarbon materials					
ITO/uSWNTs/PEDOT:PSS	spin coating	PBDB-T-2F:IT-4F	0.85	23.39	73.17	14.60	[254]
ITO/FrGO (4.9)	spray casting	PM6:Y6 PSC.	0.77	24.64	69.60	13.26	[235]
L-GO:NiO/Ag	spin coating	PBDB-T:IT-M	0.91	17.81	71.00	12.13	[245]
G-MoS$_2$/Ag (4.42)	spin coating	PTB7-Th:PC71BM	0.80	17.10	67.70	9.50	[236]
ITO/G-MoS$_2$/PEDOT:PSS (5.0)	spin coating	PTB7-Th:PC71BM	0.77	17.20	72.00	9.40	[236]
ITO/FrGO (4.9)	spray casting	PTB7-Th:EH-IDTBR	1.00	14.86	61.80	9.22	[236]
ITO/FrGO (5.1)	spin coating	PTB7-Th:PC71BM	0.79	16.89	64.80	8.60	[233]
ITO/P-GO (4.70)	spin coating	PTB7:PC71BM	0.71	16.12	68.40	7.90	[232]
ITO/GO:CuCl$_2$ (5.1)	spin coating	PTB7-Th:PC71BM	0.79	15.52	63.00	7.74	[240]
ITO/G-MoO$_3$ (5.32)	spin coating	PCDTBT:PC71BM	0.86	12.83	63.67	7.07	[246]
af-MWNTs (5.22)	spin coating	PCDTBT:PC71BM	0.87	12.65	63.50	6.97	[252]
ITO/GO:NPs (4.9)	spin coating	PTB7:PC71BM	0.75	11.55	67.91	5.88	[242]
ITO/F$_5$-rGO (5.1)	spin coating	PTB7:PC71BM	0.68	14.78	57.30	5.82	[234]
ITO/GBD (4.9)	spin coating	PBDTTT-C-T:PC70BM	0.71	13.38	52.54	5.01	[241]
ITO/GO:MoO$_3$ (5.3)	spin coating	PCDTBT:PC71BM	0.66	16.16	47.11	5.10	[244]

4.4. Conducting Polymers and Their Composites

4.4.1. PEDOT

PEDOT:PSS is the most common conducting polymer used as hole-transporting material in OSCs due to its easy solution processing, suitable WF around 5.1 eV, high conductivity, good transparency, good mechanical properties, and adapted wettability on the BHJ layer [255]. For instance, a patterning interfacial PEDOT:PSS layer formed by a nanoimprinting technique using poly(dimethylsiloxane) (PDMS) stamp was employed on OSCs based on poly(3-hexylthiophene):phenyl-C61-butyric acid methyl ester (P3HT:PCBM), showing an increased PCE of 1.53% [256]. PEDOT:PSS was incorporated on an inverted OSC based on P3HT:O-IDTBR with an evaporated Ag back electrode, showing an enhanced device performance [257]. The incorporation of PEDOT:PSS into P3HTN:PEG-C_{60} based OSCs increased the V_{oc} to 1.3 V, attributed to the large collection barrier [258]. PEDOT:PSS has strong acidic nature due to the polystyrene sulfonate (PSS, pH~2), which deteriorates the anode material and the photoactive layer, affecting the performance and stability of the device. Besides, this polyelectrolyte has high affinity for environmental water (hygroscopic), making it necessary to encapsulate the OSC's before the durability test. Humidity is a major problem in this type of device. Accordingly, different modifications to the PEDOT:PSS layer have been developed to overcome these issues. Some post-treatments to the PEDOT:PSS layer have been tested using solvents, surfactants, and by exchange of PSS with less acidic dopants as well as the addition of small molecules. These modifications aim to reach a uniform morphology, increase the interface contact, and produce a neutral pH hole-transporting polymer to improve the stability and cell performance. The use of a layer composed of PEDOT and grafted sulfonated-acetone-formaldehyde lignin (GSL) instead of PSS resulted in a better photovoltaic performance than conventional PEDOT:PSS [259]. GSL is a less acidic copolymer of lignin. A homogeneous surface of the PEDOT:GSL HTL in a PTB7-Th:PC_{71}BM/poly[(9,9bis(3′-(*N*,*N*-dimethylamino)propyl)-2,7-fluorene)-alt-2,7-(9,9dioctylfluorene)] (PFN)-based OSC resulted in a PCE of 8.47%. PEDOT:PSS treatment with solvents as isopropanol (IPA) also shows a better performance, mainly due to more uniform morphology, increased J_{SC}, and improved cell-light absorption [260]. 2-Methoxyethanol (EGME) and dimethyl sulfoxide (DMSO) solvents were added to a PEDOT:PSS solution [261]. The conductivity after doping was about seven times higher and the OSCs based on P3HT:PCBM improved the PCE from 2.8% to 3.9% owing to increased J_{sc} 16.5 mA cm^{-2} and FF of 38.0%. Another approach includes the addition of commercial surfactants such as Zonyl FS-31, which improves the wettability of the interface between the hydrophobic photoactive layer and the PEDOT:PSS HTL [262]. The fluorination of PEDOT:PSS HTL by fluorinated molecules showed an increased device efficiency [263]. On the other hand, a PSS-free, stable PEDOT HTL was obtained by using solid-state polymerization resulting in robust, stable, and solution-processable OSCs based on PCDTBT:PC_{71}BM [264]. Water-soluble polyelectrolyte poly(4-(2,3-dihydrothieno[3,4-b][1,4]dioxin-2-yl-methoxy)-1-butanesulfonic acid) (PEDOT-S), which shows the same PEDOT backbone containing an ethoxyalkylsulfonate branch, showed better performance than conventional PEDOT:PSS layers [265]. PEDOT-sulfonated polyelectrolyte complexes were also tested as an anode buffer layer [266]. Different commercial grades of PEDOT:PSS and additive solvent EG were used to form a hybrid PEDOT:PSS (PH 1000:Al 4083) layer tested as an HTL and anode electrode for inverted OSCs based on P3HT:PCBM [267]. An OSCs based on PTB7-Th:PC_{71}BM was built using a hole-transport double layer made of pyridine-based tetrathiafulvalene derivative (TTF-py) on PEDOT:PSS [268]. This modification resulted in an increased short circuit current (J_{sc}) of 17.19 mA cm^{-2} and a PCE of 9.37%. The anode configuration showed a WF of 5.28 eV for the TTF-py layer, resulting in a closer valence band toward the donor material (see Figure 14). PEDOT:PSS/TTF-py had better wettability and enhanced hole mobility, resulting in charge-loss reduction and charge-recombination suppression. Furthermore, TTF-py's molecular structure allowed molecular π-π stacking and formed an orderly molecular arrangement for hole transfer. TTF-py modification also improved the device stability, retaining 96% of the initial PCE after storing for 28 days by the suppression PEDOT:PSS permeation.

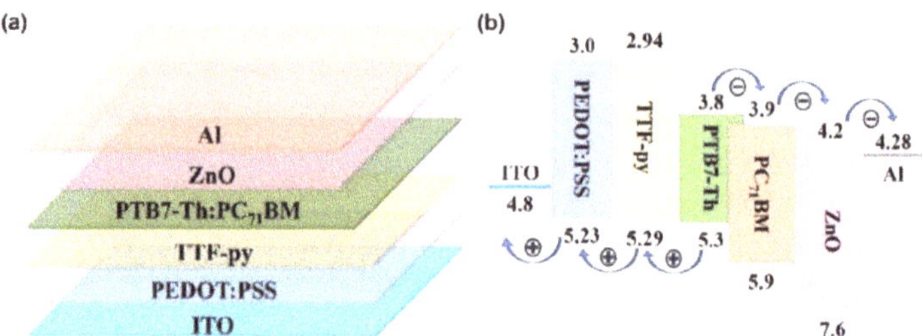

Figure 14. (**a**) Schematic illustration of device structure of ITO/PEDOT.PSS/TTF-py/PTB7-Th:PC$_{71}$BM/ZnO/Al and (**b**) energy level diagram of an OSC with the TTF-py modified PEDOT:PSS as HTL. Adapted with permission from [268]. Copyright 2019, American Chemical Society.

Other modifications on PEDOT:PSS have introduced an inorganic transition metal salt, such as nickel formate dihydrate (NFD), to tune the surface free energy (γ_s) and control the molecular orientation in the BHJ [269]. An enhanced PCE of 10.76% was achieved for the PM6:PC$_{71}$BM-based OSCs. The NFD:PEDOT:PSS HTL had a WF of 5.01 eV that well-matched the donor material and an increased γ_s of 68.96 mN m^{-1}, which led to increased FF and J$_{sc}$. Polymeric donor material PM6 preferred a face-on molecular orientation. Enhanced molecular stacking was promoted with an increased γ_s of PEDOT:PSS induced by NFD. This modification improved the molecular orientation along the charge-transport direction; thus, carrier mobility was enhanced, and the charge recombination was suppressed. The modified HTL was also tested in non-fullerene OSCs based on PM6: IT-4F, obtaining an enhanced PCE of 14.08% with FF of 78.75%. Oxoammonium salts (TEMPO$^+$ Br$^-$, 2,2,6,6-tetramethylpiperidine-1-oxoammonium) were tested as a p-type dopant of PEDOT:PSS layers, resulting in an enhanced PCE of 16.1% in OSCs based on PM6:Y6 [270]. PEDOT:PSS was further oxidized by oxoammonium salt, improving the doping level of PEDOT:PSS. Doped PEDOT:PSS (TEMPO$^+$ Br$^-$) possess higher conductivity and better energy alignment. Metallo phthalocyanines (PC) such as vapor-deposited vanadylphthalocyanine (VoPC), NiPC, and SnPC were tested as buffer layers with PEDOT:PSS, enhancing the efficiency of P3HT:PCBM-based OSCs [271]. PEDOT:PSS:In$_2$S$_3$ was also employed as HTL material for OSCs based on PBDB-T:ITIC and PM6:Y6; these showed an enhanced PCE of 11.22% and 15.89%, respectively [272]. Improved device performance was observed because of increased J$_{sc}$ and FF, and reduced R$_s$ with bimolecular recombination suppression due to partial removal of PSS from the surface. PEDOT also suffered a benzoic-quinoid transition (coil-linear structure) which delocalized charge carriers, enhancing the layer conductivity. Furthermore, device performance stability showed a retained 36% PCE for modified HTLs after 48 h compared with a non-modified PEDOT:PSS HTL, which showed a retained 10% PCE. OSCs based on ITO/PEDOT:PSS-Dopamine (DA)/PM6:Y6/poly[[2,7-bis(2-ethylhexyl)-1,2,3,6,7,8-hexahydro-1,3,6,8-tetraoxobenzo[lmn] [3,8]phenanthroline-4,9-diyl]-2,5-thiophenediyl[9,9-bis[3'((N,N-dimethyl)-N-ethylammonium)]propyl]-9H-fluorene-2,7-diyl]-2,5-thiophenediyl] (PNDIT-F3N)/Ag showed an increased PCE from 16.01% to 16.55% [273]. The DA-doped PEDOT:PSS layer showed an enhanced conductivity ascribed to (i) a more regular stack by the enhanced intermolecular packing of DA:PSS, (ii) an increased WF of 5.14 eV compatible with HOMO level of PM6 donor polymer, and (iii) enhanced film uniformity. PEDOT:PSS-DA was also tested for devices based on different active layers such as PBDB-T:ITIC, PM6: IDIC, and P3HT:PCBM, resulting in improved performances as well. PEDOT:PSS was also used together with various polymers as HTLs, such as nanoimprinted poly(methylmethacrylate) (PMMA) [274], and conjugated polyelectrolytes (CPEs), e.g., poly[(9,9-bis(4-sulfonatobutyl sodium) fluorene-alt-phenylene)-ran-(4,7-di-2-thienyl-2,1,3-benzothiadiazole-alt-phenylene)] (PSFP-DTBTP), that resulted in a PCE improvement of 13% for PCDTBT:PC$_{71}$BM-based OSCs [275]. Microporous polymer networks are a class of conjugated material that shows high specific surface areas and porosity with potential application in various fields including organic photovoltaics [276–283]. For instance, a porous organic polymer, poly(carbazolyl triphenylethylene) derivative (PTPCz), obtained by electropolymerization was used in the HTL PEDOT:PSS/PTPCz for OSCs based on PTB7:PC$_{71}$BM, resulting in an smooth surface morphology, increased WF of 5.23 eV, J$_{sc}$, and FF, reduced R$_s$, and increased R$_{sh}$, reaching an improved PCE of 8.54% [284]. Electropolymerized polytriphenylcarbazole fluoranthene (p-TPCF) and PEDOT:PSS were used for OSCs based on PTB7-Th:PC$_{71}$BM, obtaining enhanced J$_{sc}$, FF, V$_{oc}$, and a PCE of 8.99% [285]. Modification of PEDOT:PSS with a neutral conjugated polymer electrolyte poly[9,9-bis(4'-sulfonatobutyl)fluorene-*alt*-thiophene] (PFT-D) composite layer improved the device performance (PCE from 7.8% to 8.2%) and the half-lives of PTB7-Th:PC$_{71}$BM-based OSCs [286]. PFT-D molecular dipole screened the attraction between PEDOT and PSS chains; additionally, the –SO$_3^-$ ions of PFT-D act as a conjugate base of PSS, improving current generation. Poly(3-hexylthiophene)-*b*-poly(p-styrenesulfonate) (P3HT$_{50}$-b-PSS$_{23}$) block polymers were incorporated between HTL PEDOT:PSS and the active-layer P3HT:PCBM [287]. The OSCs with P3HT-b-PSS interfacial layer improved PCE by 12% due to increased V$_{oc}$ and FF that compensate for the decreased J$_{sc}$ caused by the blocked light irradiance to the P3HT. The energy level matching was improved. HOMO level of P3HT-b-PSS (−4.68 eV) was higher than P3HT of active layer, which facilitates the hole transport. In addition, P3HT-b-PSS film had a smoother surface than PEDOT:PSS, enhancing the interfacial contact and thus improving the FF of the device. Modification of the commonly used PEDOT:PSS with metallic NPs contributes with some features such as an enhanced localized field and light scattering by the localized surface-plasmon resonance (LSPR) that improves the absorption of the active layer [288]. NPs also assist in the charge transport at the interface. NPs are synthesized by different methods such as chemical reduction, the polyol method, and ultrasonochemical synthesis [289]. Hao et al. reported mixed AuNPs (rod, bone-like, cube and spheres shape) doped in PEDOT:PSS HTLs in OSCs based on PTB7:PC$_{71}$BM [290]. The addition of mixed AuNPs generated wide absorption spectra covering from the visible to the near-infrared region and induced an increase of enhancement of internal field in the active layer, resulting in improved absorption and enhanced device performance up to 9.26%. AuNPs also contributed to decrease the bulk resistance of PEDOT:PSS. Periodic Ag nanodot (Ag ND) arrays were fabricated by laser-interference lithography (LIL) between ITO and PEDOT:PSS layers in OSCs based on PTB7:PC$_{70}$BM (see Figure 15) [291]. This HTL showed increased J$_{sc}$ of 23.26 mA/cm^2, enhanced EQE induced by the plasmonic and light-scattering effect, and improved PCE of 10.11%. LSPR band matched optimally with the absorption of the photoactive layer, increasing its light-absorption.

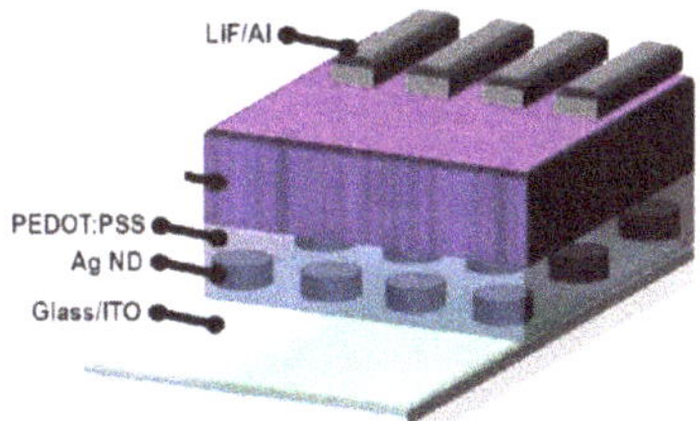

Figure 15. Schematic representation of an OPV device with Ag NDs. Adapted with permission from [291]. Copyright 2016, American Chemical Society.

AuNPs and AgNPs blended with PEDOT:PSS were used as HTLs in OSCs [292]. Both PEDOT:PSS:NPs showed an enhanced device performance in comparison with pristine PEDOT:PSS. An optimized PCE of 5.65% was obtained for PEDOT:PSS:AuNPs HTLs in rrP3HT:PC$_{71}$BM-based OSCs. Better device performances were obtained with NPs because of the surface-plasmon effect (at the visible region for AuNPs) that increased the photoabsorption length and the trapping of scattering and incident light. Segmented silver nanowires (AgNWs) were incorporated in PEDOT:PSS HTLs in OSCs with configuration

ITO/PEDOT:PSS:AgNWs/P3HT:PC$_{61}$BM/ZnO/Al [289]. These OSCs exhibited enhanced device performance with a PCE of 3.3%, increased J$_{sc}$ and FF due to LSPR, and optical-scattering properties from the AgNWs. Sah et al. reported the fabrication of OSCs based on PTB7-Th:PC$_{71}$BM with bimetallic Ag-Au-Ag nanorods (NRs) in PEDOT:PSS HTLs [293]. These devices showed improved performances up to 7.36% owing to an increased FF and J$_{sc}$. The enhancement was ascribed to an improved charge transport, broad absorption region covering visible to the near-infrared region, light-scattering-induced absorption enhancement, electric field enhancement, and improved EQE by the LSPR effect. Incorporation of copper, a cheaper and more abundant metal, as Cu-Au NPs in PEDOT:PSS for OSC based on P3HT:PC$_{61}$BM and PTB7-Th:PC$_{71}$BM showed improved PCEs of 3.63% and 8.48%, respectively [294]. Cu-AuNPs:PEDOT:PSS presented an absorption enhancement by LSPR and light-scattering effect. The improved PCE in the device was attributed to an increased J$_{sc}$, higher hole mobility, and reduced R$_s$. However, the FF decreased compared with pristine PEDOT:PSS HTLs, mainly due to an induced charge recombination by the NP doping. Adedeji et al. employed copper sulfide NPs in PEDOT:PSS HTLs to fabricate OSCs based on P3HT:PC$_{61}$BM [295]. These devices showed an enhanced PCE of 4.51% (an increase of 115% over the pristine PEDOT:PSS HTLs) and good stability, retaining up to 40% of their initial PCEs after 48 h. CuNPs exhibited surface-plasmon resonance absorption near-infrared region and induced an electric field beneficial to exciton dissociation and photon harvesting. The incorporation of nickel sulphide NPs in PEDOT:PSS layers exhibited an enhanced device performance of 6.03% in OSC based on P3HT:PC$_{61}$BM [288]. An improved photogenerated current was attributed to the effective trapping of light through scattering and improved charge collection. OSCs based on NiS NPs:PEDOT:PSS HTLs showed reduced R$_s$, indicative of an improved conductivity at the interface, and improved optical transparency enhancing the internal quantum efficiency. Furthermore, the device showed higher hole mobility, and thus reduced carrier recombination and enhanced charge transport. PCE enhancement was ascribed to the LSPR absorption (in the visible and infrared region) and light-scattering process. ZnSTe quantum dots (QDs) incorporated into PEDOT:PSS HTL showed an improved device performance in OSC based on P3HT:PC$_{71}$BM due to improved mobility and enhanced light absorption, attributed to surface-plasmon resonance [296]. Zhang et al. reported OSCs based on PTB7-Th:PC$_{71}$BM and PM6:IT-4F with high PCEs of 9.11% and 12.81%, respectively, by the insertion of black phosphorous quantum dots (BPQDs) on PEDOT:PSS [297]. BPQDs is a 2D p-type semiconductor, which, inside the device, formed a cascade band structure between the anode and the active layer. The valance band of BPQDs (−4.92 eV) was higher than the valance band of donor polymers PTB7-Th (−5.24 eV) and PM6 (−5.5 eV), providing enough driving force for the hole injection from the active layer to the BPQDs layer. The increased efficiency in devices with BPQDs interfacial layer was attributed to an increased J$_{sc}$ and FF due to excellent hole mobility in BPQDs and better energy alignment in the device, indicating improved charge extraction and exciton dissociation. Other metal NPs incorporated in PEDOT:PSS involve Al micro-stars [298], Al NPs [299], and Au NPs, gold nanorods (Au NRs) [300], and Au QDs [301]. Up-conversion NP process converts low-energy photons into high-energy photons (in the absorption region of organic polymers) to enhance the optical-to-electrical conversion performance [302]. Mei et al. incorporated sodium yttrium fluoride (β-NaYF$_4$):Er^{3+}, Yb^{3+} up-conversion NPs into PEDOT:PSS HTLs in OSCs based on P3HT:PC$_{61}$BM, resulting in enhanced J$_{sc}$ and PCE of 3.02% [303]. These results were ascribed to light scattering and photoluminescence (PL) emission from up-conversion NPs. Silver-zinc bimetallic NPs were incorporated between the PEDOT:PSS HTL and the photoactive layer of P3HT:PCBM [304]. OSCs exhibited an improved PCE of 3.6%, which was 90% higher than the reference device. This effect was attributed to LSPR of the Ag:Zn NPs, which enhanced the optical absorption and charge-carrier collection. Another modification to PEDOT:PSS was reported by Michalska et al. using wet ultra-sonochemical synthesized titanium dioxide TiO$_2$ anatase decorated with Ag NPs [305]. TiO$_2$/Ag solution was added to PEDOT:PSS HTLs showing an improved PCE of 2.07%. Au NPs on PEDOT:PSS in vacuum-free OSCs improved the device performance resulting from the increased J_{sc} [306]. The absorption of the active layer and the device PCE were enhanced, especially by Au nanorods' presence. Many studies have been focused on improving the performance of PEDOT:PSS HTL by addressing the acidic and hygroscopic nature of PEDOT:PSS that affects the stability and efficiency of the photovoltaic devices. Incorporating metal oxides (MO) can enhance the stability, efficiency, and electron-blocking properties of the HTL. Among these MOs that have been incorporated in PEDOT:PSS are vanadium oxides (V_2O_5), sol-gel synthesized VO_x, continuous-spray pyrolyzed-synthesized molybdenum oxide (MoO_3), and tungsten oxide (WO) [307]. Spin-coated V_2O_5 NWs prepared by the hydrothermal method on PEDOT:PSS HTL (see Figure 16a) showed improved V_{oc} and FF in OSCs based on P3HT:PCBM [308]. The PCE improved a 15.58% in comparison with the pristine PEDOT:PSS reference cell. The LUMO level of V_2O_5 (2.4 eV) is higher than P3HT LUMO, and thus better electron blocking properties than pristine PEDOT:PSS are expected (see Figure 16b). Modified HTL had increased incident light paths by reflection and refraction caused by V_2O_5 NWs.

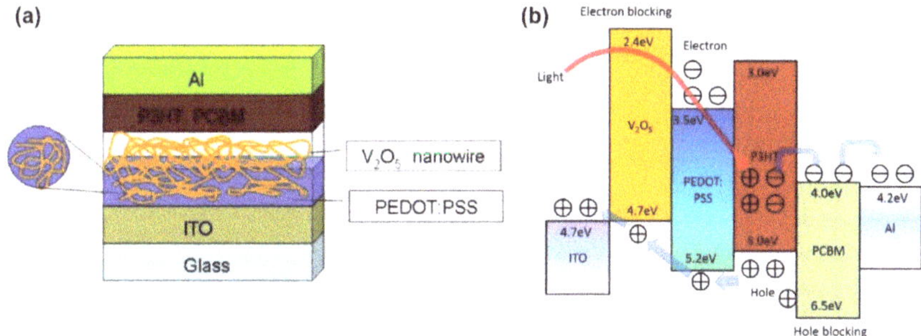

Figure 16. (**a**) Schematic representation of polymer photovoltaic cell based on PEDOT:PSS/V_2O_5 HTL and (**b**) energy level alignment for the cell components. Adapted with permission from [308]. Copyright 2016, Elsevier.

Li et al. reported a PCE of 9.44% for OSCs with V_2O_5.PEDOT:PSS as HTL due to an enhanced J_{sc} and FF, smaller R_s, and larger R_{sh} [309]. The incorporation of V_2O_5 offered an effective path for exciton extraction and suppressed charge recombination, reflected by a larger hole mobility and a higher conductivity. The composite HTL surface was uniform and smooth, related to V_2O_5 NPs filling the pinholes in PEDOT:PSS. Furthermore, better wetting and physical contact were obtained between the photoactive layer and the HTL as well as enhanced crystallinity of the active layer. Molybdenum oxide (MoO_x) NPs/PEDOT:PSS HTLs were blade coated in inverted OSCs based on PTB7-Th:$PC_{60}BM$, resulting in an increased FF and enhanced PCE of 7.4% [310]. The modification of PEDOT:PSS with MoO_3 mitigated the degradation of non-fullerene OSCs based on PM6:IT-4F by suppressing the interfacial reaction between PEDOT:PSS and IT-4F [311]. MoO_3-PEDOT:PSS hybrid HTL improved the device's operational stability, which was five times longer than reference devices. The hybrid HTL also improved the hole mobility favoring the charge extraction. Zinc oxide-doped single-carbon nanotubes (CNT) were incorporated in PEDOT:PSS as an anode buffer layer (ZnO:CNT/PEDOT:PSS), showing excellent transmittance and a smooth morphology [312]. P3HT:PCBM-based OSCs with an HTL containing 2.5% ZnO:CNT showed an improved PCE of 4.1%, enhanced J_{sc} and FF, and reduced R_s. CNT provided surface homogeneity, and ZnO prevented humidity uptake. The device parameters decreased at a slower rate than PEDOT:PSS devices under a nitrogen environment. Zheng et al. fabricated fullerene-free OSCs with tungsten oxide WO_x NPs in PEDOT:PSS as HTL [313]. The system architecture ITO/WO_x:PEDOT:PSS/PM6:IT-4F/PFN-Br/Al achieved a high FF of 80.79% and enhanced PCE of 14.57%. A more balanced hole and electron mobility was obtained for WO_x:PEDOT:PSS based of BHJ OSCs measured as a ratio μ_e/μ_h

of 0.88, which contributed to increase the FF. The longer lifetime of carriers and faster extract time of WO_x:PEDOT:PSS also benefited the device parameters. WO_3/PEDOT:PSS bilayer was used as HTL in inverted SMD2: ITIC-Th-based OSCs [314]. An optimized cell achieved a high PCE of 10.3%, with enhanced J_{sc}, V_{oc}, and FF. The WO_3/PEDOT:PSS device presented increased R_{sh} and decreased R_s by a well-matched energy level alignment, high hole mobility, a more balanced charge-carrier transport, and increased photostability. Furthermore, flexible inverted OSC modules were fabricated by slot-die coating achieving a PCE of 5.25% and a power output of 419.6 mW. A layer of hydrogen molybdenum bronze (H_xMoO_3) with PEDOT:PSS layer was also used in all solution-processed non-fullerene OSCs based on PM6:IDIC:Y6 [315]. Phosphomolybdic acid (PMA) in PEDOT:PSS layers were tested in different organic fullerene-based OSCs, showing good performances [316]. GO, a two-dimensional carbon material, has also been investigated to modified PEDOT:PSS as hole-transport materials in different OSCs. The Oleyamine-functionalized GO/PEDOT:PSS layer on PBDB-T:ITIC [317], PEDOT:PSS treated with GO layers on PTB7:PC_{71}BM devices [318], on P3HT:PC_{60}BM devices [319], on P3HT:PCBM devices [320], on inverted P3HT:PCBM OSCs [321], on inverted P3HT:PC_{71}BM OSCs [322], and on reduced GO-germanium QDs modified PEDOT:PSS on P3HT:PCBM [323]. Raj et al. reported the fabrication of PTB7:PC_{70}BM-based OSCs with PEDOT:PSS:GO, resulting in enhanced PCE of 7.68% [324]. The modified HTL showed a fine fiber-like structure that improved the conductivity. GO showed to increase the device resistance degradation. GO is generally prepared by variations of the Hummers method using graphite powder as the starting material [325,326]. PEDOT:PSS:GO was also tested on P3HT:PC_{61}BM-based OSCs, showing an increased J_{sc}, FF, and a 14% higher PCE than a reference device [327]. GO in the HTL reduced the HOMO-LUMO gap and the R_s, improving the hole mobility and the energy level matching. A double-decked GO/PEDOT:PSS HTL in PCDTBT:PC_{71}BM-based OSCs was reported by Rafique et al. [328]. The modified HTL provided a better hole extraction and transportation by a suitable WF of GO (4.9 eV) and PEDOT:PSS (5.1 eV) that well-matched energy levels. This device showed an improved PCE of 4.28% ascribed to an increased charge-carrier mobility, J_{sc}, V_{oc}, and FF, and a reduced R_s. Besides, better stability than PEDOT:PSS was reached, since GO served as a barrier that protected ITO corrosion due to the acidic nature of PEDOT:PSS (see Figure 17). Similarly, improved photovoltaic stability was achieved with GO/PEDOT:PSS HTLs in P3HT:PC_{60}BM devices [329]. This device showed an increased R_{sh} and a decreased R_s, which facilitate the hole transportation. The composite HTL was smooth and uniform, contributing to an improved device performance with a PCE of 4.82%. Nitrogen-doped graphene quantum dots (nGQDs) were blended with PEDOT:PSS HTLs in PTB7:PC_{71}BM-based OSCs, resulting in an enhanced PCE of 8.5% [330]. The modified HTL improved the charge-carrier transport, increased the hole mobility, and suppressed charge recombination. The nitrogen doping led to a high content of quaternary nitrogen, enhancing the electrical conductivity of GQDs. UV-ozone (UVO)-treated GO/PEDOT:PSS bilayer in OSCs based on PCDTBT:PC_{71}BM presented an improved PCE of 5.24% [331]. An increased J_{sc}, V_{oc}, and FF were obtained in the cells using the modified HTL and improved ambient stability, retaining above 90% of initial PCE after 240 h. The enhanced conductivity was ascribed to the reduction of oxygen content in GO after UVO treatment.

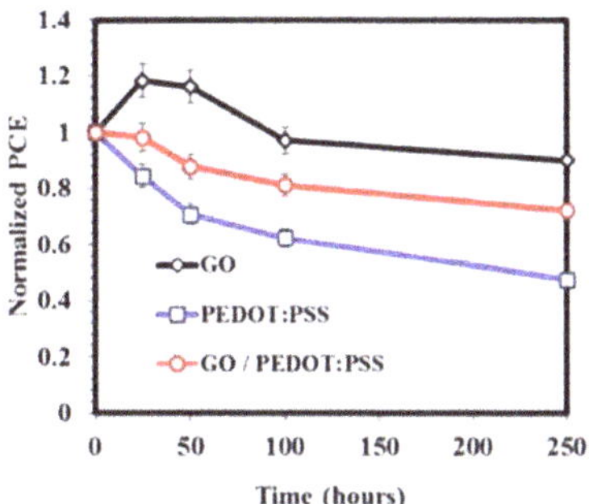

Figure 17. Stability test of various OSCs over 250 h. Adapted with permission from [328].

Other modifications on PEDOT:PSS have also been reported with a graphene analog, the two-dimensional transition metal dichalcogenides. For instance, hybrid PEDOT:PSS/WS_2 was incorporated as HTL in OSCs [332]. PEDOT:PSS worked as an effective exfoliating agent to the WS_2 2D structure. The photovoltaic device based on P3HT:PC_{61}BM and PTB7-Th:PC_{71}BM exhibited enhanced PCE of 3.07% and 7.24%, respectively, attributed to increased J_{sc} and FF as well as to enhanced hole mobility and enhanced conductivity of the PEDOT:PSS/WS_2 layer. Besides, PEDOT:PSS/WS_2-based OSCs had improved stability, retaining 77.3% of initial PCE after 36 days. Koo et al. fabricated PTB7:PC_{71}BM-based OSCs with tungsten diselenide (WSe_2)/PEDOT:PSS HTLs (see Figure 18a), showing an enhanced PCE of 8.5% [333]. The composite HTL exhibited a homogeneous film formation. WSe_2 negative surface induced the segregation of PEDOT and PSS, which enhanced the layer conductivity. Furthermore, photoluminescence peak intensity decreased, indicating diminished recombination (see Figure 18b). Thus, PEDOT:PSS-WSe_2 showed improved hole-transport ability and a better charge extraction than the reference device. Oleyamine-functionalized molybdenum disulfide MoS_2 has also been reported in the modification of PEDOT:PSS HTLs [334].

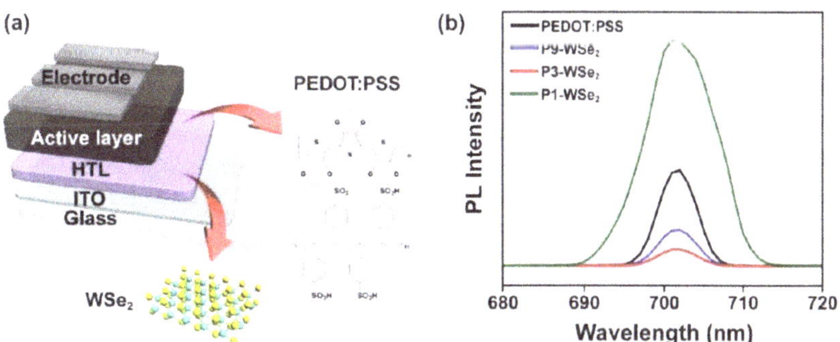

Figure 18. (**a**) Schematic representation of a WSe_2-PEDOT:PSS HTL-based OSC and (**b**) PL spectra measured from structure of glass/PEDOT:PSS and P-WSe_2-based devices. Adapted with permission from [333]. Copyright 2018, IOP Publishing Ltd.

Boronic acid functionalized multi-walled CNs (bf-MWCNTs)-doped PEDOT:PSS HTLs showed excellent hole mobility and electrical conductivity [335]. The OSC with PEDOT:PSS:bf-MWCNTs showed reduced R_s and increased R_{sh}, exhibiting excellent hole collectivity. A 28% increased PCE for PCDTBT:PC_{71}BM based OSC was attributed to an enhanced J_{sc} and FF. PL intensity of HTL doped with bf-MWCNTs was reduced, indicating an enhancement in charge transport from the active layer. The WF increased to 5.39 eV that well-matched with the HOMO energy level of PCDTBT. Graphitic carbon nitrile (g-C_3N_4) was used as a secondary dopant for PEDOT:PSS in OSCs based on PM6:Y6, leading to an improved PCE of 16.38% [336]. The g-C$_3$N$_4$:PEDOT:PSS HTL showed a higher conductivity, an improved charge transport, and a suppressed charge recombination. This modified HTL had increased hole mobility, leading to more balanced charge transport. The g-C$_3$N$_4$ insulated the PSS moiety, so the conducting PEDOT chain was exposed. A two-dimensional titanium carbide (Ti$_3$C$_2$T$_x$) bilayer was incorporated into PEDOT:PSS HTLs in non-fullerene PBDB-T:ITIC and PM6:Y6 OSCs [337]. This bilayer enhanced the conductivity of PEDOT:PSS by a reduced coulombic attraction between PEDOT and PSS, causing the conformational transition of PEDOT from coil to linear structures. The HTL roughness increased upon Ti$_3$C$_2$T$_x$ incorporation, which enlarged the contact area between HTL and the photoactive layer. The hole mobility increased because of the interconnected conducting network between PEDOT and Ti$_3$C$_2$T$_x$. The PL peak was reduced, indicating improved hole transmission. As a consequence, the PCE of devices improved to 11.02% and 14.55% for PBDB-T:ITIC- and PM6:Y6-based OSCs, respectively. Moreover, PEDOT:PSS/Ti$_3$C$_2$T$_x$ HTLs enhanced the nitrogen atmosphere's long-term stability, retaining 79.67% of the initial PCE after 300 h.

4.4.2. Other Conjugated Polymers

Another approach aims to replace the use of PEDOT:PSS with different conjugated polymers. The chemical structure of a series of conjugated polymers used as HTL for OSCs are shown in Scheme 1.

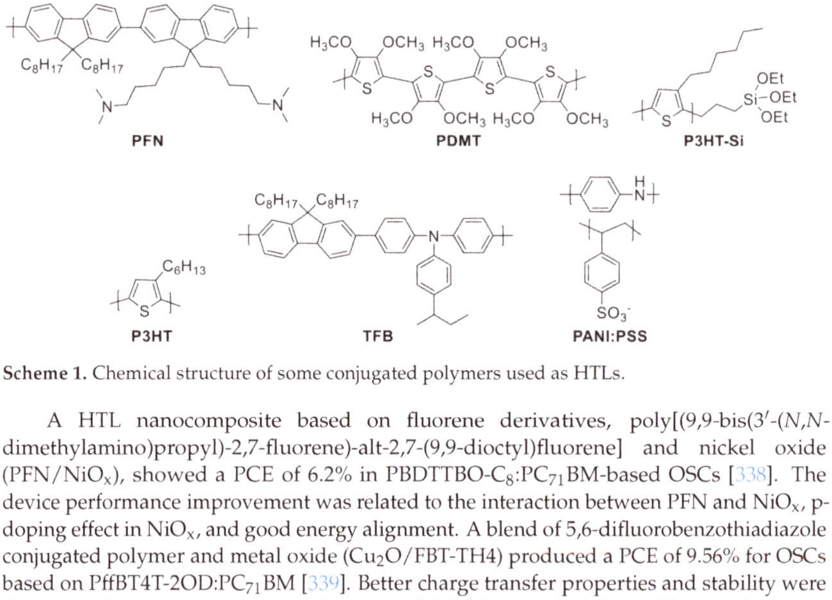

Scheme 1. Chemical structure of some conjugated polymers used as HTLs.

A HTL nanocomposite based on fluorene derivatives, poly[(9,9-bis(3′-(*N*,*N*-dimethylamino)propyl)-2,7-fluorene)-alt-2,7-(9,9-dioctyl)fluorene] and nickel oxide (PFN/NiO$_x$), showed a PCE of 6.2% in PBDTTBO-C$_8$:PC$_{71}$BM-based OSCs [338]. The device performance improvement was related to the interaction between PFN and NiO$_x$, p-doping effect in NiO$_x$, and good energy alignment. A blend of 5,6-difluorobenzothiadiazole conjugated polymer and metal oxide (Cu$_2$O/FBT-TH4) produced a PCE of 9.56% for OSCs based on PffBT4T-2OD:PC$_{71}$BM [339]. Better charge transfer properties and stability were determined, maintaining 75% of the original PCE for up to 30 days. This result was attributed to the hydrophobic character of the HTL. Poly(3,4-dimethoxythiophene) (PDMT) deposited via oxidative chemical vapor deposition were also used as hole-transport materials in OSCs [340]. Awada et al. fabricated OSCs based on hydrophobic triethoxysilane-terminated poly(3-hexylthiophene) (P3HT-Si) HTLs exhibiting a slightly enhanced device stability [341]. Some other polymers and composites used as hole-transport layers include P3HT:SWCNTs [342] and polyaniline/gold and silver NPs composites (Au$_{10}$Ag$_{10}$PANI) [343]. The interconnected network of grafted CNTs, polythiophenic agents, and conjugated PANI bottlebrushes (CNT-g-PDDT:P3ThEt-g-PANI) were used as HTL in OSCs based on PBDT-DTNT:PC$_{61}$BM, showing smooth morphology, low sheet resistance, and a PCE of 5.65% [344]. A network of CNTs and polythiophene/polyaniline bottlebrushes (CNT:P3ThEt-g-PANI) was tested as HTL in OSCs based on P3HT:PC$_{71}$BM, reaching an improved PCE of

5.30% [345]. Another approach involves the use of low acidic water-stable PSS-doped PANI as HTL based on P3HT:ICBA OSCs [346]. The PANI:PSS layer presented a well-matched WF, high conductivity, and transmittance around 90% that resulted in OSCs with PCE of 4.5%. Additionally, PANI:PSS HTL has also been tested for indoor photovoltaics [347,348]. The OSCs based on P3HT:ICBA showed a lower PCE than a device using PEDOT:PSS, but possessed better stability over 1176 h, retaining 39% of its initial PCE. PANI was also tested with GO as an acid-free composite HTL in OSCs based on P3HT:PCBM and PCDTBT:PC$_{71}$BM, resulting in optimized performance for the nanocomposite with a GO loading of 7.3 wt% [349]. A hole-transporting bilayer of copper(I) thiocyanate and poly[(9,9-dioctylfluorenyl-2,7-diyl)-alt-(4,4′-(*N*-(4-butylphenyl)))] (CuSCN/TFB) was tested in OSCs based on non-fullerene PM6:Y6 and fullerene PTB7-Th:PC$_{71}$BM by Dong et al. [350]. Better photovoltaic performance with the CuSCN/TFB bilayer than with pristine CuSCN HTL was related to enhanced J$_{sc}$ and FF. The decreased roughness and increased contact angle of the bilayer favored the interfacial contact of the HTL and the active layer, leading to better energy matching and device performance (up to 15.10%). Furthermore, the CuSCN/TFB-based device presented improved hole mobility, higher exciton dissociation efficiency, and lower recombination loss, which contributed to its enhanced exciton dissociation and charge transportation and extraction.

CPEs composed of conjugated backbone and side chains containing ionic groups are attractive materials due to their intrinsic dual electronic and ionic conductivity, and good solubility in polar solvents [351–356]. Some examples of the chemical structure of CPEs used as HTLs are shown in Scheme 2.

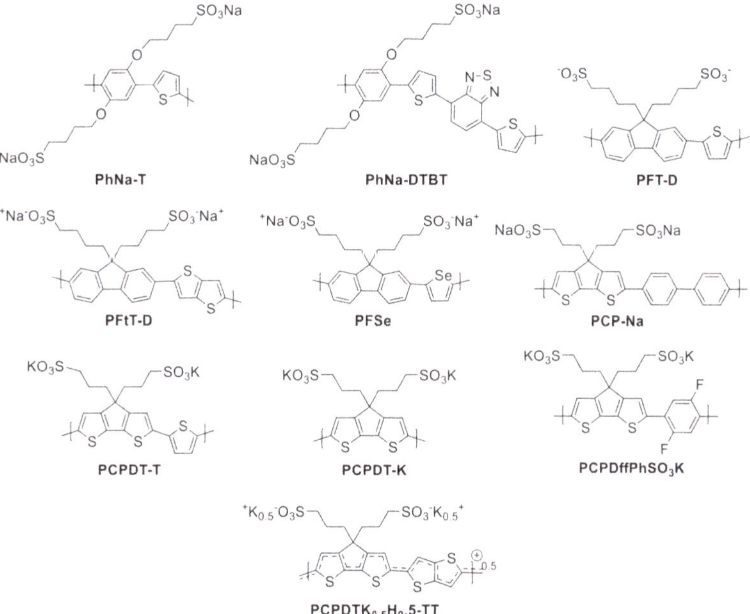

Scheme 2. Chemical structure of some CPEs used as HTLs.

Poly[1,4-bis(4-sulfonatobutoxy)benzene-thiophene] (PhNa-1T) self-doped in a neutral state achieved an increased WF of 5.21 eV, resulting in PCEs of 9.89% and 8.38% for ITO/PhNa-1T/PTB7-Th:PC$_{71}$BM/fullerene derivative (bis-C$_{60}$)/Ag and ITO/PhNa-1T/PTB7:PC$_{71}$BM/TiO$_2$/Al cells, respectively [357]. The enhanced device performance was ascribed to improved interfacial properties, a high WF, and a smoother surface resulting in a favorable contact, improved charge extraction, and an efficient hole collection. PhNa-DTBT CPE, which is based on a weakly electron-donating 2-phenyl thiophene, an electron-acceptor, 2,1,3-benzothiadiazole, and sulfonate sodium salt as an ionic functional group were used in PTB7-Th:PC71BM OSCs, reaching a PCE of 9.29% [358]. PhNa-DTBT showed a high electrical conductivity, improved J_{sc} and FF, and high WF (5.39 eV). The device also showed improved stability with a retained PCE of ca. 40% after 96 h. A pH-neutral self-doped polymer based on phenyl and thienyl units, poly[2,6-(4,4-bis-(propane-1-sulfonate sodium)-4H-cyclopenta[2,1-b;3,4-b′]dithiophene)-alt-(4,4′-biphenyl)] (PCP-Na), was used as HTL for PBDT-TS1:PC$_{71}$BM-based OSCs [359]. PCP-Na had a suitable HOMO level, smooth surface, and a high electrical conductivity due to the presence of polaronic states (radical cations). PCP-Na exhibited appreciable hole collection and charge-transport properties. The photovoltaic device using PCP-Na showed a PCE of 9.89% that resulted mainly from an enhanced FF. Following the same line, pH neutral poly[9,9-bis(4′-sulfonatobutyl)fluorene-alt-selenophene] (PFSe) was used as HTL in OSCs with architecture ITO/PFSe/PTB7:PC$_{71}$BM/PFN/Al, exhibiting a PCE of 7.2% [360]. The increased J_{sc} and FF were ascribed to a strong dipole moment at the interface. A WF of 5.15 eV of PFSe assured a good ohmic contact and a better matching energy level. Moreover, the air stability of the cell was improved by the neutral nature of the HTL polymer. Xu et al. reported a pH-neutral CPE, 3,4-dithia-7H-cyclopenta[a]pentalene and thienyl units (PCPDT) used in OSCs based on PTB7-Th:PC$_{71}$BM with a PCE of 9.3% [361]. Improved device performance by using PCPDT HTL was attributed mainly to a reduced leakage current and R_s. A tuned WF of –4.87 eV, enhanced transmittance, and improved and homogeneous mobility of HTL were related to the strong p-type self-doped nature of this HTL. Moreover, the hole layer showed improved interface compatibility, evidenced by the reduced surface energy (30.7 mN m^{-1}). The use of PCPDT-K HTL in OSC based on P3HT:PCBM showed improved device performance, increased J_{sc}, reduced R_s, a smooth surface, and better stability than a PEDOT:PSS reference device [362]. PCPDffPhSO$_3$K, a neutral CPE based on 3,4-dithia-7H-cyclopenta[a]pentalene and 1,4-difluorobenzene units, was used as HTL for ITO/HTL/PTB7-Th:PC$_{71}$BM/PFN/Al OSCs, resulting in PCE of 9.5% [363]. The self-doping effect in PCPDffPhSO$_3$K improved its conductivity. A WF around -5.18 eV ensured a better energy level alignment, achieving a higher V_{oc}, J_{sc}, and hole mobility. Lee et al. utilized poly[9,9-bis(4′-sulfonatobutyl)fluorene-alt-thieno[3,2-b]thiophene] (PFtT-D) HTLs showing a PCE of 8.3% for OSCs based on PTB7-Th:PC$_{71}$BM [364]. The WF of the modified electrode with PFtT-D was 5.19 eV, which resulted in superior ohmic contact due to well-matched energy levels facilitating the hole transportation. The modification of WF was attributed to the molecular dipole orientations. The device showed an improved lifetime because of the neutral nature of CPE; the PCE slowly decreased with a half-life of 153 h. PCPDTK$_{0.50}$H$_{0.50}$-TT, a neutral self-doped CPE, was used as HTL for the OSCs based on PM6:Y6:PC$_{71}$BM [365]. Potassium ions were exchanged to protons through ion-exchange chromatography using acid-sulfonated polystyrene resin. PCPDTK$_{0.50}$H$_{0.50}$-TT HTL had a higher WF and increased mobility, and it also exhibited a higher hole-extraction efficiency. The device performance with this HTL was improved with a PCE of 16.3%. The V_{oc} increased due to the improved hole mobility, and the J_{sc} and FF were also improved, ascribed to reduced carrier recombination and reduced bulk resistance. The device showed improved stability, and the PCE was retained by a longer time than PEDOT:PSS-based reference cells. OSCs with an area of 1.0 cm^2 prepared by wire-bar coating achieved a PCE greater than 10%, showing potential for large-area printing techniques (see Figure 19).

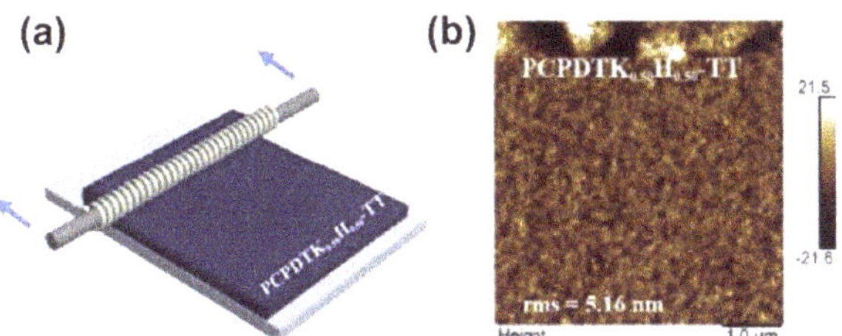

Figure 19. (a) Wire-bar coating process and (**b**) AFM image of the PCPDTK0.50H0.50-TT film prepared by the wire-bar coating process. Adapted with permission from [365]. Copyright 2021, American Chemical Society.

4.5. Small Organic Molecules

As an alternative to conjugated polymers, small organic molecules can be used as HTLs for photovoltaic applications [36]. Polymeric materials shown in the previous section have several drawbacks e.g., complicated synthesis, costly purification processes, and precise control of their molecular weight. In general, polymeric HTL materials used in OSCs usually have molar mass over 10,000 g mol^{-1}, which makes them expensive [366]. Moreover, the hole mobilities of polymers such as PTAA are sensitive to molecular weights, polydispersity indices, and purities [367]. HTLs based on small organic molecules, compared with inorganic and polymeric materials, present a variety of benefits such as simple synthesis, structural versatility, high purity, and tunable energy levels [368]. Some examples of the molecular structure of small molecules used as HTL are shown in Scheme 3.

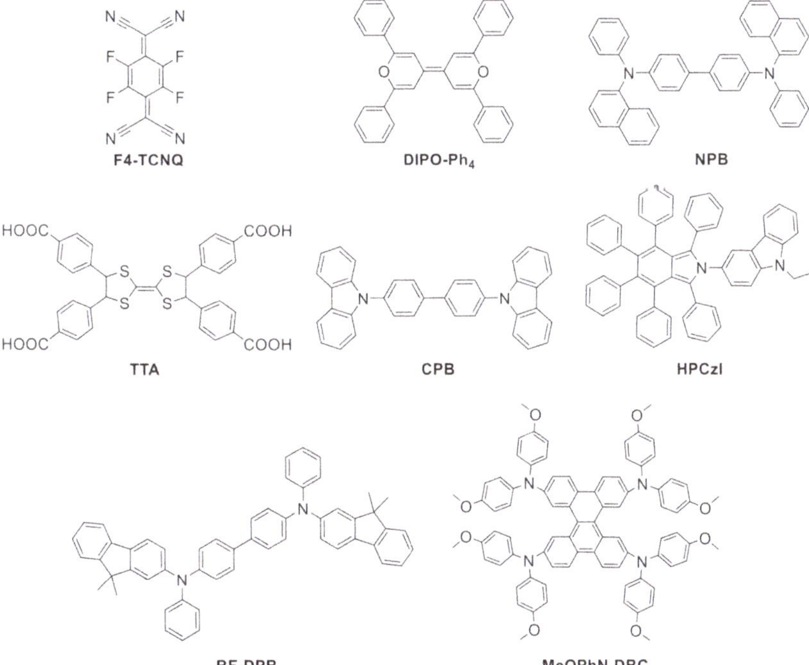

Scheme 3. Chemical structure of some small organic molecules used as HTLs.

NDP9 doped *N,N'*-((diphenyl-*N,N'*-bis)9,9,-dimethyl-fluoren-2-yl)-benzidine (BF-DPB) was used as a hole-transport material in OSCs based on zinc phtalocyanine (ZnPC):fullerene

C_{60} [369]. Spin-coated BF-DPB HTLs over AgNWs electrodes exhibited a PCE of 4.4%. BF-DPB smoothed the AgNWs topography. Cheng et al. fabricated NiOx/2,3,5,6-tetrafluoro-7,7,8,8-tetracyanoquinodimethane (F4-TCNQ) composite HTLs for the fabrication of OSCs without pre-treatment of ITO nor post-treatment on the HTL [370]. The device performance of one-step ethanol-processed NiOx:F4-TCNQ in P3HT:PC_{61}BM-based OSCs was 15.8% better than one-step PEDOT:PSS-based OSCs. NiOx:F4-TCNQ HTL was also used on PTB7-Th:PC_{71}BM-based OSCs, resulting in an enhanced PCE of 8.59%. A planar quinoid molecule, 2,2′,6,6′-tetraphenyl-dipyranylidene (DIPO-Ph_4), was tested as an anodic interfacial layer with PEDOT:PSS in P3HT:PCBM OSCs [371]. Vacuum-deposited DIPO-Ph_4 (10 nm thickness) on spin-coated PEDOT:PSS (5 nm thickness) increased OSCs' current and enhanced efficiency to 4.6%. DIPO-Ph_4's needle-like morphology increased the contact area between the active layer and the anode with high hole conductivity. 1,3,4,5,6,7-Hexaphenyl-2-{3′-(9-ethylcarbazolyl)}-isoindole (HPCzI) HTLs exhibited improved performance in comparison with MoO_3-based OSCs, reaching a PCE of 1.69% for CuPC:C60-based OSCs, due to larger FF and J_{sc} [372]. *N,N′*-bis(1-naphthalenyl)*N,N′*-bis-phenyl-(1,1′-biphenyl)-4,4′-diamine (NPB) was incorporated as HTL on inverted P3HT:PC_{71}BM OSCs, resulting in a PCE of 2.63%, a J_{sc} of 9.49 mA cm^{-2}, and low R_s [373]. These results suggested the formation of an ohmic contact between the photoactive layer and anode, which contributed to the hole extraction efficiency. Alternatively, the NPB layer was inserted between the MoO_3 layer and the photoactive layer in inverted OSCs based on P3HT:PC_{61}BM [374]. The PCE was enhanced from 3.20% to 3.94%, owing to the increased J_{sc} and reduced R_s by an improved charge transportation and reduced recombination at the interface. MoO_3 p-doped 4,4′-*N,N′*-dicarbazole-biphenyl (CBP:MoO_3) was also utilized as HTLs in inverted P3HT:PC_{61}BM-based OSCs [375]. A 3,6,11,14-Tetramethoxyphenylamine-dibenzo[g,p]chrysene (MeOPhN-DBC) layer was incorporated between MoO_3 and active-layer P3HT:PC_{61}BM-inverted OSCs, showing an enhanced PCE of 3.68%, attributed to improved J_{sc} and FF, and reduced leakage current [376]. Liu et al. reported a tetrathiafulvalene derivative with four carboxyl groups (TTA) as an HTL in OSCs (see Figure 20a) [377]. This HTL displayed a well-matched energy level (see Figure 20b) and an enhanced PCE of 9.09% in comparison with a PEDOT:PSS-based OSC (see Figure 20c). The improved FF and J_{SC} were related to the smooth surface of the TTA layer, improved charge transfer and hole mobility, and reduced charge recombination.

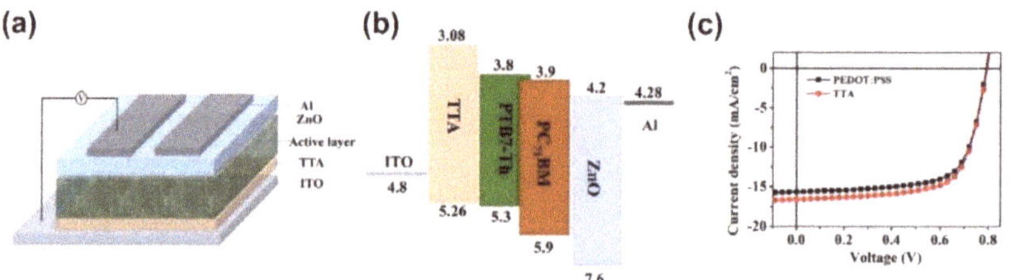

Figure 20. (**a**) Illustration of OSCs structure based on TTA, (**b**) energy level diagram of different layers, and (**c**) J-V curves of devices with TTA and PEDOT:PSS as HTLs. Adapted with permission from [377]. Copyright 2020, Elsevier.

Large-scale printing processes, like roll-to roll, require that the prepared small molecules present good adhesion and compatibility between the anode and the active layer, high carrier transport, good stability, and high solubility in non-pollutant solvents. Flexible devices will additionally require having mechanical flexibility of the layer. Convenient modification of certain functional groups allows the tuning of the electronic properties of the material to match the work function and conductivity. The research activity in these fields is very active, representing the main challenges to pave the road towards the wide commercialization of the OSCs [78].

Finally, Table 2 enlists a series of HTLs based on organic conjugated polymers and small molecules. The anode configuration with its work function, deposition technique for the HTL, active layer composition, and the OSCs performance parameters are provided as well as the reference where the information was taken from.

Table 2. Performance parameters of some representative OSCs with different organic conjugated polymers and small molecules as HTLs.

Anode Configuration and WF (eV)	Deposition Technique	Active Layer	VOC (V)	JSC (mA cm-2)	FF (%)	PCE (%)	Ref.
		PEDOT:PSS					
ITO/PEDOT:PSS-DA (5.14)	spin coating	PM6:Y6	0.84	25.52	77.1	16.55	[273]
ITO/g-C$_3$N$_4$:PEDOT:PSS (4.89)	spin coating	PM6:Y6	0.84	26.71	73.0	16.38	[336]
ITO/PEDOT:PSS:TEMPO$^+$Br$^-$ (4.95)	spin coating	PM6:Y6	0.82	27.18	72.6	16.10	[270]
ITO/PEDOT:PSS:a-In$_2$Se$_3$ (5.06)	spin coating	PM6:Y6	0.84	25.47	74.5	15.89	[272]
ITO/WO$_x$:PEDOT:PSS (4.7)	spin coating	PM6:IT-4F	0.87	20.73	80.8	14.57	[313]
ITO/PEDOT:PSS/Ti$_3$C$_2$T$_x$ (5.0)	selectively etching/spin coating	PM6:Y6	0.83	25.63	68.4	14.55	[337]
ITO/PEDOT:PSS-MoO$_3$ (5.22)	spin coating	PBDB-T-2F:IT-4F	0.86	21.71	70.6	13.19	[311]
ITO/PEDOT:PSS/BPQD (4.92)	spin coating	PM6:IT-4F	0.85	21.14	71.3	12.81	[297]
AgNWs/PEDOT:PSS/H$_x$MoO$_3$ (5.44)	transfer printing	PM6:IDIC:Y6	0.83	21.00	68.0	11.90	[315]
ITO/PEDOT:PSS:a-In$_2$Se$_3$ (5.06)	spin coating	PBDB-T:ITIC	0.91	17.31	71.1	11.22	[272]
ITO/PEDOT:PSS/Ti$_3$C$_2$T$_x$ (5.0)	selectively etching/spin coating	PBDB-T:ITIC	0.91	17.08	70.9	11.02	[337]
ITO/NiFD:PEDOT:PSS (5.01)	spin coating	PM6:PC$_{71}$BM	0.98	13.82	79.4	10.76	[269]
WO$_3$/PEDOT:PSS/Ag (5.27)	spin coating	SMD2:ITIC-Th	0.90	17.30	66.0	10.30	[314]
ITO/Ag ND/PEDOT:PSS	LIL/spin coating	PTB7:PC$_{70}$BM	0.73	23.26	61.0	10.11	[291]
ITO/PEDOT:PSS-AuNRs	spin coating	PTB7-Th:PC$_{71}$BM-Au NRs	0.80	17.90	68.8	9.89	[300]
ITO/V$_2$O$_5$: PEDOT:PSS	spin coating	PTB7-Th:PC$_{71}$BM	0.80	16.83	70.1	9.44	[309]
ITO/PEDOT:PSS/TTF-py (5.29)	spin coating	PTB7-Th:PC$_{71}$BM	0.79	17.19	70.6	9.37	[268]
ITO/PEDOT:PSS + Au NPs (5.4)	spin coating	PTB7:PC$_{71}$BM	0.74	18.30	68.0	9.26	[290]
ITO/PEDOT:PSS/BPQD (4.92)	spin coating	PTB7-Th:PC$_{71}$BM	0.80	16.40	69.4	9.11	[297]
ITO/PEDOT:PSS/p-TPCF (5.28)	electrochemical cyclic voltammetry	PTB7-Th:PC$_{71}$BM	0.80	16.98	66.2	8.99	[285]
ITO/GOs/PEDOT:PSS (4.55)	spin coating	PBDB-T:ITIC	0.90	15.10	65.7	8.93	[317]
PMA:PEDOT:PSS/Al (5.02)	spin coating	PTB7-Th:PC$_{71}$BM	0.79	17.10	68.0	8.88	[316]
PMA:PEDOT:PSS/Al (5.02)	spin coating	PffBT4T-2OD:PC$_{71}$BM	0.77	18.44	64.0	8.75	[316]
ITO/PEDOT:PSS/PTPCz (5.23)	spin coating/electrodeposition	PTB7:PC$_{71}$BM	0.74	16.23	71.1	8.54	[284]
ITO/PEDOT:PSS-WSe$_2$	spin coating	PTB7:PC$_{71}$BM	0.78	16.60	65.5	8.50	[333]
ITO/PEDOT:PSS:Cu-Au NPs	spin coating	PTB7-Th:PC$_{71}$BM	0.79	17.78	60.1	8.48	[294]
ITO/PEDOT:GSL (5.05)	spin coating	PTB7-Th:PC$_{71}$BM	0.77	15.82	68.7	8.47	[259]
ITO/GO/PEDOT:PSS (4.9)	chemical vapor deposition/drop casting	PTB7:PC$_{71}$BM	0.75	16.10	69.5	8.40	[318]
ITO/PEDOT:PSS:FOS (4.90)	spin coating	PTB7:PC$_{70}$BM	0.70	16.94	69.3	8.26	[263]
ITO/PEDOT:PSS+PFT-D (5.0)	spin coating	PTB7-Th:PC$_{71}$BM	0.77	14.90	71.3	8.20	[286]
FTO/PMMA/PEDOT:PSS	nanoimprinting/spin coating	PTB7:PC$_{70}$BM	0.73	16.30	68.2	8.12	[274]
ITO/PSS:PEDOT:PSS (4.80)	spin coating	PDCBT:PC$_{71}$BM	0.83	12.44	77.2	7.97	[269]
ITO/PEDOT:PSS:GO (5.1)	spin coating	PTB7:PCBM	0.75	14.90	67.5	7.68	[324]
PEDOT:PSS+MoO$_3$ NPs/Ag (5.0)	spin coating/blade coated	PTB7-Th:PC$_{60}$BM	0.78	14.99	63.0	7.39	[310]
ITO/PEDOT:PSS:Ag-Au-Au NRs	spin coating	PTB7:PC$_{71}$BM	0.73	16.87	60.0	7.36	[293]
ITO/PEDOT:PSS/WS$_2$	spin coating	PTB7-Th:PC$_{71}$BM	0.79	15.67	58.6	7.24	[332]
ITO/PEDOT:PSS:Cu-Au NPs	spin coating	PTB7-Th:PC$_{61}$BM	0.80	15.50	57.9	7.13	[294]
ITO/PEDOT:PSS:bf-MWCNTs (5.39)	spin coating	PCDTBT:PC$_{71}$BM	0.88	12.51	63.1	6.95	[335]
ITO/PEDOT-S (5.2)	spin coating	P3TI:PC$_{71}$BM	0.73	12.80	72.0	6.70	[265]
ITO/PEDOT:PSS-NiS	spin coating	P3HT:PC$_{61}$BM	0.58	18.65	55.9	6.03	[288]
ITO/PEDOT:PSS + Au NPs (5.0)	spin coating	rrP3HT:PC$_{71}$BM	0.58	16.10	61.0	5.65	[292]
ITO/PEDOT:PSS + Au NPs (5.0)	spin coating	rrP3HT:PC$_{71}$BM	0.58	14.70	61.0	5.29	[292]
ITO/PEDOT:PSS:PSFP-DTBTP (5.14)	spin coating	PCDTBT:PC$_{71}$BM	0.88	9.46	66.3	5.26	[275]
ITO/GO/PEDOT:PSS (4.9)	spin coating	PCDTBT:PC$_{71}$BM	0.85	10.82	57.0	5.24	[331]
ITO/PEDOT:PSS:GO (5.52)	spin casting	PTB7:PC$_{71}$BM	0.65	15.17	53.0	5.22	[327]
ITO/PEDOT:PSS-MoO$_3$ (5.3)	spray deposition	PTB7:PC$_{71}$BM	0.69	15.20	48.3	5.11	[307]
PMA:PEDOT:PSS/Ag NWs (5.02)	doctor-blade coating	PTB7-Th:PC$_{71}$BM	0.78	11.28	57.0	5.01	[316]

Table 2. *Cont.*

Anode Configuration and WF (eV)	Deposition Technique	Active Layer	VOC (V)	JSC (mA cm−2)	FF (%)	PCE (%)	Ref.
		Other conjugated polymers					
ITO/PCPDTKH-TT (5.24)	wire-bar coating	PM6:Y6:PC$_{71}$BM	0.85	25.10	75.9	16.30	[365]
ITO/CuSCN/TFB (5.32)	spin coating	PM6:Y6	0.85	24.45	72.7	15.10	[350]
ITO/PhNa-1T (5.21)	spin coating	PTB7-Th:PC$_{71}$BM	0.79	16.98	71.1	9.89	[357]
ITO/PCP-Na (5.22)	spin coating	PBDT-TS1:PC$_{71}$BM	0.80	17.46	70.6	9.89	[359]
ITO/Cu$_2$O/FBT-TH4 (5.08)	sputtered method	PffBT4T-2OD:PC$_{71}$BM	0.77	17.50	70.7	9.56	[339]
ITO/PCPDffPhSO3K (5.18)	spin coating	PTB7-Th:PC$_{71}$BM	0.79	18.08	67.0	9.50	[363]
ITO/PCPDT-T (4.87)	spin casting	PTB7-Th:PC$_{71}$BM	0.77	18.92	63.5	9.30	[361]
ITO/PhNa-DTBT (5.3)	spin coating	PTB7-Th:PC$_{71}$BM	0.79	16.92	69.5	9.29	[358]
ITO/CuSCN/TFB (5.32)	spin coating	PTB7-Th:PC$_{71}$BM	0.79	16.42	66.3	8.56	[350]
ITO/PhNa-1T (5.21)	spin coating	PTB7:PC$_{71}$BM	0.75	16.17	68.6	8.38	[357]
ITO/PFtT-D (5.19)	spin coated	PTB7-Th:PC$_{71}$BM	0.76	16.00	68.4	8.30	[364]
ITO/PFSe (5.1)	spin casting	PTB7:PC$_{71}$BM	0.68	14.40	69.0	7.20	[360]
ITO/NiO$_x$:PFN (5.34)	spin casting	PBDTTBO-C8:PC$_{71}$BM	0.71	13.75	63.7	6.20	[338]
ITO/CNT-g-PDDT:P3ThEt-g-PANI	spin coating	PBDT-DTNT:PC$_{61}$BM	0.71	12.84	62.0	5.65	[344]
ITO/CNt:P3ThEt-g-PANI	spin coating	P3HT:PC$_{71}$BM	0.68	12.85	60.7	5.30	[345]
		Small organic molecules					
ITO/TTA (5.26)	spin coating	PTB7-Th:PC71BM	0.80	16.56	69.04	9.09	[377]
ITO/NiO$_x$:F$_4$-TCNQ (5.30)	spin coating	PTB7-Th:PC71BM	0.78	16.80	65.20	8.59	[370]
ITO/DIPO-Ph$_4$/PEDOT:PSS (4.7)	vacuum deposition/spin coating	P3HT:PC61BM/	0.60	11.50	47.00	4.60	[371]
NPB/MoO$_3$/Ag (5.4)	thermal evaporation	P3HT:PC61BM	0.60	10.04	63.00	3.94	[374]
MeOPhN-DBC/MoO3/Al (5.0)	vacuum deposition	P3HT:PC61BM	0.63	12.44	47.00	3.68	[376]
ITO/NiO$_x$:F$_4$-TCNQ (5.30)	spin coating	P3HT:PC61BM	0.59	9.89	61.60	3.59	[370]
NPB/Ag (5.4)	vacuum deposition	P3HT:PC71BM	0.57	9.49	48.90	2.63	[373]
ITO/BF-DPB:NDP9 (5.23)	spin coating	ZnPC:C60	0.51	7.50	55.00	2.10	[369]
Ag NWs/BF-DPB:NDP9 (5.23)	spin coating	ZnPC:C60	0.49	7.60	55.00	2.10	[369]
ITO/MoO3:HPCzI (5.3/5.1)	thermal evaporation	CuPC:C60	0.49	6.63	53.00	1.71	[372]
ITO/HPCzI (5.1)	thermal evaporation	CuPC:C60	0.49	6.22	53.00	1.62	[372]

5. Conclusions

In summary, HTLs are fundamental to assure the high performance and stability of OSCs. Inorganic and nanocarbon materials including MoO_3, WO_3, V_2O_5, NiO_x, CuO_x, CoO_x, $CuCrO_x$, CuSCN, MoS_2, WS_2, NiS, CuS and GO, QCDs, and CNTs have shown great potential as HTLs in conventional and inverted OSCs. These hole-extracting materials can form an ohmic contact between the active and electrodes depending on their optical and electrical properties. Their high transparency enables them to absorb high light into the active layer to afford the hole-electron pairs generation, and the tuning of the Fermi levels with the donor allows the hole collection. Usually, the hole transport takes place in the HTL valence band, but in *n*-type metals such as MoO_3, it has been found that the conduction band facilitates the hole transport. Thus, the type of hole-transport path will vary with the WF and energy levels of the inorganic and nanocarbon materials as HTLs. Modification in the particles' size or addition of metal NPs results in the LSPR effect, which increases the light absorption. Inorganic materials such as Mo and Ni were doped with V and Cu to tune the WF and increase conductivity and transparency, resulting in a high V_{oc}, FF, and J_{sc}. Hybrid layers, such as MoS_2:MoO_3, improved the electron-blocking properties and increased conductivity of the layer. Nanocarbon materials such as GO were doped with F_4TCNQ to induce a change in the WF by shifting the Fermi levels, resulting in an enhanced hole transport. CNTs were functionalized with amino groups to increase the charge-carrier properties and reduce R_s, improving FF and J_{sc}. HTLs can also be subjected to ultraviolet ozone (UVO), annealing, and microwave-annealing post-treatments to increase V_{oc}, FF, and J_{sc} due to the reduction of oxygen defects in the surface morphology. The most-used conjugated polymer is the PEDOT:PSS and its composites such as PEDOT:PSS with NPs, MOs, and GO either in bilayer or composite monolayer. PANI was the second choice of conducting polymers as HTLs. The different modification to PEDOT:PSS in many cases increases the J_{SC}, improves the conductivity, and decreases the recombination loss of the device. In general, the addition of metallic NPs to PEDOT:PSS enhances the absorption ability of the photoactive layer, increases the conductivity, and improves the charge carrier collection by increasing the device performance. The incorporation of metal oxides mainly helps to (i) increase the stability of the device (e.g., MoO_3) by mitigating the degradation, (ii) serve as an electron blocking layer (e.g., V_2O_5), and (iii) suppress the charge recombination. On the other hand, GO addition to PEDOT:PSS improves the conductivity and increases the device resistance to degradation. Furthermore, CPEs were also used as HTL for OSCs; in general, these materials were pH-neutral layers that improved the stability of the devices and showed high electrical conductivity and good interface compatibility. In the last decade, the standard HTL small molecule has been spiro-OMeTAD; nevertheless, this review pointed out that research to optimize these materials is growing in activity and importance. Important points like shorter and more efficient chemical synthesis, access from cheaper starting materials, analysis of the active layer (perovskite or organic) interactions with the HTL small molecules, as well as better understanding of the charge transport process (carrier diffusion, recombination process) makes this field very important for optimization of the solar cells.

Additionally, large-area deposition techniques are mandatory to facilitate the commercialization of organic photovoltaics. Compared with the conventional spin-coating technique, laser-assisted and electrospray techniques allow the control of the surface morphology and thickness at low temperatures and short-time processing. The roll-to-roll technique is also attractive for large industrial-scale manufacturing of metal oxides, such as the inkjet printing of NiO_x. Overall, inorganic and nanocarbon HTLs are very favorable for OSCs, mainly because of their high stability, improved electrical properties, and transparency in the visible range. Solution processing is a great advantage of using small organic molecules as HTL. The continuous investigation of a vast number of new inorganic and organic HTLs, which can assure high efficiency, high stability, low costs, facile preparation, and improved film-forming properties over large areas, is essential for the future commercialization of OSCs. Therefore, not only the chemical or electrochemical properties of the prepared HTL materials are important to study, but also it is required to develop materials that fulfill the technological requirements to apply them at large scales. The future in this direction looks very promising.

Author Contributions: The manuscript was written through contributions of all authors. C.A.-C. and K.P.-I. wrote the draft manuscript. B.A.F.-U. and A.P.-C. revised the manuscript. A.P.-C. supervised the writing process and edited the manuscript. All authors have read and agreed to the published version of the manuscript.

Funding: This research was funded by German Academic Exchange Service (DAAD), grant number 57552347 and PAPIIT-DGAPA-UNAM project IN208919.

Acknowledgments: We kindly thank Prof. Dr. Ullrich Scherf for his sponsorship. A.P.-C. thanks the German Academic Exchange Service (DAAD) for the funding through the Re-invitation Programme for Former Scholarship Holders 2021 No. 57552347 and the University Yachay Tech for the internal grant No. Chem19-17. B.A.F.-U. recognizes the economic support through PAPIIT-DGAPA-UNAM project IN208919. Citlalit Martinez Soto is acknowledged for the technical support.

Conflicts of Interest: The authors declare no conflict of interest.

Abbreviations

Y6	(2,2′-((2Z,2Z)-((12,13-bis(2-ethylhexyl)-3,9-diundecyl-12,13-dihydro-[1,2,5] thiadiazolo[3,4-e]thieno[2″,3″:4′,50]thieno[2′,3′:4,5]pyrrolo[3,2-g] thieno[2′,3′:4,5] thieno[3,2-b]indole-2,10-diyl)bis(methanylylidene))bis(5,6- difluoro-3-oxo-2,3-dihydro-1H-indene-2,1-diylidene))dimalononitrile)
O-IDTBR	(5Z,5′Z)-5,5′-{[7,7′-(4,4,9,9-tetraoctyl-4,9-dihydro-s-indaceno[1,2-b:5,6-b′] dithiophene-2,7-diyl)bis(benzo[c][1,2,5]thiadiazole-7,4-diyl)] bis(methanylylidene)}bis(3-ethyl-2-thioxothiazolidin-4-one)
PC$_{71}$BM	(6,6)-phenyl-C71-butyric acid methyl ester
HPCzI	1,3,4,5,6,7-hexaphenyl-2-{3′-(9-ethylcarbazolyl)}-isoindole
ICBA	1′,1″,4′,4″-Tetrahydro-di[1,4]methanonaphthaleno[1,2:2′,3′,56,60:2″,3″][5,6] fullerene-C60
TEMPO	2,2,6,6-tetramethylpiperidine-1-oxoammonium
IDIC	2,2′-[(4,4,9,9-tetrahexyl-4,9-dihydro-s-indaceno[1,2-b:5,6-b′]dithiophene-2,7-diyl) bis[methylidyne(3-oxo-1H-indene-2,1(3H)-diylidene)]]bis-propanedinitrile
DIPO-Ph$_4$	2,2′,6,6′-tetraphenyl-dipyranylidene
F4-TCNQ	2,3,5,6-tetrafluoro-7,7,8,8-tetracyanoquinodimethane
EGME	2-methoxyethanol
PTCDA	3,4,9,10-perylenetetracarboxylic dianhydride
IT-4F	3,9-bis(2-methylene-((3-(1,1-dicyanomethylene)-6,7-difluoro)-indanone))-5,5,11, 11-tetrakis(4-hexylphenyl)-dithieno[2,3-d:2′,3′-d′]-s-indaceno [1,2-b:5,6-b′]dithiophene
ITIC	3,9-bis(2-methylene-(3-(1,1-dicyanomethylene)-indanone))-5,5,11, 11-tetrakis(4-hexylphenyl)-dithieno-[2,3-d:2′,3′-d′]-s-indaceno [1,2-b:5,6-b′]dithiophene
ITIC-Th	3,9-bis(2-methylene-(3-(1,1-dicyanomethylene)-indanone))-5,5,11, 11-tetrakis(5-hexylthienyl)-dithieno-[2,3-d:2′,3′-d′]-s-indaceno [1,2-b:5,6-b′]dithiophene
CBP	4,4′-N,N′-dicarbazole-biphenyl
a-MWNTs	Amino-functionalized multi-walled carbon nanotubes
AHM	Ammonium heptamolybdate
AIL	Anode interfacial layer
AFM	Atomic force microscope
BPQDs	Black phosphorous quantum dots
bf-MWCNTs	Boronic acid functionalized multi-walled carbon nanotubes
BHJ	Bulk heterojunction
CNT	Carbon nanotubes
CTAB	Cetyltrimethylammonium bromide
CoO$_x$	Cobalt oxide
CPEs	Conjugated polyelectrolytes
CIGS	Copper indium gallium diselenide
CuO$_x$	Copper oxide
CuS$_x$	Copper sulfide
MeOPhN-DBC	Dibenzo[g,p]chrysene derivative, 3,6,11,14-tetramethoxyphenylamine-dibenzo[g,p]chrysene
DMSO	Dimethyl sulfoxide
DMF	Dimethylformamide
DC	Direct current
DA	Dopamine
ETL	Electron transport layer
EQE	External quantum efficiency
FF	Fill factor
GSL	Grafted sulfonated-acetone-formaldehyde lignin
GO	Graphene oxide
GQDs	Graphene quantum dots
g-C$_3$N$_4$	Graphitic carbon nitrile
HOMO	High occupied molecular orbital

HTL	Hole transport layer
H_yMoO_{3-x}	Hydrogenated molybdenum oxide
ITO	Indium tin oxide
P_{in}	Input power
IPA	Isopropanol
LIL	Laser interference lithography
LSPR	Localized surface plasmon resonance
LUMO	Lowest unoccupied molecular orbital
P_{max}	Maximum power output
MO	Metal oxides
MW	Microwave
NPB	N,N'-bis(1-naphthalenyl)N,N'-bis-phenyl-(1,1'-biphenyl)-4,4'-diamine
BF-DPB	N,N'-((diphenyl-N,N'-bis)9,9,-dimethyl-fluoren-2-yl)-benzidine
NDs	Nanodots
NPs	Nanoparticles
NRs	Nanorods
NWs	Nanowires
NFD	Nickel formate dihydrate
NiO_x	Nickel oxide
NiS_x	Nickel sulfide
nGQDs	Nitrogen doped graphene quantum dots
V_{oc}	Open circuit voltage
OLEDs	Organic light-emiting diodes
OSCs	Organic solar cells
P_{out}	Output power
PCBM	Phenyl-C61-butyric acid methyl ester
PMA	Phosphomolybdic acid
PL	Photoluminescence
PV	Photovoltaic
PC	Phthalocyanine
PTB7	Poly [[4,8-bis[(2-ethylhexyl)oxy]benzo[1,2-b:4,5-b']dithiophene-2, 6-diyl][3-fluoro-2-[(2-ethylhexyl)carbonyl]thieno[3,4-b]thiophenediyl]]
PCPDT	Poly 3,4-dithia-7H-cyclopenta[a]pentalene
FBT-TH4	Poly 5,6 difluorobenzothiadiazole
PDMT	Poly(3,4-dimethoxythiophene)
PEDOT	Poly(3,4-ethylenedioxythiophene)
PEDOT:PSS	Poly(3,4-ethylenedioxythiophene)-poly(styrene sulfonate)
P3HT	Poly(3-hexylthiophene)
$P3HT_{50}$-b-PSS_{23}	Poly(3-hexylthiophene)-b-poly(p-styrenesulfonate)
PEDOT-S	Poly(4-(2,3-dihydrothieno[3,4-b][1,4]dioxin-2-yl-methoxy)-1-butanesulfonic acid)
PTPCz	Poly(carbazolyl triphenylethylene) derivative
PDMS	Poly(dimethylsiloxane)
PMMA	Poly(methylmethacrylate)
PSS	Poly(styrenesulfonate)
PM6	Poly[(2,6-(4,8-bis(5-(2-ethylhexyl)-4-fluorothiophen-2-yl)benzo [1,2-b:4,5-b']dithiophene))-co-(1,3-di(5-thiophene-2-yl)-5,7-bis (2-ethylhexyl)-benzo[1,2-c:4,5-c']dithiophene-4,8-dione))]
PBDB-T	Poly[(2,6-(4,8-bis(5-(2-ethylhexyl)thiophen-2-yl)benzo[1,2-b:4,5-b']dithiophene)-co-(1,3-di(5-thiophene-2-yl)-5,7-bis(2-ethylhexyl)benzo[1,2-c:4,5-c']dithiophene-4,8-dione)]
PffBT4T-2OD	Poly[(5,6-difluoro-2,1,3-benzothiadiazol-4,7-diyl)-alt-(3,3'''-di(2-octyldodecyl)-2,2';5',2'';5'',2'''-quaterthiophen-5,5'''-diyl)]
PFN	Poly[(9,9bis(3'-(N,N-dimethylamino)propyl)-2,7-fluorene)-alt-2, 7-(9,9dioctylfluorene)]
PSFP-DTBTP	Poly[(9,9-bis(4-sulfonatobutyl sodium) fluorene-alt-phenylene)-ran-(4,7-di-2-thienyl-2,1,3-benzothiadiazole-alt-phenylene)]
TFB	Poly[(9,9-dioctylfluorenyl-2,7-diyl)-alt-(4,4'-(N-(4-butylphenyl)))]

PNDIT-F3N	Poly[[2,7-bis(2-ethylhexyl)-1,2,3,6,7,8-hexahydro-1,3,6,8-tetraoxobenzo[lmn][3,8]phenanthroline-4,9-diyl]-2,5-thiophenediyl[9,9-bis[3'((*N*,*N*-dimethyl)-*N*-ethylammonium)]propyl]-9*H*-fluorene-2,7-diyl]-2,5-thiophenediyl]
PhNa-1T	Poly[1,4-bis(4-sulfonatobutoxy)benzene-thiophene]
PCP-Na	Poly[2,6-(4,4-bis-(propane-1-sulfonate sodium)-4*H*-cyclopenta[2,1-b;3,4-b']dithiophene)-alt-(4,4'-biphenyl)]
P3HTN	Poly[3-(6'-*N*,*N*,*N*-trimethyl ammonium)-hexylthiophene] bromide
PTB7-Th	Poly[4,8-bis(5-(2-ethylhexyl)thiophen-2-yl)benzo[1,2-b;4,5-b']dithiophene-2,6-diyl-alt-(4-(2-ethylhexyl)-3-fluorothieno[3,4-b]thiophene-)-2-carboxylate-2-6-diyl)]
PBDT-TS1	Poly[4,8-bis(5-(octylthio)thiophen-2-yl)benzo[1,2-b;4,5-b']dithiophene-2,6-diyl–alt–(4-(2- ethylhexyl)-3-fluorothieno[3,4-b]thiophene-)-2-carboxylate-2-6-diyl]
PFtT-D	Poly[9,9-bis(4'-sulfonatobutyl)fluorene-alt-thieno[3,2-b]thiophene]
PFSe	Poly[9,9-bis(4'-sulfonatobutyl)fluorene-alt-selenophene]
PFT-D	Poly[9,9-bis(4'-sulfonatobutyl)fluorene-*alt*-thiophene]
PBDT-DTNT	Poly[benzodithiophene-bis(decyltetradecyl-thien)naphthothiadiazole]
PCDTBT	Poly[*N*-9'-heptadecanyl-2,7-carbazole-alt-5,5(4',7'-di-2-thienyl-2',1',3'-benzothiadiazole)]
PANI	Polyaniline
PEG	Polyethylene glycol
PSC	Polymer solar cell
PMC	Polynuclear metal-oxo clusters
p-TPCF	Polytriphenylcarbazole fluoranthene
PCE	Power conversion efficiency
QDs	Quantum dots
rrP3HT	Regioregular poly(3-hexylthiophene-2,5-diyl)
RMS	Root meand square
SEM	Scanning electron microscopy
R_s	Series resistance
J_{sc}	Short circuit current
R_{sh}	Shunt resistance
SWCNTs	Single-walled carbon nanotubes
SILAR	Successive ionic layer adsorption and reaction
F_4TCNQ	Tetrafluorotetracyanoquino-dimethane
TTA	Tetrathiafulvalene
TTF-py	Tetrathiafulvalene pyridine derivative
P3HT-Si	Triethoxysilane terminated poly(3-hexylthiophene)
WO_x	Tungsten oxide
WS_x	Tungsten sulfide
UVO	Ultraviolet ozone
uSWNT	Unzipped single-walled carbon nanotubes
VO_x	Vanadium oxide
VoPC	Vanadylphthalocyanine
WF	Work function

References

1. Owusu, P.A.; Asumadu-Sarkodie, S. A review of renewable energy sources, sustainability issues and climate change mitigation. *Cogent Eng.* **2016**, *3*, 1167990. [CrossRef]
2. Ellabban, O.; Abu-Rub, H.; Blaabjerg, F. Renewable energy resources: Current status, future prospects and their enabling technology. *Renew. Sustain. Energy Rev.* **2014**, *39*, 748–764. [CrossRef]
3. Luqman, M.; Ahmad, S.R.; Khan, S.; Ahmad, U.; Raza, A.; Akmal, F. Estimation of Solar Energy Potential from Rooftop of Punjab Government Servants Cooperative Housing Society Lahore Using GIS. *Smart Grid Renew. Energy* **2015**, *6*, 128–139. [CrossRef]
4. Kabir, E.; Kumar, P.; Kumar, S.; Adelodun, A.A.; Kim, K.-H. Solar energy: Potential and future prospects. *Renew. Sustain. Energy Rev.* **2018**, *82*, 894–900. [CrossRef]
5. Hosenuzzaman, M.; Rahim, N.A.; Selvaraj, J.; Hasanuzzaman, M.; Malek, A.B.M.A.; Nahar, A. Global prospects, progress, policies, and environmental impact of solar photovoltaic power generation. *Renew. Sustain. Energy Rev.* **2015**, *41*, 284–297. [CrossRef]

6. Kannan, N.; Vakeesan, D. Solar energy for future world:—A review. *Renew. Sustain. Energy Rev.* **2016**, *62*, 1092–1105. [CrossRef]
7. Goswami, D.Y.; Vijayaraghavan, S.; Lu, S.; Tamm, G. New and emerging developments in solar energy. *Sol. Energy* **2004**, *76*, 33–43. [CrossRef]
8. Li, G.; Zhu, R.; Yang, Y. Polymer solar cells. *Nat. Photonics* **2012**, *6*, 153–161. [CrossRef]
9. Lin, Y.; Shao, Y.; Dai, J.; Li, T.; Liu, Y.; Dai, X.; Xiao, X.; Deng, Y.; Gruverman, A.; Zeng, X.C.; et al. Metallic surface doping of metal halide perovskites. *Nat. Commun.* **2021**, *12*, 7. [CrossRef]
10. Ur Rehman, S.; Noman, M.; Khan, A.D.; Saboor, A.; Ahmad, M.S.; Khan, H.U. Synthesis of polyvinyl acetate /graphene nanocomposite and its application as an electrolyte in dye sensitized solar cells. *Optik* **2020**, *202*, 163591. [CrossRef]
11. Liu, M.; Johnston, M.B.; Snaith, H.J. Efficient planar heterojunction perovskite solar cells by vapour deposition. *Nature* **2013**, *501*, 395–398. [CrossRef]
12. Mathew, S.; Yella, A.; Gao, P.; Humphry-Baker, R.; Curchod, B.F.E.; Ashari-Astani, N.; Tavernelli, I.; Rothlisberger, U.; Nazeeruddin, M.K.; Grätzel, M. Dye-sensitized solar cells with 13% efficiency achieved through the molecular engineering of porphyrin sensitizers. *Nat. Chem.* **2014**, *6*, 242–247. [CrossRef] [PubMed]
13. Liu, S.; Yuan, J.; Deng, W.; Luo, M.; Xie, Y.; Liang, Q.; Zou, Y.; He, Z.; Wu, H.; Cao, Y. High-efficiency organic solar cells with low non-radiative recombination loss and low energetic disorder. *Nat. Photonics* **2020**, *14*, 300–305. [CrossRef]
14. Chuang, C.-H.M.; Brown, P.R.; Bulović, V.; Bawendi, M.G. Improved performance and stability in quantum dot solar cells through band alignment engineering. *Nat. Mater.* **2014**, *13*, 796–801. [CrossRef] [PubMed]
15. Bernède, J.C. Organic photovoltaic cells: History, principle and techniques. *J. Chil. Chem. Soc.* **2008**, *53*. [CrossRef]
16. Kearns, D.; Calvin, M. Photovoltaic Effect and Photoconductivity in Laminated Organic Systems. *J. Chem. Phys.* **1958**, *29*, 950–951. [CrossRef]
17. Tang, C.W. Two-layer organic photovoltaic cell. *Appl. Phys. Lett.* **1986**, *48*, 183–185. [CrossRef]
18. Sariciftci, N.S.; Smilowitz, L.; Heeger, A.J.; Wudl, F. Semiconducting polymers (as donors) and buckminsterfullerene (as acceptor): Photoinduced electron transfer and heterojunction devices. *Synth. Met.* **1993**, *59*, 333–352. [CrossRef]
19. Yu, G.; Gao, J.; Hummelen, J.C.; Wudl, F.; Heeger, A.J. Polymer Photovoltaic Cells: Enhanced Efficiencies via a Network of Internal Donor-Acceptor Heterojunctions. *Science* **1995**, *270*, 1789–1791. [CrossRef]
20. Yin, Z.; Wei, J.; Zheng, Q. Interfacial Materials for Organic Solar Cells: Recent Advances and Perspectives. *Adv. Sci.* **2016**, *3*, 1500362. [CrossRef]
21. Green, M.A. Recent developments in photovoltaics. *Sol. Energy* **2004**, *76*, 3–8. [CrossRef]
22. Darling, S.B.; You, F. The case for organic photovoltaics. *RSC Adv.* **2013**, *3*, 17633. [CrossRef]
23. Liu, Q.; Jiang, Y.; Jin, K.; Qin, J.; Xu, J.; Li, W.; Xiong, J.; Liu, J.; Xiao, Z.; Sun, K.; et al. 18% Efficiency organic solar cells. *Sci. Bull.* **2020**, *65*, 272–275. [CrossRef]
24. Green, M.; Dunlop, E.; Hohl-Ebinger, J.; Yoshita, M.; Kopidakis, N.; Hao, X. Solar cell efficiency tables (version 57). *Prog. Photovolt. Res. Appl.* **2021**, *29*, 3–15. [CrossRef]
25. Salim, M.B.; Nekovei, R.; Jeyakumar, R. Organic tandem solar cells with 18.6% efficiency. *Sol. Energy* **2020**, *198*, 160–166. [CrossRef]
26. Dennler, G.; Scharber, M.C.; Brabec, C.J. Polymer-Fullerene Bulk-Heterojunction Solar Cells. *Adv. Mater.* **2009**, *21*, 1323–1338. [CrossRef]
27. Zhao, W.; Li, S.; Yao, H.; Zhang, S.; Zhang, Y.; Yang, B.; Hou, J. Molecular Optimization Enables over 13% Efficiency in Organic Solar Cells. *J. Am. Chem. Soc.* **2017**, *139*, 7148–7151. [CrossRef]
28. Hou, J.; Inganas, O.; Friend, R.H.; Gao, F. Organic solar cells based on non-fullerene acceptors. *Nat. Mater.* **2018**, *17*, 119–128. [CrossRef]
29. Jacoby, M. The future of low-cost solar cells. *Chem. Eng. News* **2016**, *94*, 30–35.
30. Riede, M.; Spoltore, D.; Leo, K. Organic Solar Cells—The Path to Commercial Success. *Adv. Energy Mater.* **2021**, *11*, 2002653. [CrossRef]
31. Krebs, F.C.; Espinosa, N.; Hösel, M.; Søndergaard, R.R.; Jørgensen, M. 25th anniversary article: Rise to power—OPV-based solar parks. *Adv. Mater.* **2014**, *26*, 29–39. [CrossRef] [PubMed]
32. Yan, C.; Barlow, S.; Wang, Z.; Yan, H.; Jen, A.K.Y.; Marder, S.R.; Zhan, X. Non-fullerene acceptors for organic solar cells. *Nat. Rev. Mater.* **2018**, *3*, 18003. [CrossRef]
33. Duan, C.; Huang, F.; Cao, Y. Solution processed thick film organic solar cells. *Polym. Chem.* **2015**, *6*, 8081–8098. [CrossRef]
34. Scharber, M.C.; Sariciftci, N.S. Efficiency of bulk-heterojunction organic solar cells. *Prog. Polym. Sci.* **2013**, *38*, 1929–1940. [CrossRef]
35. Yeh, N.; Yeh, P. Organic solar cells: Their developments and potentials. *Renew. Sustain. Energy Rev.* **2013**, *21*, 421–431. [CrossRef]
36. Park, J.H.; Lee, T.-W.; Chin, B.-D.; Wang, D.H.; Park, O.O. Roles of Interlayers in Efficient Organic Photovoltaic Devices. *Macromol. Rapid Commun.* **2010**, *31*, 2095–2108. [CrossRef]
37. Lai, T.-H.; Tsang, S.-W.; Manders, J.R.; Chen, S.; So, F. Properties of interlayer for organic photovoltaics. *Mater. Today* **2013**, *16*, 424–432. [CrossRef]
38. Steim, R.; Kogler, F.R.; Brabec, C.J. Interface materials for organic solar cells. *J. Mater. Chem.* **2010**, *20*, 2499. [CrossRef]
39. Po, R.; Carbonera, C.; Bernardi, A.; Camaioni, N. The role of buffer layers in polymer solar cells. *Energy Environ. Sci.* **2011**, *4*, 285–310. [CrossRef]

40. Roman, L.S.; Mammo, W.; Pettersson, L.A.A.; Andersson, M.R.; Inganäs, O. High Quantum Efficiency Polythiophene. *Adv. Mater.* **1998**, *10*, 774–777. [CrossRef]
41. Tian, L.; Xue, Q.; Hu, Z.; Huang, F. Recent advances of interface engineering for non-fullerene organic solar cells. *Org. Electron.* **2021**, *93*, 106141. [CrossRef]
42. Palilis, L.C.; Vasilopoulou, M.; Verykios, A.; Soultati, A.; Polydorou, E.; Argitis, P.; Davazoglou, D.; Mohd Yusoff, B.A.R.; Nazeeruddin, M.K. Inorganic and Hybrid Interfacial Materials for Organic and Perovskite Solar Cells. *Adv. Energy Mater.* **2020**, *10*, 2000910. [CrossRef]
43. Gusain, A.; Faria, R.M.; Miranda, P.B. Polymer Solar Cells—Interfacial Processes Related to Performance Issues. *Front. Chem.* **2019**, *7*, 61. [CrossRef]
44. Amollo, T.A.; Mola, G.T.; Nyamori, V.O. Organic solar cells: Materials and prospects of graphene for active and interfacial layers. *Crit. Rev. Solid State Mater. Sci.* **2020**, *45*, 261–288. [CrossRef]
45. Wu, C.; Wang, K.; Batmunkh, M.; Bati, A.S.R.; Yang, D.; Jiang, Y.; Hou, Y.; Shapter, J.G.; Priya, S. Multifunctional nanostructured materials for next generation photovoltaics. *Nano Energy* **2020**, *70*, 104480. [CrossRef]
46. Huang, Z.; Ouyang, D.; Shih, C.; Yang, B.; Choy, W.C.H. Solution-Processed Ternary Oxides as Carrier Transport/Injection Layers in Optoelectronics. *Adv. Energy Mater.* **2020**, *10*, 1900903. [CrossRef]
47. Li, Y. Molecular Design of Photovoltaic Materials for Polymer Solar Cells: Toward Suitable Electronic Energy Levels and Broad Absorption. *Acc. Chem. Res.* **2012**, *45*, 723–733. [CrossRef] [PubMed]
48. Yao, H.; Ye, L.; Zhang, H.; Li, S.; Zhang, S.; Hou, J. Molecular Design of Benzodithiophene-Based Organic Photovoltaic Materials. *Chem. Rev.* **2016**, *116*, 7397–7457. [CrossRef]
49. Kim, J.Y.; Lee, K.; Coates, N.E.; Moses, D.; Nguyen, T.-Q.; Dante, M.; Heeger, A.J. Efficient Tandem Polymer Solar Cells Fabricated by All-Solution Processing. *Science* **2007**, *317*, 222–225. [CrossRef]
50. Meillaud, F.; Shah, A.; Droz, C.; Vallat-Sauvain, E.; Miazza, C. Efficiency limits for single-junction and tandem solar cells. *Sol. Energy Mater. Sol. Cells* **2006**, *90*, 2952–2959. [CrossRef]
51. Lee, Y.-J.; Adkison, B.L.; Xu, L.; Kramer, A.A.; Hsu, J.W.P. Comparison of conventional and inverted organic photovoltaic devices with controlled illumination area and extraction layers. *Sol. Energy Mater. Sol. Cells* **2016**, *144*, 592–599. [CrossRef]
52. Nunzi, J.-M. Organic photovoltaic materials and devices. *Comptes Rendus Phys.* **2002**, *3*, 523–542. [CrossRef]
53. Shrotriya, V.; Li, G.; Yao, Y.; Moriarty, T.; Emery, K.; Yang, Y. Accurate Measurement and Characterization of Organic Solar Cells. *Adv. Funct. Mater.* **2006**, *16*, 2016–2023. [CrossRef]
54. Qi, B.; Wang, J. Fill factor in organic solar cells. *Phys. Chem. Chem. Phys.* **2013**, *15*, 8972. [CrossRef] [PubMed]
55. Parker, I.D. Carrier tunneling and device characteristics in polymer light-emitting diodes. *J. Appl. Phys.* **1994**, *75*, 1656–1666. [CrossRef]
56. Elumalai, N.K.; Uddin, A. Open circuit voltage of organic solar cells: An in-depth review. *Energy Environ. Sci.* **2016**, *9*, 391–410. [CrossRef]
57. Ratcliff, E.L.; Zacher, B.; Armstrong, N.R. Selective Interlayers and Contacts in Organic Photovoltaic Cells. *J. Phys. Chem. Lett.* **2011**, *2*, 1337–1350. [CrossRef] [PubMed]
58. Mihailetchi, V.D.; Blom, P.W.M.; Hummelen, J.C.; Rispens, M.T. Cathode dependence of the open-circuit voltage of polymer:fullerene bulk heterojunction solar cells. *J. Appl. Phys.* **2003**, *94*, 6849–6854. [CrossRef]
59. Günes, S.; Neugebauer, H.; Sariciftci, N.S. Conjugated Polymer-Based Organic Solar Cells. *Chem. Rev.* **2007**, *107*, 1324–1338. [CrossRef]
60. Street, R.A.; Song, K.W.; Cowan, S. Influence of series resistance on the photocurrent analysis of organic solar cells. *Org. Electron.* **2011**, *12*, 244–248. [CrossRef]
61. Servaites, J.D.; Yeganeh, S.; Marks, T.J.; Ratner, M.A. Efficiency Enhancement in Organic Photovoltaic Cells: Consequences of Optimizing Series Resistance. *Adv. Funct. Mater.* **2010**, *20*, 97–104. [CrossRef]
62. Antohe, S.; Iftimie, S.; Hrostea, L.; Antohe, V.A.; Girtan, M. A critical review of photovoltaic cells based on organic monomeric and polymeric thin film heterojunctions. *Thin Solid Films* **2017**, *642*, 219–231. [CrossRef]
63. Antohe, S.; Ion, L.; Tomozeiu, N.; Stoica, T.; Barna, E. Electrical and photovoltaic properties of photosensitised ITO/a-Si:H p–i–n/TPyP/Au cells. *Sol. Energy Mater. Sol. Cells* **2000**, *62*, 207–216. [CrossRef]
64. Antohe, S.; Ruxandra, V.; Tugulea, L.; Gheorghe, V.; Ionascu, D. Three-Layered Photovoltaic Cell with an Enlarged Photoactive Region of Codeposited Dyes. *J. Phys. III* **1996**, *6*, 1133–1144. [CrossRef]
65. Proctor, C.M.; Nguyen, T.-Q. Effect of leakage current and shunt resistance on the light intensity dependence of organic solar cells. *Appl. Phys. Lett.* **2015**, *106*, 083301. [CrossRef]
66. Wurfel, U.; Cuevas, A.; Wurfel, P. Charge Carrier Separation in Solar Cells. *IEEE J. Photovolt.* **2015**, *5*, 461–469. [CrossRef]
67. Yip, H.-L.; Jen, A.K.Y. Recent advances in solution-processed interfacial materials for efficient and stable polymer solar cells. *Energy Environ. Sci.* **2012**, *5*, 5994. [CrossRef]
68. Braun, S.; Salaneck, W.R.; Fahlman, M. Energy-Level Alignment at Organic/Metal and Organic/Organic Interfaces. *Adv. Mater.* **2009**, *21*, 1450–1472. [CrossRef]
69. Tokito, S.; Noda, K.; Taga, Y. Metal oxides as a hole-injecting layer for an organic electroluminescent device. *J. Phys. D Appl. Phys.* **1996**, *29*, 2750–2753. [CrossRef]

70. Meyer, J.; Hamwi, S.; Kröger, M.; Kowalsky, W.; Riedl, T.; Kahn, A. Transition Metal Oxides for Organic Electronics: Energetics, Device Physics and Applications. *Adv. Mater.* **2012**, *24*, 5408–5427. [CrossRef]
71. Chen, L.-M.; Xu, Z.; Hong, Z.; Yang, Y. Interface investigation and engineering—Achieving high performance polymer photovoltaic devices. *J. Mater. Chem.* **2010**, *20*, 2575. [CrossRef]
72. Curiel, D.; Más-Montoya, M. Hole Transporting Layers in Printable Solar Cells. In *Printable Solar Cells*; John Wiley & Sons, Inc.: Hoboken, NJ, USA, 2017; pp. 93–161.
73. Shrotriya, V.; Li, G.; Yao, Y.; Chu, C.-W.; Yang, Y. Transition metal oxides as the buffer layer for polymer photovoltaic cells. *Appl. Phys. Lett.* **2006**, *88*, 073508. [CrossRef]
74. Han, S.; Shin, W.S.; Seo, M.; Gupta, D.; Moon, S.-J.; Yoo, S. Improving performance of organic solar cells using amorphous tungsten oxides as an interfacial buffer layer on transparent anodes. *Org. Electron.* **2009**, *10*, 791–797. [CrossRef]
75. Irwin, M.D.; Buchholz, D.B.; Hains, A.W.; Chang, R.P.H.; Marks, T.J. p-Type semiconducting nickel oxide as an efficiency-enhancing anode interfacial layer in polymer bulk-heterojunction solar cells. *Proc. Natl. Acad. Sci. USA* **2008**, *105*, 2783–2787. [CrossRef]
76. Wang, K.; Ren, H.; Yi, C.; Liu, C.; Wang, H.; Huang, L.; Zhang, H.; Karim, A.; Gong, X. Solution-Processed Fe_3O_4 Magnetic Nanoparticle Thin Film Aligned by an External Magnetostatic Field as a Hole Extraction Layer for Polymer Solar Cells. *ACS Appl. Mater. Interfaces* **2013**, *5*, 10325–10330. [CrossRef]
77. Gu, X.; Cui, W.; Li, H.; Wu, Z.; Zeng, Z.; Lee, S.-T.; Zhang, H.; Sun, B. A Solution-Processed Hole Extraction Layer Made from Ultrathin MoS_2 Nanosheets for Efficient Organic Solar Cells. *Adv. Energy Mater.* **2013**, *3*, 1262–1268. [CrossRef]
78. Xu, H.; Xu, H.; Yuan, F.; Zhou, D.; Liao, X.; Chen, L.; Chen, Y.; Chen, Y. Hole transport layers for organic solar cells: Recent progress and prospects. *J. Mater. Chem. A* **2020**, *8*, 11478–11492. [CrossRef]
79. Kawano, K.; Pacios, R.; Poplavskyy, D.; Nelson, J.; Bradley, D.D.C.; Durrant, J.R. Degradation of organic solar cells due to air exposure. *Sol. Energy Mater. Sol. Cells* **2006**, *90*, 3520–3530. [CrossRef]
80. Wang, T.; Scarratt, N.W.; Yi, H.; Dunbar, A.D.F.; Pearson, A.J.; Watters, D.C.; Glen, T.S.; Brook, A.C.; Kingsley, J.; Buckley, A.R.; et al. Fabricating High Performance, Donor-Acceptor Copolymer Solar Cells by Spray-Coating in Air. *Adv. Energy Mater.* **2013**, *3*, 505–512. [CrossRef]
81. Krebs, F.C. Fabrication and processing of polymer solar cells: A review of printing and coating techniques. *Sol. Energy Mater. Sol. Cells* **2009**, *93*, 394–412. [CrossRef]
82. Terán-Alcocer, Á.; Bravo-Plascencia, F.; Cevallos-Morillo, C.; Palma-Cando, A. Electrochemical Sensors Based on Conducting Polymers for the Aqueous Detection of Biologically Relevant Molecules. *Nanomaterials* **2021**, *11*, 252. [CrossRef] [PubMed]
83. Zhitomirsky, I.; Niewczas, M.; Petric, A. Electrodeposition of hybrid organic–inorganic films containing iron oxide. *Mater. Lett.* **2003**, *57*, 1045–1050. [CrossRef]
84. Krebs, F.C. Polymer solar cell modules prepared using roll-to-roll methods: Knife-over-edge coating, slot-die coating and screen printing. *Sol. Energy Mater. Sol. Cells* **2009**, *93*, 465–475. [CrossRef]
85. Jung, J.W.; Jo, W.H. Annealing-Free High Efficiency and Large Area Polymer Solar Cells Fabricated by a Roller Painting Process. *Adv. Funct. Mater.* **2010**, *20*, 2355–2363. [CrossRef]
86. Yang, B.; Chen, Y.; Cui, Y.; Liu, D.; Xu, B.; Hou, J. Over 100-nm-Thick MoO x Films with Superior Hole Collection and Transport Properties for Organic Solar Cells. *Adv. Energy Mater.* **2018**, *8*, 1800698. [CrossRef]
87. Lou, X.W.; Zeng, H.C. Complex α-MoO_3 Nanostructures with External Bonding Capacity for Self-Assembly. *J. Am. Chem. Soc.* **2003**, *125*, 2697–2704. [CrossRef]
88. Pan, W.; Tian, R.; Jin, H.; Guo, Y.; Zhang, L.; Wu, X.; Zhang, L.; Han, Z.; Liu, G.; Li, J.; et al. Structure, Optical, and Catalytic Properties of Novel Hexagonal Metastable h-MoO_3 Nano-and Microrods Synthesized with Modified Liquid-Phase Processes. *Chem. Mater.* **2010**, *22*, 6202–6208. [CrossRef]
89. Kyaw, A.K.K.; Sun, X.W.; Jiang, C.Y.; Lo, G.Q.; Zhao, D.W.; Kwong, D.L. An inverted organic solar cell employing a sol-gel derived ZnO electron selective layer and thermal evaporated MoO_3 hole selective layer. *Appl. Phys. Lett.* **2008**, *93*, 221107. [CrossRef]
90. Lee, D.; Kim, J.; Park, G.; Bae, H.W.; An, M.; Kim, J.Y. Enhanced Operating Temperature Stability of Organic Solar Cells with Metal Oxide Hole Extraction Layer. *Polymers* **2020**, *12*, 992. [CrossRef]
91. Liao, X.; Jeong, A.R.; Wilks, R.G.; Wiesner, S.; Rusu, M.; Félix, R.; Xiao, T.; Hartmann, C.; Bär, M. Tunability of MoO_3 Thin-Film Properties Due to Annealing in Situ Monitored by Hard X-ray Photoemission. *ACS Omega* **2019**, *4*, 10985–10990. [CrossRef]
92. Boudaoud, L.; Benramdane, N.; Desfeux, R.; Khelifa, B.; Mathieu, C. Structural and optical properties of MoO_3 and V_2O_5 thin films prepared by Spray Pyrolysis. *Catal. Today* **2006**, *113*, 230–234. [CrossRef]
93. Inzani, K.; Nematollahi, M.; Selbach, S.M.; Grande, T.; Waalekalv, M.L.; Brakstad, T.; Reenaas, T.W.; Kildemo, M.; Vullum-Bruer, F. Tailoring properties of nanostructured MoO_3-x thin films by aqueous solution deposition. *Appl. Surf. Sci.* **2018**, *459*, 822–829. [CrossRef]
94. Bortoti, A.A.; de Gavanski, A.F.; Velazquez, Y.R.; Galli, A.; de Castro, E.G. Facile and low cost oxidative conversion of MoS_2 in α-MoO_3: Synthesis, characterization and application. *J. Solid State Chem.* **2017**, *252*, 111–118. [CrossRef]
95. Ji, R.; Cheng, J.; Yang, X.; Yu, J.; Li, L. Enhanced charge carrier transport in spray-cast organic solar cells using solution processed MoO_3 micro arrays. *RSC Adv.* **2017**, *7*, 3059–3065. [CrossRef]
96. Kim, J.-H.; Park, E.-K.; Kim, J.-H.; Cho, H.J.; Lee, D.-H.; Kim, Y.-S. Improving charge transport of P3HT:PCBM organic solar cell using MoO_3 nanoparticles as an interfacial buffer layer. *Electron. Mater. Lett.* **2016**, *12*, 383–387. [CrossRef]

97. Choy, W.C.H.; Ren, X. Plasmon-Electrical Effects on Organic Solar Cells by Incorporation of Metal Nanostructures. *IEEE J. Sel. Top. Quantum Electron.* **2016**, *22*, 1–9. [CrossRef]

98. Bobeico, E.; Mercaldo, L.V.; Morvillo, P.; Usatii, I.; Della Noce, M.; Lancellotti, L.; Sasso, C.; Ricciardi, R.; Delli Veneri, P. Evaporated MoO_x as General Back-Side Hole Collector for Solar Cells. *Coatings* **2020**, *10*, 763. [CrossRef]

99. Jagadamma, L.K.; Hu, H.; Kim, T.; Ndjawa, G.O.N.; Mansour, A.E.; El Labban, A.; Faria, J.C.D.; Munir, R.; Anjum, D.H.; McLachlan, M.A.; et al. Solution-processable MoO_x nanocrystals enable highly efficient reflective and semitransparent polymer solar cells. *Nano Energy* **2016**, *28*, 277–287. [CrossRef]

100. Wang, J.; Jia, X.; Zhou, J.; Pan, L.; Huang, S.; Chen, X. Improved Performance of Polymer Solar Cells by Thermal Evaporation of AgAl Alloy Nanostructures into the Hole-Transport Layer. *ACS Appl. Mater. Interfaces* **2016**, *8*, 26098–26104. [CrossRef]

101. Cong, S.; Hadipour, A.; Sugahara, T.; Wei, T.; Jiu, J.; Ranjbar, S.; Hirose, Y.; Karakawa, M.; Nagao, S.; Aernouts, T.; et al. Modifying the valence state of molybdenum in the efficient oxide buffer layer of organic solar cells via a mild hydrogen peroxide treatment. *J. Mater. Chem. C* **2017**, *5*, 889–895. [CrossRef]

102. Jung, S.; Lee, J.; Kim, U.; Park, H. Solution-Processed Molybdenum Oxide with Hydroxyl Radical-Induced Oxygen Vacancy as an Efficient and Stable Interfacial Layer for Organic Solar Cells. *Sol. RRL* **2020**, *4*, 1900420. [CrossRef]

103. Kobori, T.; Kamata, N.; Fukuda, T. Effect of annealing-induced oxidation of molybdenum oxide on organic photovoltaic device performance. *Org. Electron.* **2016**, *37*, 126–133. [CrossRef]

104. Li, Y.; Yu, H.; Huang, X.; Wu, Z.; Chen, M. A simple synthesis method to prepare a molybdenum oxide hole-transporting layer for efficient polymer solar cells. *RSC Adv.* **2017**, *7*, 7890–7900. [CrossRef]

105. Cheng, F.; Wu, Y.; Shen, Y.; Cai, X.; Li, L. Enhancing the performance and stability of organic solar cells using solution processed MoO_3 as hole transport layer. *RSC Adv.* **2017**, *7*, 37952–37958. [CrossRef]

106. Cai, P.; Ren, P.; Huang, X.; Zhang, X.; Zhan, T.; Xiong, J.; Xue, X.; Wang, Z.; Zhang, J.; Chen, J. An Ultraviolet-Deposited MoO_3 Film as Anode Interlayer for High-Performance Polymer Solar Cells. *Adv. Mater. Interfaces* **2020**, *7*, 1901912. [CrossRef]

107. Tran, H.N.; Park, S.; Wibowo, F.T.A.; Krishna, N.V.; Kang, J.H.; Seo, J.H.; Nguyen-Phu, H.; Jang, S.; Cho, S. 17% Non-Fullerene Organic Solar Cells with Annealing-Free Aqueous MoOx. *Adv. Sci.* **2020**, *7*, 2002395. [CrossRef]

108. Irfan; Ding, H.; Gao, Y.; Small, C.; Kim, D.Y.; Subbiah, J.; So, F. Energy level evolution of air and oxygen exposed molybdenum trioxide films. *Appl. Phys. Lett.* **2010**, *96*, 243307. [CrossRef]

109. Vasilopoulou, M.; Soultati, A.; Argitis, P.; Stergiopoulos, T.; Davazoglou, D. Fast Recovery of the High Work Function of Tungsten and Molybdenum Oxides via Microwave Exposure for Efficient Organic Photovoltaics. *J. Phys. Chem. Lett.* **2014**, *5*, 1871–1879. [CrossRef]

110. Soultati, A.; Kostis, I.; Argitis, P.; Dimotikali, D.; Kennou, S.; Gardelis, S.; Speliotis, T.; Kontos, A.G.; Davazoglou, D.; Vasilopoulou, M. Dehydration of molybdenum oxide hole extraction layers via microwave annealing for the improvement of efficiency and lifetime in organic solar cells. *J. Mater. Chem. C* **2016**, *4*, 7683–7694. [CrossRef]

111. Chang, F.-K.; Huang, Y.-C.; Jeng, J.-S.; Chen, J.-S. Band offset of vanadium-doped molybdenum oxide hole transport layer in organic photovoltaics. *Solid-State. Electron.* **2016**, *122*, 18–22. [CrossRef]

112. Marchal, W.; Verboven, I.; Kesters, J.; Moeremans, B.; De Dobbelaere, C.; Bonneux, G.; Elen, K.; Conings, B.; Maes, W.; Boyen, H.G.; et al. Steering the properties of MoO_x hole transporting layers in OPVs and OLEDs: Interface morphology vs. electronic structure. *Materials* **2017**, *10*, 123. [CrossRef] [PubMed]

113. Bai, Y.; Zhao, C.; Guo, Q.; Zhang, J.; Hu, S.; Liu, J.; Hayat, T.; Alsaedi, A.; Tan, Z. Enhancing the electron blocking ability of n-type MoO_3 by doping with p-type NiO for efficient nonfullerene polymer solar cells. *Org. Electron.* **2019**, *68*, 168–175. [CrossRef]

114. Li, X.; Xie, F.; Zhang, S.; Hou, J.; Choy, W.C.H. MoO_x and V_2O_x as hole and electron transport layers through functionalized intercalation in normal and inverted organic optoelectronic devices. *Light Sci. Appl.* **2015**, *4*, e273. [CrossRef]

115. Li, X.; Xie, F.; Zhang, S.; Hou, J.; Choy, W.C.H. Over 1.1 eV Workfunction Tuning of Cesium Intercalated Metal Oxides for Functioning as Both Electron and Hole Transport Layers in Organic Optoelectronic Devices. *Adv. Funct. Mater.* **2014**, *24*, 7348–7356. [CrossRef]

116. Yoon, S.; Kim, H.; Shin, E.-Y.; Bae, I.-G.; Park, B.; Noh, Y.-Y.; Hwang, I. Enhanced hole extraction by interaction between CuI and MoO_3 in the hole transport layer of organic photovoltaic devices. *Org. Electron.* **2016**, *32*, 200–207. [CrossRef]

117. Li, Z.; Guo, W.; Liu, C.; Zhang, X.; Li, S.; Guo, J.; Zhang, L. Impedance investigation of the highly efficient polymer solar cells with composite $CuBr_2$ /MoO_3 hole transport layer. *Phys. Chem. Chem. Phys.* **2017**, *19*, 20839–20846. [CrossRef] [PubMed]

118. Li, Y.; Yu, H.; Huang, X.; Wu, Z.; Xu, H. Improved performance for polymer solar cells using CTAB-modified MoO_3 as an anode buffer layer. *Sol. Energy Mater. Sol. Cells* **2017**, *171*, 72–84. [CrossRef]

119. Kang, Q.; Zu, Y.; Liao, Q.; Zheng, Z.; Yao, H.; Zhang, S.; He, C.; Xu, B.; Hou, J. An inorganic molecule-induced electron transfer complex for highly efficient organic solar cells. *J. Mater. Chem. A* **2020**, *8*, 5580–5586. [CrossRef]

120. Kwon, K.C.; Lee, T.H.; Choi, S.; Choi, K.S.; Gim, S.O.; Bae, S.-R.; Lee, J.-L.; Jang, H.W.; Kim, S.Y. Synthesis of atomically thin alloyed molybdenum-tungsten disulfides thin films as hole transport layers in organic light-emitting diodes. *Appl. Surf. Sci.* **2021**, *541*, 148529. [CrossRef]

121. Cheng, P.-P.; Zhou, L.; Li, J.-A.; Li, Y.-Q.; Lee, S.-T.; Tang, J.-X. Light trapping enhancement of inverted polymer solar cells with a nanostructured scattering rear electrode. *Org. Electron.* **2013**, *14*, 2158–2163. [CrossRef]

122. Zhang, W.; Lan, W.; Lee, M.H.; Singh, J.; Zhu, F. A versatile solution-processed MoO_3/Au nanoparticles/MoO_3 hole contact for high performing PEDOT:PSS-free organic solar cells. *Org. Electron.* **2018**, *52*, 1–6. [CrossRef]

123. Hofmann, M.; Hofmann, H.; Hagelüken, C.; Hool, A. Critical raw materials: A perspective from the materials science community. *Sustain. Mater. Technol.* **2018**, *17*, e00074. [CrossRef]
124. Peñafiel-Vicuña, S.; Rendón-Enríquez, I.; Vega-Poot, A.; López-Téllez, G.; Palma-Cando, A.; Frontana-Uribe, B.A. Recovering Indium Tin Oxide Electrodes for the Fabrication of Hematite Photoelectrodes. *J. Electrochem. Soc.* **2020**, *167*, 126512. [CrossRef]
125. Chen, S.; Dai, Y.; Zhao, D.; Zhang, H. ITO-free flexible organic photovoltaics with multilayer MoO_3 /LiF/MoO_3 /Ag/MoO_3 as the transparent electrode. *Semicond. Sci. Technol.* **2016**, *31*, 055013. [CrossRef]
126. Lee, H.-M.; Kim, S.-S.; Kim, H.-K. Artificially MoO_3 graded ITO anodes for acidic buffer layer free organic photovoltaics. *Appl. Surf. Sci.* **2016**, *364*, 340–348. [CrossRef]
127. Sun, J.; Zheng, Q.; Cheng, S.; Zhou, H.; Lai, Y.; Yu, J. Comparing molybdenum oxide thin films prepared by magnetron sputtering and thermal evaporation applied in organic solar cells. *J. Mater. Sci. Mater. Electron.* **2016**, *27*, 3245–3249. [CrossRef]
128. Chaturvedi, N.; Swami, S.K.; Dutta, V. Electric field assisted spray deposited MoO_3 thin films as a hole transport layer for organic solar cells. *Sol. Energy* **2016**, *137*, 379–384. [CrossRef]
129. Dong, W.J.; Ham, J.; Jung, G.H.; Son, J.H.; Lee, J.-L. Ultrafast laser-assisted synthesis of hydrogenated molybdenum oxides for flexible organic solar cells. *J. Mater. Chem. A* **2016**, *4*, 4755–4762. [CrossRef]
130. Meyer, J.; Hamwi, S.; Bülow, T.; Johannes, H.-H.; Riedl, T.; Kowalsky, W. Highly efficient simplified organic light emitting diodes. *Appl. Phys. Lett.* **2007**, *91*, 113506. [CrossRef]
131. Tao, C.; Ruan, S.; Xie, G.; Kong, X.; Shen, L.; Meng, F.; Liu, C.; Zhang, X.; Dong, W.; Chen, W. Role of tungsten oxide in inverted polymer solar cells. *Appl. Phys. Lett.* **2009**, *94*, 043311. [CrossRef]
132. Miyake, K.; Kaneko, H.; Sano, M.; Suedomi, N. Physical and electrochromic properties of the amorphous and crystalline tungsten oxide thick films prepared under reducing atmosphere. *J. Appl. Phys.* **1984**, *55*, 2747–2753. [CrossRef]
133. Vida, G.; Josepovits, V.K.; Győr, M.; Deák, P. Characterization of Tungsten Surfaces by Simultaneous Work Function and Secondary Electron Emission Measurements. *Microsc. Microanal.* **2003**, *9*, 337–342. [CrossRef] [PubMed]
134. Stubhan, T.; Li, N.; Luechinger, N.A.; Halim, S.C.; Matt, G.J.; Brabec, C.J. High Fill Factor Polymer Solar Cells Incorporating a Low Temperature Solution Processed WO 3 Hole Extraction Layer. *Adv. Energy Mater.* **2012**, *2*, 1433–1438. [CrossRef]
135. Brabec, C.J. Organic photovoltaics: Technology and market. *Sol. Energy Mater. Sol. Cells* **2004**, *83*, 273–292. [CrossRef]
136. Lee, Y.H.; Abdu, H.A.E.; Kim, D.H.; Kim, T.W. Enhancement of the power conversion efficiency of organic photovoltaic cells due to $Au@SiO_2$ core shell nanoparticles embedded into a WO_3 hole transport layer. *Org. Electron.* **2019**, *68*, 182–186. [CrossRef]
137. Li, X.; Choy, W.C.H.; Ren, X.; Xin, J.; Lin, P.; Leung, D.C.W. Polarization-independent efficiency enhancement of organic solar cells by using 3-dimensional plasmonic electrode. *Appl. Phys. Lett.* **2013**, *102*, 153304. [CrossRef]
138. Li, X.; Choy, W.C.H.; Huo, L.; Xie, F.; Sha, W.E.I.; Ding, B.; Guo, X.; Li, Y.; Hou, J.; You, J.; et al. Dual Plasmonic Nanostructures for High Performance Inverted Organic Solar Cells. *Adv. Mater.* **2012**, *24*, 3046–3052. [CrossRef]
139. Li, X.H.; Sha, W.E.I.; Choy, W.C.H.; Fung, D.D.S.; Xie, F.X. Efficient Inverted Polymer Solar Cells with Directly Patterned Active Layer and Silver Back Grating. *J. Phys. Chem. C* **2012**, *116*, 7200–7206. [CrossRef]
140. Lee, Y.H.; Kim, D.H.; Kim, T.W. Enhanced power conversion efficiency of organic photovoltaic devices due to the surface plasmonic resonance effect generated utilizing Au-WO_3 nanocomposites. *Org. Electron.* **2017**, *45*, 256–262. [CrossRef]
141. Shen, W.; Tang, J.; Wang, D.; Yang, R.; Chen, W.; Bao, X.; Wang, Y.; Jiao, J.; Wang, Y.; Huang, Z.; et al. Enhanced efficiency of polymer solar cells by structure-differentiated silver nano-dopants in solution-processed tungsten oxide layer. *Mater. Sci. Eng. B* **2016**, *206*, 61–68. [CrossRef]
142. Wang, C.C.D.; Choy, W.C.H.; Duan, C.; Fung, D.D.S.; Sha, W.E.I.; Xie, F.-X.; Huang, F.; Cao, Y. Optical and electrical effects of gold nanoparticles in the active layer of polymer solar cells. *J. Mater. Chem.* **2012**, *22*, 1206–1211. [CrossRef]
143. Zhang, D.; Choy, W.C.H.; Xie, F.; Sha, W.E.I.; Li, X.; Ding, B.; Zhang, K.; Huang, F.; Cao, Y. Plasmonic Electrically Functionalized TiO_2 for High-Performance Organic Solar Cells. *Adv. Funct. Mater.* **2013**, *23*, 4255–4261. [CrossRef]
144. Li, X.; Choy, W.C.H.; Xie, F.; Zhang, S.; Hou, J. Room-temperature solution-processed molybdenum oxide as a hole transport layer with Ag nanoparticles for highly efficient inverted organic solar cells. *J. Mater. Chem. A* **2013**, *1*, 6614. [CrossRef]
145. Xie, F.; Choy, W.C.H.; Sha, W.E.I.; Zhang, D.; Zhang, S.; Li, X.; Leung, C.; Hou, J. Enhanced charge extraction in organic solar cells through electron accumulation effects induced by metal nanoparticles. *Energy Environ. Sci.* **2013**, *6*, 3372. [CrossRef]
146. Ren, X.; Cheng, J.; Zhang, S.; Li, X.; Rao, T.; Huo, L.; Hou, J.; Choy, W.C.H. High Efficiency Organic Solar Cells Achieved by the Simultaneous Plasmon-Optical and Plasmon-Electrical Effects from Plasmonic Asymmetric Modes of Gold Nanostars. *Small* **2016**, *12*, 5200–5207. [CrossRef] [PubMed]
147. Sha, W.E.I.; Zhu, H.L.; Chen, L.; Chew, W.C.; Choy, W.C.H. A General Design Rule to Manipulate Photocarrier Transport Path in Solar Cells and Its Realization by the Plasmonic-Electrical Effect. *Sci. Rep.* **2015**, *5*, 8525. [CrossRef] [PubMed]
148. Sha, W.E.I.; Li, X.; Choy, W.C.H. Breaking the Space Charge Limit in Organic Solar Cells by a Novel Plasmonic-Electrical Concept. *Sci. Rep.* **2015**, *4*, 6236. [CrossRef]
149. Li, X.; Choy, W.C.H.; Lu, H.; Sha, W.E.I.; Ho, A.H.P. Efficiency Enhancement of Organic Solar Cells by Using Shape-Dependent Broadband Plasmonic Absorption in Metallic Nanoparticles. *Adv. Funct. Mater.* **2013**, *23*, 2728–2735. [CrossRef]
150. Remya, R.; Gayathri, P.T.G.; Deb, B. Studies on solution-processed tungsten oxide nanostructures for efficient hole transport in the inverted polymer solar cells. *Mater. Chem. Phys.* **2020**, *255*, 123584. [CrossRef]
151. Liao, H.-H.; Chen, L.-M.; Xu, Z.; Li, G.; Yang, Y. Highly efficient inverted polymer solar cell by low temperature annealing of Cs_2CO_3 interlayer. *Appl. Phys. Lett.* **2008**, *92*, 173303. [CrossRef]

152. Huang, J.-S.; Chou, C.-Y.; Liu, M.-Y.; Tsai, K.-H.; Lin, W.-H.; Lin, C.-F. Solution-processed vanadium oxide as an anode interlayer for inverted polymer solar cells hybridized with ZnO nanorods. *Org. Electron.* **2009**, *10*, 1060–1065. [CrossRef]

153. Meyer, J.; Zilberberg, K.; Riedl, T.; Kahn, A. Electronic structure of Vanadium pentoxide: An efficient hole injector for organic electronic materials. *J. Appl. Phys.* **2011**, *110*, 033710. [CrossRef]

154. Xu, M.-F.; Xu, T.; Wang, C.-N.; Zhai, Z.-C.; Jin, Y.-L.; Yang, X.-H. Low-temperature, aqueous solution-processed V_2O_5 as the hole-transport layer for high performance organic solar cells. *J. Mater. Sci. Mater. Electron.* **2018**, *29*, 14783–14787. [CrossRef]

155. Jiang, Y.; Xiao, S.; Xu, B.; Zhan, C.; Mai, L.; Lu, X.; You, W. Enhancement of Photovoltaic Performance by Utilizing Readily Accessible Hole Transporting Layer of Vanadium(V) Oxide Hydrate in a Polymer-Fullerene Blend Solar Cell. *ACS Appl. Mater. Interfaces* **2016**, *8*, 11658–11666. [CrossRef] [PubMed]

156. Cong, H.; Han, D.; Sun, B.; Zhou, D.; Wang, C.; Liu, P.; Feng, L. Facile Approach to Preparing a Vanadium Oxide Hydrate Layer as a Hole-Transport Layer for High-Performance Polymer Solar Cells. *ACS Appl. Mater. Interfaces* **2017**, *9*, 18087–18094. [CrossRef]

157. Xie, F.; Choy, W.C.H.; Wang, C.; Li, X.; Zhang, S.; Hou, J. Low-Temperature Solution-Processed Hydrogen Molybdenum and Vanadium Bronzes for an Efficient Hole-Transport Layer in Organic Electronics. *Adv. Mater.* **2013**, *25*, 2051–2055. [CrossRef]

158. Vishnumurthy, K.A.; Kesavan, A.V.; Swathi, S.K.; Ramamurthy, P.C. Low band gap thienothiophene-diketopyrrolopyrole copolymers with V2O5 as hole transport layer for photovoltaic application. *Opt. Mater.* **2020**, *109*, 110303. [CrossRef]

159. Ravi, R.; Deb, B. Studies on One-Step-Synthesized Hydrated Vanadium Pentoxide for Efficient Hole Transport in Organic Photovoltaics. *Energy Technol.* **2020**, *8*, 1901323. [CrossRef]

160. Xu, W.; Liu, Y.; Huang, X.; Jiang, L.; Li, Q.; Hu, X.; Huang, F.; Gong, X.; Cao, Y. Solution-processed VO_x prepared using a novel synthetic method as the hole extraction layer for polymer solar cells. *J. Mater. Chem. C* **2016**, *4*, 1953–1958. [CrossRef]

161. Shafeeq, K.M.; Athira, V.P.; Kishor, C.H.R.; Aneesh, P.M. Structural and optical properties of V2O5 nanostructures grown by thermal decomposition technique. *Appl. Phys. A* **2020**, *126*, 586. [CrossRef]

162. Alsulami, A.; Griffin, J.; Alqurashi, R.; Yi, H.; Iraqi, A.; Lidzey, D.; Buckley, A. Thermally Stable Solution Processed Vanadium Oxide as a Hole Extraction Layer in Organic Solar Cells. *Materials* **2016**, *9*, 235. [CrossRef] [PubMed]

163. Zafar, M.; Yun, J.-Y.; Kim, D.-H. Highly stable inverted organic photovoltaic cells with a V_2O_5 hole transport layer. *Korean J. Chem. Eng.* **2017**, *34*, 1504–1508. [CrossRef]

164. Teng, N.-W.; Li, C.-H.; Lo, W.-C.; Tsai, Y.-S.; Liao, C.-Y.; You, Y.-W.; Ho, H.-L.; Li, W.-L.; Lee, C.-C.; Lin, W.-C.; et al. Highly Efficient Nonfullerene Organic Photovoltaic Devices with 10% Power Conversion Efficiency Enabled by a Fine-Tuned and Solution-Processed Hole-Transporting Layer. *Sol. RRL* **2020**, *4*, 2000223. [CrossRef]

165. Xia, C.; Hong, W.T.; Kim, Y.E.; Choe, W.-S.; Kim, D.-H.; Kim, J.K. Metal-Organic Decomposition-Mediated Nanoparticulate Vanadium Oxide Hole Transporting Buffer Layer for Polymer Bulk-Heterojunction Solar Cells. *Polymers* **2020**, *12*, 1791. [CrossRef] [PubMed]

166. Beliatis, M.J.; Helgesen, M.; García-Valverde, R.; Corazza, M.; Roth, B.; Carlé, J.E.; Jørgensen, M.; Krebs, F.C.; Gevorgyan, S.A. Slot-Die-Coated V_2O_5 as Hole Transport Layer for Flexible Organic Solar Cells and Optoelectronic Devices. *Adv. Eng. Mater.* **2016**, *18*, 1494–1503. [CrossRef]

167. Arbab, E.A.A.; Mola, G.T. V_2O_5 thin film deposition for application in organic solar cells. *Appl. Phys. A Mater. Sci. Process.* **2016**, *122*, 405. [CrossRef]

168. Kavuri, H.A.; Fukuda, T.; Takahira, K.; Takahashi, A.; Kihara, S.; McGillivray, D.J.; Willmott, G. Electrospray-deposited vanadium oxide anode interlayers for high-efficiency organic solar cells. *Org. Electron.* **2018**, *57*, 239–246. [CrossRef]

169. Lee, D.-Y.; Cho, S.-P.; Na, S.-I.; Kim, S.-S. ITO-free polymer solar cells with vanadium oxide hole transport layer. *J. Ind. Eng. Chem.* **2017**, *45*, 1–4. [CrossRef]

170. Ko, E.-H.; Kim, H.-K. Highly transparent vanadium oxide-graded indium zinc oxide electrodes for flexible organic solar cells. *Thin Solid Films* **2016**, *601*, 2–6. [CrossRef]

171. Ai, L.; Fang, G.; Yuan, L.; Liu, N.; Wang, M.; Li, C.; Zhang, Q.; Li, J.; Zhao, X. Influence of substrate temperature on electrical and optical properties of p-type semitransparent conductive nickel oxide thin films deposited by radio frequency sputtering. *Appl. Surf. Sci.* **2008**, *254*, 2401–2405. [CrossRef]

172. Zhang, H.; Cheng, J.; Lin, F.; He, H.; Mao, J.; Wong, K.S.; Jen, A.K.Y.; Choy, W.C.H. Pinhole-Free and Surface-Nanostructured NiO_x Film by Room-Temperature Solution Process for High-Performance Flexible Perovskite Solar Cells with Good Stability and Reproducibility. *ACS Nano* **2016**, *10*, 1503–1511. [CrossRef]

173. Manders, J.R.; Tsang, S.-W.; Hartel, M.J.; Lai, T.-H.; Chen, S.; Amb, C.M.; Reynolds, J.R.; So, F. Solution-Processed Nickel Oxide Hole Transport Layers in High Efficiency Polymer Photovoltaic Cells. *Adv. Funct. Mater.* **2013**, *23*, 2993–3001. [CrossRef]

174. Parthiban, S.; Kim, S.; Tamilavan, V.; Lee, J.; Shin, I.; Yuvaraj, D.; Jung, Y.K.; Hyun, M.H.; Jeong, J.H.; Park, S.H. Enhanced efficiency and stability of polymer solar cells using solution-processed nickel oxide as hole transport material. *Curr. Appl. Phys.* **2017**, *17*, 1232–1237. [CrossRef]

175. Chavhan, S.D.; Hansson, R.; Ericsson, L.K.E.; Beyer, P.; Hofmann, A.; Brütting, W.; Opitz, A.; Moons, E. Low temperature processed NiO_x hole transport layers for efficient polymer solar cells. *Org. Electron.* **2017**, *44*, 59–66. [CrossRef]

176. Jiang, F.; Choy, W.C.H.; Li, X.; Zhang, D.; Cheng, J. Post-treatment-Free Solution-Processed Non-stoichiometric NiO_x Nanoparticles for Efficient Hole-Transport Layers of Organic Optoelectronic Devices. *Adv. Mater.* **2015**, *27*, 2930–2937. [CrossRef] [PubMed]

177. Huang, Z.; Ouyang, D.; Ma, R.; Wu, W.; Roy, V.A.L.; Choy, W.C.H. A General Method: Designing a Hypocrystalline Hydroxide Intermediate to Achieve Ultrasmall and Well-Dispersed Ternary Metal Oxide for Efficient Photovoltaic Devices. *Adv. Funct. Mater.* **2019**, *29*, 1904684. [CrossRef]

178. Kim, J.K. PEG-assisted Sol-gel Synthesis of Compact Nickel Oxide Hole-Selective Layer with Modified Interfacial Properties for Organic Solar Cells. *Polymers* **2019**, *11*, 120. [CrossRef]

179. Rho, Y.; Kang, K.-T.; Lee, D. Highly crystalline Ni/NiO hybrid electrodes processed by inkjet printing and laser-induced reductive sintering under ambient conditions. *Nanoscale* **2016**, *8*, 8976–8985. [CrossRef]

180. Singh, A.; Gupta, S.K.; Garg, A. Inkjet printing of NiO films and integration as hole transporting layers in polymer solar cells. *Sci. Rep.* **2017**, *7*. [CrossRef]

181. Huang, S.; Wang, Y.; Shen, S.; Tang, Y.; Yu, A.; Kang, B.; Silva, S.R.P.; Lu, G. Enhancing the performance of polymer solar cells using solution-processed copper doped nickel oxide nanoparticles as hole transport layer. *J. Colloid Interface Sci.* **2019**, *535*, 308–317. [CrossRef]

182. Ouyang, D.; Zheng, J.; Huang, Z.; Zhu, L.; Choy, W.C.H. An efficacious multifunction codoping strategy on a room-temperature solution-processed hole transport layer for realizing high-performance perovskite solar cells. *J. Mater. Chem. A* **2021**, *9*, 371–379. [CrossRef]

183. Al-Jawhari, H.A.; Caraveo-Frescas, J.A.; Hedhili, M.N.; Alshareef, H.N. P-Type Cu_2O/SnO Bilayer Thin Film Transistors Processed at Low Temperatures. *ACS Appl. Mater. Interfaces* **2013**, *5*, 9615–9619. [CrossRef] [PubMed]

184. Murali, D.S.; Kumar, S.; Choudhary, R.J.; Wadikar, A.D.; Jain, M.K.; Subrahmanyam, A. Synthesis of Cu_2O from CuO thin films: Optical and electrical properties. *AIP Adv.* **2015**, *5*, 047143. [CrossRef]

185. Fortunato, E.; Figueiredo, V.; Barquinha, P.; Elamurugu, E.; Barros, R.; Gonçalves, G.; Park, S.-H.K.; Hwang, C.-S.; Martins, R. Thin-film transistors based on p-type Cu_2O thin films produced at room temperature. *Appl. Phys. Lett.* **2010**, *96*, 192102. [CrossRef]

186. Zuo, C.; Ding, L. Solution-Processed Cu_2O and CuO as Hole Transport Materials for Efficient Perovskite Solar Cells. *Small* **2015**, *11*, 5528–5532. [CrossRef]

187. Yu, Z.; Liu, W.; Fu, W.; Zhang, Z.; Yang, W.; Wang, S.; Li, H.; Xu, M.; Chen, H. An aqueous solution-processed CuO X film as an anode buffer layer for efficient and stable organic solar cells. *J. Mater. Chem. A* **2016**, *4*, 5130–5136. [CrossRef]

188. Yu, R.-S.; Tasi, C.-P. Structure, composition and properties of p-type $CuCrO_2$ thin films. *Ceram. Int.* **2014**, *40*, 8211–8217. [CrossRef]

189. Han, M.; Jiang, K.; Zhang, J.; Yu, W.; Li, Y.; Hu, Z.; Chu, J. Structural, electronic band transition and optoelectronic properties of delafossite $CuGa1-_xCr_xO_2$ ($0 \leq x \leq 1$) solid solution films grown by the sol–gel method. *J. Mater. Chem.* **2012**, *22*, 18463. [CrossRef]

190. Xiong, D.; Xu, Z.; Zeng, X.; Zhang, W.; Chen, W.; Xu, X.; Wang, M.; Cheng, Y.-B. Hydrothermal synthesis of ultrasmall $CuCrO_2$ nanocrystal alternatives to NiO nanoparticles in efficient p-type dye-sensitized solar cells. *J. Mater. Chem.* **2012**, *22*, 24760. [CrossRef]

191. Marquardt, M.A.; Ashmore, N.A.; Cann, D.P. Crystal chemistry and electrical properties of the delafossite structure. *Thin Solid Films* **2006**, *496*, 146–156. [CrossRef]

192. Yang, B.; Ouyang, D.; Huang, Z.; Ren, X.; Zhang, H.; Choy, W.C.H. Multifunctional Synthesis Approach of In:$CuCrO_2$ Nanoparticles for Hole Transport Layer in High-Performance Perovskite Solar Cells. *Adv. Funct. Mater.* **2019**, *29*, 1902600. [CrossRef]

193. Wang, J.; Lee, Y.-J.; Hsu, J.W.P. Sub-10 nm copper chromium oxide nanocrystals as a solution processed p-type hole transport layer for organic photovoltaics. *J. Mater. Chem. C* **2016**, *4*, 3607–3613. [CrossRef]

194. Wang, J.; Daunis, T.B.; Cheng, L.; Zhang, B.; Kim, J.; Hsu, J.W.P. Combustion Synthesis of p-Type Transparent Conducting $CuCrO_{2+x}$ and Cu:CrO_x Thin Films at 180 °C. *ACS Appl. Mater. Interfaces* **2018**, *10*, 3732–3738. [CrossRef] [PubMed]

195. Wahl, T.; Zellmer, S.; Hanisch, J.; Garnweitner, G.; Ahlswede, E. Thin indium tin oxide nanoparticle films as hole transport layer in inverted organic solar cells. *Thin Solid Films* **2016**, *616*, 419–424. [CrossRef]

196. Bhargav, R.; Gairola, S.P.; Patra, A.; Naqvi, S.; Dhawan, S.K. Improved performance of organic solar cells with solution processed hole transport layer. *Opt. Mater.* **2018**, *80*, 138–142. [CrossRef]

197. Kan, M.; Wang, J.Y.; Li, X.W.; Zhang, S.H.; Li, Y.W.; Kawazoe, Y.; Sun, Q.; Jena, P. Structures and Phase Transition of a MoS_2 Monolayer. *J. Phys. Chem. C* **2014**, *118*, 1515–1522. [CrossRef]

198. Barrera, D.; Jawaid, A.; Daunis, T.B.; Cheng, L.; Wang, Q.; Lee, Y.-J.; Kim, M.J.; Kim, J.; Vaia, R.A.; Hsu, J.W.P. Inverted OPVs with MoS_2 hole transport layer deposited by spray coating. *Mater. Today Energy* **2017**, *5*, 107–111. [CrossRef]

199. Martinez-Rojas, F.; Hssein, M.; El Jouad, Z.; Armijo, F.; Cattin, L.; Louarn, G.; Stephant, N.; del Valle, M.A.; Addou, M.; Soto, J.P.; et al. $Mo(S_xO_y)$ thin films deposited by electrochemistry for application in organic photovoltaic cells. *Mater. Chem. Phys.* **2017**, *201*, 331–338. [CrossRef]

200. Xing, W.; Chen, Y.; Wang, X.; Lv, L.; Ouyang, X.; Ge, Z.; Huang, H. MoS_2 Quantum Dots with a Tunable Work Function for High-Performance Organic Solar Cells. *ACS Appl. Mater. Interfaces* **2016**, *8*, 26916–26923. [CrossRef]

201. Wei, J.; Yin, Z.; Chen, S.-C.; Cai, D.; Zheng, Q. Solution-processed MoS_x thin-films as hole-transport layers for efficient polymer solar cells. *RSC Adv.* **2016**, *6*, 39137–39143. [CrossRef]

202. Adilbekova, B.; Lin, Y.; Yengel, E.; Faber, H.; Harrison, G.; Firdaus, Y.; El-Labban, A.; Anjum, D.H.; Tung, V.; Anthopoulos, T.D. Liquid phase exfoliation of MoS_2 and WS_2 in aqueous ammonia and their application in highly efficient organic solar cells. *J. Mater. Chem. C* **2020**, *8*, 5259–5264. [CrossRef]

203. Lin, Y.; Adilbekova, B.; Firdaus, Y.; Yengel, E.; Faber, H.; Sajjad, M.; Zheng, X.; Yarali, E.; Seitkhan, A.; Bakr, O.M.; et al. 17% Efficient Organic Solar Cells Based on Liquid Exfoliated WS_2 as a Replacement for PEDOT:PSS. *Adv. Mater.* **2019**, *31*, 1902965. [CrossRef] [PubMed]

204. Ram, K.S.; Singh, J. Over 20% Efficient and Stable Non-Fullerene-Based Ternary Bulk-Heterojunction Organic Solar Cell with WS_2 Hole-Transport Layer and Graded Refractive Index Antireflection Coating. *Adv. Theory Simul.* **2020**, *3*, 2000047. [CrossRef]

205. Hilal, M.; Han, J.I. Preparation of hierarchical flower-like nickel sulfide as hole transporting material for organic solar cells via a one-step solvothermal method. *Sol. Energy* **2019**, *188*, 403–413. [CrossRef]

206. Bhargav, R.; Patra, A.; Dhawan, S.K.; Gairola, S.P. Solution processed hole transport layer towards efficient and cost effective organic solar cells. *Sol. Energy* **2018**, *165*, 131–135. [CrossRef]

207. Jose, E.; Mohan, M.; Namboothiry, M.A.G.; Kumar, M.C.S. Room temperature deposition of high figure of merit p-type transparent conducting Cu–Zn–S thin films and their application in organic solar cells as an efficient hole transport layer. *J. Alloys Compd.* **2020**, *829*, 154507. [CrossRef]

208. Wang, Z.; Gao, L.; Wei, X.; Zhao, M.; Miao, Y.; Zhang, X.; Zhang, H.; Wang, H.; Hao, Y.; Xu, B.; et al. Energy level engineering of PEDOT:PSS by antimonene quantum sheet doping for highly efficient OLEDs. *J. Mater. Chem. C* **2020**, *8*, 1796–1802. [CrossRef]

209. Wang, Z.; Wang, Z.; Wang, Z.; Zhao, M.; Zhou, Y.; Zhao, B.; Miao, Y.; Liu, P.; Hao, Y.; Wang, H.; et al. Novel 2D material from AMQS-based defect engineering for efficient and stable organic solar cells. *2D Mater.* **2019**, *6*, 045017. [CrossRef]

210. Gao, Z.; Wang, Y.; Ouyang, D.; Liu, H.; Huang, Z.; Kim, J.; Choy, W.C.H. Triple Interface Passivation Strategy-Enabled Efficient and Stable Inverted Perovskite Solar Cells. *Small Methods* **2020**, *4*, 2000478. [CrossRef]

211. Ma, R.; Ren, Z.; Li, C.; Wang, Y.; Huang, Z.; Zhao, Y.; Yang, T.; Liang, Y.; Sun, X.W.; Choy, W.C.H. Establishing Multifunctional Interface Layer of Perovskite Ligand Modified Lead Sulfide Quantum Dots for Improving the Performance and Stability of Perovskite Solar Cells. *Small* **2020**, *16*, 2002628. [CrossRef]

212. Ma, R.; Zheng, J.; Tian, Y.; Li, C.; Lyu, B.; Lu, L.; Su, Z.; Chen, L.; Gao, X.; Tang, J.; et al. Self-Polymerization of Monomer and Induced Interactions with Perovskite for Highly Performed and Stable Perovskite Solar Cells. *Adv. Funct. Mater.* **2022**, *32*, 2105290. [CrossRef]

213. Wijeyasinghe, N.; Eisner, F.; Tsetseris, L.; Lin, Y.-H.; Seitkhan, A.; Li, J.; Yan, F.; Solomeshch, O.; Tessler, N.; Patsalas, P.; et al. p-Doping of Copper(I) Thiocyanate (CuSCN) Hole-Transport Layers for High-Performance Transistors and Organic Solar Cells. *Adv. Funct. Mater.* **2018**, *28*, 1802055. [CrossRef]

214. Wang, Z.; Dong, J.; Guo, J.; Wang, Z.; Yan, L.; Hao, Y.; Wang, H.; Xu, B.; Yin, S. Hybrid Hole Extraction Layer Enabled High Efficiency in Polymer Solar Cells. *ACS Appl. Mater. Interfaces* **2020**, *12*, 55342–55348. [CrossRef]

215. Worakajit, P.; Hamada, F.; Sahu, D.; Kidkhunthod, P.; Sudyoadsuk, T.; Promarak, V.; Harding, D.J.; Packwood, D.M.; Saeki, A.; Pattanasattayavong, P. Elucidating the Coordination of Diethyl Sulfide Molecules in Copper(I) Thiocyanate (CuSCN) Thin Films and Improving Hole Transport by Antisolvent Treatment. *Adv. Funct. Mater.* **2020**, *30*, 2002355. [CrossRef]

216. Suresh Kumar, M.; Mohanta, K.; Batabyal, S.K. Solution processed Cu_2CdSnS_4 as a low-cost inorganic hole transport material for polymer solar cells. *Sol. Energy Mater. Sol. Cells* **2017**, *161*, 157–161. [CrossRef]

217. Xu, Z. Fundamental Properties of Graphene. In *Graphene*; Elsevier: Amsterdam, The Netherlands, 2018; pp. 73–102.

218. Konkena, B.; Vasudevan, S. Understanding Aqueous Dispersibility of Graphene Oxide and Reduced Graphene Oxide through p K a Measurements. *J. Phys. Chem. Lett.* **2012**, *3*, 867–872. [CrossRef] [PubMed]

219. Mattevi, C.; Eda, G.; Agnoli, S.; Miller, S.; Mkhoyan, K.A.; Celik, O.; Mastrogiovanni, D.; Granozzi, G.; Garfunkel, E.; Chhowalla, M. Evolution of Electrical, Chemical, and Structural Properties of Transparent and Conducting Chemically Derived Graphene Thin Films. *Adv. Funct. Mater.* **2009**, *19*, 2577–2583. [CrossRef]

220. Sun, L. Structure and synthesis of graphene oxide. *Chin. J. Chem. Eng.* **2019**, *27*, 2251–2260. [CrossRef]

221. Rafique, S.; Abdullah, S.M.; Alhummiany, H.; Abdel-wahab, M.S.; Iqbal, J.; Sulaiman, K. Bulk Heterojunction Organic Solar Cells with Graphene Oxide Hole Transport Layer: Effect of Varied Concentration on Photovoltaic Performance. *J. Phys. Chem. C* **2017**, *121*, 140–146. [CrossRef]

222. Allen, M.J.; Tung, V.C.; Kaner, R.B. Honeycomb Carbon: A Review of Graphene. *Chem. Rev.* **2010**, *110*, 132–145. [CrossRef]

223. Huang, X.; Yu, H.; Wu, Z.; Li, Y. Improving the performance of polymer solar cells by efficient optimizing the hole transport layer-graphene oxide. *J. Solid-State Electrochem.* **2018**, *22*, 317–329. [CrossRef]

224. Lee, S.; Yeo, J.-S.; Yun, J.-M.; Kim, D.-Y. Water dispersion of reduced graphene oxide stabilized via fullerenol semiconductor for organic solar cells. *Opt. Mater. Express* **2017**, *7*, 2487. [CrossRef]

225. Kwon, S.-N.; Jung, C.-H.; Na, S.-I. Electron-beam-induced reduced graphene oxide as an alternative hole-transporting interfacial layer for high-performance and reliable polymer solar cells. *Org. Electron.* **2016**, *34*, 67–74. [CrossRef]

226. Fakharan, Z.; Naji, L.; Madanipour, K. Nd:YAG pulsed laser production of reduced-graphene oxide as hole transporting layer in polymer solar cells and the influences of solvent type. *Org. Electron.* **2020**, *76*, 105459. [CrossRef]

227. Municiler, K.; Alishah, H.M.; Bozar, S.; Güneş, S.; Kaya, F. A novel method for graphene synthesis via electrochemical process and its utilization in organic photovoltaic devices. *Appl. Phys. A* **2020**, *126*, 904. [CrossRef]

228. Xia, Y.; Pan, Y.; Zhang, H.; Qiu, J.; Zheng, Y.; Chen, Y.; Huang, W. Graphene Oxide by UV-Ozone Treatment as an Efficient Hole Extraction Layer for Highly Efficient and Stable Polymer Solar Cells. *ACS Appl. Mater. Interfaces* **2017**, *9*, 26252–26256. [CrossRef]

229. Rafique, S.; Abdullah, S.M.; Iqbal, J.; Jilani, A.; Vattamkandathil, S.; Iwamoto, M. Moderately reduced graphene oxide via UV-ozone treatment as hole transport layer for high efficiency organic solar cells. *Org. Electron.* **2018**, *59*, 140–148. [CrossRef]

230. Zhao, F.-G.; Hu, C.-M.; Kong, Y.-T.; Pan, B.; Yao, X.; Chu, J.; Xu, Z.-W.; Zuo, B.; Li, W.-S. Sulfanilic Acid Pending on a Graphene Scaffold: Novel, Efficient Synthesis and Much Enhanced Polymer Solar Cell Efficiency and Stability Using It as a Hole Extraction Layer. *ACS Appl. Mater. Interfaces* **2018**, *10*, 24679–24688. [CrossRef]
231. Ali, A.; Khan, Z.S.; Jamil, M.; Khan, Y.; Ahmad, N.; Ahmed, S. Simultaneous reduction and sulfonation of graphene oxide for efficient hole selectivity in polymer solar cells. *Curr. Appl. Phys.* **2018**, *18*, 599–610. [CrossRef]
232. Chen, X.; Liu, Q.; Zhang, M.; Ju, H.; Zhu, J.; Qiao, Q.; Wang, M.; Yang, S. Noncovalent phosphorylation of graphene oxide with improved hole transport in high-efficiency polymer solar cells. *Nanoscale* **2018**, *10*, 14840–14846. [CrossRef]
233. Cheng, X.; Long, J.; Wu, R.; Huang, L.; Tan, L.; Chen, L.; Chen, Y. Fluorinated Reduced Graphene Oxide as an Efficient Hole-Transport Layer for Efficient and Stable Polymer Solar Cells. *ACS Omega* **2017**, *2*, 2010–2016. [CrossRef] [PubMed]
234. Nicasio-Collazo, J.; Maldonado, J.-L.; Salinas-Cruz, J.; Barreiro-Argüelles, D.; Caballero-Quintana, I.; Vázquez-Espinosa, C.; Romero-Borja, D. Functionalized and reduced graphene oxide as hole transport layer and for use in ternary organic solar cell. *Opt. Mater.* **2019**, *98*, 109434. [CrossRef]
235. Park, J.-J.; Heo, Y.-J.; Yun, J.-M.; Kim, Y.; Yoon, S.C.; Lee, S.-H.; Kim, D.-Y. Orthogonal Printable Reduced Graphene Oxide 2D Materials as Hole Transport Layers for High-Performance Inverted Polymer Solar Cells: Sheet Size Effect on Photovoltaic Properties. *ACS Appl. Mater. Interfaces* **2020**, *12*, 42811–42820. [CrossRef] [PubMed]
236. Zheng, X.; Zhang, H.; Yang, Q.; Xiong, C.; Li, W.; Yan, Y.; Gurney, R.S.; Wang, T. Solution-processed Graphene-MoS$_2$ heterostructure for efficient hole extraction in organic solar cells. *Carbon* **2019**, *142*, 156–163. [CrossRef]
237. Shoyiga, H.O.; Martincigh, B.S.; Nyamori, V.O. Hydrothermal synthesis of reduced graphene oxide-anatase titania nanocomposites for dual application in organic solar cells. *Int. J. Energy Res.* **2020**, *45*, 7293–7314. [CrossRef]
238. Lee, J.H.; Yoon, S.; Ko, M.S.; Lee, N.; Hwang, I.; Lee, M.J. Improved performance of organic photovoltaic devices by doping F$_4$TCNQ onto solution-processed graphene as a hole transport layer. *Org. Electron.* **2016**, *30*, 302–311. [CrossRef]
239. Lee, C.K.; Seo, J.G.; Kim, H.J.; Hong, S.J.; Song, G.; Ahn, C.; Lee, D.J.; Song, S.H. Versatile and Tunable Electrical Properties of Doped Nonoxidized Graphene Using Alkali Metal Chlorides. *ACS Appl. Mater. Interfaces* **2019**, *11*, 42520–42527. [CrossRef]
240. Sun, B.; Zhou, D.; Wang, C.; Liu, P.; Hao, Y.; Han, D.; Feng, L.; Zhou, Y. Copper(II) chloride doped graphene oxides as efficient hole transport layer for high-performance polymer solar cells. *Org. Electron.* **2017**, *44*, 176–182. [CrossRef]
241. Capasso, A.; Salamandra, L.; Faggio, G.; Dikonimos, T.; Buonocore, F.; Morandi, V.; Ortolani, L.; Lisi, N. Chemical Vapor Deposited Graphene-Based Derivative As High-Performance Hole Transport Material for Organic Photovoltaics. *ACS Appl. Mater. Interfaces* **2016**, *8*, 23844–23853. [CrossRef]
242. Sarkar, A.S.; Rao, A.D.; Jagdish, A.K.; Gupta, A.; Nandi, C.K.; Ramamurthy, P.C.; Pal, S.K. Facile embedding of gold nanostructures in the hole transporting layer for efficient polymer solar cells. *Org. Electron.* **2018**, *54*, 148–153. [CrossRef]
243. Iakobson, O.D.; Gribkova, O.L.; Tameev, A.R.; Nekrasov, A.A.; Saranin, D.S.; Di Carlo, A. Graphene nanosheet/polyaniline composite for transparent hole transporting layer. *J. Ind. Eng. Chem.* **2018**, *65*, 309–317. [CrossRef]
244. Aatif, M.; Patel, J.; Sharma, A.; Chauhan, M.; Kumar, G.; Pal, P.; Chand, S.; Tripathi, B.; Pandey, M.K.; Tiwari, J.P. Graphene oxide-molybdenum oxide composite with improved hole transport in bulk heterojunction solar cells. *AIP Adv.* **2019**, *9*, 075215. [CrossRef]
245. Cheng, J.; Zhang, H.; Zhao, Y.; Mao, J.; Li, C.; Zhang, S.; Wong, K.S.; Hou, J.; Choy, W.C.H. Self-Assembled Quasi-3D Nanocomposite: A Novel p-Type Hole Transport Layer for High Performance Inverted Organic Solar Cells. *Adv. Funct. Mater.* **2018**, *28*, 1706403. [CrossRef]
246. Dang, Y.; Wang, Y.; Shen, S.; Huang, S.; Qu, X.; Pang, Y.; Silva, S.R.P.; Kang, B.; Lu, G. Solution processed hybrid Graphene-MoO$_3$ hole transport layers for improved performance of organic solar cells. *Org. Electron.* **2019**, *67*, 95–100. [CrossRef]
247. Christopholi, L.P.; da Cunha, M.R.P.; Spada, E.R.; Gavim, A.E.X.; Hadano, F.S.; da Silva, W.J.; Rodrigues, P.C.; Macedo, A.G.; Faria, R.M.; de Deus, J.F. Reduced graphene oxide and perylene derivative nanohybrid as multifunctional interlayer for organic solar cells. *Synth. Met.* **2020**, *269*, 116552. [CrossRef]
248. Lu, J.; Yeo, P.S.E.; Gan, C.K.; Wu, P.; Loh, K.P. Transforming C60 molecules into graphene quantum dots. *Nat. Nanotechnol.* **2011**, *6*, 247–252. [CrossRef]
249. Peng, J.; Gao, W.; Gupta, B.K.; Liu, Z.; Romero-Aburto, R.; Ge, L.; Song, L.; Alemany, L.B.; Zhan, X.; Gao, G.; et al. Graphene Quantum Dots Derived from Carbon Fibers. *Nano Lett.* **2012**, *12*, 844–849. [CrossRef]
250. Ho, N.T.; Van Tam, T.; Tien, H.N.; Jang, S.-J.; Nguyen, T.K.; Choi, W.M.; Cho, S.; Kim, Y.S. Solution-Processed Transparent Intermediate Layer for Organic Tandem Solar Cell Using Nitrogen-Doped Graphene Quantum Dots. *J. Nanosci. Nanotechnol.* **2017**, *17*, 5686–5692. [CrossRef]
251. Hoang, T.T.; Pham, H.P.; Tran, Q.T. A Facile Microwave-Assisted Hydrothermal Synthesis of Graphene Quantum Dots for Organic Solar Cell Efficiency Improvement. *J. Nanomater.* **2020**, *2020*, 3207909. [CrossRef]
252. Zhang, X.; Sun, S.; Liu, X. Amino functionalized carbon nanotubes as hole transport layer for high performance polymer solar cells. *Inorg. Chem. Commun.* **2019**, *103*, 142–148. [CrossRef]
253. Rajanna, P.M.; Meddeb, H.; Sergeev, O.; Tsapenko, A.P.; Bereznev, S.; Vehse, M.; Volobujeva, O.; Danilson, M.; Lund, P.D.; Nasibulin, A.G. Rational design of highly efficient flexible and transparent p-type composite electrode based on single-walled carbon nanotubes. *Nano Energy* **2020**, *67*, 104183. [CrossRef]

254. Zhang, W.; Bu, F.; Shen, W.; Qi, X.; Yang, N.; Chen, M.; Yang, D.; Wang, Y.; Zhang, M.; Jiang, H.; et al. Strongly enhanced efficiency of polymer solar cells through unzipped SWNT hybridization in the hole transport layer. *RSC Adv.* **2020**, *10*, 24847–24854. [CrossRef]

255. Ibanez, J.G.; Rincón, M.E.; Gutierrez-Granados, S.; Chahma, M.; Jaramillo-Quintero, O.A.; Frontana-Uribe, B.A. Conducting Polymers in the Fields of Energy, Environmental Remediation, and Chemical-Chiral Sensors. *Chem. Rev.* **2018**, *118*, 4731–4816. [CrossRef] [PubMed]

256. Emah, J.B.; George, N.J.; Akpan, U.B. Interfacial Surface Modification via Nanoimprinting to Increase Open-Circuit Voltage of Organic Solar Cells. *J. Electron. Mater.* **2017**, *46*, 4989–4998. [CrossRef]

257. Fernández Castro, M.; Mazzolini, E.; Sondergaard, R.R.; Espindola-Rodriguez, M.; Andreasen, J.W. Flexible ITO-Free Roll-Processed Large-Area Nonfullerene Organic Solar Cells Based on P3HT:O-IDTBR. *Phys. Rev. Appl.* **2020**, *14*, 034067. [CrossRef]

258. Vohra, V.; Shimizu, S.; Takeoka, Y. Water-Processed Organic Solar Cells with Open-Circuit Voltages Exceeding 1.3V. *Coatings* **2020**, *10*, 421. [CrossRef]

259. Hong, N.; Xiao, J.; Li, Y.; Li, Y.; Wu, Y.; Yu, W.; Qiu, X.; Chen, R.; Yip, H.-L.; Huang, W.; et al. Unexpected fluorescent emission of graft sulfonated-acetone-formaldehyde lignin and its application as a dopant of PEDOT for high performance photovoltaic and light-emitting devices. *J. Mater. Chem. C* **2016**, *4*, 5297–5306. [CrossRef]

260. Abdel-Fattah, T.M.; Younes, E.M.; Namkoong, G.; El-Maghraby, E.M.; Elsayed, A.H.; Abo Elazm, A.H. Solvents effects on the hole transport layer in organic solar cells performance. *Sol. Energy* **2016**, *137*, 337–343. [CrossRef]

261. Kurukavak, Ç.K.; Polat, S. Influence of the volume of EGME-DMSO mixed co-solvent doping on the characteristics of PEDOT:PSS and their application in polymer solar cells. *Polym. Polym. Compos.* **2020**, *29*, 1222–1228. [CrossRef]

262. Yi, M.; Hong, S.; Kim, J.-R.; Kang, H.; Lee, J.; Yu, K.; Kee, S.; Lee, W.; Lee, K. Modification of a PEDOT:PSS hole transport layer for printed polymer solar cells. *Sol. Energy Mater. Sol. Cells* **2016**, *153*, 117–123. [CrossRef]

263. Howells, C.T.; Marbou, K.; Kim, H.; Lee, K.J.; Heinrich, B.; Kim, S.J.; Nakao, A.; Aoyama, T.; Furukawa, S.; Kim, J.-H.; et al. Enhanced organic solar cells efficiency through electronic and electro-optic effects resulting from charge transfers in polymer hole transport blends. *J. Mater. Chem. A* **2016**, *4*, 4252–4263. [CrossRef]

264. Bhargav, R.; Bhardwaj, D.; Shahjad, A.P.; Chand, S. Poly(Styrene Sulfonate) Free Poly(3,4-Ethylenedioxythiophene) as a Robust and Solution-Processable Hole Transport Layer for Organic Solar Cells. *ChemistrySelect* **2016**, *1*, 1347–1352. [CrossRef]

265. Cai, W.; Musumeci, C.; Ajjan, F.N.; Bao, Q.; Ma, Z.; Tang, Z.; Inganäs, O. Self-doped conjugated polyelectrolyte with tuneable work function for effective hole transport in polymer solar cells. *J. Mater. Chem. A* **2016**, *4*, 15670–15675. [CrossRef]

266. Gribkova, O.L.; Iakobson, O.D.; Nekrasov, A.A.; Cabanova, V.A.; Tverskoy, V.A.; Tameev, A.R.; Vannikov, A.V. Ultraviolet-Visible-Near Infrared and Raman spectroelectrochemistry of poly(3,4-ethylenedioxythiophene) complexes with sulfonated polyelectrolytes. The role of inter- and intra-molecular interactions in polyelectrolyte. *Electrochim. Acta* **2016**, *222*, 409–420. [CrossRef]

267. Kim, D.H.; Lee, D.J.; Kim, B.; Yun, C.; Kang, M.H. Tailoring PEDOT:PSS polymer electrode for solution-processed inverted organic solar cells. *Solid-State Electron.* **2020**, *169*, 107808. [CrossRef]

268. Liu, L.; Li, F.; Zhao, C.; Bi, F.; Jiu, T.; Zhao, M.; Li, J.; Zhang, G.; Wang, L.; Fang, J.; et al. Performance Enhancement of Conventional Polymer Solar Cells with TTF-py-Modified PEDOT:PSS Film as the Hole Transport Layer. *ACS Appl. Energy Mater.* **2019**, *2*, 6577–6583. [CrossRef]

269. Wang, J.; Zheng, Z.; Zhang, D.; Zhang, J.; Zhou, J.; Liu, J.; Xie, S.; Zhao, Y.; Zhang, Y.; Wei, Z.; et al. Regulating Bulk-Heterojunction Molecular Orientations through Surface Free Energy Control of Hole-Transporting Layers for High-Performance Organic Solar Cells. *Adv. Mater.* **2019**, *31*, 1806921. [CrossRef]

270. Tang, H.; Liu, Z.; Hu, Z.; Liang, Y.; Huang, F.; Cao, Y. Oxoammonium enabled secondary doping of hole transporting material PEDOT:PSS for high-performance organic solar cells. *Sci. China Chem.* **2020**, *63*, 802–809. [CrossRef]

271. Khan, M.A.; Farva, U. Elucidation of hierarchical metallophthalocyanine buffer layers in bulk heterojunction solar cells. *RSC Adv.* **2017**, *7*, 11304–11311. [CrossRef]

272. Wang, J.; Yu, H.; Hou, C.; Zhang, J. Solution-Processable PEDOT:PSS:α-In$_2$Se$_3$ with Enhanced Conductivity as a Hole Transport Layer for High-Performance Polymer Solar Cells. *ACS Appl. Mater. Interfaces* **2020**, *12*, 26543–26554. [CrossRef]

273. Zeng, M.; Wang, X.; Ma, R.; Zhu, W.; Li, Y.; Chen, Z.; Zhou, J.; Li, W.; Liu, T.; He, Z.; et al. Dopamine Semiquinone Radical Doped PEDOT:PSS: Enhanced Conductivity, Work Function and Performance in Organic Solar Cells. *Adv. Energy Mater.* **2020**, *10*, 2000743. [CrossRef]

274. Roh, S.H.; Kim, J.K. Hexagonal Array Patterned PMMA Buffer Layer for Efficient Hole Transport and Tailored Interfacial Properties of FTO-Based Organic Solar Cells. *Macromol. Res.* **2018**, *26*, 1173–1178. [CrossRef]

275. Kwak, C.K.; Pérez, G.E.; Freestone, B.G.; Al-Isaee, S.A.; Iraqi, A.; Lidzey, D.G.; Dunbar, A.D.F. Improved efficiency in organic solar cells via conjugated polyelectrolyte additive in the hole transporting layer. *J. Mater. Chem. C* **2016**, *4*, 10722–10730. [CrossRef]

276. Palma-Cando, A.; Scherf, U. Electrochemically Generated Thin Films of Microporous Polymer Networks: Synthesis, Properties, and Applications. *Macromol. Chem. Phys.* **2016**, *217*, 827–841. [CrossRef]

277. Palma-Cando, A.; Scherf, U. Electrogenerated Thin Films of Microporous Polymer Networks with Remarkably Increased Electrochemical Response to Nitroaromatic Analytes. *ACS Appl. Mater. Interfaces* **2015**, *7*, 11127–11133. [CrossRef] [PubMed]

278. Räupke, A.; Palma-Cando, A.; Shkura, E.; Teckhausen, P.; Polywka, A.; Görrn, P.; Scherf, U.; Riedl, T. Highly sensitive gas-phase explosive detection by luminescent microporous polymer networks. *Sci. Rep.* **2016**, *6*, 29118. [CrossRef]

279. Palma-Cando, A.; Woitassek, D.; Brunklaus, G.; Scherf, U. Luminescent tetraphenylethene-cored, carbazole- and thiophene-based microporous polymer films for the chemosensing of nitroaromatic analytes. *Mater. Chem. Front.* **2017**, *1*, 1118–1124. [CrossRef]
280. Palma-Cando, A.; Preis, E.; Scherf, U. Silicon-or Carbon-Cored Multifunctional Carbazolyl Monomers for the Electrochemical Generation of Microporous Polymer Films. *Macromolecules* **2016**, *49*, 8041–8047. [CrossRef]
281. Palma-Cando, A.; Rendón-Enríquez, I.; Tausch, M.; Scherf, U. Thin Functional Polymer Films by Electropolymerization. *Nanomaterials* **2019**, *9*, 1125. [CrossRef]
282. Palma-Cando, A.; Brunklaus, G.; Scherf, U. Thiophene-Based Microporous Polymer Networks via Chemical or Electrochemical Oxidative Coupling. *Macromolecules* **2015**, *48*, 6816–6824. [CrossRef]
283. Gu, C.; Huang, N.; Chen, Y.; Qin, L.; Xu, H.; Zhang, S.; Li, F.; Ma, Y.; Jiang, D. π-Conjugated Microporous Polymer Films: Designed Synthesis, Conducting Properties, and Photoenergy Conversions. *Angew. Chem.* **2015**, *127*, 13798–13802. [CrossRef]
284. Liu, C.; Luo, H.; Shi, G.; Nian, L.; Chi, Z.; Ma, Y. Electrochemical route to fabricate porous organic polymers film and its application for polymer solar cells. *Dyes Pigments* **2017**, *142*, 132–138. [CrossRef]
285. Li, Y.; Sun, X.; Zhang, X.; Zhang, D.; Xia, H.; Zhou, J.; Ahmad, N.; Leng, X.; Yang, S.; Zhang, Y.; et al. Built-in voltage enhanced by in situ electrochemical polymerized undoped conjugated hole-transporting modifiers in organic solar cells. *J. Mater. Chem. C* **2020**, *8*, 2676–2681. [CrossRef]
286. Lee, E.J.; Han, J.P.; Jung, S.E.; Choi, M.H.; Moon, D.K. Improvement in Half-Life of Organic Solar Cells by Using a Blended Hole Extraction Layer Consisting of PEDOT:PSS and Conjugated Polymer Electrolyte. *ACS Appl. Mater. Interfaces* **2016**, *8*, 31791–31798. [CrossRef] [PubMed]
287. Pérez, G.E.; Erothu, H.; Topham, P.D.; Bastianini, F.; Alanazi, T.I.; Bernardo, G.; Parnell, A.J.; King, S.M.; Dunbar, A.D.F. Improved Performance and Stability of Organic Solar Cells by the Incorporation of a Block Copolymer Interfacial Layer. *Adv. Mater. Interfaces* **2020**, *7*, 2000918. [CrossRef]
288. Hamed, M.S.G.; Oseni, S.O.; Kumar, A.; Sharma, G.; Mola, G.T. Nickel sulphide nano-composite assisted hole transport in thin film polymer solar cells. *Sol. Energy* **2020**, *195*, 310–317. [CrossRef]
289. Wei, Y.; Zhang, Q.; Wan, H.; Zhang, Y.; Zheng, S.; Zhang, Y. A facile synthesis of segmented silver nanowires and enhancement of the performance of polymer solar cells. *Phys. Chem. Chem. Phys.* **2018**, *20*, 18837–18843. [CrossRef]
290. Hao, J.; Xu, Y.; Chen, S.; Zhang, Y.; Mai, J.; Lau, T.-K.; Zhang, R.; Mei, Y.; Wang, L.; Lu, X.; et al. Broadband plasmon-enhanced polymer solar cells with power conversion efficiency of 9.26% using mixed Au nanoparticles. *Opt. Commun.* **2016**, *362*, 50–58. [CrossRef]
291. Oh, Y.; Lim, J.W.; Kim, J.G.; Wang, H.; Kang, B.-H.; Park, Y.W.; Kim, H.; Jang, Y.J.; Kim, J.; Kim, D.H.; et al. Plasmonic Periodic Nanodot Arrays via Laser Interference Lithography for Organic Photovoltaic Cells with >10% Efficiency. *ACS Nano* **2016**, *10*, 10143–10151. [CrossRef]
292. Singh, A.; Dey, A.; Das, D.; Iyer, P.K. Combined influence of plasmonic metal nanoparticles and dual cathode buffer layers for highly efficient rrP3HT:PCBM-based bulk heterojunction solar cells. *J. Mater. Chem. C* **2017**, *5*, 6578–6587. [CrossRef]
293. Sah, P.-T.; Chang, W.-C.; Chen, J.-H.; Wang, H.-H.; Chan, L.-H. Bimetallic Ag-Au-Ag nanorods used to enhance efficiency of polymer solar cells. *Electrochim. Acta* **2018**, *259*, 293–302. [CrossRef]
294. Tang, M.; Sun, B.; Zhou, D.; Gu, Z.; Chen, K.; Guo, J.; Feng, L.; Zhou, Y. Broad-band plasmonic Cu-Au bimetallic nanoparticles for organic bulk heterojunction solar cells. *Org. Electron.* **2016**, *38*, 213–221. [CrossRef]
295. Adedeji, M.A.; Hamed, M.S.G.; Mola, G.T. Light trapping using copper decorated nano-composite in the hole transport layer of organic solar cell. *Sol. Energy* **2020**, *203*, 83–90. [CrossRef]
296. Najeeb, M.A.; Abdullah, S.M.; Aziz, F.; Ahmad, Z.; Shakoor, R.A.; Mohamed, A.M.A.; Khalil, U.; Swelm, W.; Al-Ghamdi, A.A.; Sulaiman, K. A comparative Study on the performance of hybrid solar cells containing ZnSTe QDs in hole transporting layer and photoactive layer. *J. Nanopart. Res.* **2016**, *18*, 384. [CrossRef]
297. Zhang, X.; Jiang, Q.; Wang, J.; Tang, J. Black phosphorous quantum dots as an effective interlayer modifier in polymer solar cells. *Sol. Energy* **2020**, *206*, 670–676. [CrossRef]
298. Ismail, Y.A.M.; Kishi, N.; Soga, T. Improvement of organic solar cells using aluminium microstructures prepared in PEDOT:PSS buffer layer by using ultrasonic ablation technique. *Thin Solid Films* **2016**, *616*, 73–79. [CrossRef]
299. Kesavan, A.V.; Jagdish, A.K.; Ramamurthy, P.C. Organic Inorganic Hybrid Hole Transport Layer for Light Management in Inverted Organic Photovoltaic. *IEEE J. Photovolt.* **2017**, *7*, 787–791. [CrossRef]
300. Liu, Z.; Park, J.; Li, B.; Chan, H.P.; Yi, D.K.; Lee, E.-C. Performance improvement of organic bulk-heterojunction solar cells using complementary plasmonic gold nanorods. *Org. Electron.* **2020**, *84*, 105802. [CrossRef]
301. Phetsang, S.; Nootchanat, S.; Lertvachirapaiboon, C.; Ishikawa, R.; Shinbo, K.; Kato, K.; Mungkornasawakul, P.; Ounnunkad, K.; Baba, A. Enhancement of organic solar cell performance by incorporating gold quantum dots (AuQDs) on a plasmonic grating. *Nanoscale Adv.* **2020**, *2*, 2950–2957. [CrossRef]
302. Haase, M.; Schäfer, H. Upconverting Nanoparticles. *Angew. Chemie Int. Ed.* **2011**, *50*, 5808–5829. [CrossRef]
303. Mei, Y.; Ma, X.; Fan, Y.; Bai, Z.; Cheng, F.; Hu, Y.; Fan, Q.; Chen, S.; Huang, W. Application of β-NaYF$_4$:Er^{3+} (2%), Yb^{3+} (18%) Up-Conversion Nanoparticles in Polymer Solar Cells and its Working Mechanism. *J. Nanosci. Nanotechnol.* **2016**, *16*, 7380–7387. [CrossRef]
304. Mola, G.T.; Arbab, E.A.A. Bimetallic nanocomposite as hole transport co-buffer layer in organic solar cell. *Appl. Phys. A* **2017**, *123*, 772. [CrossRef]

305. Michalska, M.; Iwan, A.; Andrzejczuk, M.; Roguska, A.; Sikora, A.; Boharewicz, B.; Tazbir, I.; Hreniak, A.; Popłoński, S.; Korona, K.P. Analysis of the surface decoration of TiO_2 grains using silver nanoparticles obtained by ultrasonochemical synthesis towards organic photovoltaics. *New J. Chem.* **2018**, *42*, 7340–7354. [CrossRef]

306. Truong, N.T.N.; Kim, C.D.; Minnam Reddy, V.R.; Thai, V.H.; Jeon, H.J.; Park, C. Shape control of plasmonic gold nanoparticles and its application to vacuum-free bulk hetero-junction solar cells. *J. Mater. Sci. Mater. Electron.* **2020**, *31*, 22957–22965. [CrossRef]

307. Mohammad, T.; Dwivedi, C.; Kumar, V.; Dutta, V. Role of Continuous Spray Pyrolyzed synthesized MoO_3 nanorods in PEDOT:PSS matrix by electric field assisted spray deposition for organic photovoltaics. *Org. Electron.* **2020**, *77*, 105525. [CrossRef]

308. Pan, J.; Li, P.; Cai, L.; Hu, Y.; Zhang, Y. All-solution processed double-decked PEDOT:PSS/V_2O_5 nanowires as buffer layer of high performance polymer photovoltaic cells. *Sol. Energy Mater. Sol. Cells* **2016**, *144*, 616–622. [CrossRef]

309. Li, J.; Qin, J.; Liu, X.; Ren, M.; Tong, J.; Zheng, N.; Chen, W.; Xia, Y. Enhanced organic photovoltaic performance through promoting crystallinity of photoactive layer and conductivity of hole-transporting layer by V_2O_5 doped PEDOT:PSS hole-transporting layers. *Sol. Energy* **2020**, *211*, 1102–1109. [CrossRef]

310. Du, X.; Lytken, O.; Killian, M.S.; Cao, J.; Stubhan, T.; Turbiez, M.; Schmuki, P.; Steinrück, H.-P.; Ding, L.; Fink, R.H.; et al. Overcoming Interfacial Losses in Solution-Processed Organic Multi-Junction Solar Cells. *Adv. Energy Mater.* **2017**, *7*, 1601959. [CrossRef]

311. Wang, Y.; Han, J.; Cai, L.; Li, N.; Li, Z.; Zhu, F. Efficient and stable operation of nonfullerene organic solar cells: Retaining a high built-in potential. *J. Mater. Chem. A* **2020**, *8*, 21255–21264. [CrossRef]

312. Mbuyise, X.G.; Arbab, E.A.A.; Kaviyarasu, K.; Pellicane, G.; Maaza, M.; Mola, G.T. Zinc oxide doped single wall carbon nanotubes in hole transport buffer layer. *J. Alloys Compd.* **2017**, *706*, 344–350. [CrossRef]

313. Zheng, Z.; Hu, Q.; Zhang, S.; Zhang, D.; Wang, J.; Xie, S.; Wang, R.; Qin, Y.; Li, W.; Hong, L.; et al. A Highly Efficient Non-Fullerene Organic Solar Cell with a Fill Factor over 0.80 Enabled by a Fine-Tuned Hole-Transporting Layer. *Adv. Mater.* **2018**, *30*, 1801801. [CrossRef] [PubMed]

314. Han, Y.W.; Jeon, S.J.; Lee, H.S.; Park, H.; Kim, K.S.; Lee, H.; Moon, D.K. Evaporation-Free Nonfullerene Flexible Organic Solar Cell Modules Manufactured by An All-Solution Process. *Adv. Energy Mater.* **2019**, *9*, 1902065. [CrossRef]

315. Sun, L.; Zeng, W.; Xie, C.; Hu, L.; Dong, X.; Qin, F.; Wang, W.; Liu, T.; Jiang, X.; Jiang, Y.; et al. Flexible All-Solution-Processed Organic Solar Cells with High-Performance Nonfullerene Active Layers. *Adv. Mater.* **2020**, *32*, 1907840. [CrossRef] [PubMed]

316. Ji, G.; Wang, Y.; Luo, Q.; Han, K.; Xie, M.; Zhang, L.; Wu, N.; Lin, J.; Xiao, S.; Li, Y.-Q.; et al. Fully Coated Semitransparent Organic Solar Cells with a Doctor-Blade-Coated Composite Anode Buffer Layer of Phosphomolybdic Acid and PEDOT:PSS and a Spray-Coated Silver Nanowire Top Electrode. *ACS Appl. Mater. Interfaces* **2018**, *10*, 943–954. [CrossRef]

317. Liu, Z.; Niu, S.; Wang, N. Oleylamine-functionalized graphene oxide as an electron block layer towards high-performance and photostable fullerene-free polymer solar cells. *Nanoscale* **2017**, *9*, 16293–16304. [CrossRef]

318. Maity, S.; Thomas, T. Hole-Collecting Treated Graphene Layer and PTB7:PC 71 BM-Based Bulk-Heterojunction OPV With Improved Carrier Collection and Photovoltaic Efficiency. *IEEE Trans. Electron Devices* **2018**, *65*, 4548–4554. [CrossRef]

319. Hilal, M.; Han, J.I. Enhancing the photovoltaic characteristics of organic solar cells by introducing highly conductive graphene as a conductive platform for a PEDOT:PSS anode interfacial layer. *J. Mater. Sci. Mater. Electron.* **2019**, *30*, 6187–6200. [CrossRef]

320. Amollo, T.A.; Mola, G.T.; Nyamori, V.O. High-performance organic solar cells utilizing graphene oxide in the active and hole transport layers. *Sol. Energy* **2018**, *171*, 83–91. [CrossRef]

321. Goumri, M.; Lucas, B.; Ratier, B.; Baitoul, M. Inverted Polymer Solar Cells with a Reduced Graphene Oxide/Poly (3,4-Ethylene Dioxythiophene):Poly(4-Styrene Sulfonate) (PEDOT:PSS) Hole Transport Layer. *J. Electron. Mater.* **2019**, *48*, 1097–1105. [CrossRef]

322. Ozcan, S.; Erer, M.C.; Vempati, S.; Uyar, T.; Toppare, L.; Çırpan, A. Graphene oxide-doped PEDOT:PSS as hole transport layer in inverted bulk heterojunction solar cell. *J. Mater. Sci. Mater. Electron.* **2020**, *31*, 3576–3584. [CrossRef]

323. Amollo, T.A.; Mola, G.T.; Nyamori, V.O. Polymer solar cells with reduced graphene oxide–germanium quantum dots nanocomposite in the hole transport layer. *J. Mater. Sci. Mater. Electron.* **2018**, *29*, 7820–7831. [CrossRef]

324. Raj, P.G.; Rani, V.S.; Kanwat, A.; Jang, J. Enhanced organic photovoltaic properties via structural modifications in PEDOT:PSS due to graphene oxide doping. *Mater. Res. Bull.* **2016**, *74*, 346–352. [CrossRef]

325. Srivastava, S.K.; Pionteck, J. Recent Advances in Preparation, Structure, Properties and Applications of Graphite Oxide. *J. Nanosci. Nanotechnol.* **2015**, *15*, 1984–2000. [CrossRef] [PubMed]

326. Toh, S.Y.; Loh, K.S.; Kamarudin, S.K.; Daud, W.R.W. Graphene production via electrochemical reduction of graphene oxide: Synthesis and characterisation. *Chem. Eng. J.* **2014**, *251*, 422–434. [CrossRef]

327. Iwan, A.; Caballero-Briones, F.; Filapek, M.; Boharewicz, B.; Tazbir, I.; Hreniak, A.; Guerrero-Contreras, J. Electrochemical and photocurrent characterization of polymer solar cells with improved performance after GO addition to the PEDOT:PSS hole transporting layer. *Sol. Energy* **2017**, *146*, 230–242. [CrossRef]

328. Rafique, S.; Abdullah, S.M.; Shahid, M.M.; Ansari, M.O.; Sulaiman, K. Significantly improved photovoltaic performance in polymer bulk heterojunction solar cells with graphene oxide /PEDOT:PSS double decked hole transport layer. *Sci. Rep.* **2017**, *7*, 39555. [CrossRef]

329. Hilal, M.; Han, J.I. Significant improvement in the photovoltaic stability of bulk heterojunction organic solar cells by the molecular level interaction of graphene oxide with a PEDOT: PSS composite hole transport layer. *Sol. Energy* **2018**, *167*, 24–34. [CrossRef]

330. Moon, B.J.; Jang, D.; Yi, Y.; Lee, H.; Kim, S.J.; Oh, Y.; Lee, S.H.; Park, M.; Lee, S.; Bae, S. Multi-functional nitrogen self-doped graphene quantum dots for boosting the photovoltaic performance of BHJ solar cells. *Nano Energy* **2017**, *34*, 36–46. [CrossRef]

331. Rafique, S.; Roslan, N.A.; Abdullah, S.M.; Li, L.; Supangat, A.; Jilani, A.; Iwamoto, M. UV-ozone treated graphene oxide/PEDOT:PSS bilayer as a novel hole transport layer in highly efficient and stable organic solar cells. *Org. Electron.* **2019**, *66*, 32–42. [CrossRef]

332. Xing, W.; Chen, Y.; Wu, X.; Xu, X.; Ye, P.; Zhu, T.; Guo, Q.; Yang, L.; Li, W.; Huang, H. PEDOT:PSS-Assisted Exfoliation and Functionalization of 2D Nanosheets for High-Performance Organic Solar Cells. *Adv. Funct. Mater.* **2017**, *27*, 1701622. [CrossRef]

333. Koo, D.; Jung, S.; Oh, N.K.; Choi, Y.; Seo, J.; Lee, J.; Kim, U.; Park, H. Improved charge transport via WSe$_2$-mediated hole transporting layer toward efficient organic solar cells. *Semicond. Sci. Technol.* **2018**, *33*, 125020. [CrossRef]

334. Ramasamy, M.S.; Ryu, K.Y.; Lim, J.W.; Bibi, A.; Kwon, H.; Lee, J.-E.; Kim, D.H.; Kim, K. Solution-Processed PEDOT:PSS/MoS$_2$ Nanocomposites as Efficient Hole-Transporting Layers for Organic Solar Cells. *Nanomaterials* **2019**, *9*, 1328. [CrossRef] [PubMed]

335. Dang, Y.; Shen, S.; Wang, Y.; Qu, X.; Huang, S.; Dong, Q.; Silva, S.R.P.; Kang, B. Hole Extraction Enhancement for Efficient Polymer Solar Cells with Boronic Acid Functionalized Carbon Nanotubes doped Hole Transport Layers. *ACS Sustain. Chem. Eng.* **2018**, *6*, 5122–5131. [CrossRef]

336. Yang, Q.; Yu, S.; Fu, P.; Yu, W.; Liu, Y.; Liu, X.; Feng, Z.; Guo, X.; Li, C. Boosting Performance of Non-Fullerene Organic Solar Cells by 2D g-C$_3$N$_4$ Doped PEDOT:PSS. *Adv. Funct. Mater.* **2020**, *30*, 1910205. [CrossRef]

337. Hou, C.; Yu, H. Modifying the nanostructures of PEDOT:PSS/Ti$_3$C$_2$TX composite hole transport layers for highly efficient polymer solar cells. *J. Mater. Chem. C* **2020**, *8*, 4169–4180. [CrossRef]

338. Chiou, G.-C.; Lin, M.-W.; Lai, Y.-L.; Chang, C.-K.; Jiang, J.-M.; Su, Y.-W.; Wei, K.-H.; Hsu, Y.-J. Fluorene Conjugated Polymer/Nickel Oxide Nanocomposite Hole Transport Layer Enhances the Efficiency of Organic Photovoltaic Devices. *ACS Appl. Mater. Interfaces* **2017**, *9*, 2232–2239. [CrossRef] [PubMed]

339. Guo, Y.; Lei, H.; Xiong, L.; Li, B.; Fang, G. Organic solar cells based on a Cu$_2$O/FBT-TH4 anode buffer layer with enhanced power conversion efficiency and ambient stability. *J. Mater. Chem. C* **2017**, *5*, 8033–8040. [CrossRef]

340. Jo, W.J.; Nelson, J.T.; Chang, S.; Bulović, V.; Gradečak, S.; Strano, M.S.; Gleason, K.K. Oxidative Chemical Vapor Deposition of Neutral Hole Transporting Polymer for Enhanced Solar Cell Efficiency and Lifetime. *Adv. Mater.* **2016**, *28*, 6399–6404. [CrossRef]

341. Awada, H.; Mattana, G.; Tournebize, A.; Rodriguez, L.; Flahaut, D.; Vellutini, L.; Lartigau-Dagron, C.; Billon, L.; Bousquet, A.; Chambon, S. Surface engineering of ITO electrode with a functional polymer for PEDOT:PSS-free organic solar cells. *Org. Electron.* **2018**, *57*, 186–193. [CrossRef]

342. Kadem, B.Y.; Kadhim, R.G.; Banimuslem, H. Efficient P$_3$HT:SWCNTs hybrids as hole transport layer in P$_3$HT:PCBM organic solar cells. *J. Mater. Sci. Mater. Electron.* **2018**, *29*, 9418–9426. [CrossRef]

343. Babaei, Z.; Rezaei, B.; Pisheh, M.K.; Afshar-Taromi, F. In situ synthesis of gold/silver nanoparticles and polyaniline as buffer layer in polymer solar cells. *Mater. Chem. Phys.* **2020**, *248*, 122879. [CrossRef]

344. Deng, L.; Li, D.; Agbolaghi, S. Networked Conductive Polythiophene/Polyaniline Bottlebrushes with Modified Carbon Nanotubes As Hole Transport Layer in Organic Photovoltaics. *J. Electron. Mater.* **2020**, *49*, 937–948. [CrossRef]

345. Sorkhishams, N.; Massoumi, B.; Saraei, M.; Agbolaghi, S. Electrode buffer layers via networks of polythiophene/polyaniline bottlebrushes and carbon nanotubes in organic solar cells. *J. Mater. Sci. Mater. Electron.* **2019**, *30*, 21117–21125. [CrossRef]

346. Biswas, S.; You, Y.-J.; Kim, J.; Ha, S.R.; Choi, H.; Kwon, S.-H.; Kim, K.-K.; Shim, J.W.; Kim, H. Decent efficiency improvement of organic photovoltaic cell with low acidic hole transport material by controlling doping concentration. *Appl. Surf. Sci.* **2020**, *512*, 145700. [CrossRef]

347. Biswas, S.; You, Y.-J.; Lee, Y.; Shim, J.W.; Kim, H. Efficiency improvement of indoor organic solar cell by optimization of the doping level of the hole extraction layer. *Dyes Pigments* **2020**, *183*, 108719. [CrossRef]

348. Biswas, S.; You, Y.-J.; Shim, J.W.; Kim, H. Utilization of poly (4-styrenesulfonic acid) doped polyaniline as a hole transport layer of organic solar cell for indoor applications. *Thin Solid Films* **2020**, *700*, 137921. [CrossRef]

349. Abdulrazzaq, O.A.; Bourdo, S.E.; Saini, V.; Biris, A.S. Acid-free polyaniline:graphene-oxide hole transport layer in organic solar cells. *J. Mater. Sci. Mater. Electron.* **2020**, *31*, 21640–21650. [CrossRef]

350. Dong, J.; Guo, J.; Wang, X.; Dong, P.; Wang, Z.; Zhou, Y.; Miao, Y.; Zhao, B.; Hao, Y.; Wang, H.; et al. A Low-Temperature Solution-Processed CuSCN/Polymer Hole Transporting Layer Enables High Efficiency for Organic Solar Cells. *ACS Appl. Mater. Interfaces* **2020**, *12*, 46373–46380. [CrossRef]

351. Houston, J.E.; Kraft, M.; Scherf, U.; Evans, R.C. Sequential detection of multiple phase transitions in model biological membranes using a red-emitting conjugated polyelectrolyte. *Phys. Chem. Chem. Phys.* **2016**, *18*, 12423–12427. [CrossRef]

352. Meazzini, I.; Blayo, C.; Arlt, J.; Marques, A.-T.; Scherf, U.; Burrows, H.D.; Evans, R.C. Ureasil organic–inorganic hybrids as photoactive waveguides for conjugated polyelectrolyte luminescent solar concentrators. *Mater. Chem. Front.* **2017**, *1*, 2271–2282. [CrossRef]

353. Willis-Fox, N.; Gutacker, A.; Browne, M.P.; Khan, A.R.; Lyons, M.E.G.; Scherf, U.; Evans, R.C. Selective recognition of biologically important anions using a diblock polyfluorene–polythiophene conjugated polyelectrolyte. *Polym. Chem.* **2017**, *8*, 7151–7159. [CrossRef]

354. Balcı Leinen, M.; Berger, F.J.; Klein, P.; Mühlinghaus, M.; Zorn, N.F.; Settele, S.; Allard, S.; Scherf, U.; Zaumseil, J. Doping-Dependent Energy Transfer from Conjugated Polyelectrolytes to (6,5) Single-Walled Carbon Nanotubes. *J. Phys. Chem. C* **2019**, *123*, 22680–22689. [CrossRef]

355. Pipertzis, A.; Papamokos, G.; Sachnik, O.; Allard, S.; Scherf, U.; Floudas, G. Ionic Conductivity in Polyfluorene-Based Diblock Copolymers Comprising Nanodomains of a Polymerized Ionic Liquid and a Solid Polymer Electrolyte Doped with LiTFSI. *Macromolecules* **2021**, *54*, 4257–4268. [CrossRef]

356. Martelo, L.; Fonseca, S.; Marques, A.; Burrows, H.; Valente, A.; Justino, L.; Scherf, U.; Pradhan, S.; Song, Q. Effects of Charge Density on Photophysics and Aggregation Behavior of Anionic Fluorene-Arylene Conjugated Polyelectrolytes. *Polymers* **2018**, *10*, 258. [CrossRef]

357. Jo, J.W.; Jung, J.W.; Bae, S.; Ko, M.J.; Kim, H.; Jo, W.H.; Jen, A.K.-Y.; Son, H.J. Development of Self-Doped Conjugated Polyelectrolytes with Controlled Work Functions and Application to Hole Transport Layer Materials for High-Performance Organic Solar Cells. *Adv. Mater. Interfaces* **2016**, *3*, 1500703. [CrossRef]

358. Jo, J.W.; Yun, J.H.; Bae, S.; Ko, M.J.; Son, H.J. Development of a conjugated donor-acceptor polyelectrolyte with high work function and conductivity for organic solar cells. *Org. Electron.* **2017**, *50*, 1–6. [CrossRef]

359. Cui, Y.; Xu, B.; Yang, B.; Yao, H.; Li, S.; Hou, J. A Novel pH Neutral Self-Doped Polymer for Anode Interfacial Layer in Efficient Polymer Solar Cells. *Macromolecules* **2016**, *49*, 8126–8133. [CrossRef]

360. Choi, M.-H.; Lee, E.J.; Han, J.P.; Moon, D.K. Solution-processed pH-neutral conjugated polyelectrolytes with one-atom variation (O, S, Se) as a novel hole-collecting layer in organic photovoltaics. *Sol. Energy Mater. Sol. Cells* **2016**, *155*, 243–252. [CrossRef]

361. Xu, H.; Fu, X.; Cheng, X.; Huang, L.; Zhou, D.; Chen, L.; Chen, Y. Highly and homogeneously conductive conjugated polyelectrolyte hole transport layers for efficient organic solar cells. *J. Mater. Chem. A* **2017**, *5*, 14689–14696. [CrossRef]

362. Moon, S.; Khadtare, S.; Wong, M.; Han, S.-H.; Bazan, G.C.; Choi, H. Hole transport layer based on conjugated polyelectrolytes for polymer solar cells. *J. Colloid Interface Sci.* **2018**, *518*, 21–26. [CrossRef]

363. Xie, Q.; Zhang, J.; Xu, H.; Liao, X.; Chen, Y.; Li, Y.; Chen, L. Self-doped polymer with fluorinated phenylene as hole transport layer for efficient polymer solar cells. *Org. Electron.* **2018**, *61*, 207–214. [CrossRef]

364. Lee, E.J.; Choi, M.H.; Han, Y.W.; Moon, D.K. Effect on Electrode Work Function by Changing Molecular Geometry of Conjugated Polymer Electrolytes and Application for Hole-Transporting Layer of Organic Optoelectronic Devices. *ACS Appl. Mater. Interfaces* **2017**, *9*, 44060–44069. [CrossRef] [PubMed]

365. Xu, H.; Zou, H.; Zhou, D.; Zeng, G.; Chen, L.; Liao, X.; Chen, Y. Printable Hole Transport Layer for 1.0 cm^2 Organic Solar Cells. *ACS Appl. Mater. Interfaces* **2020**, *12*, 52028–52037. [CrossRef] [PubMed]

366. Xu, X.-P.; Li, S.-Y.; Li, Y.; Peng, Q. Recent progress in organic hole-transporting materials with 4-anisylamino-based end caps for efficient perovskite solar cells. *Rare Met.* **2021**, *40*, 1669–1690. [CrossRef]

367. Rombach, F.M.; Haque, S.A.; Macdonald, T.J. Lessons learned from spiro-OMeTAD and PTAA in perovskite solar cells. *Energy Environ. Sci.* **2021**, *14*, 5161–5190. [CrossRef]

368. Li, H.; Fu, K.; Boix, P.P.; Wong, L.H.; Hagfeldt, A.; Grätzel, M.; Mhaisalkar, S.G.; Grimsdale, A.C. Hole-Transporting Small Molecules Based on Thiophene Cores for High Efficiency Perovskite Solar Cells. *ChemSusChem* **2014**, *7*, 3420–3425. [CrossRef]

369. Bormann, L.; Selzer, F.; Weiß, N.; Kneppe, D.; Leo, K.; Müller-Meskamp, L. Doped hole transport layers processed from solution: Planarization and bridging the voids in noncontinuous silver nanowire electrodes. *Org. Electron.* **2016**, *28*, 163–171. [CrossRef]

370. Cheng, J.; Ren, X.; Zhu, H.L.; Mao, J.; Liang, C.; Zhuang, J.; Roy, V.A.L.; Choy, W.C.H. Pre- and post-treatments free nanocomposite based hole transport layer for high performance organic solar cells with considerably enhanced reproducibility. *Nano Energy* **2017**, *34*, 76–85. [CrossRef]

371. Courté, M.; Alaaeddine, M.; Barth, V.; Tortech, L.; Fichou, D. Structural and electronic properties of 2,2′,6,6′-tetraphenyl-dipyranylidene and its use as a hole-collecting interfacial layer in organic solar cells. *Dye. Pigment.* **2017**, *141*, 487–492. [CrossRef]

372. Wang, L.; Li, X.; Tan, T.W.; Shi, Y.; Zhao, X.Y.; Mi, B.X.; Gao, Z.Q. An organic semiconductor as an anode-buffer for the improvement of small molecular photovoltaic cells. *RSC Adv.* **2017**, *7*, 38204–38209. [CrossRef]

373. Li, J.; Zheng, Y.; Zheng, D.; Yu, J. Effect of organic small-molecule hole injection materials on the performance of inverted organic solar cells. *J. Photonics Energy* **2016**, *6*, 035502. [CrossRef]

374. Li, P.; Wu, B.; Yang, Y.C.; Huang, H.S.; De Yang, X.; Zhou, G.D.; Song, Q.L. Improved charge transport ability of polymer solar cells by using NPB/MoO$_3$ as anode buffer layer. *Sol. Energy* **2018**, *170*, 212–216. [CrossRef]

375. Shi, Z.; Deng, H.; Zhao, W.; Cao, H.; Yang, Q.; Zhang, J.; Ban, D.; Qin, D. An alternative hole extraction layer for inverted organic solar cells. *Appl. Phys. A* **2018**, *124*, 676. [CrossRef]

376. Wang, S.; Yang, P.; Chang, K.; Lv, W.; Mi, B.; Song, J.; Zhao, X.; Gao, Z. A new dibenzo[g.p]chrysene derivative as an efficient anode buffer for inverted polymer solar cells. *Org. Electron.* **2019**, *74*, 269–275. [CrossRef]

377. Liu, L.; Zhou, S.; Zhao, C.; Jiu, T.; Bi, F.; Jian, H.; Zhao, M.; Zhang, G.; Wang, L.; Li, F.; et al. TTA as a potential hole transport layer for application in conventional polymer solar cells. *J. Energy Chem.* **2020**, *42*, 210–216. [CrossRef]

nanomaterials

MDPI

Article

Ultra-Flexible Organic Solar Cell Based on Indium-Zinc-Tin Oxide Transparent Electrode for Power Source of Wearable Devices

Jun Young Choi, In Pyo Park and Soo Won Heo *

Nanomaterials and Nanotechnology Center, Electronic Convergence Division, Korea Institute of Ceramic Engineering and Technology (KICET), 101, Soho-ro, Jinju-si 52851, Gyeongsangnam-do, Korea; ichoijy@naver.com (J.Y.C.); pyop4869@naver.com (I.P.P.)
* Correspondence: soowon.heo@kicet.re.kr; Tel.: +82-55-792-2679

Abstract: We have developed a novel structure of ultra-flexible organic photovoltaics (UFOPVs) for application as a power source for wearable devices with excellent biocompatibility and flexibility. Parylene was applied as an ultra-flexible substrate through chemical vapor deposition. Indium-zinc-tin oxide (IZTO) thin film was used as a transparent electrode. The sputtering target composed of 70 at.% In_2O_3-15 at.% ZnO-15 at.% SnO_2 was used. It was fabricated at room temperature, using pulsed DC magnetron sputtering, with an amorphous structure. UFOPVs, in which a 1D grating pattern was introduced into the hole-transport and photoactive layers were fabricated, showed a 13.6% improvement (maximum power conversion efficiency (PCE): 8.35%) compared to the reference device, thereby minimizing reliance on the incident angle of the light. In addition, after 1000 compression/relaxation tests with a compression strain of 33%, the PCE of the UFOPVs maintained a maximum of 93.3% of their initial value.

Keywords: ultra-flexible organic solar cells; IZTO; transparent electrode; mechanical stability; 1D grating pattern

Citation: Choi, J.Y.; Park, I.P.; Heo, S.W. Ultra-Flexible Organic Solar Cell Based on Indium-Zinc-Tin Oxide Transparent Electrode for Power Source of Wearable Devices. *Nanomaterials* **2021**, *11*, 2633. https://doi.org/10.3390/nano11102633

Academic Editor: Vlad Andrei Antohe

Received: 9 September 2021
Accepted: 3 October 2021
Published: 7 October 2021

Publisher's Note: MDPI stays neutral with regard to jurisdictional claims in published maps and institutional affiliations.

1. Introduction

Recently, interest in human implantable sensors, skin-attached sensors, and wearable display devices that display acquired information in real time for fast and accurate monitoring of biosignals has been increasing [1–4]. When these devices are implanted or attached to the human body, they must be ultra-light-weight and flexible so that there is no heterogeneity on the skin or other human tissues [5,6]. Furthermore, it is also necessary to develop new device structures, such as ones with the sensor and power-supply units combined, as well as the development of power sources for the sensors [1,4,7]. Ultra-flexible organic photovoltaics (UFOPVs) are devices that convert light-energy into electrical energy. They are one of the important candidates for use as a power source for wearable devices such as biosensors [8–11]. UFOPVs exhibit a high signal-to-noise ratio because they have a lower noise current than batteries or conventional power supplies. Therefore, when applied to biosensors, improving the accuracy of the collection and analysis of biosignals is possible [1]. In addition, due to the thickness of UFOPVs not exceeding 5 μm, they have excellent adaptability to skin and other tissues given their flexibility and adhesion to 3D surfaces [12]. Therefore, in order to apply UFOPVs as power sources for wearable devices, such as implantable or attachable biosensors, the following conditions must be satisfied. First, the used transparent electrode must have excellent transmittance and conductivity, as well as mechanical flexibility [13]. Second, for the stable power supply, the dependence on incident light should be low [14]. Third, since various types of organic materials (substrate, photoactive layer, electrode, etc.) are used, the device manufacturing process is better at room temperature. Fourth, the fabricated device must have excellent mechanical stability [15,16]. Currently, indium-tin oxide (ITO) transparent electrodes are being applied to various optoelectronic devices, such as liquid crystal displays (LCDs), organic light-emitting diodes (OLEDs), organic photovoltaics (OPVs), and perovskite solar cells due to their excellent transmittance and conductivity [17–20]. However, in order to improve the electro-optical properties of ITO, a thermal treatment process of 300 °C or higher is required, which makes it difficult to apply to organic-based ultra-flexible substrates [21]. In addition, due to the brittle properties of ITO, cracks or fractures occur with only 2–3% deformation [22,23]. Therefore, the use of organic material-based transparent electrodes with excellent mechanical stability—such as poly (3,4-ethylenedioxythiophene): polystyrene sulfonate (PEDOT:PSS), metal nanowires or meshes, carbon nanotubes and graphene—is increasing [24–27]. However, due to their low conductivity, poor surface roughness, and complicated manufacturing process compared to ITO, further research on alternative materials to replace ITO is required. Therefore, in this study, we used parylene as an ultra-flexible substrate and amorphous indium-zinc-tin oxide (IZTO) as a transparent electrode, respectively, to fabricate UFOPVs as a power source that satisfies the various aforementioned conditions. Although the results of using IZTO for various purposes has been explored by other researchers, no study evaluating its mechanical properties using transparent electrodes for UFOPVs [28–30] had been done. Parylene has excellent optical transmittance (>90%), thermal stability (>290 °C), and chemical stability, making it suitable as an ultra-flexible substrate material [1,12,31]. The IZTO transparent electrode was fabricated using a sputtering target composed of 70 at.% In_2O_3-15 at.% ZnO-15 at.% SnO_2 and deposited by pulsed DC magnetron sputtering system. The IZTO transparent electrode fabricated at room temperature showed an amorphous structure, a transmittance of 88.5% (including parylene substrate) in the visible wavelength region, and a mobility of 35 cm^2/Vs. In addition, to reduce the dependence on the incident angle of light, soft imprinting lithography was applied to introduce a 1D grating pattern into the hole transport layer and the photoactive layer, respectively. All UFOPVs were fabricated at room temperature with conventional structures. After introducing the 1D grating pattern, the dependence of the incident angle of light in UFOPVs decreased by 2.2 times at 80°. In addition, to measure the mechanical stability of UFOPVs, they were attached to an elastomer film stretched by 200%, with compression/relaxation repeated 1000 times at 33% compressive strain. After the mechanical property test, the power conversion efficiency (PCE) of the UFOPVs using the IZTO transparent electrode showed a 30% improvement compared to a device using the ITO transparent electrode.

2. Materials and Methods

2.1. Materials

Parylene, an ultra-flexible substrate material, was purchased from KISCO Ltd. (diX-SR, Tokyo, Japan) and used without further purification. The composition of the IZTO target for transparent electrode was 70 at.% In_2O_3-15 at.% ZnO-15 at.% SnO_2 and was purchased from Kojundo Chemical Laboratory Co., Ltd. (Saitama, Japan). For the PEDOT:PSS used as the hole transport layer, AI4083 (Ossila) was chosen. The electron donor polymer (PTB7) and electron acceptor material ($PC_{71}BM$, >99%) were purchased from 1-material (Tokyo, Japan) and Solenne (Groningen, the Netherlands), respectively. Sylgard 184, a polydimethylsiloxane (PDMS) produced by Dow Corning (Midland, MI, USA), was used for the soft-patterned stamp.

2.2. Fabrication of 1D Grating Patterned PDMS

To fabricate a patterned PDMS stamp with a 1D grating structure, a blank DVD was used. We discussed in detail how to fabricate a patterned PDMS stamp in a previous study [8]. The polycarbonate protective layer of the blank DVD was removed using a surgical knife. When the pattern was exposed, the dye layer was removed with methanol, then dried out using a N_2 gun. For the PDMS blend solution, a base material and curing agent were mixed in a weight ratio of 10:1, then degassed in a vacuum chamber. The prepared PDMS solution was poured onto the DVD and cured for 3 h on a hot plate at 90 °C. After the PDMS had solidified, it was peeled off from the DVD and used as the 1D grating patterned PDMS stamp. The period of the 1D grating pattern was 760 nm; the depth was 100 nm.

2.3. Preparation of Ultra-Flexible Substrate and IZTO Transparent Electrode

In order to easily delaminate parylene, an ultra-flexible substrate, from the sacrificial substrate, a self-assembly monolayer (SAM) for lowering the surface energy of the sacrificial substrate was formed. The sacrificial substrate was treated with UV/Ozone (Ahtech LTS Co., Ltd., Gyeonggi-do, Korea) for 30 min and a small Petri dish was placed in a container with a capacity of 300 mL. After that, 30 µL of trichloro(perfluorooctyl)silane (FOTS, 98%, Merck, Darmstadt, Germany) was dropped onto the Petri dish and the container lid was closed. It was then placed in a vacuum chamber of 5×10^{-3} Torr for 1 h. To fabricate a parylene ultra-flexible substrate, a SAM-treated sacrificial substrate was placed in a CVD chamber and filled with 10 g of parylene. A 1 µm ultra-flexible substrate was fabricated by its deposition in a CVD chamber of 5×10^{-4} Torr for 1 h. UVO treatment was performed for 10 min to improve the adhesion between the parylene and IZTO transparent electrode. The prepared substrate was loaded in a sputtering chamber. Sputtering was performed with 125 W input power, 30 kHz frequency, and 6 mTorr oxygen partial pressure, with a flow ratio of 3% $[O_2/(O_2 + Ar)]$. The deposition temperature was performed at room temperature. The thickness of the deposited IZTO film was 100 nm.

2.4. Fabrication of Ultra-Flexible OPVs

UFOPVs were fabricated with a 1D grating pattern applied to the hole transport layer (PEDOT:PSS), the photoactive layer, and again on both layers. For the device without patterns, 150 µL of PEDOT:PSS was applied and spin-coated at 3000 rpm for 30 s. Thermal annealing was performed on a hot plate at 140 °C for 20 min, with a fabricated thickness of 40 nm. In order to introduce a pattern to the PEDOT:PSS layer, 100 µL of PEDOT:PSS was applied and spin-coated at 5000 rpm for 10 s. The patterned PDMS was then quickly but gently placed on the coated PEDOT:PSS layer. After 60 s, the patterned PDMS stamp was removed and thermal annealed on a hot plate at 140 °C for 20 min. The period of the fabricated 1D grating pattern was 760 nm and the depth 35 nm. To form the photoactive layer, PTB7, as an electron donor material, and PC$_{71}$BM, as an electron acceptor material, were used, dissolved in 1 mL of o-dichlorobenzene (DCB) at a weight ratio of 1:1.5 (10 mg:15 mg). Three volume percent of 1,8-diiodooctane (DIO, 98%, Merck) was added as a processing additive. The photoactive solution was spin-coated on a PEDOT:PSS layer, with or without a pattern, at 800 rpm for 7 s, then put in a N_2-filled glove box. To form a nanostructure, the patterned PDMS was gently placed on the spin-coated photoactive layer, then removed after 60 s. Like the PEDOT:PSS layer, no pressure was applied to the patterned PDMS; even after the patterning process, the photoactive layer was not thermally treated. The period of the fabricated 1D grating pattern was 760 nm and the depth 55 nm. For the metal electrode, a thermal evaporator was used. A Ca electron transport layer (0.5 Å s^{-1}, 20 nm) and an Al electrode (0.5 Å s^{-1}, 100 nm) were deposited under a vacuum of 10^{-7} Torr, using a metal mask.

2.5. Measurements

The thicknesses of the prepared parylene thin film, IZTO, PEDOT:PSS, and photoactive layers were measured using a surface profiler (DEKTAK 6M, Bruker, Billerica, MA, USA). The transmittance of the IZTO transparent electrode was measured using UV-Vis spectroscopy (V-650, JASCO, Tokyo, Japan). The X-ray diffraction (XRD) pattern of the IZTO film was analyzed with a Rigaku SmartLab diffractometer. Monochromatic Cu Kα radiation (λ = 0.154 nm) was generated at 45 kV, 200 mA. The surface morphologies of the patterned PDMS, PEDOT:PSS, photoactive layer, and IZTO were observed using an AFM (5400 scanning probe microscope, Agilent Technologies, Santa Clara, CA, USA) in the tapping mode. The surface morphology of the IZTO film was characterized using a field emission scanning electron microscope (FESEM S-4800, Hitachi, Tokyo, Japan). Electrical properties, such as resistivity, carrier concentration, and mobility of the IZTO film, were measured using the van der Pauw method, together with a Hall measuring system (HMS-3000, Ecopia, Gyeonggi-do, Korea).

The UFOVPs' current density-voltage characteristics were measured using a Keithley 2400 source meter unit under simulated solar illumination (AM 1.5, 100 mW cm^{-2}) in a 150 W Xe lamp -based solar simulator (PEC-L11, Peccell Technologies, Yokohama, Japan). The light intensity was calibrated with a standard silicon solar cell (BS520, Bunkoh-Keiki, Tokyo, Japan). The active area of each device was defined, using a metal photomask, to be 0.12 cm^2. The external quantum efficiency (EQE) of each device was measured with monochromatic light (SM-250F, Bunkoh-Keiki, Tokyo, Japan). For the mechanical stability test of the UFOPVs, they were removed from a sacrificial substrate and attached to a 200% pre-stretched acrylic–elastomer film (VHB Y-4905 J, 3M, Saint Paul, MN, USA). The degree of stretching was controlled using a home-made uniaxial screw machine and ruler, and the measurements were conducted under simulated solar illumination (AM 1.5, 100 mW cm^{-2}) and ambient conditions. The pre-stretched acrylic elastomer film, to which the ultra-flexible OPV devices were attached, was slowly compressed in the uniaxial center direction, while reducing the stretching force. This mechanical test was conducted for 1000 repeated compression cycles, with a compression of 33%. The data obtained from the stretching test were recorded using a Keithley 2400 source meter unit.

3. Results and Discussion

Figure 1 shows the UV-visible spectroscopy data of bare glass substrate, parylene ultra-flexible substrate, and IZTO transparent electrodes deposited at room temperature. IZTO was deposited on parylene at room temperature. The transmittance of the bare glass substrate in the visible wavelength region (380–780 nm) was 92%, whereas the transmittance of the parylene ultra-flexible substrate was 92.4%, which was slightly higher than bare glass substrate [32]. As can be seen in Figure 1, the spectra of the parylene substrate with the IZTO transparent electrode has a sinusoidal shape. This is because the parylene substrate is very thin (1 μm) and also because of the optical interference caused by the refractive index difference between the parylene substrate and air interface, or that of the parylene substrate and IZTO interface. An interesting point is that even without thermal treatment of the substrate, the IZTO/parylene transparent electrode showed a slightly higher transmittance of 88.5%, which was improved compared to the ITO/parylene transparent electrode (88%).

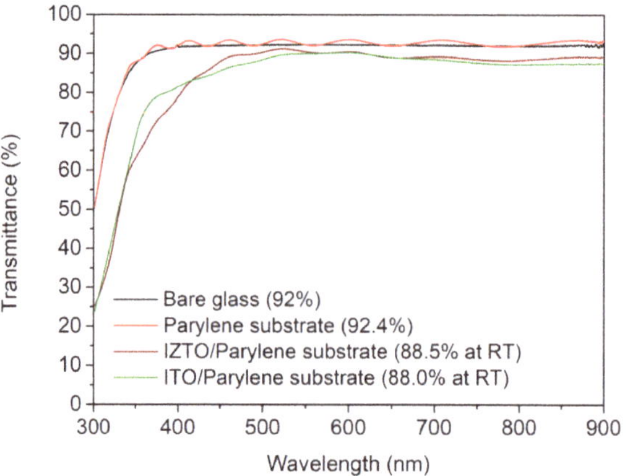

Figure 1. Transmission spectra of the bare glass, parylene free-standing film, the IZTO, and ITO films grown at room temperatures on the parylene substrates, respectively.

Figure 2 shows the XRD pattern of IZTO deposited at room temperature. The inset shows the surface morphologies of the IZTO thin film observed by FESEM and AFM. Combining the data in Figure 2, it can be seen that IZTO deposited at room temperature has an amorphous structure and a very low surface roughness of 0.521 nm. The resistivity, carrier concentration, and mobility of the IZTO thin film measured by the Hall measurement system were 6.2×10^{-4} Ω cm, 7.0×10^{20} cm^{-3}, and 45 cm^2 V^{-1} s^{-1}, respectively, showing superior properties than the electrical properties of ITO fabricated at room temperature (resistivity: 8.3×10^{-4} Ω cm, carrier concentration: 5.3×10^{20} cm^{-3} and mobility: 35 cm^2 V^{-1} s^{-1}). Therefore, it was possible to conclude that the ultra-flexible transparent electrode based on IZTO can be used as the transparent electrode of a wearable device.

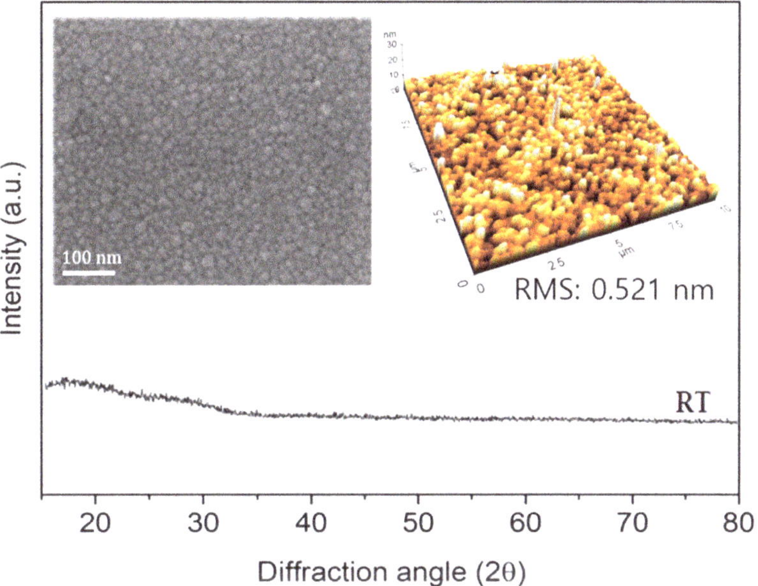

Figure 2. X-ray diffraction patterns of IZTO films with room temperature deposition on a parylene substrate. Insets are SEM image (**left**) and surface roughness (**right**) of IZTO film, respectively.

The fabrication process of the UFOPVs—with the structure of IZTO/PEDOT:PSS (with pattern (p-PEDOT) and without (f-PEDOT), 40 nm)/PTB7:PC$_7$1BM (with pattern (p-Al) and without (f-Al), 100 nm)/Ca (20 nm)/Al electrode (100 nm)—is shown in Figure 3. Nanostructures were introduced using a blank DVD template by applying soft imprinting lithography at room temperature. The surface morphology of the UFOPVs with nanostructures was characterized by AFM and is summarized in the inset of Figure 3a–c. The detailed IZTO-based UFOPVs' fabrication method is described in Section 2.4. The 1D grating structure in the patterned PDMS stamp had a period of about 760 nm and a depth of 100 nm (Figure 3a). In PEDOT:PSS introduced through soft nanoimprinting lithography, the period of the 1D grating structure was the same as that of the PDMS stamp, and the depth was 35 nm (Figure 3b). Similarly, the period of the 1D grating structure in the photoactive layer was 760 nm and the depth 55 nm (Figure 3c). Figure 3d,e shows the chemical structures of the electron donor material and the electron acceptor material used in this study, respectively.

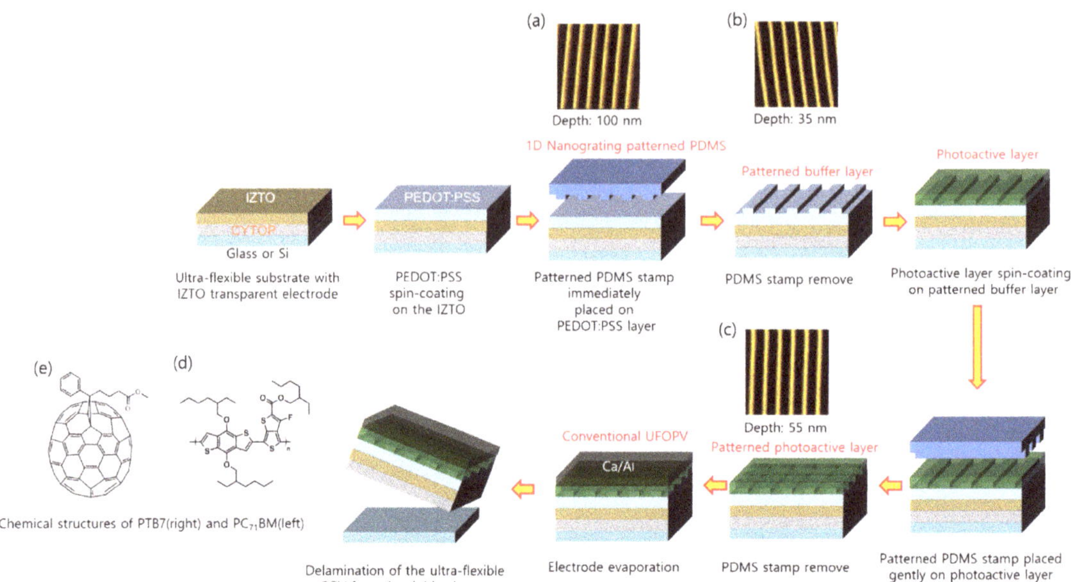

Figure 3. Schematic illustration of the fabrication of ultra-flexible OPVs with a 1D grating pattern by soft imprinting lithography using a PDMS mold. Inset figures present: (**a**) 1D grating patterned PDMS mold (period: 760 nm, depth: 100 nm); (**b**) 1D grating patterned PEDOT:PSS layer (period: 760 nm, depth: 35 nm); (**c**) 1D grating patterned photoactive layer; mixture of PTB7 and PC$_{71}$BM (pitch: 760 nm, depth: 55 nm); (**d,e**), chemical structures of PTB7 and PC$_{71}$BM, respectively.

Figure 4a shows the current density-voltage (J-V) characteristics of the IZTO-based UFOPVs. Detailed electrical properties, such as short circuit current density (J_{SC}), open circuit voltage (V_{OC}), fill factor (FF), and power conversion efficiency (PCE) are summarized in Table 1. The J_{SC} of the device in which the nanostructure was introduced into PEDOT:PSS (p-PEDOT/f-Al) was 15.08 mA cm^{-2}, which was better than the reference device's (f-PEDOT/f-Al) 14.83 mA cm^{-2}. In addition, the FF of the p-PEDOT/f-Al structure device was 69.3%, slightly better than the reference device (68.0%). All V_{OC} were 0.72 V, with no significant change. With the increase of the J_{SC} and FF, the PCE of the device of the p-PEDOT/f-Al structure was improved by 3.6% from 7.26% (reference device) to 7.52%. When a 1D grating pattern was introduced into the photoactive layer (f-PEDOT/p-Al), the J_{SC} value increased to 16.05 mA cm^{-2}, with ignorable change in the V_{OC} and FF, and the calculated PCE showing 7.95%. In addition, when 1D grating patterns were introduced in both the PEDOT:PSS and photoactive layer (p-PEDOT/p-Al), V_{OC} stayed at 0.72 V, while J_{SC} increased to 16.81 mA cm^{-2}, showing the highest PCE of 8.35%.

Table 1. Device characteristics of various device structures. The average values were calculated from eight or more cells. R_{SH} and R_S were obtained from the slope of the J-V curves in the dark at 0 and 1 V, respectively.

	Device Structure	J_{SC} (mA cm^{-2})	V_{OC} (V)	FF (%)	PCE (%)	Increment in PCE (%)	R_{SH} (kΩ cm^2)	R_S (Ω cm^2)
IZTO	Reference (f-PEDOT/f-Al)	14.83 ± 0.02	0.72	68.0 ± 0.2	7.26 ± 0.3	-	4.1	3.51
	p-PEDOT/f-Al	15.08 ± 0.06	0.72	69.3 ± 0.3	7.52 ± 0.2	3.6	9.9	2.78
	f-PEDOT/p-Al	16.05 ± 0.04	0.72	68.8 ± 0.2	7.95 ± 0.4	8.1	9.0	2.74
	p-PEDOT/p-Al	16.81 ± 0.06	0.72	69.0 ± 0.2	8.34 ± 0.3	13.6	15.3	2.32
ITO	p-PEDOT/p-Al	17.07 ± 0.03	0.72	68.9 ± 0.3	8.58 ± 0.2	-	15.8	2.22

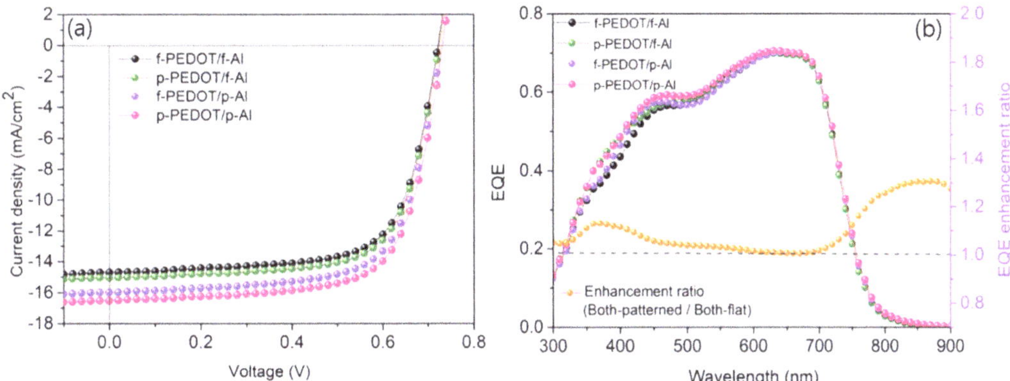

Figure 4. J-V characteristics of ultra-flexible OPVs based on PTB7:PC$_{71}$BM BHJ with (**a**) f-PEDOT/f-Al, p-PEDOT/f-Al, f-PEDOT/p-Al, and p-PEDOT/p-Al. J-V characteristics of UFOPVs measured under AM 1.5G illumination at 100 mW cm^{-2}. (**b**) EQE spectra and EQE enhancement ratios of the patterned devices relative to the reference.

As can be seen in Table 1, when the 1D grating pattern was introduced into both the PEDOT:PSS and the photoactive layer, the series resistance (R_S) and shunt resistance (R_{SH}) values both improved compared to the reference device. The FF and J_{SC} increased by 1.5% and 13.4%, respectively. In addition, when the 1D grating pattern was introduced into the PEDOT:PSS layer, transmittance in the 300–450 nm wavelength region was improved due to the anti-reflection effect. Figure 4b shows that the EQE value in the 300–450 nm wavelength region was better than the reference device. Therefore, it can be seen that the introduction of nanostructures in the PEDOT:PSS layer leads to the improvement of J_{SC}. Furthermore, the introduction of a 1D grating pattern into the photoactive layer significantly increased the J_{SC} to 16.05 mA cm^{-2} (p-PEDOT/f-Al) and 16.81 mA cm^{-2} (p-PEDOT/p-Al), respectively. In addition, propagating surface plasmon polariton (p-SPP), generated at the patterned photoactive layer and metal interface, also induces the cumulative gain of J_{SC}. This can be confirmed by observing the wavelength region after 700 nm in Figure 4b. Finally, when a 1D grating pattern was introduced in all layers, J_{SC} increased to 16.81 mA cm^{-2}, showing the highest PCE of 8.35%, due to the combined effects of anti-reflection, scattering, and propagating surface plasmon polariton.

Figure 5 shows the data analyzing the PCE characteristics of the UFOPVs with respect to the incident angle of light. It was confirmed that the device in which the nanostructure was introduced had a reduced dependence on incident light compared to the device without the nanostructure. This is because the incident light is scattered by the 1D grating pattern formed between the PEDOT:PSS layer and the photoactive layer, thereby increasing the optical path length. Therefore, the absorption amount of photons at a low incident angle is increased compared to a device without a nanostructure. When the incident angle was 80°, the PCE of UFOPVs with nanostructures was improved by 2.2 times compared to the reference device. This demonstrates that UFOPVs with nanostructures can be used as power sources for wearable devices.

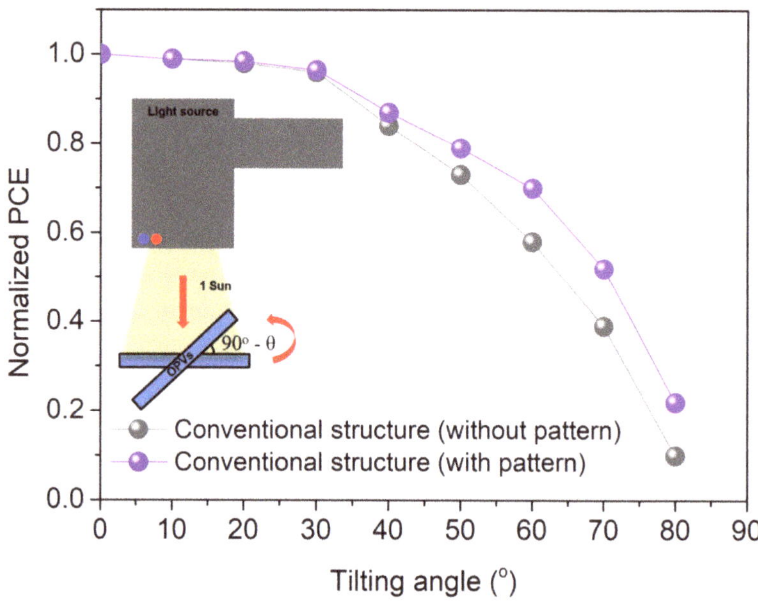

Figure 5. Incident angle dependence characteristics of with/without patterned ultra-flexible OPVs. Data shows normalized values of PCE. Inset shows the illustration of incident angle of light.

To confirm the mechanical stability of IZTO/parylene-based UFOPVs, repeated compression tests were performed. The UFOPVs delaminated from the sacrificial substrate were attached to a 200% pre-stretched acrylic elastomer film. UFOPVs fixed on an elastomer film were compressed in a uniaxial direction. Compression and relaxation were repeated 1000 times at a compression strain of 33% and their results are summarized in Table 2. In the case of IZTO-based UFOPVs, the PCE decreased by 6.7%, from 8.3% to 7.8%. On the other hand, in the case of UFOPVs using ITO as a transparent electrode, the PCE decreased by 43%, from 8.6% to 6%. As can be seen in Figure 6a–c, the PCE of the ITO-based device decreased due to the reduction of J_{SC} and FF. On the other hand, the decrease in J_{SC} and FF was not noticeable in the IZTO-based device due to the high brittleness of ITO. It was confirmed that the amorphous IZTO showed excellent mechanical stability despite the test's harsh conditions. In addition, by using flexible PEDOT:PSS as a hole transport layer, it is expected that PEDOT:PSS would resist deformation, thereby playing a role in protecting the IZTO transparent electrode.

Table 2. Device characteristics of various device structures after mechanical test.

Device Structure		J_{SC} (mA cm^{-2})	V_{OC} (V)	FF (%)	PCE (%)
After mechanical test	ITO	13.69	0.71	61.8	6.01
	IZTO	16.01	0.71	68.2	7.75

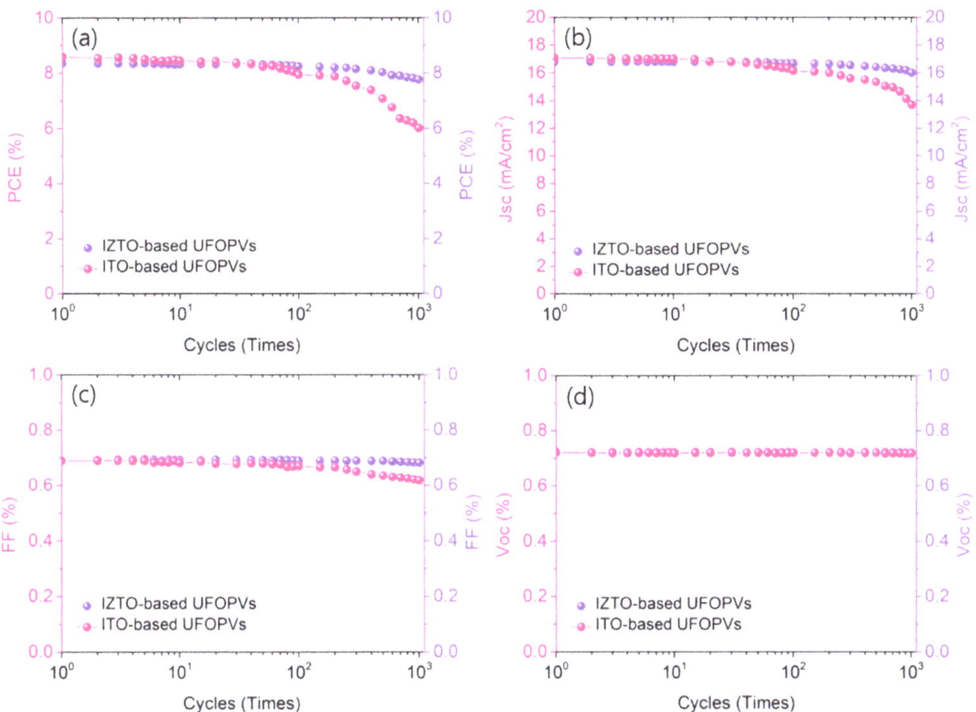

Figure 6. Behavior of photovoltaic parameters under cyclic compression of ultra-flexible OPVs with ITO, and ultra-flexible OPVs with IZTO: (**a**) PCE, (**b**) J_{SC}, (**c**) FF, and (**d**) Voc.

4. Conclusions

We successfully fabricated 1 μm thick UFOPVs that were highly compatible with the human body and capable of supplying stable energy. The substrate of UFOPVs used parylene, and the IZTO used as the transparent electrode was fabricated at room temperature through pulsed DC magnetron sputtering system using a sputtering target composed of 70 at.% In_2O_3-15 at.% ZnO-15 at.% SnO_2. The 1 μm-thick parylene, ultra-flexible substrate showed a high transmittance of 92.4%. The IZTO/parylene-based transparent electrode, to which IZTO deposited at room temperature was applied, has an amorphous structure and a demonstrated transmittance of 88.5%. In addition, the resistivity, carrier concentration, and mobility were 6.2×10^{-4} Ω cm, 5.3×10^{20} cm^{-3}, and 35 cm^2 V^{-1} s^{-1}, respectively.

In addition, we introduced a 1D grating pattern on the hole transport and photoactive layers through soft imprinting lithography at room temperature to lower the dependence on the incident angle of light and thereby obtain higher output power. The fabricated device showed a PCE of 8.35%, which is an improvement of 13.6% compared to the reference device. With the introduction of the 1D grating pattern, anti-reflection, scattering, and propagation surface plasmon polaritons were expressed in a complex manner, causing the cumulative accumulation of J_{SC}. Moreover, with the introduction of the 1D grating pattern, the dependence on the incident angle of light was reduced by 2.2 times at 80°. IZTO/parylene-based UFOPVs also showed very good mechanical stability. Although compression–relaxation was repeated 1000 times with a compression strain of 33%, there was only a 6.7% decrease compared to the initial PCE. Therefore, IZTO/parylene-based, ultra-flexible, transparent electrodes can be used for future optoelectronic devices. Furthermore, fabricated UFOPVs can be combined with various wearable devices, as well as used self-powered devices that do not require an external power source.

Author Contributions: S.W.H. conceived and designed the research. J.Y.C. and I.P.P. fabricated the ultra-thin substrate, OPVs, and performed electrical/optical measurements. J.Y.C. and S.W.H. conducted the incident angle dependence and mechanical tests. S.W.H. analyzed and interpreted the data and prepared the manuscript. All authors have read and agreed to the published version of the manuscript.

Funding: This research was supported by the National Research Foundation of Korea (NRF) funded by the Ministry of Science and ICT (NRF-2020M3H4A3081867) and (NRF-2020R1F1A1051586). This work was also supported by the Ceramic Strategic Research Program (No.KPP20004-1) through the Korea Institute of Ceramic Engineering & Technology (KICET).

Data Availability Statement: The data presented in this study are available on request from the corresponding author.

Conflicts of Interest: The authors declare no conflict of interest.

References

1. Park, S.; Heo, S.W.; Lee, W.; Inoue, D.; Jiang, Z.; Yu, K.; Jinno, H.; Hashizume, D.; Sekino, M.; Yokota, T.; et al. Self-powered ultra-flexible electronics via nano-grating-patterned organic photovoltaics. *Nature* **2018**, *561*, 516–521. [CrossRef]
2. Baik, S.; Lee, H.J.; Kim, D.W.; Kim, J.W.; Lee, Y.; Pang, C. Bioinspired Adhesive Architectures: From Skin Patch to Integrated Bioelectronics. *Adv. Mater.* **2019**, *31*, 1803309. [CrossRef]
3. Ray, T.R.; Choi, J.; Bandodkar, A.J.; Krishnan, S.; Gutruf, P.; Tian, L.; Ghaffari, R.; Rogers, J.A. Bio-Integrated Wearable Systems: A Comprehensive Review. *Chem. Rev.* **2019**, *119*, 5461–5533. [CrossRef]
4. Sultana, A.; Ghosh, S.K.; Sencadas, V.; Zheng, T.; Higgins, M.J.; Middya, T.R.; Mandal, D. Human Skin Interactive Self-powered Wearable Piezoelectric Bio-e-skin by Electrospun Poly-L-lactic Acid Nanofibers for Non-invasive Physiological Signal Monitoring. *J. Mater. Chem. B* **2017**, *5*, 7352–7359. [CrossRef]
5. Chun, S.; Kim, D.W.; Baik, S.; Lee, H.J.; Lee, J.H.; Bhang, S.H.; Pang, C. Conductive and Stretchable Adhesive Electronics with Miniaturized Octopus-Like Suckers against Dry/Wet Skin for Biosignal Monitoring. *Adv. Funct. Mater.* **2018**, *28*, 1805224. [CrossRef]
6. Ha, M.; Lim, S.; Ko, H. Wearable and flexible sensors for user-interactive health-monitoring devices. *J. Mater. Chem. B* **2018**, *6*, 4043–4064. [CrossRef]
7. Ghosh, S.K.; Adhikary, P.; Jana, S.; Biswas, A.; Sencadas, V.; Gupta, S.D.; Tudu, B.; Mandal, D. Electrospun gelatin nanofiber based self-powered bio-e-skin for health care monitoring. *Nano Energy* **2017**, *36*, 166–175. [CrossRef]
8. Heo, S.W.; Le, T.H.H.; Tanaka, T.; Osaka, I.; Takimiya, K.; Tajima, K. Cumulative gain in organic solar cells by using multiple optical nanopatterns. *J. Mater. Chem. A* **2017**, *5*, 10347–10354. [CrossRef]
9. Hashemi, S.A.; Ramakrishna, S.; Aberle, A.G. Recent Progress in Flexible-Wearable Solar Cells for Self-Powered Electronic Device. *Energy Environ. Sci.* **2020**, *13*, 685–743. [CrossRef]
10. Heo, S.W. Vacuum-Free Fabrication Strategies for Nanostructure-Embedded Ultrathin Substrate in Flexible Polymer Solar Cells. *Energies* **2020**, *13*, 5375. [CrossRef]
11. Heo, S.W. Ultra-Flexible Organic Photovoltaics with Nanograting Patterns Based on CYTOP/Ag Nanowires Substrate. *Nanomaterials* **2020**, *10*, 2185. [CrossRef] [PubMed]
12. Takakuwa, M.; Heo, S.W.; Fukuda, K.; Tajima, K.; Park, S.; Umezu, S.; Someya, T. Nanograting Structured Ultrathin Substrate for Ultraflexible Organic Photovoltaics. *Small Methods* **2020**, *4*, 1900762. [CrossRef]
13. Zhang, Y.; Ng, S.-W.; Lu, X.; Zheng, Z. Solution-Processed Transparent Electrodes for Emerging Thin-Film Solar Cells. *Chem. Rev.* **2020**, *120*, 2049–2122. [CrossRef] [PubMed]
14. Zhang, Y.-X.; Fang, J.; Li, W.; Shen, Y.; Chen, J.-D.; Li, Y.; Gu, H.; Pelivani, S.; Zhang, M.; Li, Y.; et al. Synergetic Transparent Electrode Architecture for Efficient Non-Fullerene Flexible Organic Solar Cells with >12% Efficiency. *ACS Nano* **2019**, *13*, 4686–4694. [CrossRef] [PubMed]
15. Yun, J.; Song, C.; Lee, H.; Park, H.; Jeong, Y.R.; Kim, J.W.; Jin, S.W.; Oh, S.Y.; Sun, L.; Zi, G.; et al. Stretchable array of high-performance micro-supercapacitors charged with solar cells for wireless powering of an integrated strain sensor. *Nano Energy* **2018**, *49*, 644–654. [CrossRef]
16. Varma, S.J.; Kumar, K.S.; Seal, S.; Rajaraman, S.; Thomas, J. Fiber-Type Solar Cells, Nanogenerators, Batteries, and Supercapacitors for Wearable Applications. *Adv. Sci.* **2018**, *5*, 180034. [CrossRef] [PubMed]
17. Han, Y.W.; Lee, H.S.; Moon, D.K. Printable and Semitransparent Nonfullerene Organic Solar Modules over 30 cm^2 Introducing an Energy-Level Controllable Hole Transport Layer. *ACS Appl. Mater. Interfaces* **2021**, *13*, 19085–19098. [CrossRef] [PubMed]
18. Minani, T. Present status of transparent conducting oxide thin-film development for Indium-Tin-Oxide (ITO) substitutes. *Thin Solid Films* **2008**, *516*, 5822–5828. [CrossRef]
19. Han, Y.W.; Jung, C.H.; Lee, H.S.; Jeon, S.J.; Moon, D.K. High-Performance Nonfullerene Organic Photovoltaics Applicable for Both Outdoor and Indoor Environments through Directional Photon Energy Transfer. *ACS Appl. Mater. Interfaces* **2020**, *12*, 38470–38482. [CrossRef]

20. Bett, A.J.; Winkler, K.M.; Bivour, M.; Cojocaru, L.; Kabakli, Ö.Ş.; Schulze, P.S.C.; Siefer, G.; Tutsch, L.; Hermle, M.; Glunz, S.W.; et al. Semi-Transparent Perovskite Solar Cells with ITO Directly Sputtered on Spiro-OMeTAD for Tandem Applications. *ACS Appl. Mater. Interfaces* **2019**, *11*, 45796–45804. [CrossRef]
21. Exarhos, G.J.; Zhou, X.-D. Discovery-based design of transparent conducting oxide films. *Thin Solid Films* **2007**, *515*, 7025–7052. [CrossRef]
22. Zardetto, V.; Brown, T.M.; Reale, A.; Carlo, A.D. Substrates for Flexible Electronics: A Practical Investigation on the Electrical, Film Flexibility, Optical, Temperature, and Solvent Resistance Properties. *J. Polym. Sci. Part B Polym. Phys.* **2011**, *49*, 638–648. [CrossRef]
23. Cairns, D.R.; Witte, R.P., II; Sparacin, D.K.; Sachsman, S.M.; Paine, D.C.; Crawford, G.P. Strain-dependent electrical resistance of tin-doped indium oxide on polymer substrates. *Appl. Phys. Lett.* **2000**, *76*, 1425. [CrossRef]
24. Jang, W.; Kim, B.G.; Seo, S.; Shawky, A.; Kim, M.S.; Kim, K.; Mikladal, B.; Kauppinen, E.I.; Maruyama, S.; Jeon, I.; et al. Strong dark current suppression in flexible organic photodetectors by carbon nanotube transparent electrodes. *Nanotoday* **2021**, *37*, 101081. [CrossRef]
25. Araki, T.; Uemura, T.; Yoshimoto, S.; Takemoto, A.; Noda, Y.; Izumi, S.; Sekitani, T. Wireless Monitoring Using a Stretchable and Transparent Sensor Sheet Containing Metal Nanowires. *Adv. Mater.* **2019**, *32*, 1902684. [CrossRef]
26. Kim, D.C.; Shim, H.J.; Lee, W.; Koo, J.H.; Kim, D.-H. Material-Based Approaches for the Fabrication of Stretchable Electronics. *Adv. Mater.* **2019**, *31*, 1902743. [CrossRef]
27. Kochetkov, F.M.; Neplokh, V.; Mastalieva, V.A.; Mukhangali, S.; Vorob'ev, A.A.; Uvarov, A.V.; Komissarenko, F.E.; Mitin, D.M.; Kapoor, A.; Eymery, J.; et al. Stretchable Transparent Light-Emitting Diodes Based on InGaN/GaN Quantum Well Microwires and Carbon Nanotube Films. *Nanomaterials* **2021**, *11*, 1503. [CrossRef] [PubMed]
28. Zhou, N.; Buchholz, D.B.; Zhu, G.; Yu, X.; Lin, H.; Facchetti, A.; Marks, T.J.; Chang, R.P.H. Ultraflexible Polymer Solar Cells Using Amorphous Zinc-Indium-Tin Oxide Transparent Electrodes. *Adv. Mater.* **2014**, *26*, 1098–1104. [CrossRef] [PubMed]
29. Noviyana, I.; Lestari, A.D.; Putri, M.; Won, M.-S.; Bae, J.-S.; Heo, Y.-W.; Lee, H.Y. High Mobility Thin Film Transistors Based on Amorphous Indium Zinc Tin Oxide. *Materials* **2017**, *10*, 702. [CrossRef] [PubMed]
30. Li, K.-D.; Chen, P.-W.; Chang, K.-S.; Hsu, S.-C.; Jan, D.-J. Indium-Zinc-Tin-Oxide Film Prepared by Reactive Magnetron Sputtering for Electrochromic Applications. *Materials* **2018**, *11*, 2221. [CrossRef]
31. Jean, J.; Wang, A.; Bulović, V. In situ vapor-deposited parylene substrates for ultra-thin, lightweight organic solar cells. *Org. Electron.* **2016**, *31*, 120–126. [CrossRef]
32. Darvishzadeh, A.; Alharbi, N.; Mosavi, A.; Gorji, N.E. Modeling the strain impact on refractive index and optical transmission rate. *Phys. B Condens. Matter* **2018**, *543*, 14–17. [CrossRef]

nanomaterials

Article

Studies on the Physical Properties of TiO$_2$:Nb/Ag/TiO$_2$:Nb and NiO/Ag/NiO Three-Layer Structures on Glass and Plastic Substrates as Transparent Conductive Electrodes for Solar Cells

Laura Hrostea [1], Petru Lisnic [2], Romain Mallet [3], Liviu Leontie [2] and Mihaela Girtan [4,*]

[1] Research Center on Advanced Materials and Technologies, Science Department, Institute of Interdisciplinary Research, Alexandru Ioan Cuza University of Iasi, Bldv. Carol I, No. 11, 700506 Iasi, Romania; laura.hrostea@uaic.ro

[2] Faculty of Physics, Alexandru Ioan Cuza University of Iasi, Bldv. Carol I, No. 11, 700506 Iasi, Romania; petru.lisnic@yahoo.com (P.L.); lleontie@uaic.ro (L.L.)

[3] SCIAM, SFR ICAT, Université d'Angers, 4 Rue Larrey, CEDEX 09, 49033 Angers, France; romain.mallet@univ-angers.fr

[4] Photonics Laboratory, (LPhiA) E.A. 4464, SFR Matrix, Faculté des Sciences, Université d'Angers, 2 Bd Lavoisier, 49000 Angers, France

* Correspondence: mihaela.girtan@univ-angers.fr

Citation: Hrostea, L.; Lisnic, P.; Mallet, R.; Leontie, L.; Girtan, M. Studies on the Physical Properties of TiO$_2$:Nb/Ag/TiO$_2$:Nb and NiO/Ag/NiO Three-Layer Structures on Glass and Plastic Substrates as Transparent Conductive Electrodes for Solar Cells. *Nanomaterials* **2021**, *11*, 1416. https://doi.org/10.3390/nano11061416

Academic Editors: Vlad Andrei Antohe and Henrich Frielinghaus

Received: 20 April 2021
Accepted: 25 May 2021
Published: 27 May 2021

Publisher's Note: MDPI stays neutral with regard to jurisdictional claims in published maps and institutional affiliations.

Abstract: In this paper, the physical properties of a new series of multilayer structures of oxide/metal/ oxide type deposited on glass and plastic substrates were studied in the context of their use as transparent conductive layers for solar cells. The optical properties of TiO$_2$/Ag/TiO$_2$, TiO$_2$:Nb/Ag/TiO$_2$:Nb and NiO/Ag/NiO tri-layers were investigated by spectrophotometry and ellipsometry. Optimized ellipsometric modeling was employed in order to correlate the optical and electrical properties with the ones obtained by direct measurements. The wetting surface properties of single layers (TiO$_2$, TiO$_2$:Nb and NiO) and tri-layers (TiO$_2$/Ag/TiO$_2$ TiO$_2$:Nb/Ag/TiO$_2$:Nb and NiO/Ag/NiO) were also studied and good correlations were obtained with their morphological properties.

Keywords: oxide/metal/oxide; OMO; DMD; ellipsometry; transparent conductive electrodes; plastic substrates; organic solar cells; perovskite solar cells; contact angle; wettability

1. Introduction

In the class of transparent conducting electrodes, there are few highly-doped oxides that are typically used as single layers of about 100 to 200 nm for electronics and solar cell applications [1]. Among these, the most well-known is Sn-doped In$_2$O$_3$ (ITO–indium tin oxide). Due to its intensive use and extremely limited resources on Earth, indium is one of the most economically important critical raw materials [2]. Hence, alternative solutions for ITO have been intensively looked for. A lot of studies have been done on Al, In and Ga-doped ZnO (AZO, IZO and GZO) thin films, and on F-doped SnO$_2$ (FTO) [3–13]. Besides, in the last few years, a new class of electrodes including ITO/Au/ITO, ITO/Ag/ITO, ZnO/Au/ZnO, AZO/Au/AZO and Bi$_2$O$_3$/Au/Bi$_2$O$_3$ [14–17] was developed on plastic substrates for OPV applications [18,19]. A lot of studies were also done on using TiO$_2$/Ag/TiO$_2$ as an electrode, especially for DSSC applications and perovskite solar cells, due to such electrodes' energy conversion efficiency [20–30]. These oxide/metal/oxide (O/M/O) electrodes have many advantages, due to their suitability for deposition on flexible substrates. Of their favorable mechanical properties, the metallic layer's ductility is notable. The necessary quantity of oxide materials is generally reduced by two or three times; hence, the total electrode film's thickness can be reduced. The oxide layers act as protective coatings against the oxidation and mechanical degradation of the metallic interlayer film. For solar cell applications, the surface film's properties positively influence the values of the extraction potential.

The novelty of this study consists in its comprehensive analysis of a new class of oxide/metal/oxide electrodes, including TiO_2:Nb/Ag/TiO_2:Nb and NiO_x/Ag/NiO_x (for simpler reading, we use the notation NiO/Ag/NiO for the last structure), which were deposited on plastic and glass substrates by sputtering from metallic targets.

Indeed, very few studies have been done on TiO_2:Nb/Ag/TiO_2:Nb [31,32] and NiO/Ag/NiO [33–35]. Recently, it was proved that the NiO/Ag/NiO antireflective multilayer electrodes used as top cathodes [33], bottom electrodes for $CH_3NH_3PBI_3$ perovskite solar cells [34], or bottom electrodes for PBDTTT-C:PCBM organic solar cells, have improved efficiency compared to the industry standard [35]. To further this important progress for organic and perovskite solar cells, the purpose of this paper is to give a complete and comparative overview of the physical properties of three of these new electrodes: TiO_2/Ag/TiO_2, TiO_2:Nb/Ag/TiO_2:Nb and NiO/Ag/NiO.

2. Materials and Methods

The study involved the preparation and analysis of three sets of samples, including single layers and three-layer oxide/metal/oxide structures deposited on plastic (HIFIPMX739 PET) and glass substrates. Thin oxide films and the metallic interlayer film were deposited by DC magnetron sputtering in reactive and argon atmospheres, respectively, using different metallic targets. The deposition was performed at room temperature in a vertical target–substrate configuration. The deposition parameters were the same regardless of the structures with which the layers were involved, which are mentioned in Table 1. They were chosen taking in account the optimal values in order to obtain simultaneously good optical and electrical properties.

Table 1. Deposition conditions for the samples.

Layer	Atmosphere Conditions	Target–Substrate Distance (cm)	Deposition Current (mA)	Pressure $(10^{-3}$ mbar)	Deposition Time	Target Composition (wt%)
TiO_2	Reactive atm.	7	100	9	4 min	Ti 100%
Ag	Argon atm.	7	20	9	18 s	Ag 100%
TiO_2	Reactive atm.	7	100	9	4 min	Ti 100%
TiO_2:Nb	Reactive atm.	7	100	9	4 min	Ti 94% Nb 6%
Ag	Argon atm.	7	20	9	18 s	Ag 100%
TiO_2:Nb	Reactive atm.	7	100	9	4 min	Ti 94% Nb 6%
NiO	Reactive atm.	7	100	10	4 min	Ni 100%
Ag	Argon atm.	7	20	10	18 s	Ag 100%
NiO	Reactive atm.	7	100	10	4 min	Ni 100%

The morphological properties were analyzed by electron microscopy and atomic force microscopy using a CP-R, Veeco thermo-microscope (CSInstruments, Les Ulis, France) and a JEOL JSM 6301F Electronic Microscope (JEOL, Croissy-sur-Seine, France). The wetting properties were studied via contact angle measurements performed at room temperature using distilled water droplets of equal volumes (3 μL). The optical properties were investigated on both single oxide layers and oxide/metal/oxide layers, using several techniques. Information regarding the transmission and reflection spectra was recorded using a double beam UV/VIS S9000 (Labomoderne, Gennevilliers, France) spectrophotometer and an AvaSpec-3648 Avantes optical fiber spectrophotometer (Avantes, Apeldoorn, The Netherlands), respectively. The optical properties were studied in a 300–1100 nm wavelength range. For instance, the amplitude (ψ) and phase difference (Δ) spectra were registered by spectroscopic ellipsometry using an UVISEL NIR Horiba Jobin Yvon ellipsometer (Horiba Jobin Yvon, Longjumeau, France) equipped with a 75 W high discharge Xe lamp. The chosen configurations for the modulator (M), analyzer (A) and polarizer (P) positions were: $M = 0°$ and $A = 45°$; the incidence angle was $AOI = 70°$. The experimental data were fitted by modelling using the Delta Psi 2 software from Horiba Jobin Yvon (Horiba Jobin Yvon, Longjumeau, France). The optimization of the models was done by following the procedure described in [17]. The electrical conductivity measurements were done using four-point method in planar geometry at room temperature using a Keithley 2600 source Metter (RS Components Ltd., Northants, UK), by measuring the total (sheet) resistance of the multilayer structure. The distance between the probe tips was 0.635 mm. The electrical conductivity was calculated using the estimated value of thickness obtained by ellipsometry for the three-layer structure.

3. Results and Discussion

Figure 1a–c depicts the SEM micrographs of the bottom oxide films prior to the deposition of subsequent layers, and Figure 1a–c depicts the top of the second oxide layer of the three-layer structures (oxide/metal/oxide).

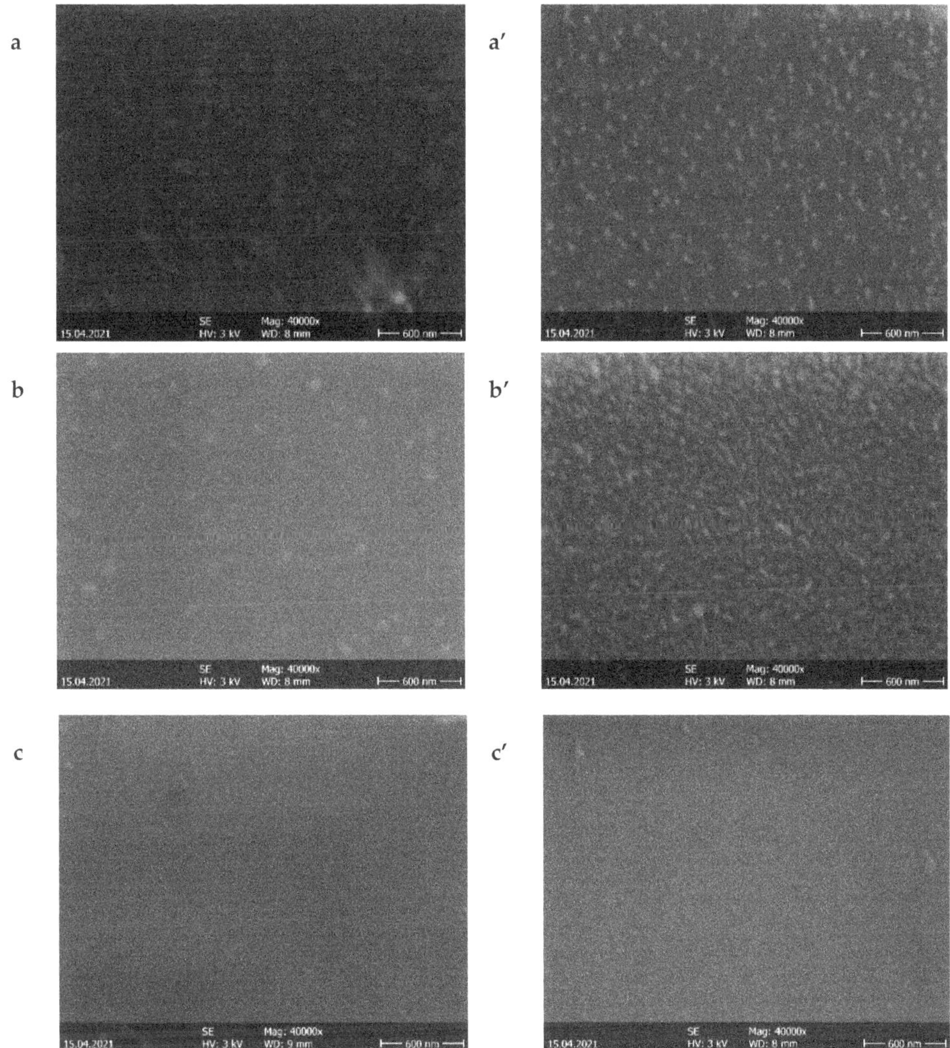

Figure 1. SEM micrographs of single oxide (bottom) layers: (**a**) TiO_2, (**b**) TiO_2:Nb and (**c**) NiO; and SEM micrographs of the top oxide layers of oxide/Ag/oxide multilayer structures: (**a'**) TiO_2, (**b'**) TiO_2:Nb and (**c'**) NiO.

Figure 2 illustrates the SEM and AFM images of the silver interlayer. The AFM analysis was done both for the surfaces of the oxide single layers deposited on glass (not shown here) and for the top oxide layers of the oxide/metal/oxide multilayer structures deposited on glass. The root mean square (RMS) and average (RA) roughness values of these layers are given in Table 2.

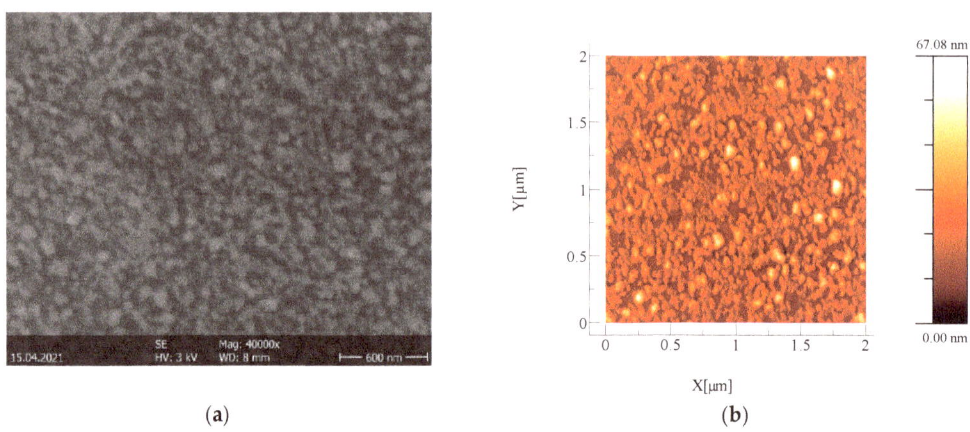

(a)   **(b)**

Figure 2. (a). SEM image of the Ag interlayer and **(b)** AFM image of the Ag interlayer.

Table 2. A summary of RMS and RA roughness values, and contact angle (CA) values of the single oxide layers and the top oxide layers of the multilayer structures (O/M/O) deposited on glass and PET substrates.

Sample	RMS (nm)	RA (nm)	Without UV Light		With UV Light	
			CA (deg) t = 0′	CA (deg) t = 10′	CA (deg) t = 0′	CA (deg) t = 10′
TiO_2	6.4	4.5	55	35	57	27
TiO_2/Ag/TiO_2	8.2	5.9	95	83	79	49
TiO_2/Ag/TiO_2 (on PET)	13.1	10.3	101	70	88	52
TiO_2:Nb	7.8	4.2	70	35	71	21
TiO_2:Nb/Ag/TiO_2:Nb	16	9.9	91	60	99	67
TiO_2:Nb/Ag/TiO_2:Nb (on PET)	26.5	20.7	90	54	93	77
NiO	3.4	2.0	93	65	98	70
NiO/Ag/NiO	2.6	1.8	96	68	100	79
NiO/Ag/NiO (on PET)	6.8	5.4	102	87	101	81
Ag	8.7	6.9	75	66	75	60

Regarding the topography and morphological properties, the SEM micrographs show that the silver layer influenced the surface morphologies of the top surfaces of the O/M/O structures with TiO_2 and TiO_2:Nb, but this influence was smaller for the NiO-based structure. This can be explained, on the one hand, by the fact that the thickness of the NiO second layer was greater than those of the other two oxides (see Table 3), and on the other hand, by the fact that NiO's roughness value was lower than those of TiO_2 and TiO_2:Nb layers.

Figure 3 reproduces the AFM images obtained by scanning the top surfaces of the oxide/metal/oxide layers deposited on glass and on plastic substrates. As was the case for films of ITO/Metal/ITO, AZO/Metal/AZO and ZnO/Metal/ZnO studied previously [15], the films deposited on plastic substrates were rougher than the films deposited on glass substrates (see Table 2). Since the morphology in this context is closely related to wettability expressed in terms of contact angles, we measured such contact angles, and the results are given in Table 2. As one can see, the existence of the metallic interlayer increased the contact angle of the oxide surface in every case. All surfaces were also sensitive to UV exposure. The changes of the contact angles as a function of time with and without exposure to UV irradiation by using a 254 UV-C a $1 \times 8W$ EF180C 1180 mW/cm^2 lamp, are given in Figure 4. For TiO$_2$ films and TiO$_2$/Ag/TiO$_2$ films on glass and plastic substrates, the contact angles decreased after exposure to UV, indicating that the surfaces became more hydrophilic. This is in agreement with the classical behavior of TiO$_2$ films [36]. For TiO$_2$:Nb/Ag/TiO$_2$:Nb and NiO/Ag/NiO, the influence of UV radiation was quite slight.

Table 3. Thicknesses obtained from ellipsometry simulations.

Sample		Thickness (nm)		X^2
TiO$_2$		28 ± 1		4.55
TiO$_2$:Nb		28 ± 1		4.90
NiO		63 ± 1		3.41
	Oxide bottom layer	Ag	Oxide top layer	
TiO$_2$/Ag/TiO$_2$	24 ± 1	8 ± 1	37 ± 1	0.85
TiO$_2$:Nb/Ag/TiO$_2$:Nb	34 ± 1	8 ± 1	42 ± 2	0.55
NiO/Ag/NiO	42 ± 9	8 ± 1	72 ± 9	7.9

Refractive index and film thickness were determined by spectroscopic ellipsometry. Thickness for single layers and the individual thicknesses in multilayer structures were determined by fitting the experimental ellipsometric spectra with those which resulted from theoretical models. For the numerical simulations and modeling, we used the Delta Psi2 software from Horiba Jobin Yvon. The global refractive indices of structures were measured and modelled using the following dispersion formulas: the new amorphous dispersion formula for TiO$_2$ and TiO$_2$:Nb; the Tauc–Lorentz formula for NiO; and the Drude and Tauc–Lorentz formulae for Ag.

Figure 5 illustrates the optical models used for the theoretical calculations and simulations for single layers (A) and O/M/O structures (B).

The films' thickness values obtained after the optimization of the models as described in [17], are given in Table 3.

In Figure 6, the experimental ellipsometric data for the global refractive index are represented as dotted lines, and the fitting curves as continuous lines. The optimal thickness of the silver interlayer, realized by ellipsometric measurements (Table 3) is of 8 nm [20]. This thickness is the lowest limit because, for lower values, the film does not completely cover the substrate, and islands of Ag might appear which are not interconnected, making the resulting layer not conductive. By increasing the thickness, the metallic interlayer is certainly conductive, but the transparency of the O/M/O electrode is reduced.

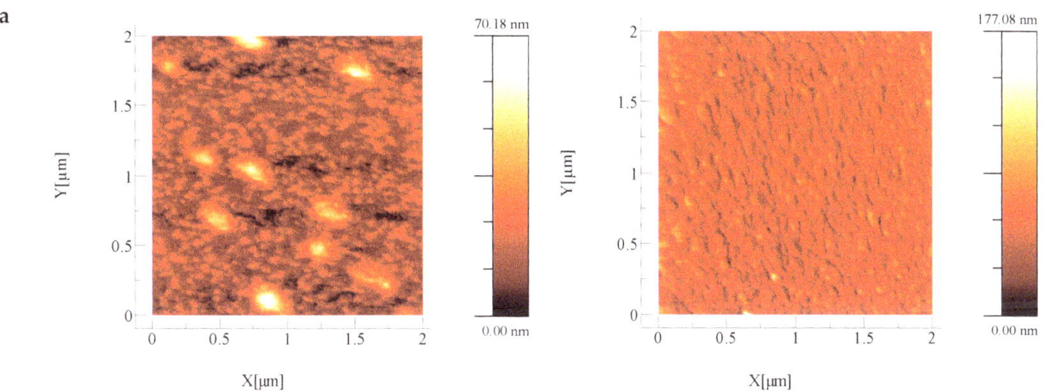

Figure 3. *Cont.*

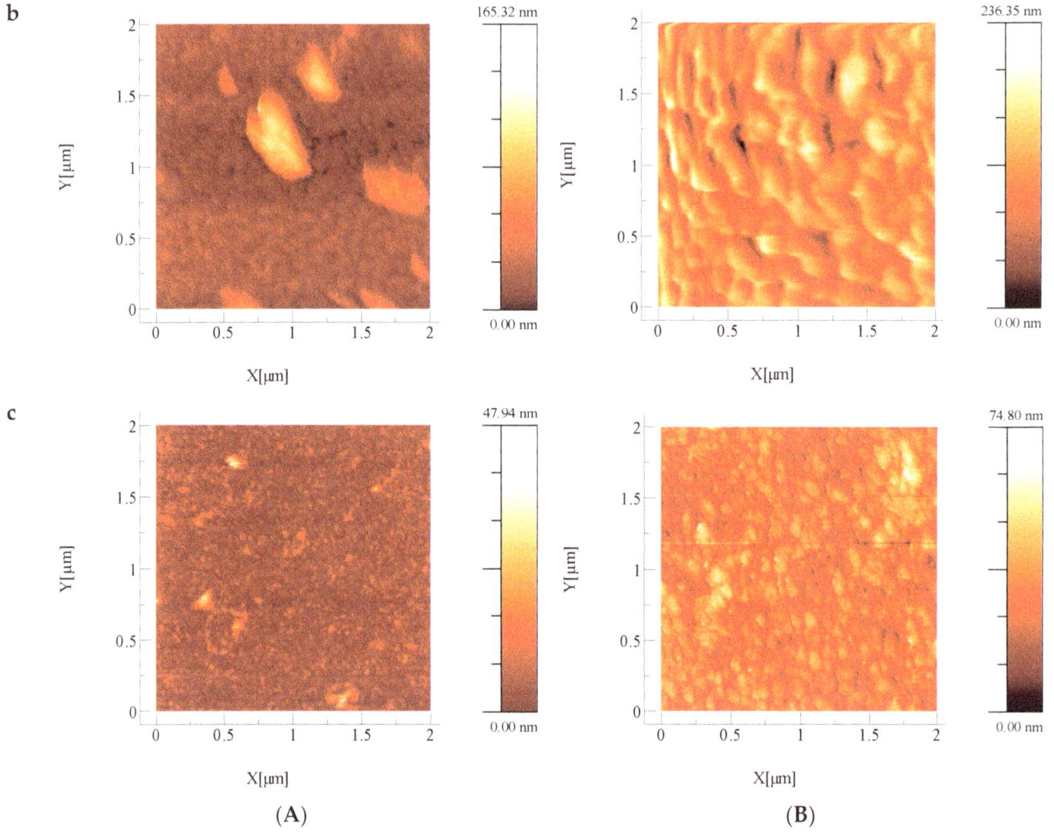

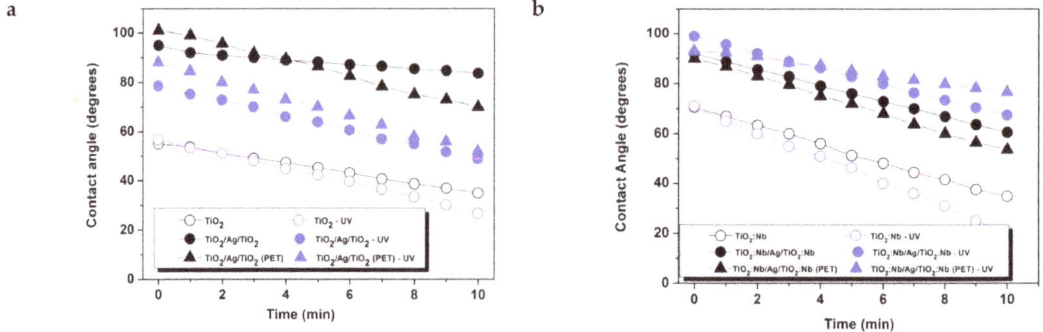

Figure 3. AFM analyses of top surfaces of oxide/Ag/oxide structures deposited on glass (**A**) and on PET (**B**): (**a**) TiO$_2$/Ag/TiO$_2$, (**b**) TiO$_2$:Nb/Ag/TiO$_2$:Nb and (**c**) NiO/Ag/NiO.

Figure 4. *Cont.*

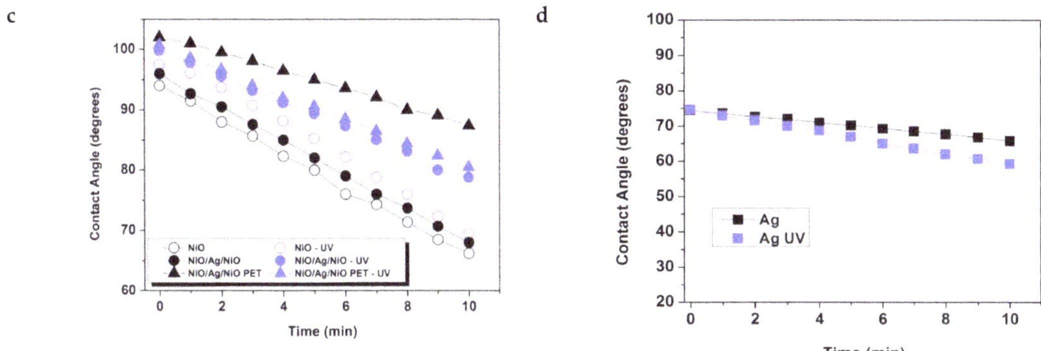

Figure 4. Contact angle measurements with and without UV irradiation for single and three-layer samples: (**a**) TiO$_2$; (**b**) TiO$_2$:Nb; (**c**) Nio; (**d**) Ag interlayer.

Figure 5. Ellipsometric models for single layers (**A**) and O/M/O three-layer structures (**B**).

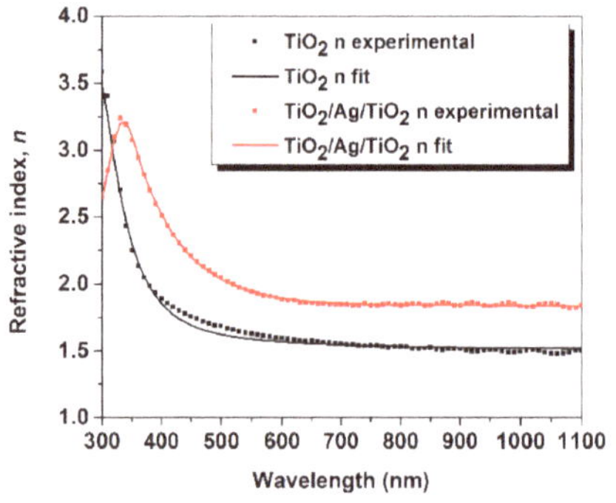

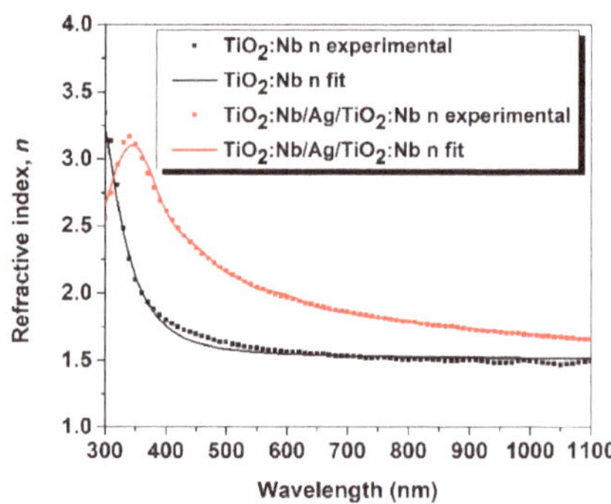

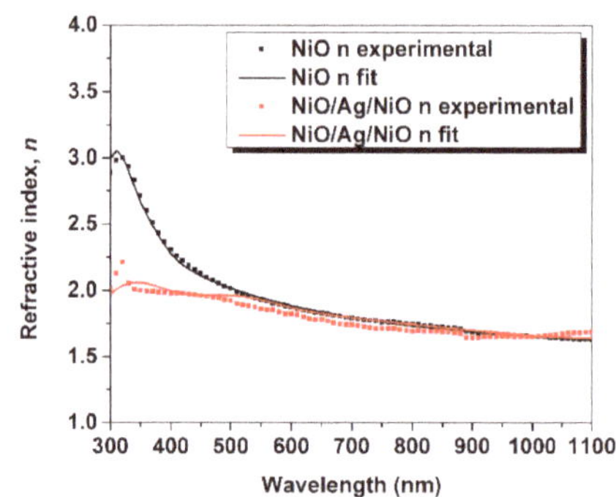

Figure 6. Ellipsometry experimental data (dotted lines) fitted by theoretically calculated curves (continuous lines) from the ellipsometric models, for single and three-layer samples.

As for single-layer films deposited by a sol–gel method—or in this case for films deposited by reactive sputtering—the refractive indices of thin Nb-doped TiO_2 films are smaller than those of an undoped TiO_2 thin films [37]. For the multilayer structures $TiO_2/Ag/TiO_2$ and TiO_2:Nb/Ag/TiO_2:Nb, the refractive indices were higher than those of single layers. On the contrary, the refractive index of NiO/Ag/NiO was smaller than the refractive index of the NiO single layer. These optical properties are important, since these films are used as electrodes for solar cells and optoelectronic devices.

The transmittance and reflectance spectra for single layers, obtained by spectrophotometry, are given in Figure 7a. From these spectra, the optical energy band gaps were calculated using the Tauc plots (Figure 7b) for indirect optical transitions. The calculated band gap values were compared with the results from the ellipsometric modelling and those given in literature (see Table 4). We can report a satisfactory correlation between the values obtained by different methods, and satisfactory correlations with those reported by others authors—this being the second verification of the validity of the ellipsometric optical models.

Table 4. Energy band gap values from spectrophotometric and ellipsometric measurements.

Sample	E_g by Ellipsometry (eV)	E_g by Spectrophotometry (eV)	E_g by Literature (eV)	Reference
TiO_2	3.36	3.59	3.28–3.32	[36]
TiO_2:Nb	3.18	3.53	3.25–3.58	[37]
NiO	3.76	3.20	3.60–4.00	[38,39]

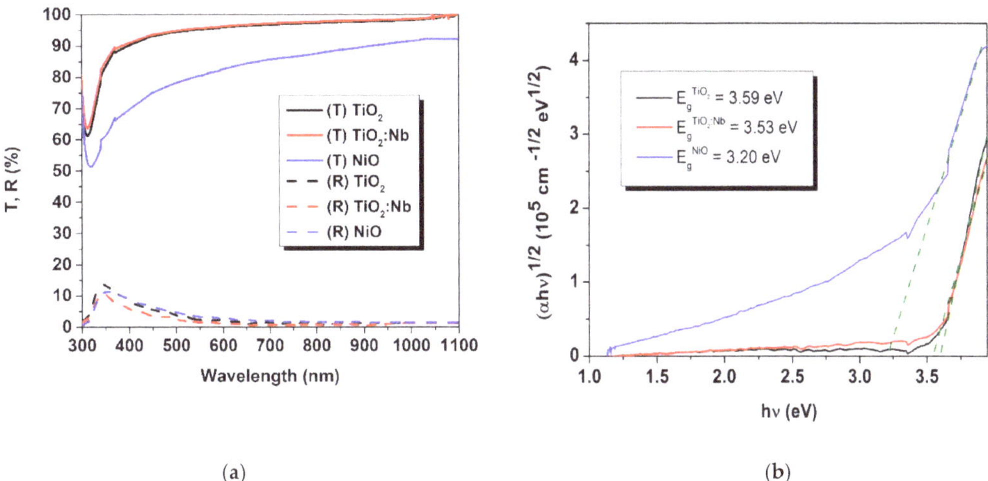

(a)   (b)

Figure 7. (**a**) Transmittance and reflectance spectra for the single-layer coatings; (**b**) energy band gap calculation using spectrophotometry data.

Similarly to Figure 7, Figure 8 gives the transmittance and reflectance spectra for O/M/O layers. Due to the increased thicknesses of these structures, the transmittance was 10% lower.

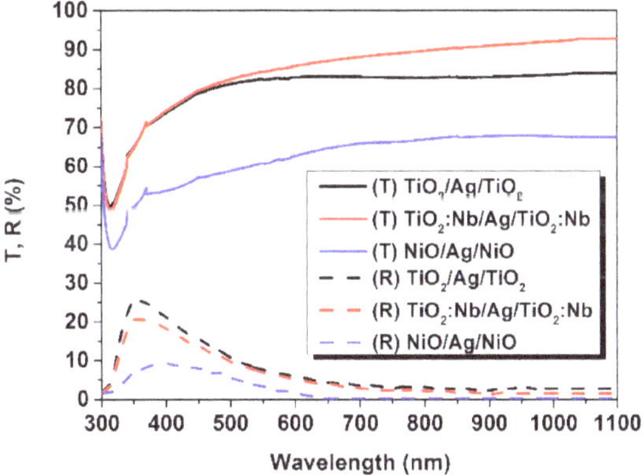

Figure 8. Transmittance and reflectance spectra for the O/M/O structures by spectrophotometry.

Due to the presence of the Ag interlayer, the reflectance increased consistently by 10%, except for the NiO/Ag/NiO three-layer structure, for which the optical features of silver were reduced by the thicker NiO top layer.

The electrical resistivity values determined from direct measurements and also from ellipsometric calculations were roughly 7×10^{-3} $\Omega{\cdot}$cm for $TiO_2/Ag/TiO_2$, 1×10^{-4} $\Omega{\cdot}$cm for $TiO_2{:}Nb/Ag/TiO_2{:}Nb$ and 2×10^{-4} $\Omega{\cdot}$cm for NiO/Ag/NiO, and are in line with the values obtained by other authors [32,33,35,40] for films deposited on oxide targets. The correlation between the electrical measurements and the ellipsometric simulations is demonstrated by equivalent values of plasma frequency. Therefore, taking into account

Drude's model describing the kinetic theory of electrons in metal, plasma frequency is defined as follows [39]:

$$\omega_p = \sqrt{\frac{4\pi\sigma}{\varepsilon_0\varepsilon_\infty\langle\tau\rangle}} \qquad (1)$$

where σ represents the electrical conductivity; ε_0 is the vacuum permittivity ($\varepsilon_0 = 8.85 \times 10^{-12} F/m$) $- \varepsilon_\infty = 1$ generally, according to the Lorentz dispersion model, on which Drude's model is based; and $\langle\tau\rangle$ is the relaxation time of electrons (for Ag electrons, $\langle\tau\rangle \cong 4 \times 10^{-14} s$ [40,41]).

The resulting values of plasma frequency based on electrical conductivity (from direct measurements) are compared with the values of plasma frequency released in the ellipsometric simulations of samples in Table 5.

Table 5. A comparison between plasma frequency values obtained from direct electrical measurements and from ellipsometric simulations.

Sample	ω_p (s^{-1}) Using Formula (1) and the Direct Measured Values of σ	ω_p (s^{-1}) From Ellipsometric Modeling
TiO$_2$/Ag/TiO$_2$	0.7×10^{15}	$(4.9 \pm 1.5) \times 10^{15}$
TiO$_2$:Nb/Ag/TiO$_2$:Nb	5.6×10^{15}	$(67.0 \pm 17.5) \times 10^{15}$
NiO/Ag/NiO	4.2×10^{15}	$(12.6 \pm 1.3) \times 10^{15}$

The differences in the values of plasma frequency calculated from electrical measurements and from ellipsometric modelling are within reasonable limits when taking into account the fact that the spectroscopic ellipsometry technique is based on reflections at one point (local measurements) and also taking into account the limits in the accuracy of the models.

By analyzing all these data, we can conclude that TiO$_2$/Ag/TiO$_2$, TiO$_2$:Nb/Ag/TiO$_2$:Nb and NiO/Ag/NiO have quite similar optical and electrical properties. However, higher values of transparency and electrical conductivity were obtained for TiO$_2$:Nb/Ag/TiO$_2$:Nb. The NiO/Ag/NiO three-layer electrodes could be slightly improved by reducing the oxide layer's thickness. The main advantage of NiO/Ag/NiO electrodes is the fact that the refractive index is lower than those of TiO$_2$/Ag/TiO$_2$ and TiO$_2$:Nb/Ag/TiO$_2$:Nb. Using ellipsometry, which is a powerful tool, the established optimal models will be used in a future work to simulate the properties of the new optimized structure.

4. Conclusions

We presented a comparative study regarding the physical properties of oxide/metal/ oxide three-layer structures which are promising alternatives to ITO electrodes in the photovoltaic field. The oxide layers (TiO$_2$, TiO$_2$:Nb, NiO) and the metallic interlayer (Ag) were laid by successive DC sputtering deposition on glass and plastic substrates. The performances of this type of electrode architecture were presented from optical and electrical points of view, and we also described the morphological features. The presence of Ag as an interlayer influences the three-layer structure. Firstly, the transmittance shows a decrease of 10%, and the reflectance an increase of 10%, the latter depending on the oxide layer's thickness. Secondly, the roughness of such a structure is directly dependent on the substrate roughness, and it too is influenced by the silver's morphological properties. Thirdly, from an electrical point of view, in terms of electrical resistivity ($\sim 10^{-3} \Omega \cdot cm$), these O/M/O structures presented huge potential for photovoltaic applications as transparent conductive electrodes. The ellipsometry optical models were validated by combining different direct measurements. These ellipsometric models can be now used to simulate the properties of new optimized structures.

Author Contributions: Conceptualization, M.G.; methodology, L.H., M.G. and R.M.; software, L.H. and M.G.; validation, M.G. and L.L.; formal analysis, L.H. and M.G.; investigation, L.H., R.M. and M.G.; resources, P.L., R.M. and M.G.; data curation, L.H. and M.G.; writing—original draft preparation, M.G. and L.H.; writing—review and editing, L.H., M.G. and L.L.; visualization, M.G.; supervision, M.G.; project administration, M.G.; funding acquisition, M.G. All authors have read and agreed to the published version of the manuscript.

Funding: The APC was funded by the European Social Fund, through Operational Programme Human Capital 2014–2020, project number POCU/380/6/13/123623, project title "PhD Students and Postdoctoral Researchers Prepared for the Labour Market".

Institutional Review Board Statement: Not applicable.

Informed Consent Statement: Not applicable.

Acknowledgments: The authors are grateful to SCIAM—Microscopy Service, for AFM and SEM micrographs, to N. Mercier and Magali Alain from Moltech Laboratory for providing the necessary facilities for XRD studies and to M. Vieira for technical support.

Conflicts of Interest: The authors declare no conflict of interest.

References

1. Loka, C.; Moon, S.W.; Choi, Y.; Lee, K.-S. High Transparent and Conductive TiO_2/Ag/TiO_2 Multilayer Electrode Films Deposited on Sapphire Substrate. *Electron. Mater. Lett.* **2018**, *14*, 125–132. [CrossRef]
2. Girtan, M.; Wittenberg, A.; Grilli, M.L.; de Oliveira, D.P.S.; Giosuè, C.; Ruello, M.L. The Critical Raw Materials Issue between Scarcity, Supply Risk, and Unique Properties. *Materials* **2021**, *14*, 1826. [CrossRef]
3. Kim, B.-G.; Kim, J.-Y.; Lee, S.-J.; Park, J.-H.; Lim, D.-G.; Park, M.-G. Structural, Electrical and Optical Properties of Ga-Doped ZnO Films on PET Substrate. *Appl. Surf. Sci.* **2010**, *257*, 1063–1067. [CrossRef]
4. Liu, C.C.; Wu, M.L.; Liu, K.C.; Hsiao, S.-H.; Chen, Y.S.; Lin, G.-R.; Huang, J. Transparent ZnO Thin-Film Transistors on Glass and Plastic Substrates Using Post-Sputtering Oxygen Passivation. *J. Disp. Technol.* **2009**, *5*, 192–197. [CrossRef]
5. Jandow, N.N.; Yam, F.K.; Thahab, S.M.; Ibrahim, K.; Abu Hassan, H. The Characteristics of ZnO Deposited on PPC Plastic Substrate. *Mater. Lett.* **2010**, *64*, 2366–2368. [CrossRef]
6. Banerjee, A.; Ghosh, C.; Chattopadhyay, K.; Minoura, H.; Sarkar, A.; Akiba, A.; Kamiya, A.; Endo, T. Low-Temperature Deposition of ZnO Thin Films on PET and Glass Substrates by DC-Sputtering Technique. *Thin Solid Film.* **2006**, *496*, 112–116. [CrossRef]
7. Girtan, M.; Kompitsas, M.; Mallet, R.; Fasaki, I. On Physical Properties of Undoped and Al and In Doped Zinc Oxide Films Deposited on PET Substrates by Reactive Pulsed Laser Deposition. *Eur. Phys. J. Appl. Phys.* **2010**, *51*. [CrossRef]
8. Girtan, M.; Rusu, G.G.; Dabos-Seignon, S.; Rusu, M. Structural and Electrical Properties of Zinc Oxides Thin Films Prepared by Thermal Oxidation. *Appl. Surf. Sci.* **2008**, *254*, 4179–4185. [CrossRef]
9. Rusu, M.; Rusu, G.G.; Girtan, M.; Seignon, S.D. Structural and Optical Properties of ZnO Thin Films Deposited onto ITO/Glass Substrates. *J. Non-Cryst. Solids* **2008**, *354*, 4461–4464. [CrossRef]
10. Ghomrani, F.-Z.; Iftimie, S.; Gabouze, N.; Serier, A.; Socol, M.; Stanculescu, A.; Sanchez, F.; Antohe, S.; Girtan, M. Influence of Al Doping Agents Nature on the Physical Properties of Al:ZnO Films Deposited by Spin-Coating Technique. *Optoelectron. Adv. Mater. Rapid Commun.* **2011**, *5*, 247–251.
11. Girtan, M.; Bouteville, A.; Rusu, G.; Rusu, M. Preparation and Properties of SnO2: F Thin Films. *J. Optoelectron. Adv. Mater.* **2006**, *8*, 27–30.
12. Fortunato, E.; Raniero, L.; Silva, L.; Goncalves, A.; Pimentel, A.; Barquinha, P.; Aguas, H.; Pereira, L.; Goncalves, G.; Ferreira, I.; et al. Highly Stable Transparent and Conducting Gallium-Doped Zinc Oxide Thin Films for Photovoltaic Applications. *Sol. Energy Mater. Sol. Cells* **2008**, *92*, 1605–1610. [CrossRef]
13. Fortunato, E.; Goncalves, A.; Pimentel, A.; Barquinha, P.; Goncalves, G.; Pereira, L.; Ferreira, I.; Martins, R. Zinc Oxide, a Multifunctional Material: From Material to Device Applications. *Appl. Phys. A Mater. Sci. Process.* **2009**, *96*, 197–205. [CrossRef]
14. Socol, M.; Preda, N.; Breazu, C.; Florica, C.; Costas, A.; Istrate, C.M.; Stanculescu, A.; Girtan, M.; Gherendi, F. Organic Heterostructures Obtained on ZnO/Ag/ZnO Electrode. *Vacuum* **2018**, *154*, 366–370. [CrossRef]
15. Girtan, M. Comparison of ITO/Metal/ITO and ZnO/Metal/ZnO Characteristics as Transparent Electrodes for Third Generation Solar Cells. *Sol. Energy Mater. Sol. Cells* **2012**, *100*, 153–161. [CrossRef]
16. Hrostea, L.; Boclinca, M.; Socol, M.; Leontie, L.; Stanculescu, A.; Girtan, M. Oxide/Metal/Oxide Electrodes for Solar Cell Applications. *Sol. Energy* **2017**, *146*, 464–469. [CrossRef]
17. Girtan, M.; Hrostea, L.; Boclinca, M.; Negulescu, B. Study of Oxide/Metal/Oxide Thin Films for Transparent Electronics and Solar Cells Applications by Spectroscopic Ellipsometry. *AIMS Mater. Sci.* **2017**, *4*, 594–613. [CrossRef]
18. Antohe, S.; Iftimie, S.; Hrostea, L.; Antohe, V.A.; Girtan, M. A Critical Review of Photovoltaic Cells Based on Organic Monomeric and Polymeric Thin Film Heterojunctions. *Thin Solid Film.* **2017**, *642*, 219–231. [CrossRef]

19. Socol, M.; Preda, N.; Breazu, C.; Stanculescu, A.; Costas, A.; Stanculescu, F.; Girtan, M.; Gherendi, F.; Popescu-Pelin, G.; Socol, G. Flexible Organic Heterostructures Obtained by MAPLE. *Appl. Phys. A Mater. Sci. Process.* **2018**, *124*, 602. [CrossRef]

20. Dhar, A.; Alford, T.L. High Quality Transparent TiO_2/Ag/TiO_2 Composite Electrode Films Deposited on Flexible Substrate at Room Temperature by Sputtering. *APL Mater.* **2013**, *1*, 012102. [CrossRef]

21. Kim, J.H.; Kim, D.-H.; Seong, T.-Y. Realization of Highly Transparent and Low Resistance TiO_2/Ag/TiO_2 Conducting Electrode for Optoelectronic Devices. *Ceram. Int.* **2015**, *41*, 3064–3068. [CrossRef]

22. Ji, L.-W.; Hsiao, Y.-J.; Young, S.-J.; Shih, W.-S.; Water, W.; Lin, S.-M. High-Efficient Ultraviolet Photodetectors Based on TiO_2/Ag/TiO_2 Multilayer Films. *IEEE Sens. J.* **2015**, *15*, 762–765. [CrossRef]

23. Kim, K.-D.; Pfadler, T.; Zimmermann, E.; Feng, Y.; Dorman, J.A.; Weickert, J.; Schmidt-Mende, L. Decoupling Optical and Electronic Optimization of Organic Solar Cells Using High-Performance Temperature-Stable TiO_2/Ag/TiO_2 Electrodes. *APL Mater.* **2015**, *3*. [CrossRef]

24. Zhu, M.-Q.; Jin, H.-D.; Bi, P.-Q.; Zong, F.-J.; Ma, J.; Hao, X.-T. Performance Improvement of TiO_2/Ag/TiO_2 Multilayer Transparent Conducting Electrode Films for Application on Photodetectors. *J. Phys. D Appl. Phys.* **2016**, *49*, 115108. [CrossRef]

25. Jia, J.H.; Zhou, P.; Xie, H.; You, H.Y.; Li, J.; Chen, L.Y. Study of Optical and Electrical Properties of TiO_2/Ag/TiO_2 Multilayers. *J. Korean Phys. Soc.* **2004**, *44*, 717–721.

26. Baskoutas, S.; Athanasiou, N. Optical Constants of TiO_2/Ag/TiO_2 Three-Layer Thin Films. *Mod. Phys. Lett. B* **1997**, *11*, 1077–1084. [CrossRef]

27. Hasan, M.M.; Haseeb, A.; Saidur, R.; Masjuki, H.H.; Johan, M.R. Structural, Optical and Morphological Properties of TiO$_2$/Ag/TiO_2 Multilayer Films. In *AIP Conference Proceedings*; American Institute of Physics: College Park, MD, USA, 2009; Volume 1136, pp. 229–233.

28. Zhao, Z.; Alford, T.L. The Optimal TiO_2/Ag/TiO_2 Electrode for Organic Solar Cell Application with High Device-Specific Haacke Figure of Merit. *Sol. Energy Mater. Sol. Cells* **2016**, *157*, 599–603. [CrossRef]

29. Zhao, Z.; Alford, T.L. The Effect of Hole Transfer Layers and Anodes on Indium-Free TiO_2/Ag/TiO_2 Electrode and ITO Electrode Based P3HT:PCBM Organic Solar Cells. *Sol. Energy Mater. Sol. Cells* **2018**, *176*, 324–330. [CrossRef]

30. Varnamkhasti, M.G.; Shahriari, E. Design and Fabrication of Nanometric TiO_2/Ag/TiO_2/Ag/TiO_2 Transparent Conductive Electrode for Inverted Organic Photovoltaic Cells Application. *Superlattices Microstruct.* **2014**, *69*, 231–238. [CrossRef]

31. Singh, S.; Sharma, V.; Asokan, K.; Sachdev, K. NTO/Ag/NTO Multilayer Transparent Conducting Electrodes for Photovoltaic Applications Tuned by Low Energy Ion Implantation. *Sol. Energy* **2018**, *173*, 651–664. [CrossRef]

32. Park, J.-H.; Kim, H.-K.; Lee, H.; Lee, H.; Yoon, S.; Kim, C.-D. Highly Transparent, Low Resistance, and Cost-Efficient Nb:TiO_2/Ag/Nb:TiO_2 Multilayer Electrode Prepared at Room Temperature Using Black Nb:TiO_2 Target. *Electrochem. Solid State Lett.* **2010**, *13*, J53–J56. [CrossRef]

33. Mardare, D.; Yildiz, A.; Girtan, M.; Manole, A.; Dobromir, M.; Irimia, M.; Adomnitei, C.; Cornei, N.; Luca, D. Surface Wettability of Titania Thin Films with Increasing Nb Content. *J. Appl. Phys.* **2012**, *112*. [CrossRef]

34. Manole, A.V.; Dobromir, M.; Girtan, M.; Mallet, R.; Rusu, G.; Luca, D. Optical Properties of Nb-Doped TiO_2 Thin Films Prepared by Sol-Gel Method. *Ceram. Int.* **2013**, *39*, 4771–4776. [CrossRef]

35. Li, H.; Chen, C.; Jin, J.; Bi, W.; Zhang, B.; Chen, X.; Xu, L.; Liu, D.; Dai, Q.; Song, H. Near-Infrared and Ultraviolet to Visible Photon Conversion for Full Spectrum Response Perovskite Solar Cells. *Nano Energy* **2018**, *50*, 699–709. [CrossRef]

36. Chang, K.-C.; Yeh, T.-H.; Lee, H.-Y.; Lee, C.-T. High Performance Perovskite Solar Cells Using Multiple Hole Transport Layer and Modulated FA(x)MA(1-x)PbI(3) Active Layer. *J. Mater. Sci. Mater. Electron.* **2020**, *31*, 4135–4141. [CrossRef]

37. Xue, Z.; Liu, X.; Zhang, N.; Chen, H.; Zheng, X.; Wang, H.; Guo, X. High-Performance NiO/Ag/NiO Transparent Electrodes for Flexible Organic Photovoltaic Cells. *Acs Appl. Mater. Interfaces* **2014**, *6*, 16403–16408. [CrossRef] [PubMed]

38. Hosny, N.M. Synthesis, Characterization and Optical Band Gap of NiO Nanoparticles Derived from Anthranilic Acid Precursors via a Thermal Decomposition Route. *Polyhedron* **2011**, *30*, 470–476. [CrossRef]

39. Ashcroft, N.W.; Mermin, N.D. *Solid State Physics*; Holt, Rinehart & Winston: New York, NY, USA; London, UK, 1976; Volume 2005.

40. Ramanathan, K.G.; Yen, S.H.; Estalote, E.A. Total Hemispherical Emissivities of Copper, Aluminum, and Silver. *Appl. Opt.* **1977**, *16*, 2810–2817. [CrossRef]

41. Yang, H.U.; D'Archangel, J.; Sundheimer, M.L.; Tucker, E.; Boreman, G.D.; Raschke, M.B. Optical Dielectric Function of Silver. *Phys. Rev. B* **2015**, *91*, 235137. [CrossRef]

MDPI

St. Alban-Anlage 66

4052 Basel

Switzerland

Tel. +41 61 683 77 34

Fax +41 61 302 89 18

www.mdpi.com

Nanomaterials Editorial Office

E-mail: nanomaterials@mdpi.com

www.mdpi.com/journal/nanomaterials